Copyright © 1991 by Saunders College Publishing, a division of Holt, Rinehart and Winston, Inc.

All rights reserved. No part of this publication may be reproduced or transmitted in any form or by any means, electronic or mechanical, including photocopy, recording, or any information storage and retrieval system, without permission in writing from the publisher.

Requests for permission to make copies of any part of the work should be mailed to Copyrights and Permissions Department, Holt, Rinehart and Winston, Inc., Orlando, Florida 32887.

Text Typeface: 10/12 Times Roman
Compositor: Progressive Typographers, Inc.
Acquisitions Editor: Robert Stern
Developmental Editor: Alexa Barnes
Managing Editor: Carol Field
Project Editor: Maureen Iannuzzi
Copy Editor: Bonnie Boehme
Manager of Art and Design: Carol Bleistine
Art Director: Christine Schueler
Art and Design Coordinator: Doris Bruey
Text Designer: NSG Design
Cover Designer: Lawrence R. Didona
Text Artwork: GRAFACON
Director of EDP: Tim Frelick
Production Manager: Bob Butler
Marketing Manager: Denise Watrobsky

Cover Credit: Colorful Knot and Ties by Dominique Sarrante/ ©1990 THE IMAGE BANK

Printed in the United States of America

College Algebra, 2/e

ISBN: 0-03-054243-X

Library of Congress Catalog Card Number:

0123   039   987654321

THIS BOOK IS PRINTED ON **ACID-FREE, RECYCLED** PAPER

Logarithm Laws

$\ln(vw) = \ln v + \ln w$ $\qquad$ $\log_b(vw) = \log_b v + \log_b w$

$\ln\left(\dfrac{v}{w}\right) = \ln v - \ln w$ $\qquad$ $\log_b\left(\dfrac{v}{w}\right) = \log_b v - \log_b w$

$\ln(v^k) = k(\ln v)$ $\qquad$ $\log_b(v^k) = k(\log_b v)$

Special Notation

$\ln e$ means $\log_e v$

$\log v$ means $\log_{10} v$

Change of Base Formula

$\log_b v = \dfrac{\ln v}{\ln b}$

GEOMETRY

The Pythagorean Theorem

$c^2 = a^2 + b^2$

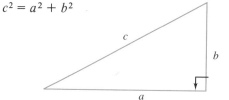

Area of a Triangle

$A = \tfrac{1}{2}bh$

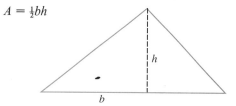

Circles

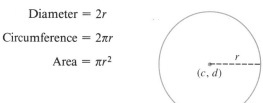

Diameter $= 2r$

Circumference $= 2\pi r$

Area $= \pi r^2$

Equation of circle with center (c, d) and radius r

$(x - c)^2 + (y - d)^2 = r^2$

Distance Formula

Length of segment $PQ =$

$\sqrt{(x_1 - x_2)^2 + (y_1 - y_2)^2}$

Midpoint Formula

Midpoint M of segment $PQ =$

$\left(\dfrac{x_1 + x_2}{2}, \dfrac{y_1 + y_2}{2}\right)$

$Q\ (x_2, y_2)$

M

$P\ (x_1, y_1)$

Lines

If $x_1 \neq x_2$, the slope of the straight line through (x_1, y_1) and (x_2, y_2) is $\dfrac{y_2 - y_1}{x_2 - x_1}$.

The equation of the straight line through (x_1, y_1) with slope m is $y - y_1 = m(x - x_1)$.

The equation of line with slope m and y-intercept b is $y = mx + b$.

College Algebra

Second Edition

Thomas W. Hungerford
Cleveland State University

Richard Mercer
Wright State University

Saunders College Publishing

Philadelphia Ft. Worth Chicago San Francisco
Montreal Toronto London Sydney Tokyo

To My Children

Anne Elizabeth Hungerford
Thomas Joseph Hungerford

To My Sister

Susan Aileen Mercer

Preface

This book is designed to provide the essentials of college algebra for students who have had two to three years of high school mathematics. As in the first edition, the goal has been to produce a text that can be understood by an average student with a minimal amount of outside assistance, and to do so without any sacrifice in rigor.

We have done our best to present sound mathematics in an informal manner that stresses meaningful motivation, detailed explanations, extensive use of pictures and diagrams, numerous examples, and emphasis on the real-world origins of basic concepts (especially functions). There are a wide variety of exercises, ranging from routine to challenging, including a large number of thought-provoking problems that are well within the reach of the average student.

The flexible design of the text (which is fully explained in the To the Instructor section on page xii) makes it suitable both for students who are preparing for calculus and for those for whom college algebra may be a terminal mathematics course.

PEDAGOGICAL CHANGES

Approximately three fourths of the book has been completely rewritten to improve clarity and convenience. Major changes include the following:

Complex numbers are introduced immediately after quadratic equations in Chapter 2.

The presentation of inverse functions has been revised and appears in Chapter 4 (Functions).

An easier method of graphing polynomial functions is now used.

The coverage of logarithms begins with a section on natural logarithms because of their central role in calculus. However, the section on logarithms to an arbitrary base is written so that it may be covered first by instructors who prefer the conventional approach.

Conic sections are treated in greater depth (including applications) in a separate chapter (Chapter 7). However, the material is arranged to allow a fast treatment of conics centered at the origin if time is short.

NEW MATERIAL

Many optional sections have been added, covering the following topics:

Variation
Descartes' Rule of Signs
Partial fractions
Matrix algebra
Higher order determinants
Systems of inequalities
Linear programming
Sequences
Infinite series
Probability theory

OTHER FEATURES

Useful features from the first edition have been retained and others have been added, including:

Geometry Review Frequently used facts from plane geometry are summarized, with examples, in an appendix at the end of the book.

Warnings Students are alerted to common errors and misconceptions by clearly marked "warning boxes."

Chapter Reviews Every chapter now concludes with a list of important concepts (referenced by section and page number), a summary of important facts and formulas, and a set of review questions. Whenever possible, these questions tie together material from several sections in the chapter or from other chapters.

Artwork All the graphs and figures have been redrawn by computer for this edition. Whenever possible, figures are placed in the body of the text at the point they are needed by the reader, rather than being stuck in the margin for the printer's convenience.

Figures are placed in the exercises only when essential. We believe that coming up with an appropriate sketch is a significant part of many problems and a skill that must be developed by students.

Exercises Each exercise set begins with routine calculation and drill problems, then proceeds to exercises that are less mechanical and require some thought. Answers for all odd-numbered problems are given at the back of the book.

Some sets include a section labeled "Unusual Problems," a few of which are quite challenging. Many of them, however, are not difficult, but simply different from what students may have seen before and are well within the reach of most students.

Calculators The use of calculators is encouraged whenever appropriate, and nonroutine calculations that may cause trouble are explained. The need to understand the underlying concepts in order to use a calculator intelligently is continually stressed and common misuses of calculators are pointed out. (Tables are supplied in appendices for those who insist upon them.)

Graphing calculators are not needed to read this text, but their use is discussed for the benefit of students who have them. For instructors who want to emphasize the use of graphing calculators, there is a Graphing Calculator Supplement available from the publisher.

SUPPLEMENTS

A Student Solutions Manual with Graphing Calculator Supplement is available for purchase. Part I, written by Dorothy Smith of the University of New Orleans, contains detailed solutions to the odd-numbered exercises, chapter summaries, and additional practice problems. Part II, written by James Angelos of Central Michigan University, explains how to use the Casio and TI graphing calculators and uses examples from the text. It also provides additional practice problems for students to try.

Instructors who adopt this text may also receive, free of charge, the following items:

Instructor's Manual with Transparency Masters Also written by Dorothy Smith, this manual contains detailed solutions to all the exercises to assist the instructor in the classroom and in grading assignments. In addition, transparency masters of important figures from the text are provided for use in preparing classroom lectures.

Test Bank This manual, written by Larry Small of Los Angeles Pierce College, contains over 1000 multiple-choice and open-ended questions arranged in five forms per chapter. There are also three final examinations. Master answer sheets and a complete answer section are included.

Computerized Test Bank (IBM version) The computerized test bank contains all the questions from the Test Bank and allows instructors to prepare quizzes and examinations easily and quickly.

A&T Software (IBM version) This software is ideal for tutorial and review purposes. It includes both algorithmic and theoretical questions, referenced by chapter and section. Hints are provided upon request, and the student's score is recorded. A demonstration disk is available.

ACKNOWLEDGMENTS

Our continued thanks go to the many people who offered assistance and advice with the first edition, without which the present edition would not have been possible. The new edition has also benefited from the comments of the following reviewers:

Paul J. Allen, University of Alabama
Daniel D. Anderson, University of Iowa
Robert Arnold, California State University—Fresno
Rick Billstein, University of Montana
Betty Detwiler, Western Kentucky University
Albert E. Filano, West Chester University of Pennsylvania
Jerrold W. Grossman, Oakland University
Catherine Hayes, Mobile College
E. John Hornsby, Jr., University of New Orleans
Michael S. Jacobson, University of Louisville
Judith Lenk, Ocean County College
Roger McCoach, County College of Morris
Mark A. Miller, Illinois State University
Mark Serebransky, Camden County College
Laurence Small, Los Angeles Pierce College
Dorothy P. Smith, University of New Orleans
Fred J. Wilson, Ferris State University
Donald J. Wright, University of Cincinnati

We are particularly grateful to our accuracy reviewers, who carefully examined the examples and exercises:

Paul Allen, University of Alabama—Tuscaloosa
Mona Choban, Portland Community College—Rock Creek Campus
Norma James, New Mexico State University
Lynne Kotrous, Central Community College—Platt Campus
Mitchel Levy, Broward Community College—Central Campus

Their sharp eyes and helpful suggestions significantly improved the book.

Finally, we want to express our appreciation to the staff of Saunders College Publishing for their assistance, with special thanks to Bob Stern, Maureen Iannuzzi, and Alexa Barnes.

Thomas W. Hungerford
Cleveland State University

Richard Mercer
Wright State University

October 1990

Contents

To the Instructor — xii
To the Student — xiv

Chapter 1 Basic Algebra — 1

- 1.1 The Real Number System — 1
- 1.1.A *Excursion:* Decimal Representation of Real Numbers — 9
- 1.2 Integral Exponents — 13
- 1.3 Radicals and Rational Exponents — 20
- 1.4 Absolute Value — 25
- 1.5 Algebraic Expressions and Polynomials — 33
- 1.5.A *Excursion:* Synthetic Division — 42
- 1.6 Factoring — 45
- 1.7 Fractional Expressions — 49

Chapter Review — 56

Chapter 2 Equations and Inequalities — 62

- 2.1 First-Degree Equations — 62
- 2.2 Applications of First-Degree Equations — 68
- 2.3 Quadratic Equations — 76
- 2.4 Applications of Quadratic Equations — 82
- 2.5 Complex Numbers — 87
- 2.6 Other Equations — 94
- 2.6.A *Excursion:* The Rational Solutions Test — 100
- 2.7 Linear Inequalities — 103
- 2.8 Polynomial and Rational Inequalities — 108

Chapter Review — 116

Chapter 3 The Coordinate Plane and Lines — 121

- 3.1 The Coordinate Plane — 121
- 3.2 Slopes of Lines — 131
- 3.3 Equations of Lines — 139

Chapter Review — 145

Chapter 4 Functions and Graphs — 150

4.1	Functions	150
4.2	Functional Notation	155
4.3	Graphs of Functions	163
4.3.A	*Excursion:* Graph Reading	171
4.4	Graphing Techniques	177
4.4.A	*Excursion:* Increasing and Decreasing Functions	187
4.5	Operations on Functions	190
4.6	Inverse Functions	197
4.7	Variation	203
	Chapter Review	207

Chapter 5 Polynomial and Rational Functions — 216

5.1	Quadratic Functions	216
5.2	Polynomial Functions	224
5.2.A	*Excursion:* What Happens for Large x?	232
5.3	Rational Functions	234
5.4	Theory of Equations	247
5.4.A	*Excursion:* Descartes' Rule of Signs	254
5.5	Approximation Techniques	257
	Chapter Review	262

Chapter 6 Exponential and Logarithmic Functions — 271

6.1	Exponential Functions	271
6.1.A	*Excursion:* Compound Interest and the Number e	278
6.2	The Natural Logarithmic Function	281
6.3	Logarithmic Functions to Other Bases	290
6.4	Exponential and Logarithmic Equations	298
	Chapter Review	305

Chapter 7 The Conic Sections — 311

7.1	Ellipses	313
7.2	Hyperbolas	323
7.3	Parabolas	333
	Chapter Review	340

Chapter 8   Systems of Equations and Inequalities   343

8.1	Systems of Linear Equations in Two Variables	343
8.1.A	*Excursion:* Nonlinear Systems in Two Variables	352
8.2	Larger Systems of Linear Equations	357
8.2.A	*Excursion:* Partial Fractions	366
8.3	Matrix Methods	371
8.3.A	*Excursion:* Matrix Algebra	377
8.4	Determinants and Cramer's Rule	388
8.4.A	*Excursion:* Higher Order Determinants	397
8.5	Systems of Inequalities	401
8.6	Introduction to Linear Programming	411
	Chapter Review	418

Chapter 9   Discrete Algebra and Probability   424

9.1	Sequences and Sums	424
9.2	Arithmetic Sequences	431
9.3	Geometric Sequences	437
9.3.A	*Excursion:* Infinite Series	443
9.4	The Binomial Theorem	446
9.5	Mathematical Induction	454
9.6	Permutations and Combinations	463
9.6.A	*Excursion:* Distinguishable Permutations	470
9.7	Introduction to Probability	473
	Chapter Review	481

Appendix 1   Geometry Review   487

Appendix 2   Logarithm Tables   492

Answers to Odd-Numbered Exercises   A-1

Index   I-1

To the Instructor

Every effort has been made to make this text as flexible as possible, so that you can easily adapt it to the needs of your own class. With minor exceptions (usually an occasional example or exercise), the interdependence of chapters is given by the chart on the next page.

Interdependence of sections within a particular chapter is indicated in three ways.

1. Sections that are not needed in the sequel (and hence can be omitted or postponed) are marked by a footnote at the beginning of the section.
2. Boldface boxes labeled **Roadmap** are scattered throughout the text at the beginnings of chapters or sections. They provide information about section interdependence and alternate orders of topics.
3. Certain sections are labeled as **Excursions.** Each Excursion is closely related to the section that precedes it and usually has that section as a prerequisite. No Excursion is a prerequisite for any other section of the text.

With rare exceptions, each Excursion is a complete discussion and includes a full complement of examples and exercises. The "Excursion" label is designed solely to make syllabus planning easier and is *not* intended as any kind of value judgment on the topic in question.

For certain audiences, a particular Excursion may be far more important than some of the "regular" sections. In other courses, however, this may not be the case. You decide what's best for your course.

INTERDEPENDENCE OF CHAPTERS

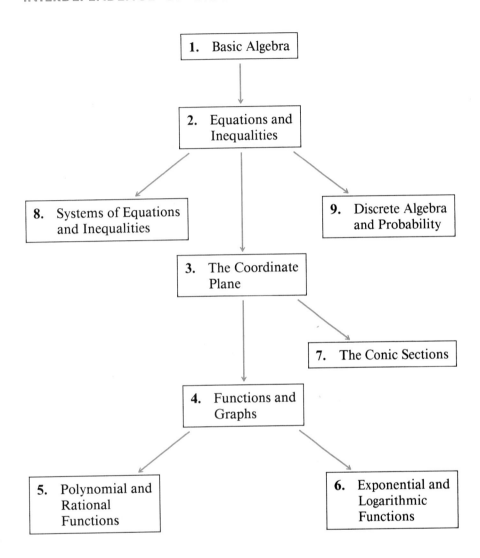

To the Student

Read this — or you will turn into a toad!

If you want to succeed in this course, remember that *mathematics is not a spectator sport*. You can't expect to learn mathematics without *doing* mathematics, any more than you could learn to swim without getting in the water. You have to take an active role, making use of your chief resources: your instructor, your fellow students, and this book.

When it comes to math textbooks, many students use their books only for finding out what the homework problems are. If they get stuck on a problem, they page back through the text until they find a similar example. If the example doesn't clarify things, they may try reading part of the text (as little as possible). Rarely, if ever, do such students read through an entire section (or subsection) from beginning to end.

If this description fits you, don't feel guilty. Some mathematics texts are unreadable. But don't use your bad past experiences as an excuse for not reading this book. It has been classroom-tested for years by students like yourself. It is understandable by an average student, with a minimal amount of outside assistance. So if you want to get the most out of this course, we strongly suggest that you follow these guidelines:

1. Read the pages assigned by your instructor from beginning to end before starting the homework problems. If you find calculations you don't understand, take pencil and paper and try to work them out. If you don't understand a particular statement, reread the preceding material to see if you missed something.

2. If you have spent a reasonable amount of time trying to figure something out, mark the place with a question mark and continue reading. Plan to ask your instructor about the material you have marked.

3. Now do the homework problems. You should be able to do all, or almost all, of the assigned problems. After you've worked at the homework for a reasonable amount of time and answered as many problems as you can, mark the exercises that are still causing trouble. Plan to ask your instructor about them.

If you follow these guidelines, you will get the most out of this book. But it won't be enough unless you actually *ask* your instructor about the things you don't understand. Unfortunately, many students are afraid to ask questions in class for fear that the questions will seem "dumb." Such students should remember this:

If you have honestly followed the guidelines above and still have unanswered questions, then there are at least six other students in your class who have the same questions.

So it's not a dumb question. Furthermore, your instructor will welcome questions that arise from a serious effort on your part. In any case, your instructor is being paid (with your tuition money) to answer questions. So do yourself a favor and get your money's worth—*ask questions.*

CHAPTER 1

Basic Algebra

This chapter is designed to refresh your algebraic and manipulative skills. Most of it is material you have seen before.

1.1 THE REAL NUMBER SYSTEM

You have been using **real numbers** most of your life. They include:

Natural numbers (or **positive integers**): 1, 2, 3, 4, ...

Integers: 0, 1, −1, 2, −2, 3, −3, 4, −4, ...

Rational numbers:* every number that can be expressed as a fraction r/s, with r and s integers and $s \neq 0$; for instance, $\frac{1}{2}, -.983 = \frac{-983}{1000}, 47 = \frac{47}{1}$, and $8\frac{3}{5} = \frac{43}{5}$. Alternatively, rational numbers may be described as numbers that can be expressed as terminating or repeating decimals; for instance, $.5 = \frac{1}{2}$ or $.3333\ldots = \frac{1}{3}$. See Excursion 1.1.A for details.

Irrational numbers:† every real number that *cannot* be expressed as a fraction, or alternatively, numbers that can be expressed as infinite *nonrepeating* decimals (see Excursion 1.1.A). For example, the number π, which is used to calculate the area of a circle, is irrational.‡

The real numbers can be pictured geometrically as points on a **number line,** as shown in Figure 1–1.

---

* The word "rational" here has no psychological implications; it refers to the "ratio" or quotient of two integers.

† Ir-rational = not rational = not a ratio.

‡ This fact is difficult to prove. In grade school you may have used 22/7 as π; a calculator might display π as 3.141592654. But these numbers are just *approximations* of π (*close to,* but not quite *equal* to π).

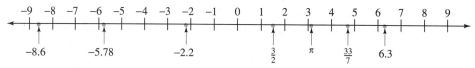

FIGURE 1-1

We shall assume that there is exactly one point on the line for every real number (and vice versa) and use phrases such as "the point 3.6"or "a number on the line." This mental identification of real numbers and points on the line is often extremely helpful.

Order

The statement "c is less than d" (written $c < d$) and the statement "d is greater than c" (written $d > c$) mean exactly the same thing:

c lies to the *left* of d on the number line.

For example, Figure 1-1 shows that $-5.78 < -2.2$ and $\pi < 4$.

The statement $c \leq d$ (or $d \geq c$) means c **is less than or equal to** d (or d **is greater than or equal to** c). The statement $b \leq c \leq d$ means

$$b \leq c \quad \text{and} \quad c \leq d.$$

Geometrically, this means that c lies between b and d on the number line (and may possibly be equal to one or both of them). Similarly, $b \leq c < d$ means $b \leq c$ and $c < d$, and so on.

The basic properties of the order relation can easily be understood by looking at examples on the number line:

PROPERTIES OF ≤

$c \leq c$ for every real number c.
If $c \leq d$ and $d \leq c$, then $c = d$.
If $b \leq c$ and $c \leq d$, then $b \leq d$.

The last property is also true if $\leq$ is replaced by $<$.

Negative Numbers and Negatives of Numbers

The **positive numbers** are those to the right of 0 on the number line, that is,

all numbers c with $c > 0$.

The **negative numbers** are those to the left of 0, that is,

all numbers c with $c < 0$.

The **nonnegative numbers** are the numbers c with $c \geq 0$. Two numbers are said to **have the same sign** if both are positive *or* both are negative. For example,

-16 and $-\frac{1}{2}$ have the same sign. Two numbers are said to have **opposite signs** if one is positive and the other is negative; for instance, -8 and 13.

The word "negative" has a second meaning in mathematics.* The **negative of a number** c is the number $-c$. For example, the negative of 5 is -5 and the negative of -3 is $-(-3) = 3$. Thus the negative of a negative number is a positive number. Zero is its own negative since $-0 = 0$. In summary,

NEGATIVES

> The negative of the number c is $-c$.
> If c is a positive number, then $-c$ is a negative number.
> If c is a negative number, then $-c$ is a positive number.

Arithmetic

A summary of the important properties of real number arithmetic is given below. Remember that a letter may represent a positive number or a negative number or zero. The names on some of these properties are convenient for reference purposes, but it is more important that you understand the *meaning* of each property and be able to use it.

PROPERTIES OF ARITHMETIC

Associative Laws

Addition: $\quad b + (c + d) = (b + c) + d$

Multiplication: $\quad b(cd) = (bc)d$

Commutative Laws

Addition: $\quad b + c = c + b$

Multiplication: $\quad bc = cb$

Identities

Addition: $\quad c + 0 = c = 0 + c$

Multiplication: $\quad c \cdot 1 = c = 1 \cdot c$

Inverses

Addition: $\quad c + (-c) = 0 = (-c) + c$

Multiplication: For $c \neq 0$, $\quad c \cdot \dfrac{1}{c} = 1 = \dfrac{1}{c} \cdot c$

Distributive Laws

$b(c + d) = bc + bd \quad$ and $\quad (c + d)b = cb + db$

$b(c - d) = bc - bd \quad$ and $\quad (c - d)b = cb - db$

---

* The use of one word for two different meanings can be confusing. But this double usage is so widespread, we're stuck with it.

Example 1 By the distributive law, $3(6 - k) = 3 \cdot 6 - 3k = 18 - 3k$. ∎

Example 2 By the commutative and associative laws for multiplication,
$$3(t \cdot 4) = 3(4t) = (3 \cdot 4)t = 12t.$$ ∎

ZERO PRODUCTS

1. For every real number c,
$$c \cdot 0 = 0 = 0 \cdot c.$$
2. If a product is zero, then at least one of the factors is zero; in other words,
$$\text{if } cd = 0, \text{ then } c = 0 \text{ or } d = 0 \text{ (or both)}.$$
Consequently, a product of nonzero factors is nonzero:
$$\text{If } c \neq 0 \text{ and } d \neq 0, \text{ then } cd \neq 0.$$

Example 3 Property 2 is often used in solving equations. If we know, for instance, that $(x - 3)(x + 5) = 0$, then we can conclude that $x - 3 = 0$ or $x + 5 = 0$. ∎

Remember that **division by 0 is *not* defined**. Expressions such as $1/0$ or $\pi/0$ or $0/0$ have no meaning; they are *not* numbers. There are good reasons for excluding division by zero; see Exercise 74.

Computational errors often occur from misuse of various rules for dealing with negatives. So be sure you understand each of the following statements.

PROPERTIES OF NEGATIVES

1. $(-1) \cdot c = -c$ and $-(-c) = c$;
2. $b + (-c) = b - c$;
3. $-(b + c) = -b - c$;
4. $b - (-c) = b + c$.

Example 4 $7 - (-x) = 7 + x$ by property 4. ∎

Example 5

$$-(4 - k) = -(4 + (-k)) \qquad [\text{property 2, with } b = 4, c = k]$$
$$= -4 - (-k) \qquad [\text{property 3, with } b = 4, c = -k]$$
$$= -4 + k \qquad [\text{property 4, with } b = -4, c = k]$$ ∎

SIGNS

Let b and c be nonzero real numbers.

1. If b and c have the same sign, then bc and b/c are positive.
2. If b and c have opposite signs, then bc and b/c are negative.

Furthermore,

3. $(-b)(-c) = bc$;
4. $(-b)c = b(-c) = -bc$;
5. $\dfrac{-b}{-c} = \dfrac{b}{c}$;
6. $\dfrac{-b}{c} = \dfrac{b}{-c} = -\dfrac{b}{c}$.

Example 6 $2(-x) = -2x$ by property 4. ■

Hierarchy of Operations and Parentheses

To avoid confusion when dealing with expressions such as $4 + 3 \cdot 2 - 12 \div 6$ mathematicians have agreed on this convention:

ORDER OF OPERATIONS

Multiplication and division are performed first, in order from left to right. Addition and subtraction are performed last, in order from left to right.

According to this convention

$$4 + 3 \cdot 2 - 12 \div 6 = 4 + 6 - 2 \quad \text{[multiplication, division first]}$$
$$= 10 - 2 = 8 \quad \text{[addition, subtraction last]}$$

WARNING

Most calculators automatically follow the convention in the box above, but some do not. Check your direction book or do a simple example like $4 + 3 \cdot 2$ to find out what priority of operations is built into your calculator. If it doesn't follow the convention, you may have to adjust your sequence of key strokes in order to get the correct answer.

Even with the convention, parentheses are often necessary. If you intend to say "subtract $2 + 4$ from 9" you must write

$$9 - (2 + 4) \text{ and } not \; 9 - 2 + 4$$

because $9 - (2 + 4) = 9 - 6 = 3$, whereas $9 - 2 + 4 = 7 + 4 = 11$. Similarly, $1 + 1/2 = 1\frac{1}{2}$ but $(1 + 1)/2 = 2/2 = 1$. So parentheses do make a difference. These examples illustrate the first of the two basic rules for dealing with parentheses:

RULES FOR PARENTHESES

1. **Do all computations inside the parentheses before doing any computations outside the parentheses.**
2. **When dealing with parentheses within parentheses, work from the inside out.**

Example 7

$$7 + (11 - (6 \div 3)) = 7 + (11 - 2) = 7 + 9 = 16.$$

Inside parentheses first

Use as many pairs of parentheses as are necessary to guarantee that operations are carried out in the correct order. To avoid confusion it may help to use brackets and braces, as in $17\{(7.2 + 4.8) - (6.4 - [7 \div 2])\}$.

WARNING

Most scientific calculators have parentheses keys. Some calculators can handle as many as eight or ten sets of nested parentheses, but others can only deal with one set of parentheses at a time. Check yours before trying complicated calculations.

Equality, Order, and Arithmetic

The operations of arithmetic and the equality relations behave pretty much as you would expect:

ARITHMETIC AND EQUALITY

1. If $a = b$, then for any number c,
$$a + c = b + c \quad \text{and} \quad ac = bc.$$
2. If $a = b \neq 0$, then $\dfrac{1}{a} = \dfrac{1}{b}$.

The interaction of the inequality relation and the operations of arithmetic is a bit trickier (especially the last two properties below):

ARITHMETIC AND INEQUALITIES

1. If $a \leq b$ and c is any number, then $a + c \leq b + c$.
2. If $a \leq b$ and $c > 0$, then $ac \leq bc$.
3. If $a \leq b$ and $c < 0$, then $ac \geq bc$. (inequality reversed!)
4. If $0 < a \leq b$, then $\dfrac{1}{a} \geq \dfrac{1}{b}$. (inequality reversed!)

All four properties are true with $<$ and $>$ in place of $\leq$ and $\geq$, respectively.

Example 8 Multiplying both sides of $5 < 7$ by -4 *reverses* the direction of the inequality by property 3: $5(-4) > 7(-4)$ because $-20 > -28$. To understand property 4, note that $3 < 9$, but $\dfrac{1}{3} > \dfrac{1}{9}$. ■

EXERCISES

1. Draw a number line and mark the location of each of these numbers: $0, -7, 8/3, 10, -1, -4.75, \frac{1}{2}, -5,$ and 2.25.

In Exercises 2–18, express the given statement in symbols.

2. 7 is less than 5.
3. -4 is greater than -8.
4. -17 is less than 14.
5. π is less than 100.
6. x is nonnegative.
7. y is less than or equal to 7.5.
8. z is greater than or equal to -4.
9. t is positive.
10. z lies strictly between -3 and -2.
11. d lies strictly between π and 7.93.
12. x is positive, but not more than 7.
13. y is negative, but not less than -8.
14. d is not greater than 2.
15. c is at most 3.
16. z is at least -17.
17. c is less than 4 and d is at least 4.
18. x is greater than -6 and y is at most -6.

In Exercises 19–28, fill the blank with = or < or >, so that the resulting statement is true.

19. $-6 \_\_ -2$
20. $5 \_\_ -3$
21. $3/4 \_\_ .75$
22. $3.1 \_\_ \pi$
23. $1/3 \_\_ .33$
24. $2 + 4 \_\_ 11 - 5$
25. $\frac{1}{2} + \frac{1}{4} \_\_ \frac{8}{11}$
26. $6(-7) + 1 \_\_ 7(-6) + 2$
27. $(-5)(-3)(-2) \_\_ (-4)7 - 3$
28. $3\{2 - 4(6 \div 2) + [-3 - 2(8 - 5)]\} \_\_ -58$

In Exercises 29–34, fill the blank with = if the resulting statement is true for *all* nonzero real numbers b, c, d; otherwise, insert $\neq$.

29. $\dfrac{bc + dc}{d} \_\_ bc + c$
30. $\dfrac{bc + dc}{c} \_\_ b + c$
31. $\dfrac{1}{b - c} \_\_ \dfrac{1}{b} - \dfrac{1}{c}$
32. $\dfrac{c + d}{b} \_\_ \dfrac{c}{b} + \dfrac{d}{b}$
33. $\dfrac{c - d}{d - c} \_\_ 1$
34. $\dfrac{c - d}{d - c} \_\_ -1$

In Exercises 35–42, find the negative of the given number.

35. 1
36. $-\pi$
37. -67.43
38. $(4 + (-6))$
39. $(-(37 - 2))$
40. $(7.53 - (6 - 2.13))$

41. $(17.77 + (-7.7 + .7))$

42. $2\pi(7 - \pi)$

In Exercises 43–56, express the given number as a single integer.

43. $14 - 10 + 3$ 44. $2 - (3 - 5) + 3 - (-7)$

45. $(1 - 8 + 7) - (3 - 5 + 8(2 - 5))$

46. $(-5)8$ 47. $7(-2)(-5)$

48. $(-1)(-2)(-3)(-4)(-5)$

49. $(-6)(5 - 1 + 4) - 3.25 + 1.25$

50. $\dfrac{-15}{2 - 7}$ 51. $\dfrac{-266}{-12 - (-26)}$

52. $\dfrac{17 + 3}{4} - 5$ 53. $(-6)\left(\dfrac{9 + 3}{9 - 3} - 2\right)$

54. $\dfrac{\dfrac{5 + 7}{5 - 7} + 10}{\dfrac{18 - 3}{5(-3)} - 3}$ 55. $\dfrac{(2 \cdot 3) + 4(2 - 11)}{2(-2) + 1}$

56. $\dfrac{(-2)\left(\dfrac{2 + 1}{3 - 5}\right) + 6}{-2 + 8\left(\dfrac{8 + 2}{7 + 9}\right)}$

57. Is the product of two rational numbers always a rational number? Justify your answer.

58. Let h be a nonzero real number. For each statement below find three different numbers that illustrate the truth of the statement.
    (a) If h is a large positive number, then $1/h$ is a small positive number.
    (b) If h is a very small positive number, then $1/h$ is a very large positive number.
    (c) If h is a very small negative number (that is, far to the left of 0), then $1/h$ is very close to 0.
    (d) If h is a negative number that is very close to 0, then $1/h$ is a very small negative number.

59. (a) Find a rational number that is strictly between $2/7$ and $5/9$ on the number line.
    (b) Let r/s and a/b be rational numbers with $r/s < a/b$. Find a rational number that is strictly between r/s and a/b on the number line.

In Exercises 60–65 fill the blank with $\leq$ or $\geq$ so that the resulting statement is true. Then give a numerical example of the statement.

60. If $a \leq b < 0$, then $1/a$ __ $1/b$.

61. If $a < 0 < b$, then $1/a$ __ $1/b$.

62. If $0 \leq a \leq b$, then a^2 __ b^2.

63. If $a \leq b \leq 0$, then a^2 __ b^2.

64. If $a \leq 0 \leq b$, then a^3 __ b^3.

65. (Sneaky) If $a \leq 0 \leq b$, then a^2 __ b^2.

In Exercises 66–72 fill the blank so as to produce two equivalent statements. For example, the arithmetic statement "a is negative" is equivalent to the geometric statement "the point a lies to the left of the point 0."

Arithmetic Statement	Geometric Statement
66. $a \geq b$	_____
67. _____	a lies c units to the right of b
68. _____	a lies between b and c
69. $a - b > 0$	_____
70. a is positive	_____
71. _____	a lies to the left of b
72. $a + b > c$ $(b > 0)$	_____

Unusual Problems

73. Let T be a right triangle in which the perpendicular sides each have length 1. If the hypotenuse of T is laid on the number line, to the right of the origin with one end on the origin, what is the coordinate of the other end of the hypotenuse?

74. Suppose that division by zero *was* defined. Then either $\frac{1}{0}$ would be a nonzero number or $\frac{1}{0}$ would be 0. This exercise shows that the first of these possibilities leads to a logical contradiction and that the second is highly unreasonable, if we want division by zero to "behave" like division by other numbers. Consequently, division by zero is not defined.
    (a) Assuming that division by zero obeys the usual rules of arithmetic of fractions, show that
    $$\dfrac{1}{\frac{1}{0}} = 0. \quad [\text{Remember: } \tfrac{0}{1} \text{ is defined and } \tfrac{0}{1} = 0.]$$

(b) Suppose $\frac{1}{0}$ is a nonzero number, say, $\frac{1}{0} = c$ with $c \neq 0$. Use part (a) to show that $1/c = 0$. Consequently,

$$c\left(\frac{1}{c}\right) = c \cdot 0$$

$$1 = 0$$

This is a logical contradiction. Therefore $\frac{1}{0}$ cannot be a nonzero number.

(c) Suppose that $\frac{1}{0} = 0$. If division by 0 behaves like division by other numbers, then all four of the following statements *should* be true:
   (i) Whenever a is very close to 5, then $1/a$ is very close to $\frac{1}{5} = .2$.
   (ii) Whenever a is very close to 2, then $1/a$ is very close to $\frac{1}{2} = .5$.
   (iii) Whenever a is very close to 1, then $1/a$ is very close to $\frac{1}{1} = 1.0$.
   (iv) Whenever a is very close to 0, then $1/a$ is very close to $\frac{1}{0} = 0$.

Use a calculator to verify that the first three statements are true but that *the last statement is false*. For example, $a = 4.999$ and $a = 5.0001$ are very close to 5 and $\frac{1}{4.999} = .200040$ and $\frac{1}{5.0001} = .199996$ are very close to $\frac{1}{5} = .2$. Since statement (iv) is false, we conclude that $\frac{1}{0} = 0$ is an unreasonable definition if we want division by 0 to behave like division by other numbers.

75. The problems noted in Exercise 74 also arise if you try to define $0/0$. Since $r/r = 1$ for every nonzero number r, it might seem reasonable to define $0/0 = 1$.
   (a) Show that the assumption $0/0 = 1$ leads to a logical contradiction. [*Hint:* What happens if you multiply both sides by 2?]
   (b) Show that the assumption $0/0 = c$, with $c \neq 0$, also leads to a contradiction.
   (c) Why would it be unreasonable to define $0/0 = 0$?

1.1.A EXCURSION: Decimal Representation of Real Numbers

Every rational number can be expressed as a terminating or repeating decimal. For instance, $3/4 = .75$. To express $15/11$ as a decimal, divide the numerator by the denominator:

```
                1.3636
          11 ) 15.0000
               11
               ──
                40
                33
                ──
                 70
                 66
                 ──
                  40
                  33
                  ──
                   70
                   66
```

Same remainder (from the first 40 to the second 40), *Repeats as above*.

Since the remainder at the first step (namely 4) occurs again at the third step, it is clear that the division process goes on forever with the two-digit block "36" repeating over and over in the quotient $\frac{15}{11} = 1.3636363636\ldots$.

The method used in the preceding example can be used to express any rational number as a decimal. During the division process some remainder *necessarily* repeats.* If the remainder at which this repetition starts is 0, the result is a repeating decimal ending in zeros—that is, a terminating decimal (for instance, .75000 . . . = .75). If the remainder at which the repetition starts is nonzero, then the result is a nonterminating repeating decimal, as in the example above.

Conversely, there is a simple method for converting any repeating decimal into a rational number.

Example 1 Write $d = .272727\ldots$ as a rational number.

Solution Assuming that the usual rules of arithmetic hold, we see that
$$100d = 27.272727\ldots \quad \text{and} \quad d = .272727\ldots$$
Now subtract d from $100d$:
$$\begin{aligned} 100d &= 27.272727\ldots \\ -d &= -.272727\ldots \\ \hline 99d &= 27 \end{aligned}$$

Dividing both sides of this last equation by 99 shows that $d = \frac{27}{99} = \frac{3}{11}$. ∎

Irrational Numbers

There are many nonterminating decimals that are *nonrepeating* (that is, no block of digits repeats forever), such as .202002000200002 . . . (where after each 2 there is one more zero than before). Although the proof is too long to give here, it is in fact true that every nonterminating and nonrepeating decimal represents an *irrational* real number.

Conversely every irrational number can be expressed as a nonterminating and nonrepeating decimal (no proof to be given here). For instance, the decimal expansion of the irrational number π begins 3.1415926535897. . . . This computation has actually been carried out to over a billion decimal places by computer.

Since every real number is either a rational number or an irrational one, the preceding discussion can be summarized as follows.

DECIMAL REPRESENTATION

1. Every real number can be expressed as a decimal.
2. Every decimal represents a real number.
3. The terminating decimals and the nonterminating repeating decimals are precisely the rational numbers.

---

* For instance, if you divide a number by 11 as in the example above, the only possible remainders at each step are the 11 numbers 0, 1, 2, 3, 4, 5, 6, 7, 8, 9, and 10. Hence after *at most* 11 steps, some remainder must occur for a second time (it happened at the third step in the example). At this point (if there are no new digits in the dividend) the division process and hence the quotient begin to repeat.

> **4. The nonterminating and nonrepeating decimals are precisely the irrational numbers.**

WARNING

> A number usually must be in decimal form in order to be entered into a calculator. The calculator can actually handle only the first 8 to 15 digits of a number, depending on its design; consequently,
>
> **All irrational numbers and many rational numbers (such as $\frac{1}{3}$, $\frac{15}{11}$, $-\frac{13}{6}$, etc.) *cannot* be entered exactly in a calculator.**

Rounding and Significant Figures

As a practical matter we must often use decimal approximations for particular numbers. For instance, since the infinite decimal expansion of π (3.1415926. . .) cannot be used for calculations, 3.14 or 3.1416 are often used as approximations. In real-life situations, approximations are needed because of limitations in measurement tools. When the length of a rod is stated as 3.27 meters, it is understood that the actual length is a number that is closer to 3.27 than to 3.28, in other words, a number between 3.265 and 3.275.

The number of decimal places shown indicates the degree of accuracy. Thus 3.789 and 45.000 are presumed accurate to three decimal places and 616.2 and 45.0 to one decimal place. If all the approximations in a calculation are accurate to a specified number of decimal places, then the answer will be accurate to the same number of decimal places. For instance, a calculator shows that $3.02/2.14 \approx 1.41121.$ . . . Since this answer is presumed accurate to two decimal places, it should be rounded to 1.41. We shall use the following:

RULES FOR ROUNDING DECIMALS

> To round a number to n decimal places, look at the portion of the number that begins in the $(n + 1)$st decimal place.
>
> **If this portion is less than 5000 . . . , delete it.**
>
> **If this portion is 5000. . .or more, delete it and add 1 to the digit in the nth decimal place.**

Rounding numbers introduces errors and it is possible for these errors to build up in a long calculation. To minimize rounding error when using a calculator, do any necessary rounding only at the end of the computation. Learn to use the memory and parentheses features to avoid rounding any intermediate results.*

---

* Most calculators can be set to display all numbers to a specified number of decimal places. However, the calculator still uses the maximum possible decimal expansions in its internal computations, rather than the shorter displayed numbers.

Another way of indicating the accuracy of an approximation is to count the number of digits in the number *from left to right,* beginning at the first *nonzero* one; this is the number of **significant figures**. For example,

Number	Number of Significant Figures
645	3
30.4	3
50.00	4
.00076	2
1.00002	6

More significant figures provide greater relative accuracy. For example, both 3.27 (with three significant figures) and 104.65 (with five significant figures) are accurate to two decimal places, so the maximum error in each is at most .005. But as a percentage of the entire number, .005 represents a larger part of 3.27 (and hence a larger error percentage) than it does of 104.65.

The result of a calculation involving approximations is only as accurate as the data being used. So the answer should contain only as many significant figures as there are in the data with the fewest significant figures. For instance, a calculator shows that $30.4/6.3 \approx 4.825396$. . . . Since 30.4 has three significant figures and 6.3 two significant figures, the answer should be given to two significant figures: 4.8.

EXERCISES

In Exercises 1–6, express the given rational number as a repeating decimal.

1. 7/9
2. 2/13
3. 23/14
4. 19/88
5. 1/19 (long)
6. 9/11

In Exercises 7–14, state whether a calculator can express the given number *exactly.*

7. 2/3
8. 7/16
9. 1/64
10. 1/22
11. $3\pi/2$
12. $\pi - 3$
13. 1/.625
14. 1/.16

In Exercises 15–21, express the given repeating decimal as a rational number.

15. .373737 . . .
16. .929292 . . .
17. 76.63424242 . . . [*Hint:* Consider $10{,}000d - 100d$, where $d = 76.63424242$. . . .]
18. 13.513513 . . . [*Hint:* Consider $1000d - d$, where $d = 13.513513$. . . .]
19. .135135135 . . . [*Hint:* See Exercise 18.]
20. .33030303 . . .
21. 52.31272727 . . .
22. If two real numbers have the same decimal expansion through three decimal places, how far apart can they be on the number line?

Unusual Problems

23. Use the methods in Exercises 15–21 to show that both .74999 . . . and .75000 . . . are decimal expansions of ¾. [Every terminating decimal can also be expressed as a decimal ending in repeated 9's. It can be proved that these are the only real numbers with more than one decimal expansion.]
24. *Finding remainders with a calculator:* If you divide 369 by 7, the quotient is 52 and the remainder is 5. Thus the answer is usually written 52⁵⁄₇, or in decimal form 52.71428571. . . . If the answer and the

divisor are known, you can find the remainder by multiplying the divisor (in this case 7) by the fractional (or decimal) part of the answer:

$$7\left(\frac{5}{7}\right) = 5, \text{ the remainder}$$

or

$$7(.71428571) = 4.99999997$$

which rounds off to the remainder 5. Use this method to find the quotient and remainder in these problems:

(a) $5683 \div 9$ (b) $1{,}000{,}000 \div 19$

(c) $53{,}000{,}000 \div 37$

In Exercises 25–30, find the decimal expansion of the given rational number. All these expansions are too long to fit in a calculator, but can be readily found by using the hint in Exercise 25.

25. 6/17 [*Hint:* The first part of dividing 6 by 17 involves working this division problem: $6{,}000{,}000 \div 17$. The method of Exercise 24 shows that the quotient is 352,941 and the remainder is 3. Thus the decimal expansion of $\frac{6}{17}$ begins .352941 and the next block of digits in the expansion will be the quotient in the problem $3{,}000{,}000 \div 17$. The remainder when 3,000,000 is divided by 17 is 10, so the next block of digits in the expansion of $\frac{6}{17}$ is the quotient in the problem $10{,}000{,}000 \div 17$. Continue in this way until the decimal expansion repeats.]

26. 3/19   27. 1/29

28. 3/43   29. 283/47

30. 768/59

31. (a) Show that there are at least as many irrational numbers (nonrepeating decimals) as there are terminating decimals. [*Hint:* With each terminating decimal associate a nonrepeating decimal.]

(b) Show that there are at least as many irrational numbers as there are repeating decimals. [*Hint:* With each repeating decimal associate a nonrepeating decimal by inserting longer and longer strings of zeros; for instance, with .11111111. . . associate the number .101001000100001. . . .]

1.2 INTEGRAL EXPONENTS

Exponents provide a convenient shorthand for certain products. If c is a real number, then c^2 denotes cc and c^3 denotes ccc. More generally, for any positive integer n

$$c^n \text{ denotes the product } ccc \cdots c \text{ (n factors)}.$$

In this notation c^1 is just c, so we usually omit the exponent 1.

Example 1 $3^4 = 3 \cdot 3 \cdot 3 \cdot 3 = 81$ and $(-2)^5 = (-2)(-2)(-2)(-2)(-2) = -32$. $0^n = 0 \cdots 0 = 0$ for every positive integer n.

Example 2 To find $(2.4)^9$, use the $\boxed{y^x}$ (or $\boxed{x^y}$) key on your calculator:*

$\boxed{2.4}$ $\boxed{y^x}$ $\boxed{9}$ $\boxed{=}$ $\boxed{2641.80754}$ †■

Because exponents are just shorthand for multiplication, it is easy to determine the rules they obey. For instance,

---

* All calculator keystroke illustrations in this book are for standard scientific calculators that use algebraic notation. Make the necessary changes if your calculator uses Reverse Polish Notation (RPN).

† Answer is approximate because of round-off.

1 BASIC ALGEBRA

$$c^3 c^5 = (ccc)(ccccc) = c^8, \text{ that is, } c^3 c^5 = c^{3+5}.$$

$$\frac{c^7}{c^4} = \frac{ccccccc}{cccc} = \frac{\cancel{cccc}ccc}{\cancel{cccc}} = ccc = c^3, \text{ that is, } \frac{c^7}{c^4} = c^{7-4}.$$

Similar arguments work in the general case:

To multiply c^m by c^n, add the exponents: $c^m c^n = c^{m+n}$.

To divide c^m by c^n, subtract the exponents: $c^m/c^n = c^{m-n}$.

Example 3 $4^2 \cdot 4^7 = 4^{2+7} = 4^9$ and $2^8/2^3 = 2^{8-3} = 2^5$. ∎

The notation c^n can be extended to the cases when n is zero or negative as follows:

If $c \neq 0$, then c^0 is defined to be the number 1.

If $c \neq 0$ and n is a positive integer, then

$$c^{-n} \text{ is defined to be the number } \frac{1}{c^n}.$$

Note that 0^0 and negative powers of 0 are *not* defined (negative powers of 0 would involve division by 0). The reason for choosing these definitions of c^{-n} for nonzero c is that the multiplication and division rules for exponents remain valid.* For instance,

$$c^5 \cdot c^0 = c^5 \cdot 1 = c^5, \text{ so that } c^5 c^0 = c^{5+0}.$$

$$c^7 c^{-7} = c^7(1/c^7) = 1 = c^0, \text{ so that } c^7 c^{-7} = c^{7-7}.$$

Example 4 $6^{-3} = 1/6^3 = 1/216$ and $(-2)^{-5} = 1/(-2)^5 = -1/32$. A calculator shows that $(.287)^{-12} \approx 3,201,969.857$.† ∎

If c and d are nonzero real numbers and m and n are integers (positive, negative, or zero), then we have these

EXPONENT LAWS

1. $c^m c^n = c^{m+n}$
2. $\dfrac{c^m}{c^n} = c^{m-n}$
3. $(c^m)^n = c^{mn}$
4. $(cd)^n = c^n d^n$
5. $\left(\dfrac{c}{d}\right)^n = \dfrac{c^n}{d^n}$
6. $\dfrac{1}{c^{-n}} = c^n$

---

*In mathematics you may define a concept any way you want, as long as it is consistent with previously established facts.

†≈ means "approximately equal to."

Example 5 Here are examples of each of the six exponent laws.

1. $\pi^{-5}\pi^2 = \pi^{-5+2} = \pi^{-3} = 1/\pi^3$.
2. $x^9/x^4 = x^{9-4} = x^5$.
3. $(5^{-3})^2 = 5^{(-3)2} = 5^{-6}$.
4. $(2x)^5 = 2^5 x^5 = 32x^5$.
5. $\left(\dfrac{7}{3}\right)^{10} = \dfrac{7^{10}}{3^{10}}$.
6. $1/x^{-5} = \dfrac{1}{\left(\dfrac{1}{x^5}\right)} = x^5$. ∎

WARNING

$(2x)^5$ is *not* the same as $2x^5$. Part 4 of the last example shows that $(2x)^5 = 32x^5$ and *not* $2x^5$.

The exponent laws can often be used to simplify complicated expressions. The examples below illustrate one way for simplifying each expression, but there are usually several different ways of arriving at the same answer.

Example 6

(a) $(2x^2 y^3 z)^4 = 2^4 (x^2)^4 (y^3)^4 z^4 = 16 x^8 y^{12} z^4$.
            ↑   ↑
        Law (4)   *Law (3)*

(b) $(r^{-3} s^2)^{-2} = (r^{-3})^{-2} (s^2)^{-2} = r^6 s^{-4} = r^6/s^4$. ∎
       ↑   ↑
   Law (4)   *Law (3)*

Example 7 To evaluate $\dfrac{2^4 \cdot 5^6}{125 \cdot 10^3}$, we first rewrite and then use the exponent laws:

$$\dfrac{2^4 \cdot 5^6}{125 \cdot 10^3} = \dfrac{2^4 \cdot 5^6}{5^3 (5 \cdot 2)^3} = \dfrac{2^4 \cdot 5^6}{5^3 (5^3 \cdot 2^3)} = \dfrac{2^4 \cdot 5^6}{5^6 \cdot 2^3} = 2^{4-3} 5^{6-6} = 2^1 5^0 = 2 \cdot 1 = 2.$$ ∎
                          ↑   ↑   ↑
                   Law (4) · *Law (1)* *Law (2)*

Example 8

$(2x^3 y^4)(5xy^2 z) = 2 \cdot 5 \cdot x^3 x y^4 y^2 z = 10 x^{3+1} y^{4+2} z = 10 x^4 y^6 z$. ∎
                ↑   ↑
      Commutative   *Law (1)*
        Law

Example 9 Simplify $\dfrac{x^5(y^2)^3}{(x^2y)^2}$.

Solution
$$\dfrac{x^5(y^2)^3}{(x^2y)^2} \underset{Law\,(3)}{=} \dfrac{x^5y^6}{(x^2y)^2} \underset{Law\,(4)}{=} \dfrac{x^5y^6}{(x^2)^2y^2} \underset{Law\,(3)}{=} \dfrac{x^5y^6}{x^4y^2} \underset{Law\,(2)}{=} x^{5-4}y^{6-2} = xy^4. \blacksquare$$

It is usually more efficient to use the exponent laws with the negative exponents rather than first converting to positive exponents. If positive exponents are required, the conversion can be made in the last step.

Example 10 Simplify and express without negative exponents $\dfrac{a^{-2}(b^2c^3)^{-2}}{(a^{-3}b^{-5})^2c}$.

Solution
$$\dfrac{a^{-2}(b^2c^3)^{-2}}{(a^{-3}b^{-5})^2c} \underset{Law\,(4)}{=} \dfrac{a^{-2}(b^2)^{-2}(c^3)^{-2}}{(a^{-3})^2(b^{-5})^2c} \underset{Law\,(3)}{=} \dfrac{a^{-2}b^{-4}c^{-6}}{a^{-6}b^{-10}c}$$

$$\underset{Law\,(2)}{=} a^{-2-(-6)}b^{-4-(-10)}c^{-6-1} = a^4b^6c^{-7} = \dfrac{a^4b^6}{c^7}. \blacksquare$$

Since $(-1)(-1) = +1$, any even power of -1, such as $(-1)^4$ or $(-1)^{12}$, will be equal to 1. Every odd power of -1 is equal to -1; for instance $(-1)^5 = (-1)^4(-1) = 1(-1) = -1$. Consequently, for every number c

$$(-c)^n = ((-1)c)^n = (-1)^n c^n = \begin{cases} c^n & \text{if } n \text{ is even} \\ -c^n & \text{if } n \text{ is odd} \end{cases}$$

Example 11 $(-3)^4 = 3^4 = 81$ and $(-5)^3 = -5^3 = -125.$ $\blacksquare$

WARNING

Some calculators will not compute negative numbers raised to a power. To compute $(-12)^7$ with such a calculator, note that $(-12)^7 = -12^7$ (odd exponent), and use the calculator to compute 12^7.

Scientific Notation

To multiply a decimal number by 10, move the decimal point one place to the right. For example, $7.83 \times 10 = 78.3$. To divide by 10 move the decimal point one place to the *left*. For instance, $89.56 \div 10 = 8.956$. Multiplying by $100 = 10^2$ means multiplying by 10 twice, so you need only move the decimal point two places to the right. Similarly, for positive n,

To multiply by 10^n, move the decimal point n places to the right.

To divide by 10^n, move the decimal point n places to the left.

For example, $56.78 \times 1000 = 56.78 \times 10^3 = 56780.0$ and $945.63 \div 10000 = 945.63 \div 10^4 = .094563$.

Using these facts we can write any real number as the product of a power of 10 and a number between 1 and 10. For example,

$$356 = 3.56 \times 100 = 3.56 \times 10^2$$

$$1{,}563{,}427 = 1.563427 \times 1{,}000{,}000 = 1.563427 \times 10^6$$

$$.072 = 7.2 \div 100 = 7.2 \times 1/100 = 7.2 \times 10^{-2}$$

$$.000862 = 8.62 \div 10{,}000 = 8.62 \times 10^{-4}$$

A number written in this form is said to be in **scientific notation.*** Scientific notation is very useful for computations with very large or very small numbers.

Example 12

$$(.00000002)(4{,}300{,}000{,}000) = (2 \times 10^{-8})(4.3 \times 10^9)$$
$$= 2(4.3)10^{-8+9} = (8.6)10^1 = 86. \blacksquare$$

Example 13

$$\frac{(50{,}000{,}000)^3(.000002)^5}{(.000008)} = \frac{(5 \times 10^7)^3(2 \times 10^{-6})^5}{8 \times 10^{-6}} = \frac{(5^3 \cdot 10^{21})(2^5 \cdot 10^{-30})}{8 \cdot 10^{-6}}$$

$$= \frac{125 \cdot 32 \cdot 10^{21-30}}{8 \cdot 10^{-6}} = 125 \cdot 4 \cdot 10^{21-30+6} = 500 \cdot 10^{-3} = \frac{500}{1000} = \frac{1}{2}. \blacksquare$$

Most scientific calculators use scientific notation. Usually they have a key labeled $\boxed{\text{EE}}$ or $\boxed{\text{EXP}}$ or $\boxed{\text{EEX}}$ which allows you to enter any exponent from -99 to 99. If you enter 7.235, then press $\boxed{\text{EXP}}$ and enter -12, the calculator display will read

$$\boxed{7.235 \qquad -12}$$

---

* The rule for determining the number of significant figures in a number written in scientific notation is different than the one given on page 12. The number of significant figures in $c \times 10^k$ is the number of digits in c. For example, 2.34×10^7 has three significant figures.

This indicates the number 7.235×10^{-12}. Many calculators automatically switch to scientific notation when a number is too large or small to be displayed in the normal way.

WARNING

Your calculator may not always obey the laws of arithmetic when dealing with very large or small numbers. For instance, the associative law shows that

$$(1 + 10^{19}) - 10^{19} = 1 + (10^{19} - 10^{19}) = 1 + 0 = 1.$$

But the calculator may round off $1 + 10^{19}$ as 10^{19} (instead of the correct number 10,000,000,000,000,000,001). So the calculator computes $(1 + 10^{19}) - 10^{19}$ as $10^{19} - 10^{19} = 0$.

EXERCISES

In Exercises 1–18, evaluate the expression.

1. $(-6)^2$
2. -6^2
3. $5 + 4(3^2 + 2^3)$
4. $(-3)2^2 + 4^2 - 1$
5. $\dfrac{(-3)^2 + (-2)^4}{-2^2 - 1}$
6. $\dfrac{(-4)^2 + 2}{(-4)^2 - 7} + 1$
7. $\left(\dfrac{-5}{4}\right)^3$
8. $-\left(\dfrac{7}{4} + \dfrac{3}{4}\right)^2$
9. $\left(\dfrac{1}{3}\right)^3 + \left(\dfrac{2}{3}\right)^3$
10. $\left(\dfrac{5}{7}\right)^2 + \left(\dfrac{2}{7}\right)^2$
11. $2^4 - 2^7$
12. $3^3 - 3^{-7}$
13. $(2^{-2} + 2)^2$
14. $(3^{-1} - 3^3)^2$
15. $2^2 \cdot 3^{-3} - 3^2 \cdot 2^{-3}$
16. $4^3 \cdot 5^{-2} + 4^2 \cdot 5^{-1}$
17. $\dfrac{1}{2^3} + \dfrac{1}{2^{-4}}$
18. $3^2 \left(\dfrac{1}{3} + \dfrac{1}{3^{-2}}\right)$

In Exercises 19–38, simplify the expression. Each letter represents a nonzero real number and should appear at most once in your answer.

19. $x^2 \cdot x^3 \cdot x^5$
20. $y \cdot y^4 \cdot y^6$
21. $(.03)y^2 \cdot y^7$
22. $(1.3)z^3 \cdot z^5$
23. $(2x^2)^3 3x$
24. $(3y^3)^4 5y^2$
25. $(3x^2 y)^2$
26. $(2xy^3)^3$
27. $(a^2)(7a)(-3a^3)$
28. $(b^3)(-b^2)(3b)$
29. $(2w)^3(3w)(4w)^2$
30. $(3d)^4(2d)^2(5d)$
31. $a^{-2} b^3 a^3$
32. $c^4 d^5 c^{-3}$
33. $(2x)^{-2}(2y)^3(4x)$
34. $(3x)^{-3}(2y)^{-2}(2x)$
35. $(-3a^4)^2(9x^3)^{-1}$
36. $(2y^3)^3(3y^2)^{-2}$
37. $(2x^2 y)^0(3xy)$
38. $(3x^2 y^4)^0$

In Exercises 39–44, express the number in scientific notation.

39. 79,327
40. 5,200,000
41. .002
42. .00000079
43. 5,963,000,000,000
44. .00000000000035

In Exercises 45–50, express the given number in decimal notation.

45. 7.4×10^5
46. 6.53×10^7
47. 3.8×10^{-12}
48. 6.02×10^{-8}
49. 3.457×10^{10}
50. 13.23×10^{13}

In Exercises 51–54, express the given number as a power of 2.

51. $(64)^2$
52. $(\tfrac{1}{8})^3$
53. $(2^4 \cdot 16^{-2})^3$
54. $(\tfrac{1}{2})^{-8}(\tfrac{1}{4})^4(\tfrac{1}{16})^{-3}$

1.2 INTEGRAL EXPONENTS

In Exercises 55–72, simplify and write the given expression without negative exponents. All letters represent nonzero real numbers.

55. $\dfrac{x^4(x^2)^3}{x^3}$

56. $\left(\dfrac{z^2}{t^3}\right)^4 \cdot \left(\dfrac{z^3}{t}\right)^5$

57. $\left(\dfrac{e^6}{c^4}\right)^2 \cdot \left(\dfrac{c^3}{e}\right)^3$

58. $\left(\dfrac{x^7}{y^6}\right)^2 \cdot \left(\dfrac{y^2}{x}\right)^4$

59. $\left(\dfrac{ab^2c^3d^4}{abc^2d}\right)^2$

60. $\dfrac{(3x)^2(y^2)^3 x^2}{(2xy^2)^3}$

61. $\left(\dfrac{a^6}{b^{-4}}\right)^2$

62. $\left(\dfrac{x^{-2}}{y^{-2}}\right)^2$

63. $\left(\dfrac{c^5}{d^{-3}}\right)^{-2}$

64. $\left(\dfrac{x^{-1}}{2y^{-1}}\right)\left(\dfrac{2y}{x}\right)^{-2}$

65. $\left(\dfrac{3x}{y^2}\right)^{-3}\left(\dfrac{-x}{2y^3}\right)^2$

66. $\left(\dfrac{5u^2 v}{2uv^2}\right)^2 \left(\dfrac{-3uv}{2u^2 v}\right)^{-3}$

67. $\dfrac{(a^{-3}b^2 c)^{-2}}{(ab^{-2}c^3)^{-1}}$

68. $\dfrac{(-2cd^2 e^{-1})^3}{(5c^{-3}de)^{-2}}$

69. $(c^{-1}d^{-2})^{-3}$

70. $((x^2 y^{-1})^2)^{-3}$

71. $a^2(a^{-1} + a^{-3})$

72. $\dfrac{a^{-2}}{b^{-2}} + \dfrac{b^2}{a^2}$

In Exercises 73–78, determine the sign of the given number without calculating the product.

73. $(-2.6)^3(-4.3)^{-2}$

74. $(4.1)^{-2}(2.5)^{-3}$

75. $(-1)^9(6.7)^5$

76. $(-4)^{12} 6^9$

77. $(-3.1)^{-3}(4.6)^{-6}(7.2)^7$

78. $(45.8)^{-7}(-7.9)^{-9}(-8.5)^{-4}$

In Exercises 79–84, r, s, and t are positive integers and a, b, and c are nonzero real numbers. Simplify and write the given expression without negative exponents.

79. $\dfrac{3^{-r}}{3^{-s-r}}$

80. $\dfrac{4^{-(t+1)}}{4^{2-t}}$

81. $\left(\dfrac{a^b}{b^{-4}}\right)^t$

82. $\dfrac{c^{-t}}{(6b)^{-s}}$

83. $\dfrac{(c^{-r}b^s)^t}{(c^t b^{-s})^r}$

84. $\dfrac{(a^r b^{-s})^{-t}}{(b^t c^r)^{-s}}$

In Exercises 85–90, perform the operations and express the result in scientific notation.

85. $(34{,}000{,}000{,}000{,}000)(.000000004)$

86. $(.00000000023)(3{,}000{,}000{,}000{,}000)$

87. $\dfrac{.00032}{160{,}000{,}000}$

88. $\dfrac{(.0000024)(.00000003)}{(18{,}000{,}000)(4{,}000{,}000)}$

89. $\dfrac{(7{,}000{,}000{,}000)^2}{(14{,}000{,}000{,}000)^{-3}}$

90. $\dfrac{(.0000000008)^{-3}}{(4{,}000)(.00000016)^{-2}}$

91. If P dollars are deposited in a savings account and interest is compounded annually at 7%, then the amount in the account after n years is $P(1.07)^n$. Determine this amount when
    (a) $P = \$500$ and $n = 10$
    (b) $P = \$1000$ and $n = 18$
    (c) By trying different values, estimate how many years it would take for $2400 to double.

In Exercises 92–96, change all numbers to scientific notation and then compute. Express your answer in scientific notation.

92. Assume the earth is a sphere with radius of 6400 kilometers. If the surface area S of a sphere of radius r is given by $S = 4\pi r^2$, what is the surface area of the earth in square kilometers?

93. If the circumference of the earth at the equator is 25,000 miles, how many inches is it around the equator?

94. If the speed of light is 186,000 miles per second, how many miles is one light year (the distance light travels in one year)?

95. If the sun is 93 million miles from the earth, how long does it take light to travel from the sun to earth? [See Exercise 94.]

96. If the national debt is $1,400,000,000,000 and there are 224 million U.S. citizens, how much money per citizen does the government owe?

Errors to Avoid

In Exercises 97–104, give an example to show that the statement may be *false* for some numbers.

97. $a^r + b^r = (a+b)^r$
98. $a^r a^s = a^{rs}$
99. $a^r b^s = (ab)^{r+s}$
100. $c^{-r} = -c^r$
101. $\dfrac{c^r}{c^s} = c^{r/s}$
102. $(a+1)(b+1) = ab + 1$
103. $(-a)^2 = -a^2$
104. $(-a)(-b) = -ab$

1.3 RADICALS AND RATIONAL EXPONENTS

A nonnegative number is one that is either positive or zero. The **square root** of a nonnegative number d is defined to be the *nonnegative* number whose square is d; it is denoted $\sqrt{d}$.* For instance,

$$\sqrt{25} = 5 \text{ because } 5^2 = 25 \quad \text{and} \quad \sqrt{1.21} = 1.1 \text{ because } 1.1^2 = 1.21.$$

In high school you may have said $\sqrt{25} = \pm 5$ since $(-5)^2$ is also 25. But in advanced mathematics, square roots are always *nonnegative*. If we want to express -5 in terms of square roots we write $-5 = -\sqrt{25}$.

> **WARNING**
>
> The square root of an integer is often an *irrational* number. In that case, a calculator provides only an *approximation*, such as $\sqrt{2} \approx 1.414213562$. If you actually square this number by hand you will find that the answer is very close to, but not equal to 2.

The **cube root** of a nonnegative number d (denoted $\sqrt[3]{d}$) is the nonnegative number whose cube is d. For example, $\sqrt[3]{8} = 2$ since $2^3 = 8$. For each positive integer n,

> The **nth root** of a nonnegative number d (denoted $\sqrt[n]{d}$) is the nonnegative number whose nth power is d, that is, $\left(\sqrt[n]{d}\right)^n = d$.

It can be proved that every nonnegative real number has a unique nth root. We shall continue to denote the square root of d by $\sqrt{d}$ rather than $\sqrt[2]{d}$.

Example 1 $\sqrt[4]{81} = 3$ because $3^4 = 81$ and $\sqrt[5]{.00032} = .2$ because $(.2)^5 = .00032$. Using the $\boxed{\sqrt[x]{}}$ key on a calculator, we see that $\sqrt[11]{225} \approx 1.63619392$; that is, $(1.63619392)^{11} \approx 225$. ∎

If c and d are nonnegative real numbers and m, n positive integers, then roots have these important properties:

---

* The symbol $\sqrt{}$ is called a **radical**.

PROPERTIES OF RADICALS

1. $(\sqrt[n]{d})^n = d$
2. $\sqrt[n]{d^n} = d$
3. $\sqrt[m]{\sqrt[n]{d}} = \sqrt[mn]{d}$
4. $\sqrt[n]{cd} = (\sqrt[n]{c})(\sqrt[n]{d})$
5. $\sqrt[n]{\dfrac{c}{d}} = \dfrac{\sqrt[n]{c}}{\sqrt[n]{d}}$ $(d \neq 0)$

Example 2 These properties may be used to simplify
(a) $\sqrt{2025}$ (b) $\sqrt[3]{\sqrt[4]{4096}}$ (c) $(3 + 5\sqrt{2})(7 - \sqrt{2})$
as follows.

(a) By property 4, $\sqrt{2025} = \sqrt{25 \cdot 81} = \sqrt{25}\sqrt{81} = 5 \cdot 9 = 45$.

(b) By properties 3 and 2, $\sqrt[3]{\sqrt[4]{4096}} = \sqrt[3 \cdot 4]{4096} = \sqrt[12]{4096} = \sqrt[12]{2^{12}} = 2$.

(c) $(3 + 5\sqrt{2})(7 - \sqrt{2}) = 3 \cdot 7 - 3\sqrt{2} + 5\sqrt{2} \cdot 7 - 5\sqrt{2}\sqrt{2}$
$= 21 - 3\sqrt{2} + 35\sqrt{2} - 5 \cdot 2 = 11 + 32\sqrt{2}$. ∎

Example 3 Assume x, y, z are positive and simplify
(a) $\sqrt[3]{54x^3y^6z}$ (b) $\dfrac{\sqrt{6xy^2}}{\sqrt{3x^5}}$
as follows.

(a) By properties 4 and 2,
$\sqrt[3]{54x^3y^6z} = \sqrt[3]{27 \cdot 2x^3y^6z} = \sqrt[3]{27}\sqrt[3]{2}\sqrt[3]{x^3}\sqrt[3]{y^6}\sqrt[3]{z} = 3\sqrt[3]{2}\,x\,\sqrt[3]{(y^2)^3}\,\sqrt[3]{z}$
$= 3\sqrt[3]{2}xy^2(\sqrt[3]{z}) = 3xy^2(\sqrt[3]{2z})$.

(b) $\dfrac{\sqrt{6xy^2}}{\sqrt{3x^5}} = \sqrt{\dfrac{6xy^2}{3x^5}} = \sqrt{\dfrac{2y^2}{x^4}} = \dfrac{\sqrt{2y^2}}{\sqrt{x^4}} = \dfrac{\sqrt{2}\sqrt{y^2}}{\sqrt{(x^2)^2}} = \dfrac{\sqrt{2}\,y}{x^2}$. ∎

Property: (5)   (5)   (4)   (2)

WARNING

In most cases $\sqrt[n]{c + d}$ is *not* equal to $\sqrt[n]{c} + \sqrt[n]{d}$. For instance, $\sqrt{9 + 16} = \sqrt{25} = 5$, but $\sqrt{9} + \sqrt{16} = 3 + 4 = 7$. So $\sqrt{9 + 16} \neq \sqrt{9} + \sqrt{16}$.

The nth root of a negative number can be defined when n is *odd* ($n = 1, 3, 5, 7,$ etc.). For example, we say that
$$\sqrt[3]{-125} = -5 \quad \text{since} \quad (-5)^3 = -125.$$
When n is *even* ($n = 2, 4, 6,$ etc.), then c^n is nonnegative, so c cannot possibly be

the nth root of a negative number. Therefore expressions such as $\sqrt{-4}$ or $\sqrt[6]{-17}$ are meaningless in the real number system. Hereafter, unless stated otherwise, *all letters in this section represent nonnegative numbers.*

Rational Exponents

Roots and radicals will now be used to define c^r when r is a rational number. These definitions are chosen so that the exponent laws will continue to hold for rational exponents. For instance, we would like $(c^{1/2})^2$ to be the number $c^{(1/2)2} = c^1 = c$. Since $(\sqrt{c})^2 = c$, it is reasonable to *define* $c^{1/2}$ to be $\sqrt{c}$. In general, if c is a nonnegative real number and k a positive integer

$$c^{1/k} \text{ denotes the number } \sqrt[k]{c}.$$

For example, $27^{1/3} = \sqrt[3]{27} = 3$ and $32^{1/5} = \sqrt[5]{32} = 2$.

We want $(c^m)^n = c^{mn}$ to hold when m and n are rational. In particular, since $\dfrac{t}{k} = \left(\dfrac{1}{k}\right)t$ and $\dfrac{t}{k} = t\left(\dfrac{1}{k}\right)$, we want to define $c^{t/k}$ in such a way that

$$c^{t/k} = (c^{1/k})^t = (\sqrt[k]{c})^t \quad \text{and} \quad c^{t/k} = (c^t)^{1/k} = \sqrt[k]{c^t}.$$

But when c, k are positive, we have

$$(\sqrt[k]{c})^t = (\sqrt[k]{c})(\sqrt[k]{c}) \cdots (\sqrt[k]{c}) = \sqrt[k]{c \cdot c \cdots c} = \sqrt[k]{c^t}.$$

A similar argument works when t is negative so that in all cases

$$(\sqrt[k]{c})^t = \sqrt[k]{c^t}.$$

Consequently, for every positive real number c and integers t, k with $k > 0$ we make this definition:

$$c^{t/k} \text{ denotes the number } (\sqrt[k]{c})^t = \sqrt[k]{c^t}.$$

Example 4 $4^{3/2} = (\sqrt{4})^3 = 2^3 = 8$ and $4^{3/2} = \sqrt{4^3} = \sqrt{64} = 8$. ∎

Example 5 To compute $27^{2/3}$, it's easier to use the first form of the definition $(27^{2/3} = (\sqrt[3]{27})^2 = 3^2 = 9)$ rather than the second $(27^{2/3} = \sqrt[3]{27^2} = \sqrt[3]{729} = 9)$. ∎

Example 6 $5^{-2/3} = \sqrt[3]{5^{-2}} = \sqrt[3]{\dfrac{1}{5^2}} = \dfrac{\sqrt[3]{1}}{\sqrt[3]{5^2}} = \dfrac{1}{\sqrt[3]{25}}$. ∎

Example 7 Every terminating decimal is a rational number, so expressions such as $5^{3.78}$ now have meaning: $5^{3.78} = 5^{378/100} = \sqrt[100]{5^{378}}$. The easiest way to evaluate such expressions is to use the $\boxed{y^x}$ key on your calculator:

$\boxed{5}\,\boxed{y^x}\,\boxed{3.78}\,\boxed{=}\,\boxed{438.6384}$ (approximately). ∎

Examples 8–12 illustrate the following fact:

All of the exponent laws listed in the box on page 14 are valid for rational exponents.

Example 8

$$(8r^{3/4}s^{-3})^{2/3} = 8^{2/3}(r^{3/4})^{2/3}(s^{-3})^{2/3} = \sqrt[3]{8^2}(r^{2/4})(s^{-2})$$

$$= \sqrt[3]{64}\, r^{1/2}s^{-2} = \frac{4r^{1/2}}{s^2}.$$

We can leave the answer in exponential form, or if it is more convenient, write it as $4\sqrt{r}/s^2$. ∎

Example 9

$$x^{1/2}(x^{3/4} - x^{3/2}) = x^{1/2}x^{3/4} - x^{1/2}x^{3/2}$$

$$= x^{1/2+3/4} - x^{1/2+3/2} = x^{5/4} - x^2. \blacksquare$$

Example 10

$$(x^{5/2}y^4)(xy^{7/4})^{-2} = (x^{5/2}y^4)x^{-2}(y^{7/4})^{-2}$$

$$= x^{5/2}y^4x^{-2}y^{(7/4)(-2)} = x^{5/2}x^{-2}y^4y^{-7/2}$$

$$= x^{(5/2)-2}y^{4-(7/2)} = x^{1/2}y^{1/2}. \blacksquare$$

Example 11 Let k be a positive rational number and express $\sqrt[10]{c^{5k}}\sqrt{(c^{-k})^{1/2}}$ without radicals, using only positive exponents.

Solution

$$\sqrt[10]{c^{5k}}\sqrt{(c^{-k})^{1/2}} = (c^{5k})^{1/10}[(c^{-k})^{1/2}]^{1/2} = c^{k/2}c^{-k/4} = c^{2k/4}c^{-k/4} = c^{k/4}. \blacksquare$$

Example 12 Express $\sqrt[5]{\dfrac{\sqrt[4]{7^3 \cdot x^5}}{\sqrt[3]{7^2 y^4}}}$ without radicals.

Solution

$$\sqrt[5]{\frac{\sqrt[4]{7^3 \cdot x^5}}{\sqrt[3]{7^2 \cdot y^4}}} = \left(\frac{(7^3 \cdot x^5)^{1/4}}{(7^2 \cdot y^4)^{1/3}}\right)^{1/5} = \frac{[(7^3 \cdot x^5)^{1/4}]^{1/5}}{[(7^2 y^4)^{1/3}]^{1/5}}$$

$$= \frac{(7^3 \cdot x^5)^{1/20}}{(7^2 \cdot y^4)^{1/15}} = \frac{7^{3/20}x^{5/20}}{7^{2/15}y^{4/15}}$$

$$= \frac{7^{3/20-2/15}x^{1/4}}{y^{4/15}} = \frac{7^{1/60}x^{1/4}}{y^{4/15}}. \blacksquare$$

Fractional exponents can be defined when the base is negative and the fraction is in *lowest terms with an odd denominator*. For example,

$$(-243)^{1/5} = \sqrt[5]{-243} = -3 \quad \text{and} \quad (-8)^{2/3} = \sqrt[3]{(-8)^2} = \sqrt[3]{64} = 4$$

However, an expression such as $(-4)^{1/2}$ is meaningless in the real number system since $\sqrt{-4}$ is not defined.

EXERCISES

Note: Unless directed otherwise, assume all letters represent positive real numbers.

In Exercises 1–20, simplify the expression without using a calculator. For example, $\sqrt{8}\sqrt{12} = \sqrt{8 \cdot 12} = \sqrt{96} = \sqrt{16 \cdot 6} = \sqrt{16}\sqrt{6} = 4\sqrt{6}$.

1. $\sqrt{.0081}$
2. $\sqrt{.000169}$
3. $\sqrt{(.08)^{12}}$
4. $\sqrt{(-11)^{28}}$
5. $\sqrt{6}\sqrt{12}$
6. $\sqrt{8}\sqrt{96}$
7. $\dfrac{\sqrt{10}}{\sqrt{8}\sqrt{5}}$
8. $\dfrac{\sqrt{6}}{\sqrt{14}\sqrt{63}}$
9. $(1+\sqrt{3})(2-\sqrt{3})$
10. $(3+\sqrt{2})(3-\sqrt{2})$
11. $(4-\sqrt{3})(5+2\sqrt{3})$
12. $(2\sqrt{5}-4)(3\sqrt{5}+2)$
13. $\sqrt[4]{10^8}$
14. $\sqrt[6]{7^{18}}$
15. $\sqrt[4]{9^2}$
16. $\sqrt[3]{54 \cdot 2^2}$
17. $(.001)^{5/3}$
18. $2^{3/2}$
19. $(4/49)^{-3/2}$
20. $16^{-5/4}$

In Exercises 21–24, use a calculator to find a decimal approximation of the given number (rounded to the nearest thousandth).

21. $6^{3.23}$
22. $8^{.0789}$
23. $12^{27/19}$
24. $26^{128/79}$

In Exercises 25–28, use a calculator and insert $<$ or $>$ or $=$ in the blank to make a true statement.

25. $18^{7/9}$ __ $18^{.77}$
26. $24^{65/32}$ __ $24^{2.032}$
27. $156^{.25}$ __ 156
28. $39^{-.5}$ __ $1/\sqrt{39}$

In Exercises 29–48, simplify the given expression. Your answer should contain no more than one radical.

29. $\sqrt{a^2 b^4}$
30. $\sqrt{25a^4 b^6}$
31. $\sqrt{(2x^2)(2y^8)}$
32. $\sqrt{(3x^4)(12y^4)}$
33. $\sqrt{18x^2 y^3}$
34. $\sqrt{28u^3 v^6}$
35. $\sqrt{3rs^3}\sqrt{3r^3 s^3}$
36. $\sqrt{2r^2 s}\sqrt{8r^4 s^3}$
37. $\sqrt[3]{(4x+2y)^3}$
38. $\sqrt[4]{(3x+7)^4}$
39. $\sqrt{c^3}\sqrt{c^2}$
40. $\sqrt{c}\sqrt{c^{10}}$
41. $\sqrt[3]{27a^6 b^3}$
42. $\sqrt[4]{81x^8 y^{16}}$
43. $\sqrt[5]{32c^{10}d^{15}}$
44. $\sqrt{\sqrt[3]{a^6 b^{12}}}$
45. $\dfrac{\sqrt{6x}}{\sqrt{24x^5}}$
46. $\dfrac{\sqrt[3]{64x^5 y^{14}}}{\sqrt[3]{8x^{-1}y^2}}$
47. $\dfrac{\sqrt[3]{27c^4 d^{-3}}}{\sqrt[3]{8c^{-7}d^9}}$
48. $\dfrac{(\sqrt{xy})^3 \sqrt{y}}{(\sqrt{x^2 y})^2 \sqrt{x}}$

In Exercises 49–54, write the given expression without using radicals.

49. $\sqrt[3]{a^2+b^2}$
50. $\sqrt[4]{a^3-b^3}$
51. $\sqrt[4]{\sqrt[4]{a^3}}$
52. $\sqrt{\sqrt[3]{a^3 b^4}}$
53. $\sqrt[5]{t}\sqrt{16t^5}$
54. $\sqrt{x}(\sqrt[3]{x^2})(\sqrt[4]{x^3})$

In Exercises 55–74, simplify the given expression.

55. $\sqrt{16a^8 b^{-2}}$
56. $\sqrt{24x^6 y^{-4}}$
57. $\dfrac{\sqrt{c^2 d^6}}{\sqrt{4c^3 d^{-4}}}$
58. $\dfrac{\sqrt{a^{-10}b^{-12}}}{\sqrt{a^{14}d^{-4}}}$
59. $5\sqrt{20} - \sqrt{45} + 2\sqrt{80}$
60. $\sqrt[3]{40} + 2(\sqrt[3]{135}) - 5(\sqrt[3]{320})$
61. $\sqrt[4]{(4x+2y)^8}$
62. $(\sqrt[3]{a+b})(\sqrt[3]{-(a+b)^2}) + \sqrt[3]{a+b}$
63. $\dfrac{2^{11/2} \cdot 2^{-7} \cdot 2^{-5}}{2^3 \cdot 2^{1/2} \cdot 2^{-10}}$
64. $\dfrac{(3^2)^{-1/2}(9^4)^{-1}}{27^{-3}}$
65. $\sqrt{x^7} \cdot x^{5/2} \cdot x^{-3/2}$
66. $(x^{1/2}y^3)(x^0 y^7)^{-2}$
67. $(c^{2/5}d^{-2/3})(c^6 d^3)^{4/3}$
68. $\left(\dfrac{r^{2/3}}{s^{1/5}}\right)^{15/9}$

69. $\dfrac{(7a)^2(5b)^{3/2}}{(5a)^{3/2}(7b)^4}$

70. $\dfrac{(6a)^{1/2}\sqrt{ab}}{a^2 b^{3/2}}$

71. $\dfrac{(2a)^{1/2}(3b)^{-2}(4a)^{3/5}}{(4a)^{-3/2}(3b)^2(2a)^{1/5}}$

72. $\dfrac{(a^{3/4}b)^2(ab^{1/4})^3}{(ab)^{1/2}(bc)^{-1/4}}$

73. $(a^{x^2})^{1/x}$

74. $\dfrac{(b^x)^{x-1}}{b^{-x}}$

In Exercises 75–78, write the given expression without radicals, using only positive exponents.

75. $(\sqrt[3]{xy^2})^{-3/5}$

76. $(\sqrt[4]{r^{14}s^{-21/5}})^{-3/7}$

77. $\dfrac{c}{(c^{5/6})^{42}(c^{51})^{-2/3}}$

78. $(c^{5/6} - c^{-5/6})^2$

In Exercises 79–84, compute and simplify.

79. $x^{1/2}(x^{2/3} - x^{4/3})$

80. $x^{1/2}(3x^{3/2} + 2x^{-1/2})$

81. $(x^{1/2} + y^{1/2})(x^{1/2} - y^{1/2})$

82. $(x^{1/3} + y^{1/2})(2x^{1/3} - y^{3/2})$

83. $(x + y)^{1/2}[(x + y)^{1/2} - (x + y)]$

84. $(x^{1/3} + y^{1/3})(x^{2/3} - x^{1/3}y^{1/3} + y^{2/3})$

In Exercises 85–90, factor the given expression. For example,

$$x - x^{1/2} - 2 = (x^{1/2} - 2)(x^{1/2} + 1).$$

85. $x^{2/3} + x^{1/3} - 6$

86. $x^{2/5} + 11x^{1/5} + 30$

87. $x + 4x^{1/2} + 3$

88. $x^{1/3} + 7x^{1/6} + 10$

89. $x^{4/5} - 81$

90. $x^{2/3} - 6x^{1/3} + 9$

91. Use a calculator to determine which is larger: 3^π or π^3.

92. Use a calculator to find a *six*-place decimal approximation of $(311)^{-4.2}$. Explain why your answer cannot possibly be the number $(311)^{-4.2}$.

93. Each of the following formulas can be used to approximate square roots:

(i) $\sqrt{a^2 + b} \approx a + \dfrac{b}{2a + 1}$

(ii) $\sqrt{a^2 + b} \approx a + \dfrac{b}{2a}$

(iii) $\sqrt{a^2 + b} \approx a + \dfrac{b}{2a} - \dfrac{b^2}{8a^3}$

(a) Use each of the formulas to find an approximation of $\sqrt{10}$. [*Hint:* $10 = 9 + 1$.] Use a calculator to see which approximation is best.
(b) Do the same for $\sqrt{19}$.

94. If c is *any* real number, then $c^2 \geq 0$, so $\sqrt{c^2}$ is defined. Think carefully and answer this: $\sqrt{c^2} = ?$

Errors to Avoid

In Exercises 95–98, give a numerical example to show that the given statement is *false* for some numbers.

95. $\sqrt[3]{a + b} = \sqrt[3]{a} + \sqrt[3]{b}$

96. $\sqrt{c^2 + d^2} = c + d$

97. $\sqrt{8a} = 4\sqrt{a}$

98. $\sqrt{a}(\sqrt[3]{a}) = \sqrt[4]{a}$

1.4 ABSOLUTE VALUE

On an informal level most students think of absolute value like this:

The absolute value of a nonnegative number is the number itself.

The absolute value of a negative number is found by "erasing the minus sign."

If $|c|$ denotes the absolute value of c, then, for example, $|5| = 5$ and $|-4| = 4$. Note that $4 = -(-4)$. Thus the absolute value of the negative number -4 is *the negative* of -4, namely, $-(-4) = 4$. This fact is the basis of the formal definition:

ABSOLUTE VALUE

The *absolute value* of a real number c is denoted $|c|$ and is defined as follows:

If $c \geq 0$, then $|c| = c$

If $c < 0$, then $|c| = -c$

Example 1

(a) $|3.5| = 3.5$ and $|-7/2| = -(-7/2) = 7/2$.

(b) To find $|\pi - 6|$ note that $\pi \approx 3.14$, so that $\pi - 6 < 0$. Hence $|\pi - 6|$ is defined to be the *negative* of $\pi - 6$, that is,

$$|\pi - 6| = -(\pi - 6) = -\pi + 6.$$

(c) $|5 - \sqrt{2}| = 5 - \sqrt{2}$ because $5 - \sqrt{2} \geq 0$.

(d) To find $|d^2 + 1|$, where d is any real number, you must determine whether $d^2 + 1$ is negative or nonnegative. But $d^2 \geq 0$ for every d, so that $d^2 + 1 > 0$. Hence $|d^2 + 1| = d^2 + 1$. ∎

Here are the important facts about absolute value:

PROPERTIES OF ABSOLUTE VALUE

1. $|c| \geq 0$ and $|c| > 0$ when $c \neq 0$.
2. $c \leq |c|$ and $-c \leq |c|$.
3. $|c| = |-c|$.
4. $|c^2| = c^2 = |c|^2$.
5. $\sqrt{c^2} = |c|$.
6. $|cd| = |c||d|$ and if $d \neq 0$, $\left|\dfrac{c}{d}\right| = \dfrac{|c|}{|d|}$.

Example 2 Here are examples of some of the properties listed in the box.

2. $|6| = 6$ and hence $6 \leq |6|$ and $-6 \leq |6|$. Similarly, $|-4| = 4$ and $-(-4) = 4$, so that $-4 \leq |-4|$ and $-(-4) \leq |-4|$.

3. $|3| = 3$ and $|-3| = 3$, so that $|3| = |-3|$.

4. If $c = -5$, then

$$|c^2| = |(-5)^2| = |25| = 25,$$
$$c^2 = (-5)^2 = 25, \quad \text{and} \quad |c|^2 = |-5|^2 = 5^2 = 25$$

so that $|(-5)^2| = (-5)^2 = |-5|^2$.

5. When c is negative, it is *not* true that $\sqrt{c^2} = c$, because square roots are always *nonnegative*. For instance, $\sqrt{(-3)^2} = \sqrt{9} = 3$. Hence $\sqrt{(-3)^2} = |-3|$, *not* -3.

6. If $c = 6$ and $d = -2$, then

$$|cd| = |6(-2)| = |-12| = 12 \quad \text{and} \quad |c||d| = |6||-2| = 6 \cdot 2 = 12$$

so that $|6(-2)| = |6||-2|$. ∎

When dealing with long expressions inside absolute value bars; do the computations inside first, and then take the absolute value.

Example 3

(a) $|5(2-4) + 7| = |5(-2) + 7| = |-10 + 7| = |-3| = 3$.

(b) $4 - |3 - 9| = 4 - |-6| = 4 - 6 = -2$.

(c) $||-1| - |-5|| = |1 - 5| = |-4| = 4$. ∎

WARNING

When c and d have opposite signs, $|c + d|$ is *not equal* to $|c| + |d|$. For example, when $c = -3$ and $d = 5$, then

$$|c + d| = |-3 + 5| = 2, \quad \text{but}$$

$$|c| + |d| = |-3| + |5| = 3 + 5 = 8.$$

The warning shows that $|c + d| < |c| + |d|$ when $c = -3$ and $d = 5$. In the general case, we have the following fact (which is proved in Exercise 105).

THE TRIANGLE INEQUALITY

$$|c + d| \leq |c| + |d| \quad \text{for any real numbers } c \text{ and } d.$$

Distance on the Number Line

Observe that the distance from -5 to 3 on the number line is 8 units:

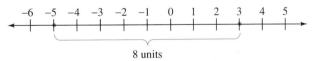

FIGURE 1-2

This distance can be expressed in terms of absolute value by noting that $|(-5) - 3| = 8$. That is, the distance is the *absolute value of the difference* of the two numbers. Furthermore, the order in which you take the difference doesn't matter; $|3 - (-5)|$ is also 8. This reflects the geometric fact that the distance from -5 to 3 is the same as the distance from 3 to -5. The same thing is true in the general case:

DISTANCE ON THE NUMBER LINE

The distance between c and d on the number line is the number
$$|c - d| = |d - c|.$$

Example 4 The distance from 4.2 to 9 is $|4.2 - 9| = |-4.8| = 4.8$ and the distance from 6 to $\sqrt{2}$ is $|6 - \sqrt{2}|$. ■

In the special case when $d = 0$, the distance formula shows that

DISTANCE TO ZERO

$|c|$ is the distance between c and 0 on the number line.

Example 5 How many real numbers have absolute value 3? In geometric terms this question is: How many real numbers are 3 units from 0 on the number line. Figure 1-3 shows that 3 and -3 are the only numbers whose distance to 0 is 3 units. ■

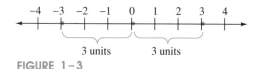

FIGURE 1-3

The procedure in Example 5 works in the general case:

If k is a positive number, then the only numbers with absolute value k are k and $-k$.

Example 5 shows that algebraic problems can sometimes be solved by translating them into equivalent geometric problems. The key is to interpret statements involving absolute value as statements about distance on the number line.

Example 6 To solve the equation $|x + 5| = 3$, we rewrite it as $|x - (-5)| = 3$. In this form it states that

*the distance from x to -5 is 3 units.**

---

* It's necessary to rewrite the equation first because the distance formula involves the *difference* of two numbers, not their sum.

Figure 1-4 shows that -8 and -2 are the only two numbers whose distance to -5 is 3 units:

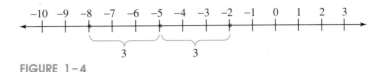

FIGURE 1-4

Thus $x = -8$ and $x = -2$ are the solutions of $|x + 5| = 3$. ∎

Example 7 If t is closer to 5 than to 0, what condition must t satisfy?

Solution The distance from t to 5, namely $|t - 5|$, must be less than the distance from t to 0, namely $|t|$. So the condition is $|t - 5| < |t|$. ∎

Example 8 The solutions of $|x - 1| \geq 2$ are all numbers x such that

the distance from x to 1 is greater than or equal to 2.

Figure 1-5 shows that the numbers 2 or more units away from 1 are the numbers x such that

$$x \leq -1 \quad \text{or} \quad x \geq 3.$$

So these numbers are the solutions of the inequality. ∎

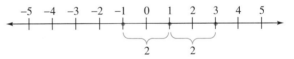

FIGURE 1-5

Example 9 The solutions of $|x - 7| < 2.5$ are all numbers x such that

the distance from x to 7 is less than 2.5

Figure 1-6 shows that the solutions of the inequality, that is, the numbers within 2.5 units of 7, are the numbers x such that $4.5 < x < 9.5$. ∎

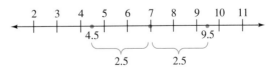

FIGURE 1-6

The set of solutions in Example 9 consists of all the numbers between 4.5 and 9.5, excluding 4.5 and 9.5. This set is usually denoted by (4.5, 9.5). In the general case, if $c < d$,

INTERVAL NOTATION

$[c, d]$ denotes the set of all real numbers x such that $c \leq x \leq d$.
(c, d) denotes the set of all real numbers x such that $c < x < d$.
$[c, d)$ denotes the set of all real numbers x such that $c \leq x < d$.
$(c, d]$ denotes the set of all real numbers x such that $c < x \leq d$.

All four of these sets are called **intervals** from c to d. The numbers c and d are the **endpoints** of the interval. $[c, d]$ is called the **closed interval** from c to d (both endpoints included and *square* brackets) and (c, d) is called the **open interval** from c to d (neither endpoint included and *round* brackets). For example,*

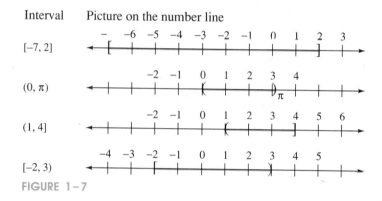

FIGURE 1-7

If b is a real number, then the half-line extending to the right or left of b is also called an interval. Depending on whether or not b is included, there are four possibilities:

INTERVAL NOTATION

$[b, \infty)$ denotes the set of all real numbers x such that $x \geq b$.
(b, ∞) denotes the set of all real numbers x such that $x > b$.
$(-\infty, b]$ denotes the set of all real numbers x such that $x \leq b$.
$(-\infty, b)$ denotes the set of all real numbers x such that $x < b$.

---

* In Figures 1–7 and 1–8 a round bracket such as) or (indicates that the endpoint is *not* included, whereas a square bracket such as] or [indicates that the endpoint is included.

1.4 ABSOLUTE VALUE

For example,

Interval	Picture on the number line
$[2, \infty)$	
$(-3.5, \infty)$	
$(-\infty, 2.75]$	
$(-\infty, -1)$	

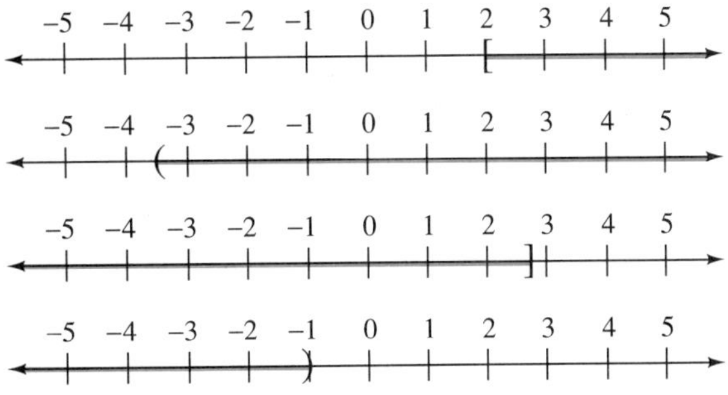

FIGURE 1-8

In a similar vein, $(-\infty, \infty)$ denotes the set of all real numbers.

NOTE The symbol ∞ is read "infinity" and we call $[b, \infty)$ "the interval from b to infinity." The symbol ∞ is part of a convenient shorthand notation for a certain set of numbers and does *not* denote a real number.

EXERCISES

In Exercises 1-14, simplify and write the given number without using absolute values.

1. $|3 - 14|$
2. $|(-2)3|$
3. $3 - |2 - 5|$
4. $-2 - |-2|$
5. $|6 - 4| + |-3 - 5|$
6. $|-6| - |6|$
7. $|(-13)^2|$
8. $-|-5|^2$
9. $|\pi - \sqrt{2}|$
10. $|\sqrt{2} - 2|$
11. $|3 - \pi| + 3$
12. $|4 - \sqrt{2}| - 5$
13. $\dfrac{|-5|}{5}$
14. $\dfrac{-|-3|}{|12|}$

In Exercises 15-22, fill the blank with $<, =,$ or $>$ so that the resulting statement is true.

15. $|-2|$ __ $|-5|$
16. 5 __ $|-2|$
17. $|3|$ __ $-|4|$
18. $|-3|$ __ 0
19. -7 __ $|-1|$
20. $-|-4|$ __ 0
21. $|-4 + |-4||$ __ 0
22. $||-7|-|-4||$ __ 2

In Exercises 23-28 draw a picture on the number line of the given interval.

23. $(0, 8]$
24. $(0, \infty)$
25. $[-2, 1]$
26. $(-1, 1)$
27. $(-\infty, 0]$
28. $[-2, 7)$

In Exercises 29-40, use interval notation to denote the set of all real numbers x that satisfy the given inequality.

29. $5 \leq x \leq 8$
30. $-2 \leq x \leq 7$
31. $-3 < x < 14$
32. $7 < x < 135$
33. $-3.7 \leq x < -2.4$
34. $-\pi < x \leq \pi$
35. $x \geq -8$
36. $x \geq 12$
37. $x > -6.7$
38. $x < -7.95$
39. $x \leq 15$
40. $x < \sqrt{2}$

In Exercises 41–46, find two pairs of numbers that make the given statement true *and* two pairs that make it false. For example, $|x| < |y|$ is true for $x = 1, y = 2$ and $x = -1, y = 7$, but it is false for $x = 3, y = 1$ and $x = -3, y = -2$.

41. $|x| + |y| = 1$
42. $|y| - |x| < 0$
43. $|x + y| = |x| + |y|$
44. $|x + y| < |x| + |y|$
45. $|x - y| = y - x$
46. $|x - y| = |x| - |y|$

In Exercises 47–54 find the distance between the given numbers.

47. -3 and 4
48. 7 and 107
49. -7 and $\frac{15}{2}$
50. $-\frac{3}{4}$ and -10
51. π and 3
52. π and -3
53. $\sqrt{2}$ and $\sqrt{3}$
54. π and $\sqrt{2}$

In Exercises 55–64, write the given expression without using absolute values.

55. $|t^2|$
56. $|u^2 + 2|$
57. $|(-3 - y)^2|$
58. $|-2 - y^2|$
59. $|b - 3|$ if $b \geq 3$
60. $|a - 5|$ if $a < 5$
61. $|c - d|$ if $c < d$
62. $|c - d|$ if $c \geq d$
63. $|u - v| - |v - u|$
64. $\frac{|u - v|}{|v - u|}$ if $u \neq v$

In Exercises 65 and 66, explain why the given statement is true for any numbers c and d. [*Hint:* Look at the properties of absolute value on page 26.]

65. $|(c - d)^2| = c^2 - 2cd + d^2$
66. $\sqrt{9c^2 - 18cd + 9d^2} = 3|c - d|$

In Exercises 67–76, express the given geometric statement about numbers on the number line algebraically, using absolute values.

67. The distance from x to 5 is less than 4.
68. x is more than 6 units from c.
69. x is at most 17 units from -4.
70. x is within 3 units of 7.
71. c is closer to 0 than b is.
72. x is closer to 1 than to 4.
73. x is farther from 0 than from -6.
74. x is closer to $-y$ than 4 is.
75. x lies either to the left of -2 or to the right of 2.
76. x lies strictly between 4 and -4.

In Exercises 77–84, translate the given algebraic statement into a geometric statement about numbers on the number line.

77. $|x - 3| < 2$
78. $|x - c| > 6$
79. $|x + 7| \leq 3$
80. $|u + v| \geq 2$
81. $|b| < |c - 3|$
82. $|b - 2| \geq |b - 5|$
83. $|b + 3| < |b - 3|$
84. $|x - 7| < |x + 5|$

85. Explain geometrically why this statement is always false:

$|c - 1| < 2$ and simultaneously $|c - 12| < 3$.

86. What is $|-c|$ when $c < 0$? What is $|-c|$ when $c > 0$?

In Exercises 87–90, determine the values of x for which the given statement is true. For example, the statement $|x| = x$ is true for all nonnegative numbers x.

87. $x \leq |x|$
88. $|x| \leq x$
89. $|x| \leq -x$
90. $-|x| \leq x$

In Exercises 91–102, use the geometric approach explained in the text to solve the given equation or inequality.

91. $|x| = 1$
92. $|x| = \frac{3}{2}$
93. $|x - 2| = 1$
94. $|x + 3| = 2$
95. $|x + \pi| = 4$
96. $|x - \frac{3}{2}| = 5$
97. $|x| < 7$
98. $|x - 5| < 2$
99. $|x + 3| < 1$
100. $|x + \frac{1}{2}| < 2$
101. $|x| \geq 5$
102. $|x - 6| > 2$

Unusual Problems

103. Explain why the statement $|a| + |b| + |c| > 0$ is algebraic shorthand for "at least one of the numbers a, b, c is different from zero."

104. Find an algebraic shorthand version of the statement "the numbers a, b, c are all different from zero."

105. Prove the triangle inequality as follows:
    (a) The box on page 26 shows that $c \le |c|$ and $d \le |d|$, and also that $-c \le |c|$ and $-d \le |d|$. Use these facts to show that $c + d \le |c| + |d|$ and $-(c + d) \le |c| + |d|$.
    (b) Use part (a) to show that $|c + d| \le |c| + |d|$. [*Hint:* What is $|c + d|$ when $c + d$ is nonnegative? When $c + d$ is negative?]

1.5 ALGEBRAIC EXPRESSIONS AND POLYNOMIALS

Expressions such as

$$b + 3c^2, \quad 3x^2 - 5x + 4, \quad \sqrt{x^3 + z}, \quad \frac{x^3 + 4xy - \pi}{x^2 + xy}$$

are called **algebraic expressions**. Each expression represents a number that is obtained by performing various algebraic operations (such as addition or taking roots) on one or more numbers, some of which may be denoted by letters.

A letter that denotes a particular real number is called a **constant**; its value remains unchanged throughout the discussion. For example, the Greek letter π has long been used to denote the number 3.14159 . . . . Sometimes a constant is a fixed but unspecified real number, as in "an angle of k degrees" or "a triangle with base of length b."

A letter that can represent *any* real number is called a **variable**. In the expression $2x + 5$, for example, the variable x can be any real number. If $x = 3$, then $2x + 5 = 2 \cdot 3 + 5 = 11$. If $x = \frac{1}{2}$, then $2x + 5 = 2 \cdot \frac{1}{2} + 5 = 6$, and so on.*

Constants are usually denoted by letters near the beginning of the alphabet and variables by letters near the end of the alphabet. Consequently, in expressions such as $cx + d$ and $cy^2 + dy$, it is understood that c and d are constants and x and y are variables.

The rules of arithmetic presented in earlier sections are valid for algebraic expressions.

Example 1 Use the distributive law to *combine like terms;* for instance,

$$3x + 5x + 4x = (3 + 5 + 4)x = 12x.$$

In practice, you do the middle part in your head and simply write $3x + 5x + 4x = 12x$. ∎

Example 2 In more complicated expressions, eliminate parentheses, use the commutative law to group like terms together, and then combine them:

---

* We assume any conditions on the constants and variables necessary to guarantee that an algebraic expression does represent a real number. For instance, in $\sqrt{z}$ we assume $z \ge 0$ and in $1/c$ we assume $c \ne 0$.

$$(a^2b - 3\sqrt{c}) + (5ab + 7\sqrt{c}) + 7a^2b = a^2b - 3\sqrt{c} + 5ab + 7\sqrt{c} + 7a^2b$$

Regroup: $\quad = a^2b + 7a^2b - 3\sqrt{c} + 7\sqrt{c} + 5ab$

Combine like terms: $\quad = 8a^2b \;+\; 4\sqrt{c} \;+\; 5ab.$ ∎

WARNING

Be careful when parentheses are preceded by a minus sign: $-(b+3) = -b - 3$ and *not* $-b + 3$. Here's the reason: $-(b+3)$ means $(-1)(b+3)$, so that by the distributive law,

$$-(b+3) = (-1)(b+3) = (-1)b + (-1)3 = -b - 3.$$

Similarly, $-(7 - y) = -7 - (-y) = -7 + y$.

The examples in the warning box illustrate the following.

RULES FOR ELIMINATING PARENTHESES

Parentheses preceded by a plus sign (or no sign) may be deleted.
Parentheses preceded by a minus sign may be deleted *if* the sign of every term within the parentheses is changed.

Example 3

$$(x^4 + 6x^2 + x - 11) - (3x^2 - 5x - 2)$$

Eliminate parentheses: $\quad = x^4 + 6x^2 + x - 11 - 3x^2 + 5x + 2$

Regroup: $\quad = x^4 + 6x^2 - 3x^2 + x + 5x - 11 + 2$

Combine like terms: $\quad = x^4 + \quad 3x^2 \;+\; 6x \;-\; 9.$ ∎

The usual method of multiplying algebraic expressions is to use the distributive laws repeatedly, as shown in the following examples. The net result is to *multiply every term in the first sum by every term in the second sum*.

Example 4 To compute $(y - 2)(3y^2 - 7y + 4)$, we first apply the distributive law, treating $(3y^2 - 7y + 4)$ as a single number:

$$(y - 2)(3y^2 - 7y + 4) = y(3y^2 - 7y + 4) - 2(3y^2 - 7y + 4)$$

Distributive law: $\quad = 3y^3 - 7y^2 + 4y - 6y^2 + 14y - 8$

Regroup: $\quad = 3y^3 - 7y^2 - 6y^2 + 4y + 14y - 8$

Combine like terms: $\quad = 3y^3 - \quad 13y^2 \;+\; 18y \;-\; 8.$ ∎

Example 5 We follow the same procedure with $(2x - 5y)(3x + 4y)$:

$$(2x - 5y)(3x + 4y) = 2x(3x + 4y) - 5y(3x + 4y)$$
$$= 2x \cdot 3x + 2x \cdot 4y + (-5y) \cdot 3x + (-5y) \cdot 4y$$
$$= 6x^2 + \underbrace{8xy - 15xy}_{} - 20y^2$$
$$= 6x^2 - 7xy - 20y^2. \blacksquare$$

Observe the pattern in the second line of Example 5 and its relationship to the terms being multiplied:

$$(2x - 5y)(3x + 4y) = 2x \cdot 3x + 2x \cdot 4y + (-5y) \cdot 3x + (-5y) \cdot 4y$$

First terms

$(2x - 5y)(3x + 4y)$ *Outside terms*

$(2x - 5y)(3x + 4y)$ *Inside terms*

$(2x - 5y)(3x + 4y)$ *Last terms*

This pattern is easy to remember by using the acronym FOIL (First, Outside, Inside, Last). The FOIL method makes it easy to find products such as this one mentally, without the necessity of writing out the intermediate steps.

Example 6

$$(3x + 2)(x + 5) = 3x^2 + 15x + 2x + 10 = 3x^2 + 17x + 10. \blacksquare$$
$$\quad\quad\quad\quad\quad\quad\; First\quad Outside\quad Inside\quad Last$$

WARNING

> The FOIL method can be used only when multiplying two expressions that each have two terms.

Certain multiplication patterns occur so often that they should be memorized. You can verify the validity of each of the statements below by multiplying out the left side. For example, statement 1 is a result of this calculation (using FOIL):
$$(u + v)(u - v) = u^2 - uv + uv - v^2 = u^2 - v^2.$$

MULTIPLICATION PATTERNS

1. $(u + v)(u - v) = u^2 - v^2$
2. $(u + v)^2 = u^2 + 2uv + v^2$
3. $(u - v)^2 = u^2 - 2uv + v^2$
4. $(u + v)^3 = u^3 + 3u^2v + 3uv^2 + v^3$
5. $(u - v)^3 = u^3 - 3u^2v + 3uv^2 - v^3$

Whenever you have the sum and/or difference of two quantities, you may be able to use these patterns. Just substitute one quantity for u and the other for v.

Example 7 Here are illustrations of the first four patterns in the box.

1. To find $(2x + 5)(2x - 5)$, let $u = 2x$ and $v = 5$. Then
$$(2x + 5)(2x - 5) = (2x)^2 - 5^2 = 4x^2 - 25.$$

2. By letting $u = y$ and $v = 3$, we see that
$$(y + 3)^2 = y^2 + 2 \cdot y \cdot 3 + 3^2 = y^2 + 6y + 9.$$

3. $(x - 3y)^2 = x^2 - 2 \cdot x \cdot 3y + (3y)^2 = x^2 - 6xy + 9y^2$ (here $u = x$ and $v = 3y$).

4. $(c + 4)^3 = c^3 + 3c^2 \cdot 4 + 3c \cdot 4^2 + 4^3 = c^3 + 12c^2 + 48c + 64.$ ∎

> **WARNING**
>
> Don't forget the middle term: $(u + v)^2 \neq u^2 + v^2$.

The multiplication patterns for $(u \pm v)^2$ and $(u \pm v)^3$ are special cases of the **Binomial Theorem**. This theorem provides a quick way to calculate higher powers, such as

$$(u + v)^4, \qquad (x - y)^5, \qquad (c + d)^6, \qquad \text{and so on.}$$

The Binomial Theorem is discussed in Section 12.4, which may be read at this time if desired.

Polynomials

Informally, a **polynomial** is an algebraic expression such as

$$x^3 - 6x^2 + \tfrac{1}{2}x \quad \text{or} \quad x^{15} + x^{10} + 7 \quad \text{or} \quad x - \pi \quad \text{or} \quad 12$$

We assume $x^0 = 1$ so that a real number, such as 12, may be thought of as $12x^0$. Consequently, a polynomial is a sum of terms of the form cx^k, where c is a constant and k is a *nonnegative integer*. Thus $x^{-2} + 6$ and $\sqrt{x^2 + 1}$ and $x^{1/4} + x - 3$ are not polynomials, but $\sqrt{5}$ and x^7 and $x + \dfrac{1}{2}$ are.

The formal definition is that a **polynomial in x** is an algebraic expression that can be written in the form

$$a_n x^n + a_{n-1} x^{n-1} + \cdots + a_3 x^3 + a_2 x^2 + a_1 x + a_0$$

where n is a nonnegative integer, x is a variable, and each of $a_0, a_1, \ldots, a_n$ is a constant.* The numbers $a_0, a_1, \ldots, a_n$ are called the **coefficients** of the polynomial. The coefficient a_0 is called the **constant term.**

We omit any term with zero coefficient, we write x^0 as 1, and we don't write a coefficient or exponent if it is the number 1; for instance, we write $x^2 + 3$ instead of $1x^2 + 0x^1 + 3x^0$. A polynomial that consists of only a constant term, such as 12 or $-1/2$, is called a **constant polynomial.** The **zero polynomial** is the constant polynomial 0.

The *exponent* of the highest power of x that appears with *nonzero* coefficient is the **degree** of the polynomial, and the nonzero coefficient of this highest power of x is the **leading coefficient.** For example,

Polynomial	Degree	Leading Coefficient	Constant Term
$6x^7 + 4x^3 + 5x^2 - 7x + 10$	7	6	10
$-x^4 + 2x^3 + \frac{1}{2}$	4	-1	½
x^3	3	1	0
12	0	12	12
$7x^5 - 3x^3 + x^2 + 4x$	5	7	0
$0x^4 + 5x^3 - 6x^2 + 2x - \frac{1}{4}$	3 (note well)	5	$-\frac{1}{4}$
$2x^6 + 3x^7 + x^8 - 2x - 4 + 3x^2$	8 (be careful)	1	-4

The degree of the zero polynomial is *not defined* since no exponent of x occurs with nonzero coefficient. First-degree polynomials are often called **linear polynomials.** Second- and third-degree polynomials are called **quadratics** and **cubics,** respectively.

Polynomial Arithmetic

Examples of addition, subtraction, and multiplication of polynomials were given above. Long division of polynomials is quite similar to long division of numbers, as we now see.

Example 8 To divide $2x^5 + 5x^4 - 4x^3 + 8x^2 + 1$ by $2x^2 - x + 1$, we first write:

$$2x^2 - x + 1 \overline{\smash{\big)}\, 2x^5 + 5x^4 - 4x^3 + 8x^2 + 1}$$

We call $2x^2 - x + 1$ the **divisor** and $2x^5 + 5x^4 - 4x^3 + 8x^2 + 1$ the **dividend.** The **quotient** of the division will be written above the horizontal line. We begin by dividing the first term of the divisor ($2x^2$) into the first term of the dividend ($2x^5$) and putting the result $\left(\text{namely, } \frac{2x^5}{2x^2} = x^3\right)$ on the top line, as shown below. Then multiply x^3 times the entire divisor, put the result on the third line, and subtract:

---

* Any letter may be used as the variable in a polynomial, but we shall usually use x.

$$
\begin{array}{r}
x^3 \\
2x^2 - x + 1 \overline{\smash{)}2x^5 + 5x^4 - 4x^3 + 8x^2 + 1} \\
\underline{2x^5 - x^4 + x^3 } \\
6x^4 - 5x^3 + 8x^2 + 1
\end{array}
$$

$\leftarrow x^3 \cdot (2x^2 - x + 1)$
$\leftarrow$ *Subtraction**

Next, divide the first term of the divisor ($2x^2$) into $6x^4$ and put the result $\left(\dfrac{6x^4}{2x^2} = 3x^2 \right)$ on the top line, as shown below. Then multiply $3x^2$ times the entire divisor, put the result on the fifth line, and subtract. Continuing this procedure, we obtain:

$$
\begin{array}{r}
x^3 + 3x^2 - x + 2 \\
2x^2 - x + 1 \overline{\smash{)}2x^5 + 5x^4 - 4x^3 + 8x^2 + 1} \\
\underline{2x^5 - x^4 + x^3 } \\
6x^4 - 5x^3 + 8x^2 + 1 \\
\underline{6x^4 - 3x^3 + 3x^2 } \\
-2x^3 + 5x^2 + 1 \\
\underline{-2x^3 + x^2 - x} \\
4x^2 + x + 1 \\
\underline{4x^2 - 2x + 2} \\
3x - 1
\end{array}
$$

$\leftarrow$ Quotient

$\leftarrow x^3 \cdot (2x^2 - x + 1)$
$\leftarrow$ Subtraction
$\leftarrow 3x^2 \cdot (2x^2 - x + 1)$
$\leftarrow$ Subtraction
$\leftarrow (-x)(2x^2 - x + 1)$
$\leftarrow$ Subtraction
$\leftarrow 2 \cdot (2x^2 - x + 1)$
$\leftarrow$ Subtraction

The polynomial $3x - 1$ is called the **remainder.** The division process always stops when the remainder is zero or has *smaller degree* than the divisor (here the divisor $2x^2 - x + 1$ has degree 2 and the remainder $3x - 1$ has degree 1). ∎

Recall how you check a long division problem with numbers:

$$
\begin{array}{r}
145 \\
31 \overline{\smash{)}4509} \\
\underline{31} \\
140 \\
\underline{124} \\
169 \\
\underline{155} \\
14
\end{array}
\qquad \text{Check:} \qquad
\begin{array}{r}
145 \\
\times 31 \\
\hline
145 \\
435 \\
\hline
4495 \\
+14 \\
\hline
4509
\end{array}
$$

$\leftarrow$ Quotient
$\leftarrow$ Divisor

$\leftarrow$ Remainder
$\leftarrow$ Dividend

We can summarize this process in one line:

$$\text{Divisor} \cdot \text{Quotient} + \text{Remainder} = \text{Dividend}.$$

The same process can be used with polynomial division. In the example above, you can easily verify that the divisor times the quotient is

---

* If this subtraction is confusing, write it out horizontally and watch the signs carefully:

$$(2x^5 + 5x^4 - 4x^3 + 8x^2 + 1) - (2x^5 - x^4 + x^3)$$
$$= 2x^5 + 5x^4 - 4x^3 + 8x^2 + 1 - 2x^5 + x^4 - x^3$$
$$= 6x^4 - 5x^3 + 8x^2 + 1$$

1.5 ALGEBRAIC EXPRESSIONS AND POLYNOMIALS

$$(2x^2 - x + 1)(x^3 + 3x^2 - x + 2) = 2x^5 + 5x^4 - 4x^3 + 8x^2 - 3x + 2$$

Adding the remainder $3x - 1$ to this result yields the original dividend:

$$(2x^5 + 5x^4 - 4x^3 + 8x^2 - 3x + 2) + (3x - 1)$$
$$= 2x^5 + 5x^4 - 4x^3 + 8x^2 + 1$$

So just as with division of numbers we have:

$$\text{Divisor} \cdot \text{Quotient} + \text{Remainder} = \text{Dividend}.$$

Because this fact is so important it is given a special name and a formal statement:

THE DIVISION ALGORITHM

> If a polynomial P is divided by a nonzero polynomial D, then there is a quotient polynomial Q and a remainder polynomial R such that
>
> $$DQ + R = P$$
>
> where either $R = 0$, or R has degree less than the degree of the divisor D.

When a polynomial P is written as the product of other polynomials, then each polynomial in the product is called a **factor** of P. For example, $x - 2$ and $3x + 1$ are factors of $3x^2 - 5x - 2$ because $(x - 2)(3x + 1) = 3x^2 - 5x - 2$. The Division Algorithm can be used to determine if one polynomial is a factor of another polynomial.

Example 9 To determine if $2x^2 + 1$ is a factor of $6x^3 - 4x^2 + 3x - 2$, we divide:

$$
\begin{array}{r}
3x - 2 \\
2x^2 + 1 \overline{\smash{)}6x^3 - 4x^2 + 3x - 2} \\
\underline{6x^3 + 3x } \\
-4x^2 - 2 \\
\underline{-4x^2 - 2} \\
0
\end{array}
$$

Since the remainder is 0, the Division Algorithm tells us that:

$$\text{Dividend} = \text{Divisor} \cdot \text{Quotient} + \text{Remainder}$$

$$6x^3 - 4x^2 + 3x - 2 = (2x^2 + 1)(3x - 2) + 0$$
$$= (2x^2 + 1)(3x - 2).$$

Therefore $2x^2 + 1$ is a factor of $6x^3 - 4x^2 + 3x - 2$ and the other factor is the quotient $3x - 2$. ∎

The same argument works in the general case:

REMAINDERS AND FACTORS

The remainder in polynomial division is 0 exactly when the divisor is a factor of the dividend. In this case the other factor is the quotient.

When the divisor in polynomial division is a first-degree polynomial such as $x - 2$ or $x + 5$, there is a convenient shorthand method of doing the division called **synthetic division**. See Excursion 1.5.A for details.

EXERCISES

In Exercises 1–54, perform the indicated operations and simplify your answer.

1. $x + 7x$
2. $5w + 7w - 3w$
3. $6a^2b + (-8b)a^2$
4. $-6x^3\sqrt{t} + 7x^3\sqrt{t} - 15x^3\sqrt{t}$
5. $(x^2 + 2x + 1) - (x^3 - 3x^2 + 4)$
6. $\left(u^4 - (-3)u^3 + \dfrac{u}{2} + 1\right) + \left(u^4 - 2u^3 + 5 - \dfrac{u}{2}\right)$
7. $\left(u^4 - (-3)u^3 + \dfrac{u}{2} + 1\right) - \left(u^4 - 2u^3 + 5 - \dfrac{u}{2}\right)$
8. $(6a^2b + 3a\sqrt{c} - 5ab\sqrt{c}) + (-6ab^2 - 3ab + 6ab\sqrt{c})$
9. $(4z - 6z^2w - (-2)z^3w^2) + (8 - 6z^2w - zw^3 + 4z^3w^2)$
10. $(x^5y - 2x + 3xy^3) - (-2x - x^5y + 2xy^3)$
11. $(9x - x^3 + 1) - (2x^3 + (-6)x + (-7))$
12. $(x - \sqrt{y} - z) - (x + \sqrt{y} + z) - (\sqrt{y} + z - x)$
13. $(x^2 - 3xy) - (x + xy) - (x^2 + xy)$
14. $2x(x^2 + 2)$
15. $(-5y)(-3y^2 + 1)$
16. $x^2y(xy - 6xy^2)$
17. $3ax(4ax - 2a^2y + 2ay)$
18. $2x(x^2 - 3xy + 2y^2)$
19. $6z^3(2z + 5)$
20. $-3x^2(12x^6 - 7x^5)$
21. $3ab(4a - 6b + 2a^2b)$
22. $(-3ay)(4ay - 5y)$
23. $(x + 1)(x - 2)$
24. $(x + 2)(2x - 5)$
25. $(-2x + 4)(-x - 3)$
26. $(y - 6)(2y + 2)$
27. $(y + 3)(y + 4)$
28. $(w - 2)(3w + 1)$
29. $(3x + 7)(-2x + 5)$
30. $(ab + 1)(a - 2)$
31. $(y - 3)(3y^2 + 4)$
32. $(y + 8)(y - 8)$
33. $(x + 4)(x - 4)$
34. $(3x - y)(3x + y)$
35. $(4a + 5b)(4a - 5b)$
36. $(x + 6)^2$
37. $(y - 11)^2$
38. $(2x + 3y)^2$
39. $(5x - b)^2$
40. $(2s^2 - 9y)(2s^2 + 9y)$
41. $(4x^3 - y^4)^2$
42. $(4x^3 - 5y^2)(4x^3 + 5y^2)$
43. $(-3x^2 + 2y^4)^2$
44. $(c - 2)(2c^2 - 3c + 1)$
45. $(2y + 3)(y^2 + 3y - 1)$
46. $(x + 2y)(2x^2 - xy + y^2)$
47. $(5w + 6)(-3w^2 + 4w - 3)$
48. $(5x - 2y)(x^2 - 2xy + 3y^2)$
49. $2x(3x + 1)(4x - 2)$
50. $3y(-y + 2)(3y + 1)$
51. $(x - 1)(x - 2)(x - 3)$
52. $(y - 2)(3y + 2)(y + 2)$
53. $(x + 4y)(2y - x)(3x - y)$
54. $(2x - y)(3x + 2y)(y - x)$

In Exercises 55–62, determine whether the given algebraic expression is a polynomial. If it is, list its leading coefficient, constant term, and degree.

55. $1 + x^3$
56. -7
57. $(x - 1)(x^2 + 1)$
58. $7^x + 2x + 1$
59. $(x + \sqrt{3})(x - \sqrt{3})$
60. $4x^2 + 3\sqrt{x} + 5$
61. $\dfrac{7}{x^2} + \dfrac{5}{x} - 15$
62. $(x - 1)^k$ (where k is a fixed positive integer)

In Exercises 63–68, divide the first polynomial by the second. Check your division by calculating (divisor)(quotient) + remainder.

63. $3x^4 + 2x^2 - 6x + 1;\ x + 1$
64. $x^5 - x^3 + x - 5;\ x - 2$
65. $x^5 + 2x^4 - 6x^3 + x^2 - 5x + 1;\ x^3 + 1$
66. $3x^4 - 3x^3 - 11x^2 + 6x - 1;\ x^3 + x^2 - 2$
67. $5x^4 + 5x^2 + 5;\ x^2 - x + 1$
68. $x^5 - 1;\ x - 1$

In Exercises 69–72, determine whether the first polynomial is a factor of the second.

69. $x^2 + 3x - 1;\ x^3 + 2x^2 - 5x - 6$
70. $x^2 + 9;\ x^5 + x^4 - 81x - 81$
71. $x^2 + 3x - 1;\ x^4 + 3x^3 - 2x^2 - 3x + 1$
72. $x^2 - 5x + 7;\ x^3 - 3x^2 - 3x + 9$

In Exercises 73–82, find the coefficient of x^2 in the given product. Avoid doing any more multiplying than necessary.

73. $(x^2 + 3x + 1)(2x - 3)$
74. $(x^2 - 1)(x + 1)$
75. $(x^3 + 2x - 6)(x^2 + 1)$
76. $(\sqrt{3} + x)(\sqrt{3} - x)$
77. $(x + 2)^3$
78. $(x^2 + x + 1)(x - 1)$
79. $(x^2 + x + 1)(x^2 - x + 1)$
80. $(2x^2 + 1)(2x^2 - 1)$
81. $(2x - 1)(x^2 + 3x + 2)$
82. $(1 - 2x)(4x^2 + x - 1)$

In Exercises 83–88, perform the indicated multiplication and simplify your answer if possible.

83. $(\sqrt{x} + 5)(\sqrt{x} - 5)$
84. $(2\sqrt{x} + \sqrt{2y})(2\sqrt{x} - \sqrt{2y})$
85. $(3 + \sqrt{y})^2$
86. $(7w - \sqrt{2x})^2$
87. $(1 + \sqrt{3}x)(x + \sqrt{3})$
88. $(2y + \sqrt{3})(\sqrt{5}y - 1)$

In Exercises 89–94, compute the product and arrange the terms of your answer according to decreasing powers of x, with each power of x appearing at most once. Example: $(ax + b)(4x - c) = 4ax^2 + (4b - ac)x - bc$.

89. $(ax + b)(3x + 2)$
90. $(4x - c)(dx + c)$
91. $(ax + b)(bx + a)$
92. $rx(3rx + 1)(4x - r)$
93. $(x - a)(x - b)(x - c)$
94. $(2dx - c)(3cx + d)$

In Exercises 95–100, assume that all exponents are nonnegative integers and find the product. Example: $2x^k(3x + x^{n+1}) = (2x^k)(3x) + (2x^k)(x^{n+1}) = 6x^{k+1} + 2x^{k+n+1}$.

95. $3^r 3^4 3^t$
96. $(2x^n)(8x^k)$
97. $(x^m + 2)(x^n - 3)$
98. $(y^r + 1)(y^s - 4)$
99. $(2x^n - 5)(x^{3n} + 4x^n + 1)$
100. $(3y^{2k} + y^k + 1)(y^k - 3)$

Errors to Avoid

In Exercises 101–110, find a numerical example to show that the given statement is *false*. Then find the mistake in the statement and correct it. Example: The statement $-(b + 2) = -b + 2$ is false when $b = 5$, since $-(5 + 2) = -7$ but $-5 + 2 = -3$. The mistake is the sign on the 2.

The correct statement is $-(b + 2) = -b - 2$.

101. $3(y + 2) = 3y + 2$
102. $x - (3y + 4) = x - 3y + 4$
103. $(x + y)^2 = x + y^2$
104. $(2x)^3 = 2x^3$
105. $(7x)(7y) = 7xy$
106. $(x + y)^2 = x^2 + y^2$
107. $y + y + y = y^3$
108. $(a - b)^2 = a^2 - b^2$
109. $(x - 3)(x - 2) = x^2 - 5x - 6$
110. $(a + b)(a^2 + b^2) = a^3 + b^3$

Unusual Problems

In Exercises 111–112, explain algebraically why each of these parlor tricks always works.

111. Write down a nonzero number. Add 1 to it and square the result. Subtract 1 from the original number and square the result. Subtract this second square from the first one. Divide by the number with which you started. The answer is 4.

112. Write down a positive number. Add 4 to it. Multiply the result by the original number. Add 4 to this result and then take the square root. Subtract the number with which you started. The answer is 2.

113. Invent a similar parlor trick in which the answer is always the number with which you started.

1.5.A EXCURSION: Synthetic Division

Synthetic division is a fast method of doing polynomial division when the divisor is a first-degree polynomial of the form $x - c$ for some real number c. To see how it works, we first consider an example of ordinary long division:

$$
\begin{array}{r}
3x^3 + 6x^2 + 4x - 3 \longleftarrow \text{Quotient}\\
x - 2 \overline{)\, 3x^4 - 8x^2 - 11x + 1} \longleftarrow \text{Dividend}\\
\underline{3x^4 - 6x^3} \\
6x^3 - 8x^2 \\
\underline{6x^3 - 12x^2} \\
4x^2 - 11x \\
\underline{4x^2 - 8x} \\
-3x + 1\\
\underline{-3x + 6}\\
-5 \longleftarrow \text{Remainder}
\end{array}
$$

with Divisor $\longrightarrow x - 2$.

This calculation obviously involves a lot of repetitions. If we insert 0 coefficients for terms that don't appear above and keep the various coefficients in the proper columns, we can eliminate the repetitions and all the x's:

$$
\begin{array}{r}
3 6 4 -3 \longleftarrow \text{Quotient}\\
1 - 2 \overline{)\, 3 0 -8 -11 1} \longleftarrow \text{Dividend}\\
\underline{-6} \\
6 \\
\underline{-12} \\
4 \\
\underline{-8} \\
-3\\
\underline{+6}\\
-5 \longleftarrow \text{Remainder}
\end{array}
$$

with Divisor $\longrightarrow 1 - 2$.

1.5.A SYNTHETIC DIVISION

We can save space by moving the lower lines upward and writing 2 in the divisor position (since that's enough to remind us that the divisor is $x - 2$):

$$
\begin{array}{r}
\;\;3\quad\;\;6\quad\;\;4\quad -3 \quad\quad\leftarrow \text{Quotient}\\
\text{Divisor}\longrightarrow 2\overline{)3\quad\;\;0\;\; -8\; -11\quad\;\;1}\leftarrow \text{Dividend}\\
\underline{-6\; -12\;\; -8\quad\;\;6}\\
6\quad\;\;4\quad -3\;\;\boxed{-5}\;\leftarrow \text{Remainder}
\end{array}
$$

Since the last line contains most of the quotient line, we can save more space and still preserve the essential information by inserting a 3 in the last line and omitting the top line:

$$
\begin{array}{r}
\text{Divisor}\longrightarrow 2\overline{)3\quad\;\;0\;\; -8\; -11\quad\;\;1}\leftarrow \text{Dividend}\\
\underline{-6\; -12\;\; -8\quad\;\;6}\\
\underbrace{3\quad\;\;6\quad\;\;4\quad -3}\;\;\boxed{-5}\;\leftarrow \text{Remainder}\\
\text{Quotient}
\end{array}
$$

Synthetic division is a quick method for obtaining the last row of this array. Here is a step-by-step explanation of the division of $3x^4 - 8x^2 - 11x + 1$ by $x - 2$:

STEP 1. In the first row list the 2 from the divisor and the coefficients of the dividend in order of decreasing powers of x (insert 0 coefficients for missing powers of x).

$$2\rfloor\quad 3\quad\;\;0\quad -8\quad -11\quad\;\;1$$

STEP 2. Bring down the first dividend coefficient (namely, 3) to the third row.

STEP 3. Multiply $2 \cdot 3$ and insert the answer 6 in the second row, in the position shown here.

STEP 4. Add $0 + 6$ and write the answer 6 in the third row.

STEP 5. Multiply $2 \cdot 6$ and insert the answer 12 in the second row.

STEP 6. Add $-8 + 12$ and write the answer 4 in the third row.

44 1 BASIC ALGEBRA

STEP 7. Multiply $2 \cdot 4$ and insert the answer 8 in the second row.

$$
\begin{array}{r|rrrrr}
2 & 3 & 0 & -8 & -11 & 1 \\
  &   & 6 & 12 & 8   &   \\ \hline
  & 3 & 6 & 4 &   &
\end{array}
$$

STEP 8. Add $-11 + 8$ and write the answer -3 in the third row.

$$
\begin{array}{r|rrrrr}
2 & 3 & 0 & -8 & -11 & 1 \\
  &   & 6 & 12 & 8   &   \\ \hline
  & 3 & 6 & 4 & -3 &
\end{array}
$$

STEP 9. Multiply $2 \cdot (-3)$ and insert the answer -6 in the second row.

$$
\begin{array}{r|rrrrr}
2 & 3 & 0 & -8 & -11 & 1 \\
  &   & 6 & 12 & 8   & -6 \\ \hline
  & 3 & 6 & 4 & -3 &
\end{array}
$$

STEP 10. Add $1 + (-6)$ and write the answer -5 in the third row.

$$
\begin{array}{r|rrrrr}
2 & 3 & 0 & -8 & -11 & 1 \\
  &   & 6 & 12 & 8   & -6 \\ \hline
  & 3 & 6 & 4 & -3 & -5
\end{array}
$$

Except for the signs in the second row, this last array is the same as the array obtained from the long division process, and we can read off the quotient and remainder:

The last number in the third row is the remainder.

The other numbers in the third row are the coefficients of the quotient (arranged in order of decreasing powers of x).

Since we are dividing the *fourth*-degree polynomial $3x^4 - 8x^2 - 11x + 1$ by the *first*-degree polynomial $x - 2$, the quotient must be a polynomial of degree *three* with coefficients $3, 6, 4, -3$, namely $3x^3 + 6x^2 + 4x - 3$. The remainder is -5.

WARNING

Synthetic division can be used *only* when the divisor is a first-degree polynomial of the form $x - c$. In the example above, $c = 2$. If you want to use synthetic division with a divisor such as $x + 3$, you must write it as $x - (-3)$, which is of the form $x - c$ with $c = -3$.

Example 1 To divide $x^5 + 5x^4 + 6x^3 - x^2 + 4x + 29$ by $x + 3$, we write the divisor as $x - (-3)$ and proceed as above:

$$
\begin{array}{r|rrrrrr}
-3 & 1 & 5 & 6 & -1 & 4 & 29 \\
   &   & -3 & -6 & 0 & 3 & -21 \\ \hline
   & 1 & 2 & 0 & -1 & 7 & 8
\end{array}
$$

The last row shows that the quotient is $x^4 + 2x^3 - x + 7$ and the remainder is 8. ∎

Example 2 Show that $x - 7$ is a factor of $8x^5 - 52x^4 + 2x^3 - 198x^2 - 86x + 14$ and find the other factor.

Solution $x - 7$ is a factor exactly when division by $x - 7$ leaves remainder 0, in which case the quotient is the other factor. Using synthetic division we have:

$$
\begin{array}{r|rrrrrr}
7 & 8 & -52 & 2 & -198 & -86 & 14 \\
  &   & 56 & 28 & 210 & 84 & -14 \\ \hline
  & 8 & 4   & 30 & 12 & -2 & 0
\end{array}
$$

Since the remainder is 0, the divisor $x - 7$ and the quotient $8x^4 + 4x^3 + 30x^2 + 12x - 2$ are factors:

$$8x^5 - 52x^4 + 2x^3 - 198x^2 - 86x + 14$$
$$= (x - 7)(8x^4 + 4x^3 + 30x^2 + 12x - 2). \blacksquare$$

EXERCISES

In Exercises 1–8, use synthetic division to find the quotient and remainder.

1. $(3x^4 - 8x^3 + 9x + 5) \div (x - 2)$
2. $(4x^3 - 3x^2 + x + 7) \div (x - 2)$
3. $(2x^4 + 5x^3 - 2x - 8) \div (x + 3)$
4. $(3x^3 - 2x^2 - 8) \div (x + 5)$
5. $(5x^4 - 3x^2 - 4x + 6) \div (x - 7)$
6. $(3x^4 - 2x^3 + 7x - 4) \div (x - 3)$
7. $(x^4 - 6x^3 + 4x^2 + 2x - 7) \div (x - 2)$
8. $(x^6 - x^5 + x^4 - x^3 + x^2 - x + 1) \div (x + 3)$

In Exercises 9–12, use synthetic division to find the quotient and the remainder. In each divisor $x - c$ the number c is not an integer, but the same technique will work.

9. $(3x^4 - 2x^2 + 2) \div (x - \frac{1}{4})$
10. $(2x^4 - 3x^2 + 1) \div (x - \frac{1}{2})$
11. $(2x^4 - 5x^3 - x^2 + 3x + 2) \div (x + \frac{1}{2})$
12. $(10x^5 - 3x^4 + 14x^3 + 13x^2 - \frac{4}{3}x + \frac{7}{3}) \div (x + \frac{1}{3})$

In Exercises 13–16, use synthetic division to show that the first polynomial is a factor of the second and find the other factor.

13. $x + 4;\ 3x^3 + 9x^2 - 11x + 4$
14. $x - 5;\ x^5 - 8x^4 + 17x^2 + 293x - 15$
15. $x - \frac{1}{2};\ 2x^5 - 7x^4 + 15x^3 - 6x^2 - 10x + 5$
16. $x + \frac{1}{3};\ 3x^6 + x^5 - 6x^4 + 7x^3 + 3x^2 - 15x - 5$

In Exercises 17–18, use a calculator and synthetic division to find the quotient and remainder.

17. $(x^3 - 5.27x^2 + 10.708x - 10.23) \div (x - 3.12)$
18. $(2.79x^4 + 4.8325x^3 - 6.73865x^2 + .9255x - 8.125) \div (x - 1.35)$

Unusual Problems

19. When $x^3 + cx + 4$ is divided by $x + 2$, the remainder is 4. Find c.
20. If $x - d$ is a factor of $2x^3 - dx^2 + (1 - d^2)x + 5$, what is d?

1.6 FACTORING

Factoring is the reverse of multiplication: We begin with a product and find the factors that multiply together to produce this product. Factoring skills are necessary to simplify expressions, to do arithmetic with fractional expressions, and to solve equations and inequalities.

The first general rule for factoring is:

COMMON FACTORS

> If there is a common factor in every term of the expression, factor out the common factor of highest degree.

Example 1 In $4x^6 - 8x$, for example, each term contains a factor of $4x$, so that $4x^6 - 8x = 4x(x^5 - 2)$. Similarly, the common factor of highest degree in $x^3y^2 + 2xy^3 - 3x^2y^4$ is xy^2 and

$$x^3y^2 + 2xy^3 - 3x^2y^4 = xy^2(x^2 + 2y - 3xy^2). \blacksquare$$

You can greatly increase your factoring proficiency by learning to recognize multiplication patterns that appear frequently. Here are some from the last section.

QUADRATIC FACTORING PATTERNS

Difference of Squares	$u^2 - v^2 = (u + v)(u - v)$
Perfect Squares	$u^2 + 2uv + v^2 = (u + v)^2$
	$u^2 - 2uv + v^2 = (u - v)^2$

Example 2

(a) $x^2 - 9y^2$ can be written $x^2 - (3y)^2$, a difference of squares. Therefore $x^2 - 9y^2 = (x + 3y)(x - 3y)$.

(b) $y^2 - 7 = y^2 - (\sqrt{7})^2 = (y + \sqrt{7})(y - \sqrt{7})$.*

(c) $36r^2 - 64s^2 = (6r)^2 - (8s)^2 = (6r + 8s)(6r - 8s)$.
$ = 2(3r + 4s)2(3r - 4s) = 4(3r + 4s)(3r - 4s). \blacksquare$

Example 3 Since the first and last terms of $4x^2 - 36x + 81$ are perfect squares, we try to use the perfect square pattern with $u = 2x$ and $v = 9$:

$$4x^2 - 36x + 81 = (2x)^2 - 36x + 9^2$$
$$= (2x)^2 - 2 \cdot 2x \cdot 9 + 9^2 = (2x - 9)^2. \blacksquare$$

CUBIC FACTORING PATTERNS

Difference of Cubes	$u^3 - v^3 = (u - v)(u^2 + uv + v^2)$
Sum of Cubes	$u^3 + v^3 = (u + v)(u^2 - uv + v^2)$
Perfect Cubes	$u^3 + 3u^2v + 3uv^2 + v^3 = (u + v)^3$
	$u^3 - 3u^2v + 3uv^2 - v^3 = (u - v)^3$

* When a polynomial has integer coefficients, we normally look only for factors with integer coefficients. But when it is easy to find other factors, as here, we shall do so.

Example 4

(a) $$x^3 - 125 = x^3 - 5^3 = (x - 5)(x^2 + 5x + 5^2)$$
$$= (x - 5)(x^2 + 5x + 25).$$

(b) $$x^3 + 8y^3 = x^3 + (2y)^3 = (x + 2y)(x^2 - x \cdot 2y + (2y)^2)$$
$$= (x + 2y)(x^2 - 2xy + 4y^2).$$

(c) $$x^3 - 12x^2 + 48x - 64 = x^3 - 12x^2 + 48x - 4^3$$
$$= x^3 - 3x^2 \cdot 4 + 3x \cdot 4^2 - 4^3$$
$$= (x - 4)^3. \blacksquare$$

When none of the multiplication patterns applies, use trial and error to factor quadratic polynomials. If a quadratic has two first-degree factors, then the factors must be of the form $ax + b$ and $cx + d$ for some constants a, b, c, d. The product of such factors is

$$(ax + b)(cx + d) = acx^2 + adx + bcx + bd$$
$$= acx^2 + (ad + bc)x + bd$$

Note that *ac is the coefficient of x^2* and *bd is the constant term* of the product polynomial. This pattern can be used to factor quadratics by reversing the FOIL Process.*

Example 5

If $x^2 + 9x + 18$ factors as $(ax + b)(cx + d)$, then we must have $ac = 1$ (coefficient of x^2) and $bd = 18$ (constant term). Thus $a = \pm 1$ and $c = \pm 1$ (the only integer factors of 1). The only possibilities for b and d are:

$$\pm 1, \pm 18 \quad \text{or} \quad \pm 2, \pm 9 \quad \text{or} \quad \pm 3, \pm 6$$

We mentally try the various possibilities, using FOIL as our guide. For example, we try $b = 2, d = 9$ and check this factorization: $(x + 2)(x + 9)$. The sum of the outside and inside terms is $9x + 2x = 11x$, so this product can't be $x^2 + 9x + 18$. By trying other possibilities we find that $b = 3, d = 6$ leads to the correct factorization: $x^2 + 9x + 18 = (x + 3)(x + 6)$. $\blacksquare$

Example 6

To factor $6x^2 + 11x + 4$ as $(ax + b)(cx + d)$, we must find numbers a and c whose product is 6, the coefficient of x^2, and numbers b and d whose product is the constant term 4. Some possibilities are:

$ac = 6$

a	± 1	± 2	± 3	± 6
c	± 6	± 3	± 2	± 1

$bd = 4$

b	± 1	± 2	± 4
d	± 4	± 2	± 1

Trial and error shows that $(2x + 1)(3x + 4) = 6x^2 + 11x + 4$. $\blacksquare$

* FOIL is explained on page 35.

Example 7 Every term of $24y^5 - 14y^4 - 5y^3$ contains a factor of y^3. Hence $24y^5 - 14y^4 - 5y^3 = y^3(24y^2 - 14y - 5)$. Now use the techniques in the preceding examples to factor $24y^2 - 14y - 5$:

$$24y^5 - 14y^4 - 5y^3 = y^3(24y^2 - 14y - 5) = y^3(6y - 5)(4y + 1). \blacksquare$$

Example 8 To factor $x^2 + 6xy - 40y^2$, rewrite it as $x^2 + (6y)x - 40y^2$ and consider it as a polynomial in x whose coefficients involve the number y. To factor it as $(ax + b)(cx + d)$ we must find a, c such that $ac = 1$ and b, d such that $bd = -40y^2$. Furthermore, the sum of the outside and inside terms must be $6xy$. Trial and error shows that $(x + 10y)(x - 4y) = x^2 + 6xy - 40y^2$. $\blacksquare$

Occasionally the patterns above can be used to factor expressions involving larger exponents than 3.

Example 9

(a) $\quad x^6 - y^6 = (x^3)^2 - (y^3)^2 = (x^3 + y^3)(x^3 - y^3)$
$\qquad\qquad = (x + y)(x^2 - xy + y^2)(x - y)(x^2 + xy + y^2).$

(b) $\quad x^8 - 1 = (x^4)^2 - 1 = (x^4 + 1)(x^4 - 1)$
$\qquad\qquad = (x^4 + 1)(x^2 + 1)(x^2 - 1)$
$\qquad\qquad = (x^4 + 1)(x^2 + 1)(x + 1)(x - 1). \blacksquare$

Example 10 To factor $x^4 - 2x^2 - 3$, let $u = x^2$. Then

$$x^4 - 2x^2 - 3 = (x^2)^2 - 2x^2 - 3$$
$$= u^2 - 2u - 3 = (u + 1)(u - 3)$$
$$= (x^2 + 1)(x^2 - 3)$$
$$= (x^2 + 1)(x + \sqrt{3})(x - \sqrt{3}). \blacksquare$$

Example 11 $3x^3 + 3x^2 + 2x + 2$ can be factored by regrouping and using the distributive law to factor out a common factor:

$$(3x^3 + 3x^2) + (2x + 2) = 3x^2(x + 1) + 2(x + 1)$$
$$= (3x^2 + 2)(x + 1). \blacksquare$$

EXERCISES

In Exercises 1–58, factor the expression.

1. $x^2 - 4$
2. $x^2 + 6x + 9$
3. $9y^2 - 25$
4. $y^2 - 4y + 4$
5. $81x^2 + 36x + 4$
6. $4x^2 - 12x + 9$
7. $5 - x^2$
8. $1 - 36u^2$
9. $49 + 28z + 4z^2$

10. $25u^2 - 20uv + 4v^2$
11. $x^4 - y^4$
12. $x^2 - \frac{1}{9}$
13. $x^2 + x - 6$
14. $y^2 + 11y + 30$
15. $z^2 + 4z + 3$
16. $x^2 - 8x + 15$
17. $y^2 + 5y - 36$
18. $z^2 - 9z + 14$
19. $x^2 - 6x + 9$
20. $4y^2 - 81$
21. $x^2 + 7x + 10$
22. $w^2 - 6w - 16$
23. $x^2 + 11x + 18$
24. $x^2 + 3xy - 28y^2$
25. $3x^2 + 4x + 1$
26. $4y^2 + 4y + 1$
27. $2z^2 + 11z + 12$
28. $10x^2 - 17x + 3$
29. $9x^2 - 72x$
30. $4x^2 - 4x - 3$
31. $10x^2 - 8x - 2$
32. $7z^2 + 23z + 6$
33. $8u^2 + 6u - 9$
34. $2y^2 - 4y + 2$
35. $4x^2 + 20xy + 25y^2$
36. $63u^2 - 46uv + 8v^2$
37. $x^3 - 125$
38. $y^3 + 64$
39. $x^3 + 6x^2 + 12x + 8$
40. $y^3 - 3y^2 + 3y - 1$
41. $8 + x^3$
42. $z^3 - 9z^2 + 27z - 27$
43. $-x^3 + 15x^2 - 75x + 125$
44. $27 - t^3$
45. $x^3 + 1$
46. $x^3 - 1$
47. $8x^3 - y^3$
48. $(x - 1)^3 + 1$
49. $x^6 - 64$
50. $x^5 - 8x^2$
51. $y^4 + 7y^2 + 10$
52. $z^4 - 5z^2 + 6$
53. $81 - y^4$
54. $x^6 + 16x^3 + 64$
55. $z^6 - 1$
56. $y^6 + 26y^3 - 27$
57. $x^4 + 2x^2y - 3y^2$
58. $x^8 - 17x^4 + 16$

In Exercises 59–68, factor by regrouping and using the distributive law (as in Example 11).

59. $x^2 - yz + xz - xy$
60. $x^6 - 2x^4 - 8x^2 + 16$
61. $a^3 - 2b^2 + 2a^2b - ab$
62. $u^2v - 2w^2 - 2uvw + uw$
63. $x^3 + 4x^2 - 8x - 32$
64. $z^8 - 5z^7 + 2z - 10$
65. $2x^2 + 5xy - 3y^2 + 6x - 3y$
66. $x^2 - 9xy + 14y^2 + 3xy - 6y^3$
67. $x^3 + x - 3y - 27y^3$
68. $8u^3 + 10u + v^3 + 5v$

In Exercises 69–74, factor the expression, as in this example:

$$x^2 - \frac{x}{4} - \frac{3}{8} = \left(x + \frac{1}{2}\right)\left(x - \frac{3}{4}\right).$$

69. $x^2 - \frac{1}{64}$
70. $x^3 - \frac{1}{8}$
71. $y^2 - \frac{2y}{3} - \frac{5}{36}$
72. $x^2 + x - \frac{3}{4}$
73. $z^2 + 3z + \frac{35}{16}$
74. $t^2 - \frac{t}{3} - \frac{2}{9}$

Unusual Problem

75. Show that there do *not* exist real numbers c and d such that $x^2 + 1 = (x + c)(x + d)$.

1.7 FRACTIONAL EXPRESSIONS

Quotients of algebraic expressions are called **fractional expressions**. A quotient of two polynomials is sometimes called a **rational expression**. The basic rules for dealing with fractional expressions are essentially the same as those for ordinary numerical fractions. For instance, $\frac{2}{4} = \frac{3}{6}$ and the "cross products" are equal: $2 \cdot 6 = 4 \cdot 3$. In the general case we have:

PROPERTIES OF FRACTIONS

1. **Equality Rule:** $\dfrac{a}{b} = \dfrac{c}{d}$ exactly when $ad = bc$.*

2. **Cancellation Property:** If $k \neq 0$, then $\dfrac{ka}{kb} = \dfrac{a}{b}$.

The cancellation property follows directly from the equality rule because $(ka)b = (kb)a$.

Example 1 Here are examples of the two properties:

1. $\dfrac{x^2 + 2x}{x^2 + x - 2} = \dfrac{x}{x - 1}$ because the cross products are equal:

$$(x^2 + 2x)(x - 1) = x^3 + x^2 - 2x = (x^2 + x - 2)x.$$

2. $\dfrac{x^4 - 1}{x^2 + 1} = \dfrac{(x^2 + 1)(x^2 - 1)}{(x^2 + 1)} = \dfrac{x^2 - 1}{1} = x^2 - 1.$ ■

A fraction is in **lowest terms** if its **numerator** (top) and **denominator** (bottom) have no common factors except ± 1. To express a fraction in lowest terms, factor numerator and denominator and cancel common factors.

Example 2 $\dfrac{x^2 + x - 6}{x^2 - 3x + 2} = \dfrac{(x - 2)(x + 3)}{(x - 2)(x - 1)} = \dfrac{x + 3}{x - 1}.$ ■

To add two fractions with the same denominator, simply add the numerators as in ordinary arithmetic: $\dfrac{a}{b} + \dfrac{c}{b} = \dfrac{a + c}{b}$. Subtraction is done similarly.

Example 3

$$\dfrac{7x^2 + 2}{x^2 + 3} - \dfrac{4x^2 + 2x - 5}{x^2 + 3} = \dfrac{(7x^2 + 2) - (4x^2 + 2x - 5)}{x^2 + 3}$$

$$= \dfrac{7x^2 + 2 - 4x^2 - 2x + 5}{x^2 + 3}$$

$$= \dfrac{3x^2 - 2x + 7}{x^2 + 3}. \ ■$$

To add or subtract fractions with different denominators, you must first find a common denominator. One common denominator for a/b and c/d is the

---

* Throughout this section we assume that all denominators are nonzero.

product of the two denominators bd because both fractions can be expressed with this denominator:

$$\frac{a}{b} = \frac{ad}{bd} \quad \text{and} \quad \frac{c}{d} = \frac{bc}{bd}.$$

Consequently,

$$\frac{a}{b} + \frac{c}{d} = \frac{ad}{bd} + \frac{bc}{bd} = \frac{ad + bc}{bd} \quad \text{and} \quad \frac{a}{b} - \frac{c}{d} = \frac{ad}{bd} - \frac{bc}{bd} = \frac{ad - bc}{bd}.$$

Example 4

$$\frac{2x + 1}{3x} - \frac{x^2 - 2}{x - 1} = \frac{(2x + 1)(x - 1)}{3x(x - 1)} - \frac{3x(x^2 - 2)}{3x(x - 1)}$$

$$= \frac{(2x + 1)(x - 1) - 3x(x^2 - 2)}{3x(x - 1)} = \frac{2x^2 - x - 1 - 3x^3 + 6x}{3x^2 - 3x}$$

$$= \frac{-3x^3 + 2x^2 + 5x - 1}{3x^2 - 3x}. \quad \blacksquare$$

Although the product of the denominators can always be used as a common denominator, it's often more efficient to use the *least common denominator*. The least common denominator can be found by factoring each denominator completely (with integer coefficients) and then taking the product of the highest power of each of the distinct factors.

Example 5 In the sum $\frac{1}{100} + \frac{1}{120}$, the denominators are $100 = 2^2 \cdot 5^2$ and $120 = 2^3 \cdot 3 \cdot 5$. The distinct factors are 2, 3, 5. The highest exponent of 2 is 3, the highest of 3 is 1, and the highest of 5 is 2. So the least common denominator is $2^3 \cdot 3 \cdot 5^2 = 600$. $\blacksquare$

Example 6 To find the least common denominator of $\frac{1}{x^2 + 2x + 1}$, $\frac{5x}{x^2 - x}$, and $\frac{3x - 7}{x^4 + x^3}$, factor each of the denominators completely:

$$(x + 1)^2, \quad x(x - 1), \quad x^3(x + 1).$$

The distinct factors are x, $x + 1$, and $x - 1$. The least common denominator is $x^3(x + 1)^2(x - 1)$. $\blacksquare$

To express one of several fractions in terms of the least common denominator, multiply its numerator and denominator by those factors in the common denominator that *don't* appear in the denominator of the fraction.

Example 7 The preceding example shows that the least common denominator of $\frac{1}{(x+1)^2}, \frac{5x}{x(x-1)},$ and $\frac{3x-7}{x^3(x+1)}$ is $x^3(x+1)^2(x-1)$. Therefore

$$\frac{1}{(x+1)^2} = \frac{1}{(x+1)^2} \cdot \frac{x^3(x-1)}{x^3(x-1)} = \frac{x^3(x-1)}{x^3(x+1)^2(x-1)}$$

$$\frac{5x}{x(x-1)} = \frac{5x}{x(x-1)} \cdot \frac{x^2(x+1)^2}{x^2(x+1)^2} = \frac{5x^3(x+1)^2}{x^3(x+1)^2(x-1)}$$

$$\frac{3x-7}{x^3(x+1)} = \frac{3x-7}{x^3(x+1)} \cdot \frac{(x+1)(x-1)}{(x+1)(x-1)} = \frac{(3x-7)(x+1)(x-1)}{x^3(x+1)^2(x-1)}. \blacksquare$$

Example 8 To find $\frac{1}{z} + \frac{3z}{z+1} - \frac{z^2}{(z+1)^2}$ we use the least common denominator $z(z+1)^2$:

$$\frac{1}{z} + \frac{3z}{z+1} - \frac{z^2}{(z+1)^2} = \frac{(z+1)^2}{z(z+1)^2} + \frac{3z^2(z+1)}{z(z+1)^2} - \frac{z^3}{z(z+1)^2}$$

$$= \frac{(z+1)^2 + 3z^2(z+1) - z^3}{z(z+1)^2}$$

$$= \frac{z^2 + 2z + 1 + 3z^3 + 3z^2 - z^3}{z(z+1)^2}$$

$$= \frac{2z^3 + 4z^2 + 2z + 1}{z(z+1)^2}. \blacksquare$$

Multiplication of fractions is easy: Multiply corresponding numerators and denominators, then simplify your answer:

Example 9

$$\frac{x^2-1}{x^2+2} \cdot \frac{3x-4}{x+1} = \frac{(x^2-1)(3x-4)}{(x^2+2)(x+1)}$$

$$= \frac{(x-1)(x+1)(3x-4)}{(x^2+2)(x+1)} = \frac{(x-1)(3x-4)}{x^2+2}. \blacksquare$$

Division of fractions is given by the rule:

Invert the divisor and multiply: $\frac{a}{b} \div \frac{c}{d} = \frac{a}{b} \cdot \frac{d}{c} = \frac{ad}{bc}.$

1.7 FRACTIONAL EXPRESSIONS

Example 10

$$\frac{x^2 + x - 2}{x^2 - 6x + 9} \div \frac{x^2 - 1}{x - 3} = \frac{x^2 + x - 2}{x^2 - 6x + 9} \cdot \frac{x - 3}{x^2 - 1}$$

$$= \frac{(x + 2)(x - 1)}{(x - 3)^2} \cdot \frac{x - 3}{(x - 1)(x + 1)}$$

$$= \frac{x + 2}{(x - 3)(x + 1)}. \blacksquare$$

Division problems can also be written as fractions. For instance, $\frac{8}{2}$ means $8 \div 2 = 4$. Similarly, the compound fraction $\dfrac{a/b}{c/d}$ means $\dfrac{a}{b} \div \dfrac{c}{d}$. So the basic rule for simplifying compound fractions is: *Invert the denominator and multiply it by the numerator.*

Example 11

(a) $\dfrac{16y^2z/8yz^2}{yz/6y^3z^3} = \dfrac{16y^2z}{8yz^2} \cdot \dfrac{6y^3z^3}{yz} = \dfrac{16 \cdot 6 \cdot y^5z^4}{8y^2z^3} = 2 \cdot 6 \cdot y^{5-2}z^{4-3} = 12y^3z.$

(b) $\dfrac{\dfrac{y^2}{y + 2}}{\dfrac{y^3 + y}{}} = \dfrac{y^2}{y + 2} \cdot \dfrac{1}{y^3 + y} = \dfrac{y^2}{(y + 2)(y^3 + y)}$

$$= \dfrac{y^2}{(y + 2)y(y^2 + 1)} = \dfrac{y}{(y + 2)(y^2 + 1)}. \blacksquare$$

Example 12 In order to simplify

$$\dfrac{\dfrac{3}{x^2 - 4} + \dfrac{1}{x + 2}}{5 - \dfrac{6}{x - 2}}$$

we first use the fact that $x^2 - 4 = (x + 2)(x - 2)$ to find a common denominator for the sum above the heavy line and then continue:

$$\dfrac{\dfrac{3}{x^2 - 4} + \dfrac{1}{x + 2}}{5 - \dfrac{6}{x - 2}} = \dfrac{\dfrac{3}{x^2 - 4} + \dfrac{x - 2}{x^2 - 4}}{\dfrac{5(x - 2)}{x - 2} - \dfrac{6}{x - 2}} = \dfrac{\dfrac{3 + x - 2}{x^2 - 4}}{\dfrac{5(x - 2) - 6}{x - 2}} = \dfrac{\dfrac{1 + x}{x^2 - 4}}{\dfrac{5x - 16}{x - 2}}$$

$$= \dfrac{1 + x}{x^2 - 4} \cdot \dfrac{x - 2}{5x - 16}$$

$$= \dfrac{(1 + x)(x - 2)}{(x + 2)(x - 2)(5x - 16)} = \dfrac{1 + x}{(x + 2)(5x - 16)}. \blacksquare$$

Example 13 $\dfrac{(2(x+h))^{-1} - (2x)^{-1}}{h}$ may be simplified as follows:

$$\dfrac{(2(x+h))^{-1} - (2x)^{-1}}{h} = \dfrac{\dfrac{1}{2(x+h)} - \dfrac{1}{2x}}{h} = \dfrac{\dfrac{x}{2x(x+h)} - \dfrac{x+h}{2x(x+h)}}{h}$$

$$= \dfrac{\dfrac{x-(x+h)}{2x(x+h)}}{h} = \dfrac{\dfrac{-h}{2x(x+h)}}{h} = \dfrac{-h}{2x(x+h)} \cdot \dfrac{1}{h}$$

$$= \dfrac{-1}{2x(x+h)}. \blacksquare$$

Rationalizing Numerators and Denominators

It is sometimes necessary to eliminate all radicals from either the numerator or denominator of a fraction. For instance, we can eliminate the radical in the denominator of $1/\sqrt{2}$ by using the fact that $\sqrt{2}/\sqrt{2} = 1$:

$$\dfrac{1}{\sqrt{2}} = \dfrac{1}{\sqrt{2}} \cdot \dfrac{\sqrt{2}}{\sqrt{2}} = \dfrac{\sqrt{2}}{2}$$

Here are some other examples of **rationalizing the denominator:**

Example 14 $\sqrt{\dfrac{5}{2x+1}} = \dfrac{\sqrt{5}}{\sqrt{2x+1}} = \dfrac{\sqrt{5}}{\sqrt{2x+1}} \cdot \dfrac{\sqrt{2x+1}}{\sqrt{2x+1}} = \dfrac{\sqrt{10x+5}}{2x+1}. \blacksquare$

Example 15 In order to rationalize the denominator of $\dfrac{7}{\sqrt{5}+\sqrt{3}}$, we must multiply both top and bottom by something which will eliminate the radicals in the denominator. Observe that $(\sqrt{5}+\sqrt{3})(\sqrt{5}-\sqrt{3}) = (\sqrt{5})^2 - (\sqrt{3})^2 = 5 - 3 = 2$. Thus

$$\dfrac{7}{\sqrt{5}+\sqrt{3}} = \left(\dfrac{7}{\sqrt{5}+\sqrt{3}}\right)\left(\dfrac{\sqrt{5}-\sqrt{3}}{\sqrt{5}-\sqrt{3}}\right) = \dfrac{7(\sqrt{5}-\sqrt{3})}{(\sqrt{5}+\sqrt{3})(\sqrt{5}-\sqrt{3})} = \dfrac{7(\sqrt{5}-\sqrt{3})}{2}. \blacksquare$$

The same techniques can be used to **rationalize the numerator** of a fraction.

Example 16

$$\dfrac{\sqrt{x+3}-\sqrt{3}}{5} = \dfrac{\sqrt{x+3}-\sqrt{3}}{5} \cdot \dfrac{\sqrt{x+3}+\sqrt{3}}{\sqrt{x+3}+\sqrt{3}}$$

$$= \frac{(\sqrt{x+3})^2 - (\sqrt{3})^2}{5(\sqrt{x+3} + \sqrt{3})} = \frac{x+3-3}{5(\sqrt{x+3} + \sqrt{3})}$$

$$= \frac{x}{5\sqrt{x+3} + 5\sqrt{3}}. \blacksquare$$

EXERCISES

In Exercises 1–10, express the fraction in lowest terms.

1. $\dfrac{63}{49}$

2. $\dfrac{121}{33}$

3. $\dfrac{13 \cdot 27 \cdot 22 \cdot 10}{6 \cdot 4 \cdot 11 \cdot 12}$

4. $\dfrac{x^2 - 4}{x + 2}$

5. $\dfrac{x^2 - x - 2}{x^2 + 2x + 1}$

6. $\dfrac{z + 1}{z^3 + 1}$

7. $\dfrac{a^2 - b^2}{a^3 - b^3}$

8. $\dfrac{x^4 - 3x^2}{x^3}$

9. $\dfrac{(x + c)(x^2 - cx + c^2)}{x^4 + c^3 x}$

10. $\dfrac{x^4 - y^4}{(x^2 + y^2)(x^2 - xy)}$

In Exercises 11–28, perform the indicated operations.

11. $\dfrac{3}{7} + \dfrac{2}{5}$

12. $\dfrac{7}{8} - \dfrac{5}{6}$

13. $\left(\dfrac{19}{7} + \dfrac{1}{2}\right) - \dfrac{1}{3}$

14. $\dfrac{1}{a} - \dfrac{2a}{b}$

15. $\dfrac{c}{d} + \dfrac{3c}{e}$

16. $\dfrac{r}{s} + \dfrac{s}{t} + \dfrac{t}{r}$

17. $\dfrac{b}{c} - \dfrac{c}{b}$

18. $\dfrac{a}{b} + \dfrac{2a}{b^2} + \dfrac{3a}{b^3}$

19. $\dfrac{1}{x+1} - \dfrac{1}{x}$

20. $\dfrac{1}{2x+1} + \dfrac{1}{2x-1}$

21. $\dfrac{1}{x+4} + \dfrac{2}{(x+4)^2} - \dfrac{3}{x^2 + 8x + 16}$

22. $\dfrac{1}{x} + \dfrac{1}{xy} + \dfrac{1}{xy^2}$

23. $\dfrac{1}{x} - \dfrac{1}{3x-4}$

24. $\dfrac{3}{x-1} + \dfrac{4}{x+1}$

25. $\dfrac{1}{x+y} + \dfrac{x+y}{x^3 + y^3}$

26. $\dfrac{6}{5(x-1)(x-2)^2} + \dfrac{x}{3(x-1)^2(x-2)}$

27. $\dfrac{1}{4x(x+1)(x+2)^3} - \dfrac{6x+2}{4(x+1)^3}$

28. $\dfrac{x+y}{(x^2 - xy)(x-y)^2} - \dfrac{2}{(x^2 - y^2)^2}$

In Exercises 29–42, express in lowest terms.

29. $\dfrac{3}{4} \cdot \dfrac{12}{5} \cdot \dfrac{10}{9}$

30. $\dfrac{10}{45} \cdot \dfrac{6}{14} \cdot \dfrac{1}{2}$

31. $\dfrac{3a^2 c}{4ac} \cdot \dfrac{8ac^3}{9a^2 c^4}$

32. $\dfrac{6x^2 y}{2x} \cdot \dfrac{y}{21xy}$

33. $\dfrac{7x}{11y} \cdot \dfrac{66y^2}{14x^3}$

34. $\dfrac{ab}{c^2} \cdot \dfrac{cd}{a^2 b} \cdot \dfrac{ad}{bc^2}$

35. $\dfrac{3x + 9}{2x} \cdot \dfrac{8x^2}{x^2 - 9}$

36. $\dfrac{4x + 16}{3x + 15} \cdot \dfrac{2x + 10}{x + 4}$

37. $\dfrac{5y - 25}{3} \cdot \dfrac{y^2}{y^2 - 25}$

38. $\dfrac{6x - 12}{6x} \cdot \dfrac{8x^2}{x - 2}$

39. $\dfrac{u}{u - 1} \cdot \dfrac{u^2 - 1}{u^2}$

40. $\dfrac{t^2 - t - 6}{t^2 - 6t + 9} \cdot \dfrac{t^2 + 4t - 5}{t^2 - 25}$

41. $\dfrac{2u^2 + uv - v^2}{4u^2 - 4uv + v^2} \cdot \dfrac{8u^2 + 6uv - 9v^2}{4u^2 - 9v^2}$

42. $\dfrac{2x^2 - 3xy - 2y^2}{6x^2 - 5xy - 4y^2} \cdot \dfrac{6x^2 + 6xy}{x^2 - xy - 2y^2}$

56 1 BASIC ALGEBRA

In Exercises 43–60, compute the quotient and express in lowest terms.

43. $\dfrac{5}{12} \div \dfrac{4}{14}$

44. $\dfrac{\frac{100}{52}}{\frac{27}{26}}$

45. $\dfrac{uv}{v^2w} \div \dfrac{uv}{u^2v}$

46. $\dfrac{3x^2y}{(xy)^2} \div \dfrac{3xyz}{x^2y}$

47. $\dfrac{\frac{x+3}{x+4}}{\frac{2x}{x+4}}$

48. $\dfrac{\frac{(x+2)^2}{(x-2)^2}}{\frac{x^2+2x}{x^2-4}}$

49. $\dfrac{x+y}{x+2y} \div \left(\dfrac{x+y}{xy}\right)^2$

50. $\dfrac{\frac{u^3+v^3}{u^2-v^2}}{\frac{u^2-uv+v^2}{u+v}}$

51. $\dfrac{\frac{(c+d)^2}{c^2-d^2}}{cd}$

52. $\dfrac{\frac{1}{x}-\frac{3}{2}}{\frac{2}{x-2}+\frac{5}{x}}$

53. $\dfrac{\frac{1}{x^2}-\frac{1}{y^2}}{\frac{1}{x}+\frac{1}{y}}$

54. $\dfrac{\frac{x}{x+1}+\frac{1}{x}}{\frac{1}{x}+\frac{1}{x+1}}$

55. $\dfrac{\frac{6}{y}-3}{1-\frac{1}{y-1}}$

56. $\dfrac{\frac{1}{3x}-\frac{1}{4y}}{\frac{5}{6x^2}+\frac{1}{y}}$

57. $\dfrac{\frac{1}{x+h}-\frac{1}{x}}{h}$

58. $\dfrac{\frac{1}{(x+h)^2}-\frac{1}{x^2}}{h}$

59. $(x^{-1}+y^{-1})^{-1}$

60. $\dfrac{(x+y)^{-1}}{x^{-1}+y^{-1}}$

In Exercises 61–69, rationalize the denominator.

61. $\dfrac{2}{\sqrt{5}}$

62. $\sqrt{\dfrac{16}{5}}$

63. $\sqrt{\dfrac{7}{10}}$

64. $\sqrt{\dfrac{9x^4}{23}}$

65. $\dfrac{1}{\sqrt{x}}$

66. $\dfrac{\sqrt{6}}{\sqrt{6}+\sqrt{2}}$

67. $\dfrac{x+1}{\sqrt{x}+1}$

68. $\dfrac{\sqrt{r}+\sqrt{s}}{\sqrt{r}-\sqrt{s}}$

69. $\dfrac{1}{\sqrt{a}-2\sqrt{b}}$

70. $\dfrac{u^2-v^2}{\sqrt{u+v}-\sqrt{u-v}}$

Errors to Avoid

In Exercises 71–77, find a numerical example to show that the given statement is *false*. Then find the mistake in the statement and correct it.

71. $\dfrac{1}{a}+\dfrac{1}{b}=\dfrac{1}{a+b}$

72. $\dfrac{x^2}{x^2+x^6}=1+x^3$

73. $\left(\dfrac{1}{\sqrt{a}+\sqrt{b}}\right)^2=\dfrac{1}{a+b}$

74. $\dfrac{r+s}{r+t}=1+\dfrac{s}{t}$

75. $\dfrac{u}{v}+\dfrac{v}{u}=1$

76. $\dfrac{\frac{1}{x}}{\frac{1}{y}}=\dfrac{1}{xy}$

77. $(\sqrt{x}+\sqrt{y})\dfrac{1}{\sqrt{x}+\sqrt{y}}=x+y$

CHAPTER REVIEW

Important Concepts

Section 1.1 Real numbers, rational numbers, irrational numbers 1
Number line 1
Order ($<, \leq, >, \geq$) 2
Negatives 2
Commutative, associative, and distributive laws 3
Hierarchy of operations 5

CHAPTER 1 REVIEW

Excursion 1.1.A	Repeating and nonrepeating decimals	9
Section 1.2	Integral exponents	9
	Exponent laws	10
	Scientific notation	17
Section 1.3	Square root	20
	nth root	20
	Rational exponents	22
Section 1.4	Absolute value	26
	Distance on the number line	28
	Intervals, open intervals, closed intervals	30
Section 1.5	Algebraic expression	33
	Multiplication patterns	35
	Polynomial	36
	Degree of a polynomial	37
	Polynomial division	37
	Division Algorithm	39
Excursion 1.5.A	Synthetic division	43
Section 1.6	Factorization	46
	Difference of Squares	46
	Perfect Squares	46
	Sum and Difference of Cubes	46
	Perfect Cubes	46
Section 1.7	Equality Rule	50
	Cancellation Property	50
	Fractions in lowest terms	50
	Arithmetic of fractions	51–53
	Rationalizing numerators and denominators	54

Important Facts and Formulas

- Laws of Exponents: $c^r c^s = c^{r+s}$ $(cd)^r = c^r d^r$

$$\frac{c^r}{c^s} = c^{r-s} \qquad \left(\frac{c}{d}\right)^r = \frac{c^r}{d^r}$$

$$(c^r)^s = c^{rs} \qquad c^{-r} = \frac{1}{c^r}$$

- $|c - d|$ = distance from c to d on the number line.

Review Questions

1. Fill the blanks with one of the symbols $<, =, >$ so that the resulting statement is true.

    (a) 141 ___ $|-51|$

    (b) $\sqrt{2}$ ___ $|-2|$

    (c) -1000 ___ $\dfrac{1}{10}$

(d) $|-2|$ ___ $-|6|$

(e) $|u - v|$ ___ $|v - u|$, where u, v are fixed real numbers.

2. List two real numbers that are *not* rational numbers.
3. Express $0.282828\ldots$ as a fraction.
4. Express $0.362362362\ldots$ as a fraction.
5. Express in symbols:

    (a) y is negative, but greater than -10;

    (b) x is nonnegative and not greater than 10.

6. Express in symbols:

    (a) $c - 7$ is nonnegative;

    (b) .6 is greater than $|5x - 2|$.

7. Express in symbols:

    (a) x is less than 3 units from -7 on the number line.

    (b) y is farther from 0 than x is from 3 on the number line.

8. Express in scientific notation: $.0000000457$
9. Express in scientific notation: $\dfrac{231 \times 10^7}{7 \times 10^{-4}}$
10. Simplify: $|b^2 - 2b + 1|$
11. Solve: $|x - 5| = 3$
12. Solve: $|x + 2| = 4$
13. Solve: $|x + 3| = \dfrac{5}{2}$
14. Solve: $|x - 5| \leq 2$
15. Solve: $|x + 2| \leq 2$
16. Solve: $|x - 1| > 4$
17. (a) $|\pi - 7| =$ _____ (b) $|\sqrt{23} - \sqrt{3}| =$ _____
18. If c and d are real numbers with $c \neq d$, what are the possible values of $\dfrac{c - d}{|c - d|}$?
19. Express in interval notation:

    (a) the set of all real numbers that are strictly greater than -8;

    (b) the set of all real numbers that are less than or equal to 5.

20. Express in interval notation:

   (a) the set of all real numbers that are strictly between -6 and 9.

   (b) the set of all real numbers that are greater than or equal to 5, but strictly less than 14.

21. Express without negative exponents: $(c^3 d^{-3} e^{-1})^5$

22. Simplify: $\sqrt{\sqrt[3]{c^{12}}}$

In questions 23–42, compute and simplify.

23. $\dfrac{|x^2 - 6x + 9|}{(3 - x)^2}$

24. $\sqrt{8x^4 y^{10}}$ (x, y positive)

25. $(-3)^{-3}$

26. $\dfrac{2a + 4a^2}{2a}$

27. $(\sqrt{2} + \sqrt{5})^2$

28. $\dfrac{x^2 - 4}{2x} \cdot \dfrac{6}{3x - 6}$

29. $(2x^2 + 5y)^3$

30. $\left(\dfrac{v}{s} - \dfrac{2s}{r}\right)^2$

31. $(u^2 + v^2 - 2w^2)^2$

32. $\dfrac{4x - 7}{8} - \dfrac{3x - 5}{12}$

33. $\dfrac{2x - 3}{x - 2} + \dfrac{x}{x + 2}$

34. $1 - \dfrac{x}{x - y}$

35. $\dfrac{\dfrac{1}{c} - \dfrac{1}{c^2}}{\dfrac{1}{c} + \dfrac{1}{c^2}}$

36. $\dfrac{\sqrt{r}}{\sqrt{r} + \sqrt{s}} + \dfrac{\sqrt{s}}{\sqrt{r} - \sqrt{s}}$

37. Simplify and use only positive exponents in your answer: $\dfrac{(8u^5)^2 2^{-4} u^{-3}}{2u^8}$

38. Simplify: $(\sqrt[3]{4}\, c^3 d^2)^3 (c\sqrt{d})^2$

39. Simplify and express without radicals: $\dfrac{\sqrt[3]{6 c^4 d^{14}}}{\sqrt[3]{48 c^{-2} d^2}}$

40. $3x(2x - 2)^2 = ?$

41. What is the remainder when $3x^4 + x^3 - x^2 + 3x + 1$ is divided by $x^2 + x$?

42. $(\sqrt{x} + 2y)(\sqrt{x} - y) = ?$

In questions 43–54, factor the given expression.

43. $x^2 + 5x + 4$ 44. $x^2 - 4x + 3$
45. $x^2 - 2x - 8$ 46. $x^2 - 10x + 25$
47. $4x^2 - 9$ 48. $49y^2 - \frac{1}{4}$
49. $4x^2 - 4x - 15$ 50. $6x^2 - x - 2$
51. $3x^3 + 5x^2 - 2x$ 52. $2x^3 + 4x^2 - 6x$
53. $x^4 - 1$ 54. $x^3 - 8$

55. Rationalize the denominator: $7/\sqrt{3}$

56. Express in lowest terms: $\dfrac{x^2 - 3x + 2}{x^2 - 4} \cdot \dfrac{x + 2}{x - 3}$

57. Let c be a positive real number. Rationalize the denominator: $\dfrac{1}{\sqrt{3} + 2\sqrt{c}}$

58. Use synthetic division to show that $x - 2$ is a factor of $x^6 - 5x^5 + 8x^4 + x^3 - 17x^2 + 16x - 4$ and find the other factor.

59. Which of the following statements is *false* for *all* real numbers t?

    (a) $t^2 < t$ (b) $t - 1 < t$
    (c) $2t < t$ (d) $t + 7 < t$
    (e) $t < t + 1$

60. A number x is one of the five numbers A, B, C, D, E on the number line shown in Figure 1–9.

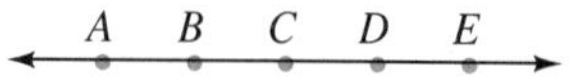

    FIGURE 1–9

    If x satisfies both $A \le x < D$ and $B < x \le E$, then
    (a) $x = A$ (b) $x = B$
    (c) $x = C$ (d) $x = D$
    (e) $x = E$

61. Let a and b be positive real numbers. Which of these statements is *false*?

    (a) $a^r a^s = a^{r+s}$ (b) $\sqrt{a}\sqrt{b} = \sqrt{ab}$
    (c) $\sqrt[3]{a^3} = a$ (d) $\sqrt{a + b} = \sqrt{a} + \sqrt{b}$
    (e) $\sqrt{\sqrt{a}} = \sqrt[4]{a}$

62. Which one of these statements is *always* true for any real numbers x, y.
    (a) $|2x| = 2x$
    (b) $\sqrt{x^2} = x$
    (c) $|x - y| = x - y$
    (d) $|x - y| = |y - x|$
    (e) $|x - y| = |x| + |y|$

63. Which one of these intervals contains *both* -3 and 2?
    (a) $(-4, 2)$
    (b) $(-3, 2]$
    (c) $[-2, 2)$
    (d) $[-3, 3]$
    (e) $[-2, 3]$

64. Which of these statements is *true*?
    (a) $(4x)^3 = 4x^3$
    (b) $(x + y)^2 = x^2 + y^2$
    (c) $(a + b)(a^2 + b^2) = a^2 + b^3$
    (d) $x - (2y + 6) = x - 2y + 6$
    (e) none of the above

CHAPTER 2

Equations and Inequalities

This chapter deals with equations and inequalities, such as

$$x^2 + 8x + 3 = 0 \qquad 4x + 3 < 9 \qquad |2x^2 + 3x - 1| \geq 3$$

A **solution** of an equation or inequality is a number that, when substituted for the variable x, produces a true statement.* For example, 5 is a solution of $3x + 2 = 17$ because $3 \cdot 5 + 2 = 17$ is a true statement. To **solve** an equation or inequality means to find all its solutions.

Two equations are said to be **equivalent** if they have the same solutions. For example, $3x + 2 = 17$ and $x - 2 = 3$ are equivalent because 5 is the only solution of each one.

BASIC PRINCIPLES FOR SOLVING EQUATIONS

> **Performing any of the following operations on an equation produces an equivalent equation:**
>
> 1. **Add or subtract the same quantity from both sides of the equation.**
> 2. **Multiply or divide both sides of the equation by the same *nonzero* quantity.**

The usual strategy in equation solving is to use these Basic Principles to transform a given equation into an equivalent one whose solutions are known. Analogous principles for solving inequalities are presented in Section 2.7.

2.1 FIRST-DEGREE EQUATIONS

Solving most first-degree equations is quite straightforward. Here are some step-by-step examples.

Example 1 To solve $3x - 6 = 7x + 4$ we use the Basic Principles to transform this equation into an equivalent one whose solution is obvious:

---

* Any letter may be used for the variable.

2.1 FIRST-DEGREE EQUATIONS

$$3x - 6 = 7x + 4$$

Add 6 to both sides: $\quad 3x = 7x + 10$

Subtract 7x from both sides: $\quad -4x = 10$

Divide both sides by -4: $\quad x = \dfrac{10}{-4} = -\dfrac{5}{2}$

Since $-5/2$ is the only solution of this last equation, $-5/2$ is the only solution of the original equation, $3x - 6 = 7x + 4$. ∎

WARNING

To guard against mistakes, check your solutions by substituting each one in the *original* equation to make sure it really *is* a solution.

Example 2 A calculator can be used to solve

$$42.19x + 121.34 = 16.83x + 19.15.$$

To avoid round-off errors in the intermediate steps, do all the algebra first without using the calculator:

Subtract 121.34 from both sides: $\quad 42.19x = 16.83x + 19.15 - 121.34$

Subtract 16.83x from both sides: $\quad 42.19x - 16.83x = 19.15 - 121.34$

Factor out x on the left side: $\quad (42.19 - 16.83)x = 19.15 - 121.34$

Divide both sides by $(42.19 - 16.83)$: $\quad x = \dfrac{19.15 - 121.34}{42.19 - 16.83}$

Now use the calculator to find $x \approx -4.03$. Because this answer is an approximation it may not check exactly when substituted in the original equation. ∎

Example 3 The first step in solving

$$\frac{(t + 7)(2t - 1)}{2} = \frac{(3t - 4)(t + 1)}{3} + t$$

is to eliminate the fractions by multiplying both sides by the common denominator 6:

$$6\left(\frac{(t + 7)(2t - 1)}{2}\right) = 6\left(\frac{(3t - 4)(t + 1)}{3} + t\right)$$

$$6\left(\frac{(t+7)(2t-1)}{2}\right) = 6\left(\frac{(3t-4)(t+1)}{3}\right) + 6t$$

Simplify: $\quad 3(t+7)(2t-1) = 2(3t-4)(t+1) + 6t$

Multiply out both sides: $\quad 3(2t^2 + 13t - 7) = 2(3t^2 - t - 4) + 6t$

$$6t^2 + 39t - 21 = 6t^2 - 2t - 8 + 6t$$

Combine like terms: $\quad 6t^2 + 39t - 21 = 6t^2 + 4t - 8$

Subtract $6t^2$ from both sides:

$$39t - 21 = 4t - 8$$

Subtract $4t$ from both sides:

$$35t - 21 = -8$$

Add 21 to both sides: $\quad 35t = 13$

Divide both sides by 35: $\quad t = 13/35.$

The only solution to this last equation, and hence the only solution to the original one, is 13/35. ∎

The second Basic Principle applies only to *nonzero* quantities. So care must be used when multiplying both sides of an equation by a quantity involving the variable, because such a quantity may be zero for some values of x.* The resulting equation may have solutions that are *not* solutions of the original one.† In such situations, *check your solutions in the original equation.*

Example 4 To solve $\dfrac{x-3}{x-7} = \dfrac{3x-17}{x-7}$ we begin by multiplying both sides by $x - 7$:

$$\frac{x-3}{x-7}(x-7) = \frac{3x-17}{x-7}(x-7)$$

Simplify both sides: $\quad x - 3 = 3x - 17$

Add 3 to both sides: $\quad x = 3x - 14$

Subtract $3x$ from both sides: $\quad -2x = -14$

Divide both sides by -2: $\quad x = 7$

Substituting 7 for x in the original equation yields:

$$\frac{7-3}{7-7} = \frac{3 \cdot 7 - 17}{7 - 7} \quad \text{or equivalently,} \quad \frac{4}{0} = \frac{4}{0}.$$

---

* For instance, $(x-2)(x+5)$ is 0 when $x = 2$ or $x = -5$.
† Such numbers are called **extraneous solutions**.

Since division by 0 is not defined, 7 is *not* a solution of the original equation. So this equation has no solutions.* ∎

Example 5 To solve $\dfrac{5}{x+6} = \dfrac{3}{x-3} + \dfrac{1}{2x-6}$ we first note that the third denominator is $2x - 6 = 2(x - 3)$. Hence $2(x - 3)(x + 6)$ is a common denominator for all three fractions. Multiplying both sides by this common denominator, we have:

$$\frac{5}{x+6} \cdot 2(x-3)(x+6) = \frac{3}{x-3} \cdot 2(x-3)(x+6) + \frac{1}{2x-6} \cdot 2(x-3)(x+6)$$

Cancel like factors: $5 \cdot 2(x - 3) = 3 \cdot 2(x + 6) + (x + 6)$

Multiply out both sides: $10(x - 3) = 6(x + 6) + x + 6$

$10x - 30 = 6x + 36 + x + 6$

Combine like terms: $10x - 30 = 7x + 42$

Subtract 7x from both sides: $3x - 30 = 42$

Add 30 to both sides: $3x = 72$

Divide both sides by 3: $x = 24$

By substituting 24 for x in the original equation, it can be seen that 24 *is* a solution:

$$\frac{5}{24+6} = \frac{5}{30} = \frac{1}{6} \quad \text{and} \quad \frac{3}{24-3} + \frac{1}{2 \cdot 24 - 6} = \frac{3}{21} + \frac{1}{42} = \frac{1}{6}. \quad \blacksquare$$

Formulas

When dealing with formulas such as

$$A = \pi r^2 \qquad S = 2\pi(r^2 + rh) \qquad I = Prt$$

it is often necessary to express one of the variables in terms of the others. This amounts to solving the equation for the desired variable, while treating the others as constants.

Example 6 To solve the equation $aw - b = cdw$ for w, we proceed as before to collect all terms involving w on one side of the equation:

$$aw - b = cdw$$

Add b to both sides: $aw = cdw + b$

Subtract cdw from both sides: $aw - cdw = b$

Factor w out of left-hand side: $(a - cd)w = b$

---

* The preceding argument shows that *if* there were a solution, it would have to be 7. Since 7 is not a solution, there is no solution.

If the quantity $(a - cd)$ is nonzero, we can divide both sides of the equation by it and obtain this solution:

$$w = \frac{b}{a - cd} \quad \text{(provided } a - cd \neq 0\text{).} \blacksquare$$

Example 7 The surface area S of a cylindrical tin can is given by the formula $S = 2\pi(r^2 + rh)$, where h is the height and r is the radius of the circular top of the can. In order to find a formula for the height in terms of the surface area and the radius, we must solve the equation $2\pi(r^2 + rh) = S$ for h:

$$2\pi(r^2 + rh) = S$$

Multiply out left-hand side: $\quad 2\pi r^2 + 2\pi rh = S$

Subtract $2\pi r^2$ from both sides: $\quad 2\pi rh = S - 2\pi r^2$

Since the radius r of the top of the can is a positive number, we can divide both sides of this last equation by the nonzero number $2\pi r$ and obtain the solution:

$$h = \frac{S - 2\pi r^2}{2\pi r} = \frac{S}{2\pi r} - r. \blacksquare$$

Example 8 Assume $b + c \neq 0$ and solve $b + c = \dfrac{c}{a + b}$ for a.

Solution

Multiply both sides by $a + b$: $\quad (b + c)(a + b) = \dfrac{c}{a + b}(a + b)$

Simplify: $\quad (b + c)(a + b) = c$

Divide both sides by $b + c$: $\quad a + b = \dfrac{c}{b + c}$

Subtract b from both sides. $\quad a = \dfrac{c}{b + c} - b. \blacksquare$

EXERCISES

In Exercises 1–42, solve the given equation and check your answer.

1. $2x = 3$
2. $4x = 0$
3. $-x = 5$
4. $2 - x = 7$
5. $3x + 2 = 26$
6. $\dfrac{y}{5} - 3 = 14$
7. $3x + 2 = 9x + 7$
8. $-7(t + 2) = 3(4t + 1)$
9. $3y/4 - 6 = y + 2$
10. $2(1 + x) = 3x + 5$
11. $\dfrac{1}{2}(2 - z) = \dfrac{3}{2} + 6z$

12. $\dfrac{4y}{5} + \dfrac{5y}{4} = 1$

13. $2x - \dfrac{x-5}{4} = \dfrac{3x-1}{2} + 1$

14. $2\left(\dfrac{x}{3}+1\right) + 5\left(\dfrac{4x}{3}-2\right) = 2$

15. $3\left(\dfrac{x}{2}+2\right) = \dfrac{5}{2} - 4\left(\dfrac{3x}{2}+1\right)$

16. $1.2z + 5.3 = 2.3 - 7.8z$

17. $\sqrt{2}x + \dfrac{1}{\sqrt{2}} = 3\sqrt{2}$

18. $\sqrt{3}x - 1 = 4\sqrt{3}x - 1 + \sqrt{3}$

19. $\dfrac{3x+2}{x-6} = 4$

20. $\dfrac{3x+2}{x-6} = \dfrac{2}{x-6} + 2$

21. $\dfrac{2x-1}{2x+1} = \dfrac{1}{4}$

22. $\dfrac{2x}{x-3} = 1 + \dfrac{6}{x-3}$

23. $\dfrac{1}{2t} - \dfrac{2}{5t} = \dfrac{1}{10t} - 1$

24. $\dfrac{1}{2} + \dfrac{2}{y} = \dfrac{1}{3} + \dfrac{3}{y}$

25. $\dfrac{2x-7}{x+4} = \dfrac{5}{x+4} - 2$

26. $\dfrac{z+4}{z+5} = \dfrac{-1}{z+5}$

27. $\dfrac{6x}{3x-1} - 5 = \dfrac{2}{3x-1}$

28. $1 + \dfrac{1}{x} = \dfrac{2+x}{x} + 1$

29. $(x+2)^2 = (x+3)^2$

30. $x^2 + 3x - 7 = (x+4)(x-4)$

31. $(x+1)^2 + 2(x-3) = (x-2)(x+4)$

32. $(y-1)^3 = (y+1)^3 - 3y(1+2y)$

33. $(z+2)^3 = (z+1)^3 + 3z^2 + 1$

34. $(x-1)^3 + 3(x+2)^2 = x^3 + 2x - 2$

35. $\dfrac{4}{3y+1} = \dfrac{1}{y}$

36. $\dfrac{-5}{t} = \dfrac{7}{t+1}$

37. $\dfrac{2}{x+4} = \dfrac{5}{x-1}$

38. $\dfrac{1+t}{1-t} - \dfrac{1-t}{1+t} = \dfrac{1}{1-t^2}$

39. $\dfrac{1}{1-x} + \dfrac{2}{1-x^2} = \dfrac{3}{1+x}$

40. $\dfrac{2z+1}{2z-5} = \dfrac{z}{z-6}$

41. $\dfrac{x-2}{x+3} = \dfrac{x+1}{x-1}$

42. $\dfrac{y-2}{y-1} + 3 = \dfrac{4y+1}{y+2}$

An equation is an *identity* if every value of the variable for which all terms of the equation are defined is a solution. For example, $x^2 - 4 = (x+2)(x-2)$ is an identity because every real number is a solution. In Exercises 43–50, determine if the equation is an identity.

43. $x^2 - 6x + 7 = (x-3)^2 - 2$

44. $(x+5)^2 - 12 = x^2 + 5x + 13$

45. $(3x+2)(2x-1) + 2 = 6x^2$

46. $(2x+1)^2 - 4x^2 = 4x + 1$

47. $\dfrac{x^2-4}{x+2} = x - 2$ [*Hint:* Factor the numerator.]

48. $\dfrac{x^2+2x-3}{x-1} = x + 3$

49. $2 + \dfrac{1}{x-3} = \dfrac{2x-6}{x-3}$

50. $5 + \dfrac{1}{x+2} = \dfrac{5x+11}{x+2}$

In Exercises 51–60, use a calculator to solve the given equation.

51. $21.31 + 41.29x = 17.51x - 8.17$

52. $2.37x + 3.1288 = 6.93x - 2.48$

53. $18.923y - 15.4228 = 10.003y + 18.161$

54. $6.31(x - 3.53) = 5.42(x + 1.07) - 21.1584$

55. $18.34x - \dfrac{14.21}{3} = \dfrac{33.41}{3} - 46.82x$

56. $\dfrac{423.1}{51.71} + \dfrac{53.19x}{21.72} = x$

57. $\dfrac{2}{476.1} + \dfrac{1}{x} = \dfrac{3}{919.7}$ 58. $\dfrac{1}{.2131} = \dfrac{1}{x} + \dfrac{1}{.7218}$ 68. $\dfrac{1}{r} = \dfrac{1}{s} + \dfrac{1}{t}$ for r 69. $S = \dfrac{b}{1 - v}$ for v

59. $\dfrac{231.2}{x} - 3.1 = \dfrac{41.7}{2x}$ 60. $\dfrac{11.2}{x} - \dfrac{1}{4} = \dfrac{15.7}{x}$ 70. $a(y - 7) + 2a\left(\dfrac{y}{2} + a\right) = 16$ for y

In Exercises 61–74, solve the given equation for the indicated variable. State any conditions that are necessary for your answer to be valid.

71. $ax + (4a - b)(x + 3b) = 2x - a$ for x

72. $at^2 - (3b + t)t - 4a + 7 = 0$ for a

73. $(a + c)(a - b)(b + z) = 14$ for z

61. $x = 3y - 5$ for y 62. $5x - 2y = 1$ for x

63. $\dfrac{1}{4}x - \dfrac{1}{3}y - 6 = 0$ for y

74. $S = \dfrac{H}{a(v - w)}$ for w

75. Find the number c such that $x = -5$ is a solution of $3x + 2 - 4c = 2x - 5$.

64. $3x + \dfrac{1}{2}y - z = 4$ for z

76. Find the number c such that $x = 10$ is a solution of $\dfrac{1}{3 - x} + \dfrac{c}{x + 3} = \dfrac{1}{x^2 - 9}$.

65. $A = \dfrac{h}{2}(b + c)$ for b

66. $V = \pi b^2 c$ for c

77. Find the number c such that $3x - 6 = c$ and $2 - x = 1$ have the same solutions.

67. $V = \dfrac{\pi d^2 h}{4}$ for h

78. If $1/3$ is a solution of $cx + d = 0$, then how are c and d related?

2.2 APPLICATIONS OF FIRST-DEGREE EQUATIONS*

Real-life problems don't usually appear as equations, but as written or verbal statements.† To solve such problems you must first translate the given verbal information into the mathematical language of equations and *then* solve the equation. There is no single method that works for all problems, but these guidelines are often helpful:

GUIDELINES FOR SOLVING APPLIED PROBLEMS

1. *Read* the problem carefully and determine quantities that are being sought.
2. *Label* the unknown quantities by letters (variables).
3. *Translate* the verbal statements and relationships between the known and unknown quantities into an equation. This may involve expressing all the unknown variables in terms of a single one. Drawing a picture may help to sort out the relationships.
4. *Solve* the equation.
5. *Interpret* the solutions in terms of the original problem. Do they make sense? Do they satisfy the stated conditions?

---

* This section is not needed in the sequel.

† Consequently, such problems are often called **word problems**.

2.2 APPLICATIONS OF FIRST-DEGREE EQUATIONS

We begin with some problems involving interest. Recall that 8% means .08 and that "8% *of* 227" means ".08 *times* 227," that is, .08(227) = 18.16. The basic rule of annual simple interest is:

Interest = rate × amount.

Example 1 (Interest) How much money must be invested at 9% to produce $128.25 interest annually?

Solution The unknown quantity is the amount to be invested; *label* it x. The relationship between x, the interest rate 9%, and the interest $128.25 is given by the interest rule above:

$$\text{Interest} = \text{rate} \cdot \text{amount}$$
$$128.25 = 9\% \text{ of } x$$
$$128.25 = .09x$$

The given information has now been *translated* into an equation, which is easily *solved* by dividing both sides by .09:

$$x = \frac{128.25}{.09} = \$1425.$$

Interpret: 9% of $1425 is .09(1425) = $128.25, so $1425 is the correct amount to invest. ∎

Example 2 (Interest) A high-risk stock pays dividends at a rate of 12% per year, while a savings account pays 6% interest per year. How much of a $9000 investment should be put in the stock and how much in the savings account in order to obtain a return of 8% per year on the total investment?

Solution *Label:* Let x be the amount invested in stock. Then the rest of the $9000, namely, $(9000 - x)$ dollars, goes in the savings account. *Translate:* We want the total return on $9000 to be 8%, so we have:

$$\begin{pmatrix}\text{Return on } x \text{ dollars} \\ \text{of stock at 12\%}\end{pmatrix} + \begin{pmatrix}\text{Return on } (9000 - x) \\ \text{dollars of savings at 6\%}\end{pmatrix} = 8\% \text{ of } \$9000$$

$$(12\% \text{ of } x \text{ dollars}) + (6\% \text{ of } (9000 - x) \text{ dollars}) = 8\% \text{ of } \$9000$$

$$.12x + .06(9000 - x) = .08(9000)$$
$$.12x + .06(9000) - .06x = .08(9000)$$
$$.12x + 540 - .06x = 720$$
$$.12x - .06x = 720 - 540$$
$$.06x = 180$$
$$x = \frac{180}{.06} = 3000.$$

Interpret: $3000 should be invested in stock and $(9000 - 3000) = \$6000$ in the savings account. If this is done, the total return will be 12% of $3000 ($360) plus 6% of $6000 ($360), a total of $720, which is precisely 8% of $9000. ∎

Example 3 (Averages) A student has scores of 66, 74, 78, and 70 on four exams. If the final exam counts double, what score must the student get in order to have: **(a)** an average of 80? **(b)** an average of 90?

Solution

(a) *Label:* let x be the score on the final exam. *Translate:* If the final exam counted the *same* as the other exams, then the average would be

$$(66 + 74 + 78 + 70 + x)$$

divided by 5. But since the final exam counts as if it were *two* exams, we have:

(∗) $$\frac{66 + 74 + 78 + 70 + x + x}{6} = \text{average}$$

Since $(66 + 74 + 78 + 70) = 288$, we must *solve* this equation:

$$\frac{288 + 2x}{6} = 80$$

$$288 + 2x = 6 \cdot 80$$

$$2x = 480 - 288 = 192$$

$$x = \frac{192}{2} = 96.$$

Interpret: The student needs a final exam score of 96—difficult but not impossible.

(b) Equation (∗) becomes:

$$\frac{288 + 2x}{6} = 90.$$

Verify that the solution is $x = 126$. *Interpret:* Assuming that 100 is a perfect score, it is impossible for this student to raise his/her average to 90. ∎

Example 4 (Mixture) How much Darjeeling tea, selling for $8.25 per pound, and how much common black tea, selling for $4.50 per pound, should be combined to make 100 pounds of a mixture that will sell for $6.00 per pound?

Solution Let x be the number of pounds of Darjeeling needed.* Then the

---

* Hereafter we omit the headings Label, Translate, etc.

remainder of the mixture, $100 - x$ pounds, will be common tea. The price of x pounds of Darjeeling is

$$(\text{price per pound}) \cdot (\text{number of pounds}) = 8.25x.$$

Similarly, the price of $(100 - x)$ pounds of common tea at \$4.50 per pound is $4.50(100 - x)$ dollars. The price of 100 pounds of mixture at \$6.00 per pound is $6(100) = 600$ dollars. We can now set up an equation by using this fact:

$$\begin{pmatrix} \text{Price of } x \text{ pounds} \\ \text{of Darjeeling} \end{pmatrix} + \begin{pmatrix} \text{Price of } (100 - x) \\ \text{pounds of common} \end{pmatrix} = \begin{pmatrix} \text{Price of 100 pounds} \\ \text{of mixture} \end{pmatrix}$$

$$8.25x + 4.50(100 - x) = 6(100)$$
$$8.25x + 450 - 4.50x = 600$$
$$8.25x - 4.50x = 600 - 450$$
$$3.75x = 150$$
$$x = \frac{150}{3.75} = 40.$$

So 40 pounds of Darjeeling should be mixed with $(100 - 40) = 60$ pounds of common tea. ∎

Example 5 (Mixture) A car radiator contains 12 quarts of fluid, 20% of which is antifreeze. How much fluid should be drained and replaced with pure antifreeze in order that the resulting mixture be 50% antifreeze?

Solution Let x be the number of quarts of fluid to be replaced by pure antifreeze. When x quarts are drained, there are $12 - x$ quarts of fluid left in the radiator, 20% of which is antifreeze. So we have:

$$\begin{pmatrix} \text{Amount of antifreeze} \\ \text{in radiator after} \\ \text{draining } x \text{ quarts} \\ \text{of fluid} \end{pmatrix} + \begin{pmatrix} x \text{ quarts of} \\ \text{antifreeze} \end{pmatrix} = \begin{pmatrix} \text{Amount of} \\ \text{antifreeze in} \\ \text{final mixture} \end{pmatrix}$$

$$(20\% \text{ of } 12 - x) + x = 50\% \text{ of } 12$$
$$.20(12 - x) + x = .50(12)$$
$$2.4 - .2x + x = 6$$
$$-.2x + x = 6 - 2.4$$
$$.8x = 3.6$$
$$x = \frac{3.6}{.8} = 4.5.$$

Therefore 4.5 quarts should be drained and replaced with pure antifreeze. ∎

The basic formula for problems involving distance and a uniform rate of speed is:

Distance = rate × time.

For instance, if you drive at a rate of 55 mph for 2 hours, you travel a distance of $55 \cdot 2 = 110$ miles.

Example 6 (Distance/Rate) A car leaves Chicago at 1 P.M. and travels at an average speed of 64 kilometers per hour toward Decatur, which is 239 kilometers away. A second car leaves Decatur at 2 P.M. and heads for Chicago on the same road at an average speed of 86 kilometers per hour. When will the cars meet?

Solution In problems like this, it's best to measure the hours elapsed after a particular time. So let t be the number of hours after 1 P.M. When the first car has been traveling for t hours, the second car will have been traveling for $t - 1$ hours, since it left an hour later. At the time the cars meet, the sum of the distances they have traveled will be exactly 239 km, as shown here:

Chicago |⟵——————————— *239 km* ———————————⟶| Decatur
| ——— *first car* ———⟶⟵— *second car* ——— |

We now translate this information into an equation:

$$\left(\begin{array}{c}\text{Distance traveled}\\ \text{by first car}\end{array}\right) + \left(\begin{array}{c}\text{Distance traveled}\\ \text{by second car}\end{array}\right) = 239$$

$$\text{rate} \cdot \text{time} \quad + \quad \text{rate} \cdot \text{time} \quad = 239$$

$$64t \quad + \quad 86(t - 1) \quad = 239$$

$$64t + 86t - 86 = 239$$

$$64t + 86t = 239 + 86$$

$$150t = 325$$

$$t = \frac{325}{150} = 2\frac{1}{6} \text{ hours}$$

$$t = 2 \text{ hours and } 10 \text{ minutes, after 1 P.M.}$$

Therefore the cars meet at 3:10 P.M. ■

Example 7 (Distance/Rate) One day a commuter drives to work at an average speed of 45 mph and arrives 1 minute before 9. The next day she leaves at the same time, drives at an average speed of 40 mph, and arrives 1 minute after 9. How far is it from home to work?

Solution On either day, we know that distance = rate·time, so that,

$$\text{Time} = \frac{\text{distance}}{\text{rate}}.$$

If d is the distance from home to work, then on the first day the time is $d/45$ and on the second day the time is $d/40$. She takes 2 minutes longer on the second day (arriving at 9:01 instead of 8:59). Since the speed is given in miles per *hour*, time should be measured in hours: 2 minutes is $2/60 = 1/30$ hour. Therefore

$$\text{Time on second day} = (\text{Time on first day}) + \frac{1}{30}$$

$$\frac{d}{40} = \frac{d}{45} + \frac{1}{30}$$

Multiplying both sides by the common denominator 360 shows that

$$\frac{d}{40} \cdot 360 = \frac{d}{45} \cdot 360 + \frac{1}{30} \cdot 360$$

$$9d = 8d + 12$$

$$d = 12 \text{ miles.}$$

Alternate Solution The distance d is the same on both days. On the first day $d = \text{rate} \cdot \text{time} = 45t$. On the second day, the time is $t + 1/30$, so $d = \text{rate} \cdot \text{time} = 40(t + 1/30)$. Therefore $45t = 40(t + 1/30)$. Solving this equation shows that $t = 4/15$ (time on first day). Hence $d = 45(4/15) = 12$. ∎

Example 8 (Work) Tom can paint the fence in 6 hours and Huck can paint it in 4 hours. If they work together, how long will it take to paint the fence?

Solution Since Tom can paint the entire fence in 6 hours, he can paint $\frac{1}{6}$ of it in 1 hour. Similarly, Huck can paint $\frac{1}{4}$ of the fence in 1 hour. If t is the number of hours it takes to paint the fence when they work together, then together they can paint $1/t$ of the fence in 1 hour. Therefore:

$$\begin{pmatrix}\text{Part of fence}\\\text{painted by Tom}\\\text{in 1 hour}\end{pmatrix} + \begin{pmatrix}\text{Part of fence}\\\text{painted by Huck}\\\text{in 1 hour}\end{pmatrix} = \begin{pmatrix}\text{Part of fence}\\\text{painted by both}\\\text{in 1 hour}\end{pmatrix}$$

$$\frac{1}{6} + \frac{1}{4} = \frac{1}{t}$$

$$12t\left(\frac{1}{6} + \frac{1}{4}\right) = 12t\left(\frac{1}{t}\right)$$

$$2t + 3t = 12$$

$$5t = 12$$

$$t = \frac{12}{5} = 2.4 \text{ hours.} \blacksquare$$

EXERCISES

1. A student has exam scores of 88, 62, and 79. What score does he need on the fourth exam in order to have an average of 80?
2. A student has an average of 72 on four exams. If the final exam is to count double, what score does she need in order to have an average of 80?
3. An item on sale costs $13.41. All merchandise was marked down 25 percent for the sale. What was the original price of the item?
4. A discount store sells a jacket for $28, which is 20% below the list price. What is the list price?
5. A worker gets an 8% pay raise and now makes $1593 per month. What was the worker's old salary?
6. Inflation caused the price of a workbook to increase by 5%. The new price is $15.12. What was the old price?
7. If Tom can do a job in 9 hours and Anne can do the same job in 6 hours, how long will it take if they work together?
8. If John can paint a room in 10 hours and Andy can paint it in 12 hours, how long will it take them working together to paint the room?
9. A student has exam grades of 64, 82, and 91. What is his average? What score must he get on the fourth exam in order to raise his average 3 points?
10. A student scored 63 and 81 on two exams, each of which accounts for 30% of the course grade. The final exam accounts for 40% of the course grade. What score must she get on the final exam in order to have an average grade of 76?
11. You have already invested $550 in a stock with an annual return of 11%. How much of an additional $1100 should be invested at 12% and how much at 6% in order that the total annual return on the entire $1650 is 9%.
12. $25,000 is placed in the bank, part in an account earning 10% interest and part in an account earning 15% interest. The annual interest paid on the $25,000 is $3,350. How much money is in the 10% account?
13. If you borrow $500 from a credit union at 12% annual interest and $250 from a bank at 18% annual interest, what is the *effective annual interest rate* (that is, what single rate of interest on $750 would result in the same total amount of interest)?
14. If you borrow $200 at 12% annual interest and another $300 at 14% annual interest, what is the effective rate of interest on your $500 loan?
15. If 9 is added to a certain number, the result is one less than three times the number. What is the number?
16. What number, when added to 30, is 11 times the number?
17. A merchant has 5 pounds of mixed nuts; they cost $30. He wants to add peanuts that cost $1.50 per pound and cashews that cost $4.50 per pound to obtain 50 pounds of a mixture that costs $2.90 per pound. How many pounds of peanuts and how many of cashews should he add?
18. Emily has 10 pounds of premium coffee which costs $7.50 per pound. She wants to mix this with ordinary coffee which costs $4 per pound in order to make a mixture which costs $5.50 per pound. How much ordinary coffee must be used? What will the weight of the final mixture be?
19. How much pure water should be added to 75 ounces of a 30% salt solution in order to obtain a 12% salt solution?
20. How many gallons of a 12% salt solution should be combined with 10 gallons of an 18% salt solution to obtain a 16% solution?
21. One alloy is one part copper to three parts tin. A second alloy is one part copper to four parts tin. How much of the second alloy should be combined with 24 pounds of the first alloy to obtain a new alloy that is two parts copper to seven parts tin?
22. A chemist has 10 ml of a 30% acid solution. How many milliliters of pure acid must be added in order to obtain a 50% acid solution?
23. A radiator contains 8 quarts of fluid, 40% of which is antifreeze. How much fluid should be drained and replaced with pure antifreeze in order that the new mixture be 60% antifreeze?
24. A radiator contains 10 quarts of fluid, 30% of which is antifreeze. How much fluid should be drained and replaced with pure antifreeze in order that the new mixture be 40% antifreeze?

25. If Lionel lost 20 pounds, he would weigh seven times as much as his pet boa constrictor. Together they weigh 200 pounds. How much does each weigh?

26. Lawnmowers are priced at $300 during April. In May the price is raised by 20%. In September the price is reduced by 20%. What is the final price? Explain why it is *not* $300.

27. A car passes a certain point at 1 P.M. and continues along at a constant speed of 64 kilometers per hour. A second car passes the point at 2 P.M. and goes on at a constant speed of 88 kilometers per hour, following the same route as the first car. At what time will the second car catch up with the first one?

28. A car leaves city C at 3 P.M. and travels at an average speed of 40 mph toward city D, 150 miles away. Thirty minutes later, a second car leaves city C and follows the same road to city D at an average speed of 55 mph. When will the second car overtake the first one?

29. Two cars leave a gas station at the same time, one traveling north and the other south. The northbound car travels at 50 mph. After 3 hours the cars are 345 miles apart. How fast is the southbound car traveling?

30. A train leaves New York for Boston, 200 miles away, at 3 P.M. and averages 75 mph. Another train leaves Boston for New York on an adjacent set of tracks at 5 P.M. and averages 45 mph. At what time will the trains meet?

31. A motorist drives from city C to city D at an average speed of 33 mph. The motorist returns to city C by the same route at an average speed of 41.25 mph. If the return trip took 40 minutes less time than the outgoing trip, how far is it from city C to city D?

32. In a motorcycle race the winner averaged 100 mph and finished 15 minutes ahead of the loser, who averaged 95 mph. How long was the race (in miles) and what was the winner's time?

33. An airplane flew with the wind for 2.5 hours and returned the same distance against the wind in 3.5 hours. If the cruising speed of the plane was a constant 360 mph, how fast was the wind blowing? [*Hint:* If the wind speed is r miles per hour, then the plane travels at $(360 + r)$ mph with the wind and at $(360 - r)$ mph against the wind.]

34. A motorboat goes 15 miles downstream at its top speed and then turns around and returns 15 miles upstream at top speed. The trip upstream took twice as long as the trip downstream. If the boat's top speed is 10 mph in still water, how fast is the current? [See Exercise 33.]

35. A rectangular garden has a perimeter of 270 feet and its length is twice its width. What are the dimensions of the garden?

36. The length of a rectangle is 15 inches more than the width. If the perimeter is 398 inches, what are the dimensions of the rectangle?

37. The perimeter of a rectangle is 160 meters. If a new rectangle is formed by halving the length of each of one pair of opposite sides and increasing the remaining pair of sides by 30 meters each, then the new rectangle also has a perimeter of 160 meters. What were the dimensions of the original rectangle?

38. The perimeter of a rectangular garden is 72 feet. If two such gardens were placed side by side they would form a square. What are the dimensions of the garden?

39. If Charlie and Nick work together they can paint the house in 20 hours. Charlie can do the job alone in 36 hours. How long would it take Nick to do the job alone?

40. Using two lawnmowers and working together Tom and Anne can mow the lawn in 36 minutes. It takes Anne 90 minutes if she mows the entire lawn herself. How long would it take Tom to do the job if he worked alone?

41. One pipe can fill a pool in 50 minutes. Another pipe takes 75 minutes. How long does it take for both pipes together to fill the pool? [*Hint:* This is a "work problem" in disguise: How much of the pool does the first pipe fill in 1 minute?]

42. One pipe can fill a tank in 4 hours, a second pipe can fill it in 10 hours, and a third pipe in 12 hours. The pipes are connected to the same tank. How long does it take to fill the tank if all three pipes are used together? [See Exercise 41.]

43. (With apologies to Mrs. Morgan of *How Green Was My Valley*) The water faucet can fill a tub in 15 minutes. But there is a hole in the tub through which a full tub of water can drain out in 18 minutes. If the tub is empty and the faucet is turned on, how long will it take to fill the tub?

44. An expert can do a certain job in 10 hours, whereas an amateur takes 16 hours to do the same job. If two experts and four amateurs work together on the job, how long will it take?

Unusual Problems

45. Insurance on overseas shipment of goods is charged at the rate of 4% of the total value of the shipment (which is the sum of the value of the goods, the freight charges, *and* the cost of the insurance). The freight charges for $30,000 worth of goods are $4,500. How much does the insurance cost?

46. A squirrel and a half eats a nut and a half in a day and a half. How many nuts do six squirrels eat in six days?

47. When Hershey raised the price of its chocolate bar from 20 to 25 cents it also increased the weight of the bar. The price increase was 9.375% per ounce. If the old bar weighed 1.05 ounces, what was the weight of the new bar?

48. Diophantus of Alexandria, a third-century mathematician, lived one sixth of his life in childhood, one twelfth in his youth, and one seventh as a bachelor. Then he married and five years later had a son. The son died four years before Diophantus at half the age Diophantus was when he himself died. How long did Diophantus live?

2.3 QUADRATIC EQUATIONS

A **quadratic equation** is one that can be written in the form

$$ax^2 + bx + c = 0$$

for some constants a, b, c with $a \neq 0$. The various techniques for solving such equations will be presented here, beginning with the **factoring method.**

Example 1 To solve $x^2 - 5x + 6 = 0$, start by factoring the left side:

$$(x - 2)(x - 3) = 0.$$

If a product of real numbers is 0, then at least one of the factors must be 0 (see the box on page 4). So this equation is equivalent to:

$$x - 2 = 0 \quad \text{or} \quad x - 3 = 0$$
$$x = 2 \quad \text{or} \quad x = 3$$

Therefore 2 and 3 are the solutions of $x^2 - 5x + 6 = 0$. ∎

Example 2 To solve $3x^2 - x = 10$, we first rearrange the terms to make one side 0 and then factor:

Subtract 10 from each side: $\quad 3x^2 - x - 10 = 0$

Factor left side: $\quad (3x + 5)(x - 2) = 0$

The last equation is equivalent to:

$$3x + 5 = 0 \quad \text{or} \quad x - 2 = 0$$
$$3x = -5 \qquad\qquad\qquad x = 2$$
$$x = -5/3$$

Therefore the solutions are $-5/3$ and 2. ∎

The solutions of $x^2 = 7$ are the numbers whose square is 7. Although 7 has just *one* square root, there are *two* numbers whose square is 7, namely, $\sqrt{7}$ and its negative, $-\sqrt{7}$. So the solutions of $x^2 = 7$ are $\sqrt{7}$ and $-\sqrt{7}$, or in abbreviated form, $\pm\sqrt{7}$. The same argument works for any positive real number d:

The solutions of $x^2 = d$ are $\sqrt{d}$ and $-\sqrt{d}$.

We now use a slight variation of this idea to develop a method for solving quadratic equations that don't readily factor. The method depends on this fact: A polynomial of the form $x^2 + bx$ can be changed into a perfect square* by adding a suitable constant. For example, $x^2 + 6x$ can be changed into a perfect square, by adding 9:

$$x^2 + 6x + 9 = (x + 3)^2.$$

This process is called **completing the square**.

To complete the square in $x^2 + 6x$, we added $9 = 3^2$. Note that 3 is one-half the coefficient of x in the original polynomial $x^2 + 6x$. The same idea works in the general case. The multiplication pattern for perfect squares shows that for any real number b:

$$\left(x + \frac{b}{2}\right)^2 = x^2 + 2\left(\frac{b}{2}\right)x + \left(\frac{b}{2}\right)^2 = x^2 + bx + \left(\frac{b}{2}\right)^2.$$

Therefore

COMPLETING THE SQUARE

> To complete the square in $x^2 + bx$, add the square of one-half the coefficient of x, namely, $\left(\dfrac{b}{2}\right)^2$. This produces a perfect square:
>
> $$x^2 + bx + \left(\frac{b}{2}\right)^2 = \left(x + \frac{b}{2}\right)^2$$

The following example shows how completing the square can be used to solve quadratic equations.

Example 3 To solve $x^2 + 6x + 1 = 0$, we first rewrite the equation as $x^2 + 6x = -1$. Next we complete the square on the left side by adding the square of half the coefficient of x, namely, $(\frac{6}{2})^2 = 9$. In order to have an equivalent equation, we must add 9 to *both* sides:

$$x^2 + 6x + 9 = -1 + 9$$

Factor left side: $\qquad (x + 3)^2 = 8$

Thus $x + 3$ is a number whose square is 8. The only numbers whose squares equal 8 are $\sqrt{8}$ and $-\sqrt{8}$. So we must have:

---

*A quadratic polynomial is a **perfect square** if it factors as $(x + d)^2$ for some constant d.

$$x + 3 = \sqrt{8} \quad \text{or} \quad x + 3 = -\sqrt{8}$$
$$x = \sqrt{8} - 3 \quad \text{or} \quad x = -\sqrt{8} - 3.$$

Therefore the solutions of the original equation are $\sqrt{8} - 3$ and $-\sqrt{8} - 3$, or in more compact notation, $\pm\sqrt{8} - 3$. ■

WARNING

This method only works when the coefficient of x^2 is 1. In an equation such as $5x^2 - x + 2 = 0$, you must first divide both sides by 5, and *then* complete the square.

We can use the completing-the-square method to solve *any* quadratic equation:*

$$ax^2 + bx + c = 0$$

Divide both sides by a:
$$x^2 + \frac{b}{a}x + \frac{c}{a} = 0$$

Subtract $\frac{c}{a}$ from both sides:
$$x^2 + \frac{b}{a}x = -\frac{c}{a}$$

Add $\left(\frac{b}{2a}\right)^2$ to both sides:†
$$x^2 + \frac{b}{a}x + \left(\frac{b}{2a}\right)^2 = \left(\frac{b}{2a}\right)^2 - \frac{c}{a}$$

Factor left side:
$$\left(x + \frac{b}{2a}\right)^2 = \left(\frac{b}{2a}\right)^2 - \frac{c}{a}$$

Find common denominator for right side:
$$\left(x + \frac{b}{2a}\right)^2 = \frac{b^2}{4a^2} - \frac{c}{a} = \frac{b^2 - 4ac}{4a^2}$$

Since the square of $x + \frac{b}{2a}$ is $\frac{b^2 - 4ac}{4a^2}$, we must have:

$$x + \frac{b}{2a} = \pm\sqrt{\frac{b^2 - 4ac}{4a^2}} = \pm\frac{\sqrt{b^2 - 4ac}}{2a}$$

Subtract $\frac{b}{2a}$ from both sides:
$$x = \frac{-b}{2a} \pm \frac{\sqrt{b^2 - 4ac}}{2a} = \frac{-b \pm \sqrt{b^2 - 4ac}}{2a}$$

We have proved:

---

* If you have trouble following any step here, do it for a numerical example, such as the case when $a = 3, b = 11, c = 5$.

† This is the square of half the coefficient of x.

THE QUADRATIC FORMULA

> The solutions of the quadratic equation $ax^2 + bx + c = 0$ are:
> $$x = \frac{-b \pm \sqrt{b^2 - 4ac}}{2a}$$

You should memorize the quadratic formula.

Example 4 To solve $x^2 + 3 = -8x$, rewrite the equation as $x^2 + 8x + 3 = 0$ and apply the quadratic formula with $a = 1$, $b = 8$, and $c = 3$:

$$x = \frac{-b \pm \sqrt{b^2 - 4ac}}{2a} = \frac{-8 \pm \sqrt{8^2 - 4 \cdot 1 \cdot 3}}{2 \cdot 1}$$

$$= \frac{-8 \pm \sqrt{52}}{2} = \frac{-8 \pm \sqrt{4 \cdot 13}}{2}$$

$$= \frac{-8 \pm 2\sqrt{13}}{2} = -4 \pm \sqrt{13}$$

Therefore the equation has two distinct real solutions, $-4 + \sqrt{13}$ and $-4 - \sqrt{13}$. ∎

Example 5 To solve $x^2 - 194x + 9409 = 0$, use a calculator and the quadratic formula with $a = 1$, $b = -194$, and $c = 9409$:

$$x = \frac{-b \pm \sqrt{b^2 - 4ac}}{2a} = \frac{-(-194) \pm \sqrt{(-194)^2 - 4 \cdot 1 \cdot 9409}}{2 \cdot 1}$$

$$= \frac{194 \pm \sqrt{37636 - 37636}}{2} = \frac{194 \pm 0}{2} = 97$$

Thus 97 is the only solution of the equation. ∎

Example 6 The solutions of $2x^2 + x + 3 = 0$ are given by the quadratic formula with $a = 2$, $b = 1$, and $c = 3$:

$$x = \frac{-b \pm \sqrt{b^2 - 4ac}}{2a} = \frac{-1 \pm \sqrt{1^2 - 4 \cdot 2 \cdot 3}}{2 \cdot 2} = \frac{-1 \pm \sqrt{1 - 24}}{4}$$

$$= \frac{-1 \pm \sqrt{-23}}{4}$$

Since $\sqrt{-23}$ is not a real number, this equation has *no real solutions* (that is, no solutions in the real number system). ∎

The expression $b^2 - 4ac$ in the quadratic formula is called the **discriminant**. As the last three examples demonstrate, the discriminant determines the *number* of real solutions of the equation $ax^2 + bx + c = 0$.

REAL SOLUTIONS OF A QUADRATIC EQUATION

Discriminant $b^2 - 4ac$	Number of Real Solutions of $ax^2 + bx + c = 0$	Example
> 0	Two distinct real solutions	$x^2 + 8x + 3 = 0$
$= 0$	One real solution	$x^2 - 194x + 9409 = 0$
< 0	No real solutions	$2x^2 + x + 3 = 0$

The quadratic formula and a calculator can be used to solve any quadratic equation with nonnegative discriminant. Experiment with your calculator to find the most efficient sequence of key strokes for doing this. It is usually advisable to compute the discriminant first and store the result in memory. It can then be recalled as needed for the rest of the computation.

Example 7 On a typical calculator with algebraic logic the equation $3.2x^2 + 15.93x - 7.1 = 0$ might be solved as follows. First compute $\sqrt{b^2 - 4ac}$ and store it in memory:

Then the solutions can be obtained by:

Remember that these answers are *approximations,* so they may not check exactly when substituted in the original equation. ■

WARNING

Although the quadratic formula can *always* be used to solve a quadratic equation, it may not be the quickest method. As a general rule, try factoring first. If $ax^2 + bx + c$ factors readily, you're done. If it doesn't, *then* use the quadratic formula.

The various techniques for solving quadratic equations work equally well for formulas involving several variables.

Example 8 To express the radius r of the top of a cylindrical can in terms of the surface area S and the height h of the can, we must solve the surface area formula $S = 2\pi(r^2 + rh)$ for r. We have:

$$2\pi(r^2 + rh) = S$$
$$2\pi r^2 + (2\pi h)r - S = 0$$

This is a quadratic equation in r. We can apply the quadratic formula, substituting r for x and $a = 2\pi$, $b = 2\pi h$, and $c = -S$:

$$r = \frac{-b \pm \sqrt{b^2 - 4ac}}{2a} = \frac{-2\pi h \pm \sqrt{(2\pi h)^2 - 4(2\pi)(-S)}}{2(2\pi)}$$

$$= \frac{-2\pi h \pm \sqrt{4\pi^2 h^2 + 8\pi S}}{4\pi} = \frac{-2\pi h \pm \sqrt{4(\pi^2 h^2 + 2\pi S)}}{4\pi}$$

$$= \frac{-2\pi h \pm 2\sqrt{\pi^2 h^2 + 2\pi S}}{4\pi} = \frac{-\pi h \pm \sqrt{\pi^2 h^2 + 2\pi S}}{2\pi}.$$

Because the radius cannot be negative, the only solution that makes sense in this problem is

$$r = \frac{-\pi h + \sqrt{\pi^2 h^2 + 2\pi S}}{2\pi}. \blacksquare$$

EXERCISES

In Exercises 1–12, solve the equation by factoring.

1. $x^2 - 8x + 15 = 0$
2. $x^2 + 5x + 6 = 0$
3. $x^2 - 5x = 14$
4. $x^2 + x = 20$
5. $2y^2 + 5y - 3 = 0$
6. $3t^2 - t - 2 = 0$
7. $4t^2 + 9t + 2 = 0$
8. $9t^2 + 2 = 11t$
9. $3u^2 + u = 4$
10. $5x^2 + 26x = -5$
11. $12x^2 + 13x = 4$
12. $18x^2 = 23x + 6$

In Exercises 13–16, solve the equation by completing the square.

13. $x^2 - 2x = 15$
14. $x^2 - 4x - 32 = 0$
15. $x^2 - x - 1 = 0$
16. $x^2 + 3x - 2 = 0$

In Exercises 17–28, use the quadratic formula to solve the equation.

17. $x^2 - 4x + 1 = 0$
18. $x^2 - 2x - 1 = 0$
19. $x^2 + 6x + 7 = 0$
20. $x^2 + 4x - 3 = 0$
21. $x^2 + 6 = 2x$
22. $x^2 + 11 = 6x$
23. $4x^2 - 4x = 7$
24. $4x^2 - 4x = 11$
25. $4x^2 - 8x + 1 = 0$
26. $2t^2 + 4t + 1 = 0$
27. $5u^2 + 8u = -2$
28. $4x^2 = 3x + 5$

In Exercises 29–34, find the *number* of real solutions of the equation by computing the discriminant.

29. $x^2 + 4x + 1 = 0$
30. $4x^2 - 4x - 3 = 0$
31. $9x^2 = 12x + 1$
32. $9t^2 + 15 = 30t$
33. $25t^2 + 49 = 70t$
34. $49t^2 + 5 = 42t$

In Exercises 35–38, solve the equation and check your answers. (*Hint:* First, eliminate fractions by multiplying both sides by a common denominator.)

35. $1 - \dfrac{3}{x} = \dfrac{40}{x^2}$

36. $\dfrac{4x^2 + 5}{3x^2 + 5x - 2} = \dfrac{4}{3x - 1} - \dfrac{3}{x + 2}$

37. $\dfrac{2}{x^2} - \dfrac{5}{x} = 4$

38. $\dfrac{x}{x - 1} + \dfrac{x + 2}{x} = 3$

In Exercises 39–48, solve the equation by any method.

39. $x^2 + 9x + 18 = 0$
40. $3t^2 - 11t - 20 = 0$
41. $4x(x + 1) = 1$
42. $25y^2 = 20y + 1$
43. $2x^2 = 7x + 15$
44. $2x^2 = 6x + 3$
45. $t^2 + 4t + 13 = 0$
46. $5x^2 + 2x = -2$
47. $\dfrac{7x^2}{3} = \dfrac{2x}{3} - 1$
48. $25x + \dfrac{4}{x} = 20$

In Exercises 49–52, use a calculator and the quadratic formula to find approximate solutions of the equation.

49. $4.42x^2 - 10.14x + 3.79 = 0$
50. $8.06x^2 + 25.8726x - 25.047256 = 0$
51. $3x^2 - 82.74x + 570.4923 = 0$
52. $7.63x^2 + 2.79x = 5.32$

In Exercises 53–60, solve the equation for the indicated letter. State any conditions that are necessary for your answer to be valid.

53. $E = mc^2$ for c
54. $V = 4\pi r^2$ for r
55. $A = \pi rh + \pi r^2$ for r
56. $d = -16t^2 + vt$ for t
57. $3x^2 + xy - 4y^2 = 9$ for x
58. $3x^2 + xy - 4y^2 = 9$ for y
59. $kx^2 + 2x = 3kx + 6$ for x
60. $2x^2 - 2ax + 4 = (a + b)^2 + 2bx$ for x

In Exercises 61–64, find a number k such that the given equation has exactly one real solution.

61. $x^2 + kx + 25 = 0$
62. $x^2 - kx + 49 = 0$
63. $kx^2 + 8x + 1 = 0$
64. $kx^2 + 24x + 16 = 0$

65. Find a number k such that 4 and 1 are the solutions of $x^2 - 5x + k = 0$.
66. Suppose a, b, c are fixed real numbers such that $b^2 - 4ac \geq 0$. Let r and s be the solutions of $ax^2 + bx + c = 0$.

   (a) Use the quadratic formula to show that $r + s = -b/a$ and $rs = c/a$.
   (b) Use part (a) to verify that $ax^2 + bx + c = (x - r)(x - s)$.
   (c) Use part (b) to factor $x^2 - 2x - 1$ and $5x^2 + 8x + 2$.

2.4 APPLICATIONS OF QUADRATIC EQUATIONS*

The guidelines for solving applied problems in Section 2.2 are also useful in situations involving quadratic equations.

Example 1 A landscaper wants to put a cement walk of uniform width around a rectangular garden that measures 24 by 40 feet. She has enough cement to cover 660 square feet. How wide should the walk be in order to use up all the cement?

Solution Let x denote the width of the walk (in feet) and draw a picture of the situation:

---

* This section is not needed in the sequel.

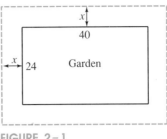

FIGURE 2-1

The length of the outer rectangle is $40 + 2x$ (the garden length plus walks on each end) and its width is $24 + 2x$.

$$\begin{pmatrix}\text{Area of outer}\\\text{rectangle}\end{pmatrix} - \begin{pmatrix}\text{Area of}\\\text{garden}\end{pmatrix} = \text{Area of walk}$$

$$\text{length} \cdot \text{width} - \text{length} \cdot \text{width} = 660$$

$$(40 + 2x)(24 + 2x) - 40 \cdot 24 = 660$$
$$960 + 128x + 4x^2 - 960 = 660$$
$$4x^2 + 128x - 660 = 0$$

Dividing both sides by 4 and applying the quadratic formula yields:

$$x^2 + 32x - 165 = 0$$
$$x = \frac{-32 \pm \sqrt{(32)^2 - 4 \cdot 1 \cdot (-165)}}{2 \cdot 1}$$
$$x = \frac{-32 \pm \sqrt{1684}}{2} \approx \begin{cases} 4.5183 \\ \text{or} \\ -36.5183 \end{cases}$$

Only the positive solution makes sense in the context of this problem.* The walk should be approximately 4.5 feet wide. ■

Example 2 A stone is dropped from the top of a cliff. Five seconds later the sound of the stone hitting the ground is heard. How high is the cliff? Assume that the speed of sound is 1 100 feet per second and that the distance traveled in t seconds by a falling object is $16t^2$ feet. Ignore wind resistance.

Solution The height of the cliff is $16t^2$, where t is the number of seconds it takes the rock to fall to the ground. The total time for the rock to fall and the sound to return to the top of the cliff is 5 seconds. So it must take the sound $5 - t$ seconds to return to the top. Therefore:

---

* When an equation has several solutions, some of them may not satisfy all the conditions in the applied problem, as here. Such meaningless solutions should be ignored.

$$\begin{pmatrix} \text{Distance rock falls} \\ \text{in } t \text{ seconds} \end{pmatrix} = \text{Height of cliff} = \begin{pmatrix} \text{Distance sound} \\ \text{travels upward} \\ \text{in } 5 - t \text{ seconds} \end{pmatrix}$$

$$16t^2 = \text{rate} \cdot \text{time}$$
$$16t^2 = 1100(5 - t)$$
$$16t^2 = 5500 - 1100t$$
$$16t^2 + 1100t - 5500 = 0$$

The quadratic formula and a calculator show that:

$$t = \frac{-1100 \pm \sqrt{(1100)^2 - 4 \cdot 16 \cdot (-5500)}}{2 \cdot 16}$$

$$= \frac{-1100 \pm \sqrt{1{,}562{,}000}}{32} \approx \begin{cases} 4.6812 \\ \text{or} \\ -73.4312 \end{cases}$$

Since the time must be a number between 0 and 5, we see that $t \approx 4.6812$ seconds. Therefore the height of the cliff is

$$16(4.6812)^2 \approx 350.6 \text{ feet.} \blacksquare$$

Example 3 A pilot wants to make an 840-mile trip from Cleveland to Peoria and back in 5 hours flying time. There will be a headwind of 30 mph going to Peoria and it is estimated that there will be a 40-mph tailwind returning to Cleveland. At what constant air speed should the plane be flown?

Solution Let x be the engine speed of the plane. On the trip to Peoria, a distance of 420 miles, the actual speed will be $x - 30$ because of the headwind. Since rate $\cdot$ time = distance, the time to Peoria will be $\dfrac{\text{distance}}{\text{rate}} = \dfrac{420}{x - 30}$. On the return trip the actual speed will be $x + 40$ because of the tailwind and the time will be $\dfrac{\text{distance}}{\text{rate}} = \dfrac{420}{x + 40}$. Therefore:

$$5 = \begin{pmatrix} \text{Time from Cleveland} \\ \text{to Peoria} \end{pmatrix} + \begin{pmatrix} \text{Time from Peoria} \\ \text{to Cleveland} \end{pmatrix}$$

$$5 = \frac{420}{x - 30} + \frac{420}{x + 40}$$

Multiplying both sides by the common denominator $(x - 30)(x + 40)$ and simplifying, we have:

$$5(x - 30)(x + 40) = \frac{420}{x - 30} \cdot (x - 30)(x + 40) + \frac{420}{x + 40} \cdot (x - 30)(x + 40)$$

$$5(x - 30)(x + 40) = 420(x + 40) + 420(x - 30)$$

$$(x - 30)(x + 40) = 84(x + 40) + 84(x - 30)$$
$$x^2 + 10x - 1200 = 84x + 3360 + 84x - 2520$$
$$x^2 - 158x - 2040 = 0$$
$$(x - 170)(x + 12) = 0$$
$$x - 170 = 0 \quad \text{or} \quad x + 12 = 0$$
$$x = 170 \quad\quad\quad x = -12.$$

Obviously, the negative solution doesn't apply. Since we multiplied both sides by a quantity involving the variable, we must check that 170 actually is a solution of the original equation. It is; so the plane should be flown at a speed of 170 mph. ∎

EXERCISES

1. Find two consecutive integers, the sum of whose squares is 313.
2. Find two consecutive odd integers, the sum of whose squares is 2050.
3. The two legs of a right triangle differ in length by 1 cm and the hypotenuse is 1 cm longer than the longer leg. How long is each side?
4. The area of a triangle is 20 square feet. If the base is 2 feet longer than the height, find the height.
5. The diameter of a circle is 16 cm. By what amount must the radius be decreased in order to decrease the area of the circle by 48π square centimeters?
6. A right triangle has a hypotenuse of length 12.054 meters. The sum of the lengths of the other two sides is 16.96 meters. How long is each side?
7. A 13-foot-long ladder leans on a wall. The bottom of the ladder is 5 feet from the wall. If the bottom is pulled out 3 feet farther from the wall, how far does the top of the ladder move down the wall? [*Hint:* The ladder, ground, and wall form a right triangle. Draw pictures of this triangle before and after the ladder is moved. Use the Pythagorean Theorem to set up an equation.]
8. A 15-foot-long pole leans against a wall. The bottom is 9 feet from the wall. How much farther should the bottom be pulled away from the wall so that the top moves the same amount down the wall? [See Exercise 7.]
9. A concrete walk of uniform width is to be built around a circular pool. The radius of the pool is 12 meters and there is enough concrete available to cover 52π square meters. If all the concrete is to be used, how wide should the walk be?
10. A decorator has 92 one-foot-square tiles that will be laid around the edges of a 12- by 15-foot room. A rectangular rug that is 3 feet longer than it is wide is to be placed over the center area where there are no tiles. To the nearest quarter foot, find the dimensions of the smallest rug that will cover the untiled part of the floor. [Assume that all the tiles are used and that none of them are split.]
11. A corner lot has dimensions of 25 by 40 yards. The city plans to take a strip of uniform width along the two sides bordering the streets in order to widen these roads. To the nearest tenth of a yard, how wide should the strip be if the remainder of the lot is to have an area of 844 square yards?
12. A box with a square base and no top is to be made from a square sheet of aluminum by cutting out 3-inch squares from each corner and folding up the sides. If the box is to have a volume of 48 cubic inches, what size should the piece of aluminum be?
13. A box with no top is to be made by taking a rectangular aluminum sheet 8 by 10 inches and cutting a square of the same size out of each corner and folding up the sides. If the area of the base is to be 24 square inches, what size squares should be cut out?
14. A box with no top is to be made by taking a rectan-

gular piece of aluminum and cutting a square of the same size out of each corner and folding up the sides. The box is to be 2 inches deep. The length of its base is to be 5 inches more than the width. If the volume is to be 352 cubic inches, what size should the piece of aluminum be?

15. A group of homeowners are to share equally in the $210 cost of repairing a bus-stop shelter near their homes. At the last moment two members of the group decide not to participate and this raises the share of each remaining person by $28. How many people were in the group at the beginning?

16. A rectangular theater seats 1620 people. If each row had six more seats in it, the number of rows would be reduced by nine. How many seats would *then* be in each row?

17. A woman and her son working together can paint a garage in 4 hours and 48 minutes. The woman working alone can paint it in 4 hours less than the son would take to do it alone. How long would it take the son to paint the garage alone?

18. When each works alone Barbara can do a job in 3 hours less time than John. When they work together it takes them 2 hours to complete the job. How long does it take each one working alone to do the job?

19. A student leaves the university at noon, bicycling south at a constant rate. At 12:30 a second student leaves the same point and heads west, bicycling 7 mph faster than the first student. At 2 P.M. they are 30 miles apart. How fast is each one going?

20. Two trains leave the same city at the same time, one going north and the other east. The northbound train travels 20 mph faster than the eastbound one. If they are 300 miles apart after 5 hours, what is the speed of each train?

21. Red Riding Hood drives the 432 miles to Grandmother's house in 1 hour less than it takes the Wolf to drive the same route. Her average speed is 6 mph faster than the Wolf's average speed. How fast does each drive?

22. To get to work Sam jogs 3 kilometers to the train, then rides the remaining 5 kilometers. If the train goes 14 kilometers per hour faster than Sam's constant rate of jogging and the entire trip takes 45 minutes, how fast does Sam jog?

23. A canoeist paddles at a constant rate and takes 2 hours longer to go 12 kilometers upstream than to go the same distance downstream. If the current runs at 3 kilometers per hour, how long would a 12-kilometer trip take in still water?

24. It takes 15 hours less to make a 40-mile boat trip downstream than it does to make the return trip upstream. If the current is 3 mph, how long would the trip take in still water?

Background for Exercises 25–28: **If an object is thrown upward, dropped, or thrown downward and travels in a vertical line subject only to gravity (with wind resistance ignored), then the height h of the object above the ground (in feet) after t seconds is given by:**

$$h = -16t^2 + v_0 t + h_0$$

where h_0 is the initial height of the object at starting time $t = 0$ and v_0 is the initial velocity (speed) of the object at time $t = 0$. The value of v_0 is taken as positive if the object starts moving upward at time $t = 0$ and negative if the object starts moving downward at $t = 0$. An object that is dropped (rather than thrown downward) has initial velocity $v_0 = 0$.

25. How long does it take an object to reach the ground if
    (a) it is dropped from the top of a 640-foot-high building?
    (b) it is thrown downward from the top of the same building, with an initial velocity of 52 feet per second?

26. You are standing on a cliff 200 feet high. How long will it take a rock to reach the ground if
    (a) you drop it?
    (b) you throw it downward at an initial velocity of 40 feet per second?
    (c) How far does the rock fall in 2 seconds if you throw it downward with an initial velocity of 40 feet per second?

27. A rocket is fired straight up from ground level with an initial velocity of 800 feet per second.
    (a) How long does it take the rocket to rise 3200 feet?
    (b) When will the rocket hit the ground?

28. A rocket loaded with fireworks is to be shot vertically upward from ground level with an initial velocity of 200 feet per second. When the rocket reaches a height of 400 feet on its upward trip the

fireworks will be detonated. How many seconds after lift-off will this take place?

Unusual Problems

29. Two ferryboats leave opposite sides of a lake at the same time. They pass each other when they are 800 meters from the nearest shore. When it reaches the opposite side each boat spends 30 minutes at the dock and then starts back. This time the boats pass each other when they are 400 meters from the nearest shore. Assuming that each boat travels at the same speed in both directions, how wide is the lake between the two ferry docks?

30. Charlie was crossing a narrow bridge. When he was halfway across he saw a truck on the opposite side of the bridge, 200 yards away and heading toward him. He turned and ran back. The truck continued at the same speed and missed him by a hair. If Charlie had tried to cross the bridge the truck would have hit him 3 yards before he reached the other end. How long is the bridge?

2.5 COMPLEX NUMBERS*

If you are restricted to nonnegative integers, you can't solve the equation $x + 5 = 0$. Enlarging the number system to include negative integers makes it possible to solve this equation ($x = -5$). Enlarging it again, to include rational numbers, makes it possible to solve equations like $3x = 7$, which have no integer solutions. Similarly, the equation $x^2 = 2$ has no solutions in the rational number system, but has $\sqrt{2}$ and $-\sqrt{2}$ as solutions in the real number system. So the idea of enlarging a number system in order to solve an equation that can't be solved in the present system is a natural one.

Equations such as $x^2 = -1$ and $x^2 = -4$ have no solutions in the real number system because $\sqrt{-1}$ and $\sqrt{-4}$ are not defined. In order to solve such equations (or equivalently, in order to find square roots of negative numbers) we must enlarge the number system again. We claim that there is a number system, called the **complex number system**, with these properties:

PROPERTIES OF THE COMPLEX NUMBER SYSTEM

1. The complex number system contains all real numbers.
2. Addition, subtraction, multiplication, and division of complex numbers obey the same rules of arithmetic that hold in the real number system (as summarized in the box at the bottom of page 3).
3. The complex number system contains a number (usually denoted by i) such that $i^2 = -1$.
4. Every complex number can be written in the *standard form* $a + bi$, where a and b are real numbers.†
5. Two complex numbers $a + bi$ and $c + di$ are equal exactly when $a = c$ and $b = d$.

---

* This section is not needed until Section 5.4; it may be postponed until then if desired.

† Hereafter whenever we write $a + bi$ or $c + di$, it is assumed that a, b, c, d are real numbers and $i^2 = -1$.

In view of our past experience with enlarging the number system, this claim *ought* to appear plausible. But the mathematicians who invented the complex numbers in the 17th century were very uneasy about a number i such that $i^2 = -1$ (that is, $i = \sqrt{-1}$). Consequently, they called numbers of the form bi (b any real number), such as $5i$ and $-\frac{1}{4}i$, **imaginary numbers**. The old familiar numbers (integers, rationals, irrationals) were called **real numbers**. Sums of real and imaginary numbers, numbers of the form $a + bi$, such as

$$5 + 2i, \qquad 7 - 4i, \qquad 18 + \frac{3}{2}i, \qquad \sqrt{3} - 12i$$

were called **complex numbers.***

Every real number is a complex number; for instance, $7 = 7 + 0i$. Similarly, every imaginary number bi is a complex number since $bi = 0 + bi$. Since the usual laws of arithmetic still hold, it's easy to add, subtract, and multiply complex numbers. As the following examples demonstrate, *all symbols can be treated as if they were real numbers, provided that i^2 is replaced by -1.* Unless directed otherwise, express your answers in the standard form $a + bi$.

Example 1
(a) $(1 + i) + (3 - 7i) = 1 + i + 3 - 7i$
$$= (1 + 3) + (i - 7i) = 4 - 6i.$$
(b) $(4 + 3i) - (8 - 6i) = 4 + 3i - 8 - (-6i)$
$$= (4 - 8) + (3i + 6i) = -4 + 9i.$$
(c) $4i(2 + \frac{1}{2}i) = 4i \cdot 2 + 4i(\frac{1}{2}i) = 8i + 4 \cdot \frac{1}{2} \cdot i^2$
$$= 8i + 2i^2 = 8i + 2(-1) = -2 + 8i.$$
(d) $(2 + i)(3 - 4i) = 2 \cdot 3 + 2(-4i) + i \cdot 3 + i(-4i)$
$$= 6 - 8i + 3i - 4i^2 = 6 - 8i + 3i - 4(-1)$$
$$= (6 + 4) + (-8i + 3i) = 10 - 5i. \blacksquare$$

The familiar multiplication patterns and exponent laws for integer exponents hold in the complex number system.

Example 2
(a) $(3 + 2i)(3 - 2i) = 3^2 - (2i)^2$
$$= 9 - 4i^2 = 9 - 4(-1) = 9 + 4 = 13.$$
(b) $(4 + i)^2 = 4^2 + 2 \cdot 4 \cdot i + i^2 = 16 + 8i + (-1) = 15 + 8i.$

---

* This terminology is still used, even though there is nothing complicated, unreal, or imaginary about complex numbers—they are just as valid mathematically as are real numbers. See Exercise 80 for a formal construction of the complex numbers and proofs of the claims made above.

(c) To find i^{54}, we first note that $i^4 = i^2 i^2 = (-1)(-1) = 1$ and that $54 = 52 + 2 = 4 \cdot 13 + 2$. Consequently,

$$i^{54} = i^{52+2} = i^{52} i^2 = i^{4 \cdot 13} i^2 = (i^4)^{13} i^2 = 1^{13}(-1) = -1. \blacksquare$$

The **conjugate** of the complex number $a + bi$ is the number $a - bi$, and the conjugate of $a - bi$ is $a + bi$. For example, the conjugate of $3 + 4i$ is $3 - 4i$ and the conjugate of $-3i = 0 - 3i$ is $0 + 3i = 3i$. Every real number is its own conjugate; for instance, the conjugate of $17 = 17 + 0i$ is $17 - 0i = 17$.

For any complex number $a + bi$, we have:

$$(a + bi)(a - bi) = a^2 - (bi)^2 = a^2 - b^2 i^2 = a^2 - b^2(-1) = a^2 + b^2$$

Since a^2 and b^2 are nonnegative real numbers, so is $a^2 + b^2$. Therefore *the product of a complex number and its conjugate is a nonnegative real number.* This fact enables us to express quotients of complex numbers in standard form.

Example 3 To express $\dfrac{3 + 4i}{1 + 2i}$ in the form $a + bi$, *multiply both numerator and denominator by the conjugate of the denominator,* namely, $1 - 2i$:

$$\frac{3 + 4i}{1 + 2i} = \frac{3 + 4i}{1 + 2i} \cdot \frac{1 - 2i}{1 - 2i} = \frac{(3 + 4i)(1 - 2i)}{(1 + 2i)(1 - 2i)}$$

$$= \frac{3 + 4i - 6i - 8i^2}{1^2 - (2i)^2} = \frac{3 + 4i - 6i - 8(-1)}{1 - 4i^2} = \frac{11 - 2i}{1 - 4(-1)}$$

$$= \frac{11 - 2i}{5} = \frac{11}{5} - \frac{2}{5} i$$

This is the form $a + bi$ with $a = 11/5$ and $b = -2/5$. $\blacksquare$

Example 4 The inverse of $1 - i$ under multiplication is $1/(1 - i)$. To express $\dfrac{1}{1 - i}$ in standard form, note that the conjugate of the denominator is $1 + i$ and therefore:

$$\frac{1}{1 - i} = \frac{1 \cdot (1 + i)}{(1 - i)(1 + i)} = \frac{1 + i}{1^2 - i^2} = \frac{1 + i}{1 - (-1)} = \frac{1 + i}{2} = \frac{1}{2} + \frac{1}{2} i$$

We can check this result by multiplying $\dfrac{1}{2} + \dfrac{1}{2} i$ by $1 - i$ to see if the product is 1 (which it should be if $\dfrac{1}{2} + \dfrac{1}{2} i = \dfrac{1}{1 - i}$):

$$\left(\frac{1}{2} + \frac{1}{2} i\right)(1 - i) = \frac{1}{2} \cdot 1 - \frac{1}{2} i + \frac{1}{2} i \cdot 1 - \frac{1}{2} i^2 = \frac{1}{2} - \frac{1}{2}(-1) = 1. \blacksquare$$

Since $i^2 = -1$, we define $\sqrt{-1}$ to be the complex number i. Similarly, since $(5i)^2 = 5^2 i^2 = 25(-1) = -25$, we define $\sqrt{-25}$ to be $5i$. In general,

SQUARE ROOTS OF NEGATIVE NUMBERS

For any positive real number b, $\sqrt{-b}$ is defined to be $\sqrt{b}\,i$.

because $(\sqrt{b}\,i)^2 = (\sqrt{b})^2 i^2 = b(-1) = -b$.

WARNING

$\sqrt{b}\,i$ is *not* the same as $\sqrt{bi}$. To avoid confusion it may help to write $\sqrt{b}\,i$ as $i\sqrt{b}$.

Example 5

(a) $\sqrt{-3} = \sqrt{3}\,i = i\sqrt{3}$.

(b) $\dfrac{1 - \sqrt{-7}}{3} = \dfrac{1 - \sqrt{7}\,i}{3} = \dfrac{1}{3} - \dfrac{\sqrt{7}}{3}\,i$. ■

WARNING

The property $\sqrt{cd} = \sqrt{c}\sqrt{d}$, which is valid for positive real numbers, *does not hold* when both c and d are negative. For example, according to the definition above

$$\sqrt{-20}\,\sqrt{-5} = \sqrt{20}\,i \cdot \sqrt{5}\,i = \sqrt{20}\,\sqrt{5}\cdot i^2 = \sqrt{20\cdot 5}\,(-1)$$
$$= \sqrt{100}\,(-1) = -10.$$

But $\sqrt{(-20)(-5)} = \sqrt{100} = 10$, so that

$$\sqrt{(-20)(-5)} \neq \sqrt{-20}\,\sqrt{-5}.$$

To avoid difficulty, *always write square roots of negative numbers in terms of i before doing any simplification.*

Example 6

$$(7 - \sqrt{-4})(5 + \sqrt{-9}) = (7 - \sqrt{4}\,i)(5 + \sqrt{9}\,i)$$
$$= (7 - 2i)(5 + 3i)$$
$$= 35 + 21i - 10i - 6i^2$$
$$= 35 + 11i - 6(-1) = 41 + 11i.\ ■$$

Since every negative real number has a square root in the complex number system, we can now find complex solutions for equations that have no real solutions. For example, the solutions of $x^2 = -25$ are $x = \pm\sqrt{-25} = \pm 5i$. In fact, *every quadratic equation with real coefficients has solutions in the complex number system.*

Example 7 To solve the equation $2x^2 + x + 3 = 0$, we apply the quadratic formula:

$$x = \frac{-1 \pm \sqrt{1^2 - 4 \cdot 2 \cdot 3}}{2 \cdot 2} = \frac{-1 \pm \sqrt{-23}}{4}$$

Since $\sqrt{-23}$ is not a real number, this equation has no real number solutions. But $\sqrt{-23}$ *is* a complex number, namely, $\sqrt{-23} = \sqrt{23}i$. Thus the equation does have solutions in the complex number system:

$$x = \frac{-1 \pm \sqrt{-23}}{4} = \frac{-1 \pm \sqrt{23}i}{4} = -\frac{1}{4} \pm \frac{\sqrt{23}}{4}i$$

Note that the two solutions, $-\frac{1}{4} + \frac{\sqrt{23}}{4}i$ and $-\frac{1}{4} - \frac{\sqrt{23}}{4}i$, are conjugates of each other. ∎

Example 8 To find *all* solutions of $x^3 = 1$, we rewrite the equation and use the Difference of Cubes pattern (page 46) to factor:

$$x^3 = 1$$
$$x^3 - 1 = 0$$
$$(x - 1)(x^2 + x + 1) = 0$$
$$x - 1 = 0 \quad \text{or} \quad x^2 + x + 1 = 0$$

The solution of the first equation is $x = 1$. The solutions of the second can be obtained from the quadratic formula:

$$x = \frac{-1 \pm \sqrt{1^2 - 4 \cdot 1 \cdot 1}}{2 \cdot 1} = \frac{-1 \pm \sqrt{-3}}{2} = \frac{-1 \pm \sqrt{3}i}{2} = -\frac{1}{2} \pm \frac{\sqrt{3}}{2}i.$$

Therefore the equation $x^3 = 1$ has one real solution ($x = 1$) and two nonreal complex solutions [$x = -\frac{1}{2} + (\sqrt{3}/2)i$ and $x = -\frac{1}{2} - (\sqrt{3}/2)i$]. Each of these solutions is said to be a **cube root of one** or a **cube root of unity**. Observe that the two nonreal complex cube roots of unity are conjugates of each other. ∎

The preceding examples illustrate this useful fact (whose proof is omitted):

CONJUGATE SOLUTIONS

If $a + bi$ is a solution of a polynomial equation with *real* coefficients, then its conjugate $a - bi$ is also a solution of this equation.

EXERCISES

In Exercises 1–54, perform the indicated operation and write the result in the form $a + bi$.

1. $(2 + 3i) + (6 - i)$
2. $(-5 + 7i) + (14 + 3i)$
3. $(2 - 8i) - (4 + 2i)$
4. $(3 + 5i) - (3 - 7i)$
5. $\frac{5}{4} - \left(\frac{7}{4} + 2i\right)$
6. $(\sqrt{3} + i) + (\sqrt{5} - 2i)$
7. $\left(\frac{\sqrt{2}}{2} + i\right) - \left(\frac{\sqrt{3}}{2} - i\right)$
8. $\left(\frac{1}{2} + \frac{\sqrt{3}i}{2}\right) + \left(\frac{3}{4} - \frac{5\sqrt{3}i}{2}\right)$
9. $(2 + i)(3 + 5i)$
10. $(2 - i)(5 + 2i)$
11. $(-3 + 2i)(4 - i)$
12. $(4 + 3i)(4 - 3i)$
13. $(2 - 5i)^2$
14. $(1 + i)(2 - i)i$
15. $(\sqrt{3} + i)(\sqrt{3} - i)$
16. $\left(\frac{1}{2} - i\right)\left(\frac{1}{4} + 2i\right)$
17. i^{15}
18. i^{26}
19. i^{33}
20. $(-i)^{53}$
21. $(-i)^{107}$
22. $(-i)^{213}$
23. $\frac{1}{5 - 2i}$
24. $\frac{1}{i}$
25. $\frac{1}{3i}$
26. $\frac{i}{2 + i}$
27. $\frac{3}{4 + 5i}$
28. $\frac{2 + 3i}{i}$
29. $\frac{1}{i(4 + 5i)}$
30. $\frac{1}{(2 - i)(2 + i)}$
31. $\frac{2 + 3i}{i(4 + i)}$
32. $\frac{2}{(2 + 3i)(4 + i)}$
33. $\frac{2 + i}{1 - i} + \frac{1}{1 + 2i}$
34. $\frac{1}{2 - i} + \frac{3 + i}{2 + 3i}$
35. $\frac{i}{3 + i} - \frac{3 + i}{4 + i}$
36. $6 + \frac{2i}{3 + i}$
37. $\sqrt{-36}$
38. $\sqrt{-81}$
39. $\sqrt{-14}$
40. $\sqrt{-50}$
41. $-\sqrt{-16}$
42. $-\sqrt{-12}$
43. $\sqrt{-16} + \sqrt{-49}$
44. $\sqrt{-25} - \sqrt{-9}$
45. $\sqrt{-15} - \sqrt{-18}$
46. $\sqrt{-12}\sqrt{-3}$
47. $\sqrt{-16}/\sqrt{-36}$
48. $-\sqrt{-64}/\sqrt{-4}$
49. $(\sqrt{-25} + 2)(\sqrt{-49} - 3)$
50. $(5 - \sqrt{-3})(-1 + \sqrt{-9})$
51. $(2 + \sqrt{-5})(1 - \sqrt{-10})$
52. $\sqrt{-3}(3 - \sqrt{-27})$
53. $1/(1 + \sqrt{-2})$
54. $(1 + \sqrt{-4})/(3 - \sqrt{-9})$

In Exercises 55–58, find x and y. Remember that $a + bi = c + di$ exactly when $a = c$ and $b = d$.

55. $3x - 4i = 6 + 2yi$
56. $8 - 2yi = 4x + 12i$
57. $3 + 4xi = 2y - 3i$
58. $8 - xi = \frac{1}{2}y + 2i$

In Exercises 59–70, solve the equation and express each solution in the form $a + bi$.

59. $3x^2 - 2x + 5 = 0$
60. $5x^2 + 2x + 1 = 0$
61. $x^2 + x + 2 = 0$
62. $5x^2 - 6x + 2 = 0$
63. $2x^2 - x = -4$
64. $x^2 + 1 = 4x$
65. $2x^2 + 3 = 6x$
66. $3x^2 + 4 = -5x$
67. $x^3 - 8 = 0$
68. $x^3 + 125 = 0$
69. $x^4 - 1 = 0$
70. $x^4 - 81 = 0$
71. Simplify: $i + i^2 + i^3 + \cdots + i^{15}$
72. Simplify: $i - i^2 + i^3 - i^4 + i^5 - \cdots + i^{15}$

Unusual Problems

If $z = a + bi$ is a complex number, then its conjugate is usually denoted $\bar{z}$, that is, $\bar{z} = a - bi$. In Exercises 73–77, prove that for any complex numbers $z = a + bi$ and $w = c + di$:

73. $\overline{z + w} = \bar{z} + \bar{w}$
74. $\overline{zw} = \bar{z} \cdot \bar{w}$

75. $\overline{\left(\dfrac{z}{w}\right)} = \dfrac{\bar{z}}{\bar{w}}$ 76. $\bar{\bar{z}} = z$

77. z is a real number exactly when $\bar{z} = z$.

78. The **real part** of the complex number $a + bi$ is defined to be the real number a. The **imaginary part** of $a + bi$ is defined to be the real number b (not bi).

   (a) Show that the real part of $z = a + bi$ is $\dfrac{z + \bar{z}}{2}$.

   (b) Show that the imaginary part of $z = a + bi$ is $\dfrac{z - \bar{z}}{2i}$.

79. If $z = a + bi$ (with a, b real numbers, not both 0), express $1/z$ in standard form.

80. **Construction of the Complex Numbers.** We assume that the real number system is known. In order to construct a new number system with the desired properties, we must do the following:
   (i) Define a set C (whose elements will be called complex numbers).
   (ii) The set C must contain the real numbers or at least a copy of them.
   (iii) Define addition and multiplication in the set C in such a way that the usual laws of arithmetic are valid.
   (iv) Show that C has the other properties listed on page 87.

   We begin by defining C to be the set of all ordered pairs of real numbers. Thus $(1, 5)$, $(-6, 0)$, $(\tfrac{4}{3}, -17)$, and $(\sqrt{2}, \tfrac{12}{5})$ are some of the elements of the set C. More generally, a complex number (= element of C) is any pair (a, b) where a and b are real numbers. By definition, two complex numbers are *equal* exactly when they have the same first and the same second coordinate.
   (a) *Addition in C* is defined by this rule:
   $$(a, b) + (c, d) = (a + c, b + d)$$
   For example,
   $$(3, 2) + (5, 4) = (3 + 5, 2 + 4) = (8, 6).$$
   Verify that this addition has the following properties. For any complex numbers (a, b), (c, d), (e, f) in C:
   (i) $(a, b) + (c, d) = (c, d) + (a, b)$
   (ii) $((a, b) + (c, d)) + (e, f) = (a, b) + ((c, d) + (e, f))$
   (iii) $(a, b) + (0, 0) = (a, b)$
   (iv) $(a, b) + (-a, -b) = (0, 0)$

   (b) *Multiplication in C* is defined by this rule:
   $$(a, b)(c, d) = (ac - bd, bc + ad)$$
   For example,
   $$(3, 2)(4, 5) = (3 \cdot 4 - 2 \cdot 5, 2 \cdot 4 + 3 \cdot 5)$$
   $$= (12 - 10, 8 + 15) = (2, 23).$$
   Verify that this multiplication has the following properties. For any complex numbers (a, b), (c, d), (e, f) in C:
   (i) $(a, b)(c, d) = (c, d)(a, b)$
   (ii) $((a, b)(c, d))(e, f) = (a, b)((c, d)(e, f))$
   (iii) $(a, b)(1, 0) = (a, b)$
   (iv) $(a, b)(0, 0) = (0, 0)$

   (c) Verify that for any two elements of C with second coordinate zero:
   (i) $(a, 0) + (c, 0) = (a + c, 0)$
   (ii) $(a, 0)(c, 0) = (ac, 0)$

   Identify $(t, 0)$ with the real number t. Statements (i) and (ii) show that when addition or multiplication in C is performed on two real numbers (that is, elements of C with second coordinate 0), the result is the usual sum or product of real numbers. Thus C contains (a copy of) the real number system.

   (d) *New Notation.* Since we are identifying the complex number $(a, 0)$ with the real number a, we shall hereafter denote $(a, 0)$ simply by the symbol a. Also, let i denote the complex number $(0, 1)$.
   (i) Show that $i^2 = -1$ [that is, $(0, 1)(0, 1) = (-1, 0)$].
   (ii) Show that for any complex number $(0, b)$, $(0, b) = bi$ [that is, $(0, b) = (b, 0)(0, 1)$].
   (iii) Show that any complex number (a, b) can be written: $(a, b) = a + bi$ [that is, $(a, b) = (a, 0) + (b, 0)(0, 1)$].

   In this new notation, every complex number is of the form $a + bi$ with a, b real and $i^2 = -1$, and our construction is finished.

2.6 OTHER EQUATIONS

We first extend the techniques for solving quadratic equations to certain polynomial equations of degree greater than 2. Although it is often difficult, **factoring** sometimes works.

Example 1 To solve $x^6 + x^5 - 9x^4 - 9x^3 = 0$, we use regrouping and the Difference of Squares pattern to factor the left side:

$$x^3[x^3 + x^2 - 9x - 9] = 0$$
$$x^3[x^2(x + 1) - 9(x + 1)] = 0$$
$$x^3[(x^2 - 9)(x + 1)] = 0$$
$$x^3(x + 3)(x - 3)(x + 1) = 0$$

If a product is 0, one of the factors must be 0. Therefore

$x^3 = 0$ or $x + 3 = 0$ or $x - 3 = 0$ or $x + 1 = 0$
$x = 0$ $x = -3$ $x = 3$ $x = -1$

So the equation has four distinct solutions: $-3, -1, 0,$ and 3. ∎

WARNING

In the last example, it is *wrong* to attempt to simplify the original equation by dividing both sides by x^3. Doing so produces the equation $x^3 + x^2 - 9x - 9 = 0$ and its solutions $-1, \pm 3$, but it *eliminates* the solution 0 of the original equation. By following the factoring procedure, as in the example, you avoid the loss of any solutions.

An equation of **quadratic type** is one that can be transformed into a quadratic equation by making a suitable substitution for the variable.

Example 2 To solve $4x^4 - 13x^2 + 3 = 0$, substitute u for x^2 and solve the resulting quadratic equation:

$$4x^4 - 13x^2 + 3 = 0$$
$$4(x^2)^2 - 13x^2 + 3 = 0$$
$$4u^2 - 13u + 3 = 0$$
$$(u - 3)(4u - 1) = 0$$

$u - 3 = 0$ or $4u - 1 = 0$
$u = 3$ $4u = 1$
                     $u = \dfrac{1}{4}$

Since $u = x^2$ we see that
$$x^2 = 3 \quad \text{or} \quad x^2 = \frac{1}{4}$$
$$x = \pm\sqrt{3} \quad\quad\quad x = \pm\frac{1}{2}$$

Hence the original equation has four solutions: $-\sqrt{3}, \sqrt{3}, -\frac{1}{2}, \frac{1}{2}$. ■

Example 3 To solve $x^4 - 4x^2 + 1 = 0$, let $u = x^2$:
$$x^4 - 4x^2 + 1 = 0$$
$$u^2 - 4u + 1 = 0$$

The quadratic formula shows that
$$u = \frac{-(-4) \pm \sqrt{(-4)^2 - 4 \cdot 1 \cdot 1}}{2 \cdot 1} = \frac{4 \pm \sqrt{12}}{2}$$
$$= \frac{4 \pm \sqrt{4 \cdot 3}}{2} = \frac{4 \pm 2\sqrt{3}}{2} = 2 \pm \sqrt{3}$$

Since $u = x^2$, we have the equivalent statements:
$$x^2 = 2 + \sqrt{3} \quad \text{or} \quad x^2 = 2 - \sqrt{3}$$
$$x = \pm\sqrt{2 + \sqrt{3}} \quad\quad x = \pm\sqrt{2 - \sqrt{3}}$$

Therefore the original equation has four solutions. ■

Other techniques for solving higher degree polynomial equations are discussed in Excursion 2.6.A.

Radical Equations

If two numbers are equal, say $a = b$, then $a^r = b^r$ for every positive integer r. The same is true of algebraic expressions. For example,
$$\text{if } x - 2 = 3, \text{ then } (x - 2)^2 = 3^2 = 9$$

Thus every solution of $x - 2 = 3$ is also a solution of $(x - 2)^2 = 9$. But *be careful:* This only works in *one* direction. For instance, -1 is a solution of $(x - 2)^2 = 9$, but not of $x - 2 = 3$. Similarly, 1 is the only solution of $x = 1$, but $x^3 = 1^3$ (that is, $x^3 - 1 = 0$) has two additional complex solutions, as shown in Example 8 on page 91. Therefore

POWER PRINCIPLE

> If both sides of an equation are raised to the same positive integer power, every solution of the original equation is also a solution of the new equation. But the new equation may have solutions that are *not* solutions of the original one.*

* Such solutions are called **extraneous solutions**.

So if you raise both sides of an equation to a power, you *must check* your solutions in the *original* equation.

Example 4 To solve $\sqrt[3]{2x^2 + 7x - 6} = 3^{2/3}$ we first cube both sides, and then solve the resulting equation:

$$(\sqrt[3]{2x^2 + 7x - 6})^3 = (3^{2/3})^3$$
$$2x^2 + 7x - 6 = 9$$
$$2x^2 + 7x - 15 = 0$$
$$(2x - 3)(x + 5) = 0$$
$$2x - 3 = 0 \quad \text{or} \quad x + 5 = 0$$
$$2x = 3 \quad\quad\quad\quad x = -5$$
$$x = \frac{3}{2}$$

Substituting 3/2 and −5 in the left side of the original equation shows that:

$$\sqrt[3]{2\left(\frac{3}{2}\right)^2 + 7\left(\frac{3}{2}\right) - 6} = \sqrt[3]{\frac{30}{2} - 6} = \sqrt[3]{9} = \sqrt[3]{3^2} = 3^{2/3}$$

$$\sqrt[3]{2(-5)^2 + 7(-5) - 6} = \sqrt[3]{9} = \sqrt[3]{3^2} = 3^{2/3}$$

Therefore both 3/2 and −5 are solutions. ∎

Example 5 To solve $5 + \sqrt{3x - 11} = x$, we first rearrange terms to get the radical expression alone on one side:

$$\sqrt{3x - 11} = x - 5$$

Then square both sides and solve the resulting equation:

$$(\sqrt{3x - 11})^2 = (x - 5)^2$$
$$3x - 11 = x^2 - 10x + 25$$
$$0 = x^2 - 13x + 36$$
$$0 = (x - 4)(x - 9)$$
$$x - 4 = 0 \quad \text{or} \quad x - 9 = 0$$
$$x = 4 \quad \text{or} \quad x = 9$$

Substituting 4 in the original equation produces

    Left side: $5 + \sqrt{3 \cdot 4 - 11} = 5 + \sqrt{1} = 6$ Right side: 4

Therefore 4 is *not* a solution. Substituting 9 shows that

    Left side: $5 + \sqrt{3 \cdot 9 - 11} = 5 + \sqrt{16} = 9$ Right side: 9

Hence 9 is the only solution of the original equation. ∎

Example 6 To solve $\sqrt{2x-3} - \sqrt{x+7} = 2$, we first rearrange terms so that one side contains only a single radical term:

$$\sqrt{2x-3} = \sqrt{x+7} + 2$$

Then square both sides and simplify:

$$(\sqrt{2x-3})^2 = (\sqrt{x+7} + 2)^2$$
$$2x - 3 = (\sqrt{x+7})^2 + 2 \cdot 2 \cdot \sqrt{x+7} + 2^2$$
$$2x - 3 = x + 7 + 4\sqrt{x+7} + 4$$
$$x - 14 = 4\sqrt{x+7}$$

Now square both sides and solve the resulting equation:

$$(x-14)^2 = (4\sqrt{x+7})^2$$
$$x^2 - 28x + 196 = 4^2 \cdot (\sqrt{x+7})^2$$
$$x^2 - 28x + 196 = 16(x+7)$$
$$x^2 - 28x + 196 = 16x + 112$$
$$x^2 - 44x + 84 = 0$$
$$(x-2)(x-42) = 0$$
$$x - 2 = 0 \quad \text{or} \quad x - 42 = 0$$
$$x = 2 \qquad\qquad x = 42$$

Substituting 2 and 42 in the left side of the original equation shows that

$$\sqrt{2 \cdot 2 - 3} - \sqrt{2+7} = \sqrt{1} - \sqrt{9} = 1 - 3 = -2$$
$$\sqrt{2 \cdot 42 - 3} - \sqrt{42+7} = \sqrt{81} - \sqrt{49} = 9 - 7 = 2$$

Therefore 42 is the only solution of the equation. ∎

Many equations involving positive or negative rational exponents are of quadratic type and can be solved by making an appropriate substitution.

Example 7 To solve $x^{2/3} - 2x^{1/3} - 15 = 0$, let $u = x^{1/3}$:

$$x^{2/3} - 2x^{1/3} - 15 = 0$$
$$(x^{1/3})^2 - 2x^{1/3} - 15 = 0$$
$$u^2 - 2u - 15 = 0$$
$$(u+3)(u-5) = 0$$
$$u + 3 = 0 \quad \text{or} \quad u - 5 = 0$$
$$u = -3 \qquad\qquad u = 5$$
$$x^{1/3} = -3 \qquad\qquad x^{1/3} = 5$$

Cubing both sides of these last equations shows that

$$(x^{1/3})^3 = (-3)^3 \quad \text{or} \quad (x^{1/3})^3 = 5^3$$
$$x = -27 \quad\quad\quad\quad x = 125$$

Since we cubed both sides, we must check these numbers in the original equation. Verify that both *are* solutions. ∎

Absolute Value Equations

As we saw on page 28, 3 and -3 are the only numbers with absolute value 3, that is, the only solutions of $|x| = 3$. A similar procedure can be used to solve other absolute value equations.

Example 8 To solve $|3x - 4| = 8$, use the fact that there are exactly two numbers whose absolute value is 8, namely, 8 and -8. Since $3x - 4$ is a number with absolute value 8, there are only two possibilities:

$$3x - 4 = 8 \quad \text{or} \quad 3x - 4 = -8$$
$$3x = 12 \quad\quad\quad\quad 3x = -4$$
$$x = 4 \quad\quad\quad\quad x = -\frac{4}{3}$$

You can readily verify that both 4 and $-4/3$ are solutions of the original equation $|3x - 4| = 8$. ∎

Example 9 The equation $|\sqrt{2x - 3} - \sqrt{x + 7}| = 2$ is equivalent to

$$\sqrt{2x - 3} - \sqrt{x + 7} = 2 \quad \text{or} \quad \sqrt{2x - 3} - \sqrt{x + 7} = -2$$

because the only numbers with absolute value 2 are 2 and -2. Example 6 above shows that 42 is the solution of the first equation. A similar argument shows that 2 is the solution of the second equation. Therefore 2 and 42 are the solutions of $|\sqrt{2x - 3} - \sqrt{x + 7}| = 2$. ∎

EXERCISES

In Exercises 1–36, find all real solutions of the equation.

1. $x^3 - 27 = 0$
2. $x^3 + 8 = 0$
3. $x^4 - 125x = 0$
4. $t^8 = t^4$
5. $x^3 - 12x^2 + 48x - 64 = 0$
6. $x^3 - 3x^2 + 3x - 1 = 0$
7. $x^5 - .0016x = 0$
8. $16x^6 = 2.0736x^2$
9. $y^4 - 7y^2 + 6 = 0$
10. $x^4 - 2x^2 + 1 = 0$
11. $x^4 + 2x^2 - 35 = 0$
12. $x^4 - 2x^2 - 24 = 0$
13. $2y^4 - 9y^2 + 4 = 0$
14. $6z^4 - 7z^2 + 2 = 0$
15. $10x^4 + 3x^2 = 1$
16. $6x^4 - 7x^2 = 3$

17. $\sqrt{x+2} = 3$
18. $\sqrt{x-7} = 4$
19. $\sqrt[3]{5-11x} = 3$
20. $\sqrt[3]{6x-10} = 2$
21. $\sqrt{x^2-x-1} = 1$
22. $\sqrt{x^2-5x+4} = 2$
23. $(x+5)^{1/3} = (x+3)^{2/3}$
24. $\sqrt[3]{x-3} = \sqrt[3]{3x^2+2x-5}$
25. $\sqrt[5]{x^2-x+123} = 5^{3/5}$
26. $(2x^2+17x+17)^{1/7} = (2x+4)^{2/7}$
27. $\sqrt{x+7} = x-5$
28. $\sqrt{x+5} = x-1$
29. $\sqrt{3x^2+7x-2} = x+1$
30. $\sqrt{4x^2-10x+5} = x-3$
31. $\sqrt[3]{x^3+x^2-4x+5} = x+1$
32. $\sqrt[3]{x^3-6x^2+2x+3} = x-1$
33. $\sqrt{5x+6} = 3+\sqrt{x+3}$
34. $\sqrt{3y+1} - 1 = \sqrt{y+4}$
35. $\sqrt{3x+5} + \sqrt{2x+3} + 1 = 0$
36. $\sqrt{20-x} = \sqrt{9-x} + 3$

In Exercises 37–40, assume that all letters represent positive numbers and solve each equation for the required variable.

37. $A = \sqrt{1+\dfrac{a^2}{b^2}}$ for b
38. $T = 2\pi\sqrt{\dfrac{m}{g}}$ for g
39. $K = \sqrt{1-\dfrac{x^2}{u^2}}$ for u
40. $R = \sqrt{d^2+k^2}$ for d

In Exercises 41–54, solve the given equation by making an appropriate substitution.

41. $x - 4x^{1/2} + 4 = 0$
42. $x - x^{1/2} - 12 = 0$
43. $2x - \sqrt{x} - 6 = 0$
44. $3x - 11\sqrt{x} - 4 = 0$
45. $x^{2/3} + 3x^{1/3} + 2 = 0$ [*Hint:* Let $u = x^{1/3}$.]
46. $x^{2/3} - 4x^{1/3} + 3 = 0$
47. $\sqrt[3]{x^2} + 2\sqrt[3]{x} - 8 = 0$
48. $2\sqrt[3]{x^2} - \sqrt[3]{x} - 6 = 0$
49. $x^{1/2} - x^{1/4} - 2 = 0$ [*Hint:* Let $u = x^{1/4}$.]
50. $x^{1/3} + x^{1/6} - 2 = 0$
51. $x^{-2} - x^{-1} - 6 = 0$ [*Hint:* Let $u = x^{-1}$.]
52. $x^{-2} - 6x^{-1} + 5 = 0$
53. $2x^{-2} + x^{-1} - 1 = 0$
54. $6x^{-2} + x^{-1} - 1 = 0$

In Exercises 55–62, solve the equation.

55. $|2x+3| = 9$
56. $|3x-5| = 7$
57. $|6x-9| = 0$
58. $|4x-5| = -9$
59. $|x^2+4x-1| = 4$
60. $|x^2+2x-9| = 6$
61. $|x^2-5x+1| = 3$
62. $|2x^2+5x-7| = 4$

63. What are the dimensions of a rectangle whose diagonal is 130 cm long and whose area is 6000 cm²?

64. Find the radius of the base of a conical container whose height is 1/3 of the radius and whose volume is 180 cubic inches. [The volume of a cone of radius r and height h is $\pi r^2 h/3$.]

65. What is the radius of the base of a cone whose surface area is 18π cm² and whose height is 4 cm? [The surface area of a cone of radius r and height h is $\pi r\sqrt{r^2+h^2}$.]

66. A power plant is located on the bank of a river that is ½ mile wide. Wiring is to be laid across the river and then along the shore to a substation 8 miles downstream, as shown in Figure 2-2.

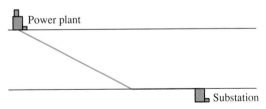

FIGURE 2-2

It costs $12,000 per mile for underwater wiring and $8,000 per mile for wiring on land. If $72,000 is to be spent on the project, how far from the substation should the wiring come to shore?

2.6.A EXCURSION: The Rational Solutions Test

There are (complicated) formulas for finding the solutions of third- and fourth-degree polynomial equations, but no such formulas exist for solving higher degree equations. When the polynomial has integer coefficients, however, you can find all the *rational* solutions by using:

THE RATIONAL SOLUTIONS TEST

> If a rational number r/s (in lowest terms) is a solution of the equation
> $$a_n x^n + \cdots + a_1 x + a_0 = 0,$$
> where the coefficients $a_n, \cdots, a_1, a_0$ are integers, then r is a factor of the constant term a_0 and s is a factor of the leading coefficient a_n.

The test states conditions that a rational solution must satisfy.* Of course, there may be numbers satisfying these conditions that are *not* solutions. By finding all the numbers that meet these conditions, we produce a list of the *possible* rational solutions. It is then relatively easy (though tedious) to test each number on the list to see if it actually *is* a solution.

Example 1 *If* the equation
$$2x^5 - x^4 - 10x^3 + 5x^2 + 12x - 6 = 0$$
has a rational number solution r/s, then by the Rational Solutions Test r must be a factor of the constant term -6. Therefore r must be one of $\pm 1, \pm 2, \pm 3, \pm 6$ (the only factors of -6). Similarly, s must be a factor of the leading coefficient 2, so s must be one of $\pm 1, \pm 2$ (the only factors of 2). Consequently, the only *possibilities* for r/s are

$$\frac{\pm 1}{\pm 1}, \frac{\pm 2}{\pm 1}, \frac{\pm 3}{\pm 1}, \frac{\pm 6}{\pm 1}, \frac{\pm 1}{\pm 2}, \frac{\pm 2}{\pm 2}, \frac{\pm 3}{\pm 2}, \frac{\pm 6}{\pm 2}$$

Eliminating duplications from this list, we see that the only *possible* rational solutions are:

$$1, -1, 2, -2, 3, -3, 6, -6, \frac{1}{2}, -\frac{1}{2}, \frac{3}{2}, -\frac{3}{2}$$

By substituting each of these numbers in the equation we can find which, if any, actually *are* solutions.† When you do it, you find that $\frac{1}{2}$ is the only number on

---

* Since the proof of the Rational Solutions Test sheds very little light on how the result is actually *used* to solve equations, it will be omitted.

† You can reduce the amount of computation if you have a graphing calculator. As we shall see in Chapter 5, the points at which the graph of $y = 2x^5 - x^4 - 10x^3 + 5x^2 + 12x - 6$ crosses the x-axis are the solutions of the equation. When you see the graph, you can eliminate any numbers from the list that aren't near a crossing point. In this case, the graph indicates that you need only check $1/2$ and $\pm 3/2$.

the list that is a solution. Since the list includes all possible rational solutions, any other real solutions must be irrational numbers. ■

Example 2 If $x^4 - 11x^2 - 2x + 12 = 0$ has a rational solution r/s, then s must be a factor of the leading coefficient 1. Hence $s = \pm 1$, so that r/s is either r or $-r$. Since r must be a factor of the constant term 12, the only *possible* rational solutions are $r = \pm 1, \pm 2, \pm 3, \pm 4, \pm 6,$ or ± 12. Substituting each of these in the equation shows that 1 and -3 are the only rational solutions. ■

Once you have found one solution of a polynomial equation, the following fact can be used to factor the polynomial, which may in turn lead to additional solutions:

THE FACTOR THEOREM

> The number c is a solution of the polynomial equation $P = 0$ exactly when $x - c$ is a factor of the polynomial P.*

Example 3 Example 1 above shows that ½ is a solution of $2x^5 - x^4 - 10x^3 + 5x^2 + 12x - 6 = 0$. By the Factor Theorem $x - \frac{1}{2}$ is a factor of $2x^5 - x^4 - 10x^3 + 5x^2 + 12x - 6$. Use synthetic or long division to verify that the other factor is $2x^4 - 10x^2 + 12$. Then use this information to solve the equation:

$$2x^5 - x^4 - 10x^3 + 5x^2 + 12x - 6 = 0$$

$$\left(x - \frac{1}{2}\right)(2x^4 - 10x^2 + 12) = 0$$

$$x - \frac{1}{2} = 0 \quad \text{or} \quad 2x^4 - 10x^2 + 12 = 0$$

The second equation is in quadratic form; by mentally substituting $u = x^2$, we can factor it:

$$x - \frac{1}{2} = 0 \quad \text{or} \quad (2x^2 - 4)(x^2 - 3) = 0$$

$$x - \frac{1}{2} = 0 \quad \text{or} \quad 2x^2 - 4 = 0 \quad \text{or} \quad x^2 - 3 = 0$$

$$x = \frac{1}{2} \qquad\qquad x^2 = 2 \qquad\qquad x^2 = 3$$

$$x = \pm\sqrt{2} \qquad\qquad x = \pm\sqrt{3}$$

Therefore the equation has one rational solution, $\frac{1}{2}$, and four irrational ones, $\sqrt{2}, -\sqrt{2}, \sqrt{3},$ and $-\sqrt{3}$. ■

---

* The Factor Theorem will be proved in Section 5.3.

Example 4 In Example 2 above we saw that 1 and -3 are solutions of $x^4 - 11x^2 - 2x + 12 = 0$. By the Factor Theorem $x - 1$ and $x - (-3) = x + 3$ are factors of $x^4 - 11x^2 - 2x + 12$. If we divide $x^4 - 11x^2 - 2x + 12$ by $x - 1$, and then divide that quotient by $x + 3$, we obtain a factorization that leads to a complete solution:

$$x^4 - 11x^2 - 2x + 12 = 0$$
$$(x - 1)(x^3 + x^2 - 10x - 12) = 0$$
$$(x - 1)(x + 3)(x^2 - 2x - 4) = 0$$

$x - 1 = 0$ or $x + 3 = 0$ or $x^2 - 2x - 4 = 0$

$x = 1$ $\qquad$ $x = -3$ $\qquad$ $x = \dfrac{-(-2) \pm \sqrt{(-2)^2 - 4 \cdot 1 \cdot (-4)}}{2 \cdot 1}$

$$= \dfrac{2 \pm \sqrt{20}}{2} = \dfrac{2 \pm 2\sqrt{5}}{2}$$

$$= 1 \pm \sqrt{5}$$

So the equation has rational solutions 1 and -3 and irrational solutions $1 + \sqrt{5}$ and $1 - \sqrt{5}$. ∎

EXERCISES

In Exercises 1–10, find all rational solutions of the equation.

1. $x^3 + 3x^2 - x - 3 = 0$
2. $x^3 - x^2 - 3x + 3 = 0$
3. $x^3 + 5x^2 - x - 5 = 0$
4. $3x^3 + 8x^2 - x - 20 = 0$
5. $2x^3 + 5x^2 = 11x - 4$
6. $2x^3 = 3x^2 + 7x + 6$
7. $\dfrac{1}{12}x^3 - \dfrac{1}{12}x^2 - \dfrac{2}{3}x + 1 = 0$ [*Hint:* First multiply both sides by 12.]
8. $\dfrac{2}{3}x^4 + \dfrac{1}{2}x^3 - \dfrac{5}{4}x^2 - x - \dfrac{1}{6} = 0$
9. $\dfrac{1}{3}x^4 - x^3 - x^2 + \dfrac{13}{3}x - 2 = 0$
10. $\dfrac{1}{3}x^3 + \dfrac{1}{6} = \dfrac{1}{2}x^2 + \dfrac{1}{6}x$

In Exercises 11–16, use the Factor Theorem to determine if the first polynomial is a factor of the second.

11. $x - 1$; $\quad x^5 + 1$
12. $x - \dfrac{1}{2}$; $\quad 2x^4 + x^3 + x - \dfrac{3}{4}$
13. $x + 2$; $\quad x^3 - 3x^2 - 4x + 12$
14. $x + 1$; $\quad x^3 - 4x^2 + 3x + 8$
15. $x + 2$; $\quad x^3 + x^2 - 4x + 4$
16. $x - 1$; $\quad 14x^{99} - 65x^{56} + 51$

In Exercises 17–24, find all real solutions of the equation. (*Hint:* First find the rational solutions.)

17. $2x^3 - 5x^2 + x + 2 = 0$

18. $t^4 - t^3 + 2t^2 - 4t - 8 = 0$

19. $6x^3 - 11x^2 + 6x = 1$

20. $z^3 + z^2 + 2z + 2 = 0$

21. $x^4 = x^3 + x^2 + x + 2$

22. $3x^5 + 2x^4 - 7x^3 = -2x^2$

23. $3y^3 + 3y = 2y^2 + 2$ 24. $x^5 + x = x^3$

In Exercises 25–28, find a number k with the stated property.

25. $x + 2$ is a factor of $x^3 + 3x^2 + kx - 2$

26. $x - 3$ is a factor of $x^4 - 5x^3 + kx^2 + 18k + 18$

27. $x - 1$ is a factor of $k^2 x^4 - 2kx^2 + 1$

28. $x + 2$ is a factor of $x^3 - kx^2 + 3x + 7k$

29. Use the Factor Theorem to show that for every real number c, $x - c$ is *not* a factor of $x^4 + x^2 + 1$.

30. Let c be a real number and n a positive integer.
    (a) Show that $x - c$ is a factor of $x^n - c^n$.
    (b) If n is even, show that $x + c$ must be a factor of $x^n - c^n$. [Remember that $x + c = x - (-c)$.]

31. (a) If c is a real number and n an odd positive integer, give an example to show that $x + c$ may not be a factor of $x^n - c^n$.
    (b) If c and n are as in part (a), show that $x + c$ is a factor of $x^n + c^n$.

32. (a) Use the Rational Solutions Test on $x^2 - 2 = 0$ to show that $\sqrt{2}$ is an irrational number.
    (b) Show that $\sqrt{3}$ is irrational.

2.7 LINEAR INEQUALITIES

Two inequalities are **equivalent** if they have the same solutions. The usual solution strategy for inequalities is to transform a given inequality into an equivalent one, whose solutions are known, by using these:

BASIC PRINCIPLES FOR SOLVING INEQUALITIES

Performing any of the following operations on an inequality produces an equivalent inequality:

1. Add or subtract the same quantity on both sides of the inequality.
2. Multiply or divide both sides of the inequality by the same *positive* quantity.
3. Multiply or divide both sides of the inequality by the same *negative* quantity and *reverse the direction of the inequality*.

Note Principle 3 carefully. It says, for example, that if you multiply both sides of $-3 < 5$ by -2, the equivalent inequality is $6 > -10$ (direction of inequality reversed).

Example 1 To solve $5x + 3 \leq 6 + 7x$ we use the Basic Principles to transform it into an inequality whose solutions are obvious:

Subtract $7x$ from both sides: $\quad -2x + 3 \leq 6$

Subtract 3 from both sides: $\quad -2x \leq 3$

Divide both sides by -2 and reverse the direction of the inequality: $\quad x \geq -\dfrac{3}{2}$

Therefore the solutions are all real numbers greater than or equal to $-3/2$, that is, the interval $[-3/2, \infty)$, as shown in Figure 2-3. ∎

FIGURE 2-3

Example 2 A solution of the inequality $2 \leq 3x + 5 < 2x + 11$ is any number that is a solution of *both* of these inequalities:

$$2 \leq 3x + 5 \quad \text{and} \quad 3x + 5 < 2x + 11$$

Each of these inequalities can be solved by the methods used above. For the first one we have:

$$2 \leq 3x + 5$$

Subtract 5 from both sides: $\quad -3 \leq 3x$

Divide both sides by 3: $\quad -1 \leq x$

The second inequality is solved similarly:

$$3x + 5 < 2x + 11$$

Subtract 5 from both sides: $\quad 3x < 2x + 6$

Subtract 2x from both sides: $\quad x < 6$

The solutions of the original inequality are the numbers x that satisfy *both* $-1 \leq x$ and $x < 6$, that is, all x with $-1 \leq x < 6$. Thus the solutions are precisely the numbers in the interval $[-1, 6)$, as shown in Figure 2-4. ∎

FIGURE 2-4

Example 3 When solving the inequality $4 < 3 - 5x < 18$, in which the variable appears only in the middle part, you can proceed as follows:

$$4 < 3 - 5x < 18$$

Subtract 3 from each part: $\quad 1 < -5x < 15$

Divide each part by -5 and reverse the direction of the inequalities: $\quad -\dfrac{1}{5} > x > -3$

Reading this last inequality from right to left we see that $-3 < x < -\frac{1}{5}$, so that the solutions are precisely the numbers in the interval $(-3, -\frac{1}{5})$. ∎

> **WARNING**
>
> All inequality signs in an inequality should point in the same direction. So *don't* write things like $4 < x > 2$ or $-3 \geq x < 5$.

Example 4 You can rent a car for $40 per day plus 5¢ per mile, or you can pay $10 per day plus 30¢ per mile. When is the first plan cheaper for one day?

Solution If x is the number of miles driven in a day, then the rental costs are

$$\text{Plan 1:} \quad \$40 + 5¢ \text{ per mile} = 40 + .05x$$
$$\text{Plan 2:} \quad \$10 + 30¢ \text{ per mile} = 10 + .30x$$

We must solve this inequality:

$$\text{Cost of Plan 1} < \text{Cost of Plan 2}$$
$$40 + .05x < 10 + .30x$$

Subtract 40 from both sides: $\qquad .05x < -30 + .30x$

Subtract .30x from both sides: $\qquad -.25x < -30$

Divide both sides by $-.25$ and reverse the direction of the inequality: $\qquad x > \dfrac{-30}{-.25}$

Simplify right side: $\qquad x > 120$

Therefore Plan 1 is cheaper when you drive more than 120 miles. ■

Absolute Value Inequalities

The inequality $|r| \leq 5$ states that the distance from r to 0 (namely $|r|$) is 5 units or less. A glance at the number line in Figure 2–5 shows that these are the numbers r with $-5 \leq r \leq 5$:

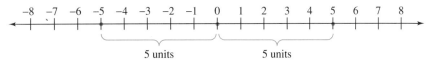

FIGURE 2–5

Similarly, the numbers r such that $|r| \geq 5$ are those whose distance to 0 is 5 or more units, that is, the numbers r with $r \leq -5$ or $r \geq 5$. This argument works with any positive number k in place of 5 and proves the following facts (which are also true with $<$ and $>$ in place of $\leq$ and $\geq$):

ABSOLUTE VALUE INEQUALITIES

> Let k be a positive number and r any real number.
>
> $|r| \leq k$ is equivalent to $-k \leq r \leq k$.
>
> $|r| \geq k$ is equivalent to $r \leq -k$ or $r \geq k$.

Example 5 To solve $|3x - 7| \leq 11$, apply the first fact in the box, with $3x - 7$ in place of r and 11 in place of k, and obtain this equivalent inequality:

$$-11 \leq 3x - 7 \leq 11$$

Add 7 to each part: $\quad -4 \leq 3x \leq 18$

Divide each part by 3: $\quad -\dfrac{4}{3} \leq x \leq 6$

Therefore the solutions of the original inequality are all numbers in the interval $[-4/3, 6]$, as shown in Figure 2-6. ∎

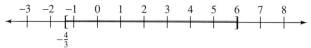

FIGURE 2-6

Example 6 To solve $|5x + 2| > 3$, apply the second fact in the box, with $5x + 2$ in place of r, 3 in place of k, and $>$ in place of $\geq$. This produces the equivalent statement:

$$5x + 2 < -3 \quad \text{or} \quad 5x + 2 > 3$$
$$5x < -5 \quad\quad\quad\quad\quad 5x > 1$$
$$x < -1 \quad \text{or} \quad\quad x > \dfrac{1}{5}$$

Therefore the solutions of the original inequality are the numbers in *either* of the intervals $(-\infty, -1)$ or $(1/5, \infty)$. ∎

Example 7 If a and δ are real numbers with δ positive, then the inequality $|x - a| < \delta$ is equivalent to $-\delta < x - a < \delta$. Adding a to each part shows that $a - \delta < x < a + \delta$. ∎

EXERCISES

In Exercises 1–22, solve the inequality and express your answer in interval notation.

1. $2x + 4 \leq 7$
2. $3x - 5 > -6$
3. $3 - 5x < 13$
4. $2 - 3x < 11$
5. $6x + 3 \leq x - 5$
6. $5x + 3 \leq 2x + 7$
7. $5 - 7x < 2x - 4$
8. $5 - 3x > 7x - 3$
9. $2 < 3x - 4 < 8$
10. $1 < 5x + 6 < 9$
11. $0 < 5 - 2x \leq 11$
12. $-4 \leq 7 - 3x < 0$
13. $2x + 7(3x - 2) < 2(x - 1)$
14. $x + 3(x - 5) \geq 3x + 2(x + 1)$
15. $\dfrac{x + 1}{2} - 3x \leq \dfrac{x + 5}{3}$
16. $\dfrac{x - 1}{4} + 2x \geq \dfrac{2x + 1}{3} + 2$
17. $2x + 3 \leq 5x + 6 < -3x + 7$
18. $4x - 2 < x + 8 < 9x + 1$
19. $3 - x < 2x + 1 \leq 3x - 4$
20. $2x + 5 \leq 4 - 3x < 1 - 4x$
21. $(x - 1)(x + 2) \leq (x + 5)(x - 2)$
22. $(x - 3)^2 + 4 > (x - 1)(x + 1) + 3$

In Exercises 23–26, use a calculator to solve the inequality. [Hint: See Example 2 on page 63.]

23. $7.35x - 6.42 > 5.37 - 12.24x$
24. $8.21 - 6.75x \leq 3.59x + 2.74$
25. $8.53(2.11x + 5.32) < 2.65(3.21 - 6.42x)$
26. $(2.57 - 3.26x)6.25 \geq 1.73(2.71x + 4.32)$

In Exercises 27–30, a, b, c, and d are positive constants. Solve the inequality for x.

27. $ax - b < c$
28. $d - cx > a$
29. $0 < x - c < a$
30. $-d < x - c < d$

In Exercises 31–44, solve the inequality.

31. $|3x + 2| \leq 2$
32. $|5x - 1| < 3$
33. $|2x + 3| - 2 < 0$
34. $3 + |1 - 2x| < 7$
35. $|2x + 3| > 1$
36. $|3x - 1| \geq 2$
37. $|5x + 2| \geq \dfrac{3}{4}$
38. $|2 - 3x| > 4$
39. $|4 - 7x| > 1$
40. $|2x - 3| \geq \dfrac{1}{2}$
41. $\left|\dfrac{12}{5} + 2x\right| > \dfrac{1}{4}$
42. $\left|\dfrac{5}{6} + 3x\right| < \dfrac{7}{6}$
43. $\left|\dfrac{3 - 4x}{2}\right| > 5$
44. $\left|\dfrac{2 - 5x}{3}\right| \geq 2$

45. The temperature F in degrees Fahrenheit is related to the temperature C in degrees Celsius by the formula $F = \dfrac{9}{5}C + 32$. Find the temperature range in degrees Celsius corresponding to the Fahrenheit temperatures from 32° (where water freezes) through 212° (where water boils).

46. A business executive leases a car for $300 per month. She decides to lease another brand for $250 per month, but has to pay a penalty of $1000 for breaking the first lease. How long must she keep the second car in order to come out ahead?

47. One freezer costs $623.95 and uses 90 kilowatt hours (kwh) of electricity each month. A second freezer costs $500 and uses 100 kwh of electricity each month. The expected life of each freezer is 12 years. What is the minimum electric rate (in *cents* per kwh) for which the 12-year total cost (purchase price + electricity costs) will be less for the first freezer?

48. A Gas Guzzler automobile has a 26-gallon gas tank and gets 12 miles per gallon. If the car travels more than 210 miles and runs out of gas, what are the possible amounts of gas that were in the tank at the beginning of the trip?

49. One salesperson is paid a salary of $1000 per month plus a commission of 2% of her total sales. A second salesperson receives no salary, but is paid a commission of 10% of her total sales. What dollar amount of sales must the second salesperson have in order to earn more per month than the first?

50. A developer subdivided 60 acres of a 100-acre tract, leaving 20% of the 60 acres as a park. Zoning laws require that at least 25% of the total tract be

set aside for parks. For financial reasons the developer wants to have no more than 30% of the tract as parks. How many one-quarter-acre lots can the developer sell in the remaining 40 acres and still meet the requirements for the whole tract?

51. If $5000 is invested at 8%, how much more should be invested at 10% in order to guarantee a total annual interest income between $800 and $940?

52. How many gallons of a 12% salt solution should be added to 10 gallons of an 18% salt solution in order to produce a solution whose salt content is between 14% and 16%?

Unusual Problems

53. Let c and d be fixed real numbers with $c < d$. Show that the solutions of $\left| x - \frac{c+d}{2} \right| < \frac{d-c}{2}$ are all x with $c < x < d$.

54. Let ϵ be a fixed positive number. Show that every solution of $|x - 3| < \epsilon/5$ is also a solution of $|(5x - 4) - 11| < \epsilon$.

2.8 POLYNOMIAL AND RATIONAL INEQUALITIES

The **value** of an algebraic expression at a number c is the number obtained by substituting c for the variable in the expression. For example, the value of $x^2 - x - 6$ at 2 is -4 because

$$2^2 - 2 - 6 = -4.$$

Similarly, the value of $\dfrac{x+3}{x-1}$ at 9 is $\dfrac{9+3}{9-1} = \dfrac{12}{8} = \dfrac{3}{2}$.

Solving an inequality such as $x^2 - x - 6 > 0$ amounts to finding those numbers for which the value of $x^2 - x - 6$ is positive. The key to solving such polynomial inequalities is the following fact.*

TEST NUMBER THEOREM

> If the equation $a_n x^n + \cdots + a_1 x + a_0 = 0$ has no real solutions in an interval,† then the value of the polynomial $a_n x^n + \cdots + a_1 x + a_0$ is either positive for *every* number in the interval or negative for *every* number in the interval.

Example 1 below shows how to use this theorem to solve inequalities (and explains its name). The footnotes to Example 1 provide a proof that the theorem is true in this particular case; similar arguments can be used to prove it in the general case.

Example 1 To solve $x^2 - x - 6 > 0$, we first solve the equation

$$x^2 - x - 6 = 0$$
$$(x + 2)(x - 3) = 0$$
$$x = -2 \quad \text{or} \quad x = 3$$

---

* The Test Number Theorem is a special case of the Intermediate Value Theorem, which is discussed in Section 5.2.

† Intervals are defined on page 30.

Thus the equation has *no* solutions when

$$x < -2 \quad \text{or} \quad -2 < x < 3 \quad \text{or} \quad x > 3$$

that is, no solutions in these intervals:

$$(-\infty, -2) \quad (-2, 3) \quad (3, \infty)$$

Choose a number in the first interval, say -4:

$$\text{Value of } x^2 - x - 6 \text{ at } -4 = \text{Value of } (x + 2)(x - 3) \text{ at } -4$$
$$= (-4 + 2)(-4 - 3) = (-2)(-7) = 14$$

Since the value of $x^2 - x - 6$ is positive for one number in the interval $(-\infty, -2)$, the Test Number Theorem shows that the value of $x^2 - x - 6$ must be positive for *every* number in the interval, that is, for every $x < -2$.*

Now choose a number in the second interval, that is, a number between -2 and 3, say $x = 0$. The value of $x^2 - x - 6$ at 0 is $0^2 - 0 - 6 = -6$, a negative number. Therefore the value of $x^2 - x - 6$ must be negative for *every* x with $-2 < x < 3$.†

Finally, choose a number in the third interval, say $x = 6$. The value of $x^2 - x - 6$ at 6 is $6^2 - 6 - 6 = 24$, a positive number. Hence the value of $x^2 - x - 6$ is positive for all $x > 3$.‡ We can summarize these facts in this chart:

Interval	$x < -2$	$-2 < x < 3$	$x > 3$
Test number in this interval	-4	0	6
Value of $x^2 - x - 6$ at test number	14	-6	24
Sign of $x^2 - x - 6$ at every number in this interval	$+$	$-$	$+$

The last line of the chart shows that the value of $x^2 - x - 6$ is positive when $x < -2$ or $x > 3$, and negative or zero everywhere else. Therefore the solutions of $x^2 - x - 6 > 0$ are all numbers x with $x < -2$ or $x > 3$, as shown in Figure 2-7 on the next page. ■

---

* The calculation with -4 is a special case of the following argument, which shows that the value of $x^2 - x - 6$ is positive for every number $c < -2$. If $c < -2$, then $c + 2 < 0$. Since $c < -2$ and $-2 < 3$, we have $c < 3$, and hence $c - 3 < 0$. Since $c + 2$ and $c - 3$ are negative, their product $(c + 2)(c - 3)$, which is the value of $x^2 - x - 6 = (x + 2)(x - 3)$ at c, is necessarily positive.

† If c is any number with $-2 < c < 3$, then $c > -2$, so that $c + 2 > 0$. Similarly, $c < 3$, so that $c - 3 < 0$. Therefore $(c + 2)(c - 3)$, the value of $x^2 - x - 6 = (x + 2)(x - 3)$ at c, is the product of a positive and a negative factor, and hence is negative.

‡ If $c > 3$, then $c + 2$ and $c - 3$ are both positive. Hence their product (the value of $x^2 - x - 6$ at c) is positive.

Example 2 The last line of the chart in Example 1 also shows that the solutions of $x^2 - x - 6 < 0$ are the numbers x such that $-2 < x < 3$. ∎

Example 3 To solve $2x^3 - 15x < x^2$, we first change it to an equivalent inequality with 0 on one side:

Subtract x^2 from both sides: $\quad 2x^3 - x^2 - 15x < 0$

Next we solve the equation $2x^3 - x^2 - 15x = 0$ by factoring:

$$x(2x^2 - x - 15) = 0$$

$$x(2x + 5)(x - 3) = 0$$

$$x = 0 \quad \text{or} \quad x = -\frac{5}{2} \quad \text{or} \quad x = 3$$

Thus the equation has no solutions in these intervals:

$$x < -\frac{5}{2} \quad -\frac{5}{2} < x < 0 \quad 0 < x < 3 \quad x > 3$$

We can now determine where $2x^3 - x^2 - 15x$ is positive or negative by finding its value at a number in each of these intervals:

Interval	$x < -\frac{5}{2}$	$-\frac{5}{2} < x < 0$	$0 < x < 3$	$x > 3$
Test number in this interval	-3	-1	1	4
Value of $2x^3 - x^2 - 15x$ at test number	-18	12	-14	52
Sign of $2x^3 - x^2 - 15x$ at every number in this interval	$-$	$+$	$-$	$+$

The last line of the chart shows that the solutions of

$$2x^3 - x^2 - 15x < 0$$

are all numbers x such that $x < -5/2$ or $0 < x < 3$, as shown in Figure 2–8. ∎

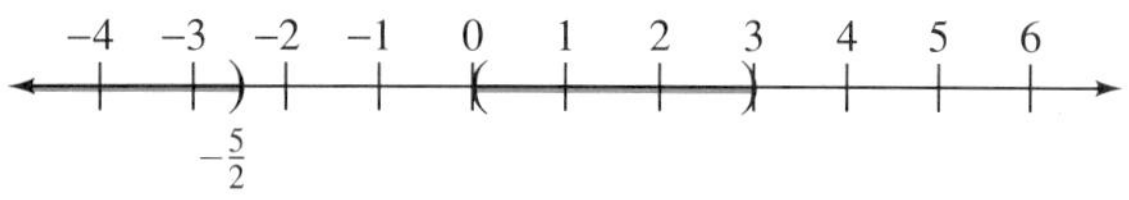

FIGURE 2-8

Example 4 To solve $2x^3 - x^2 - 15x \leq 0$, use the last line of the chart in Example 3 and the fact that $2x^3 - x^2 - 15x = 0$ when $x = -5/2$ or 0 or 3. The solutions are all numbers x such that $x \leq -5/2$ or $0 \leq x \leq 3$. ∎

Example 5 To solve $x^2 + x + 3 > 0$, we first solve the equation $x^2 + x + 3 = 0$. The quadratic formula shows that it has no real solutions, that is, no real solutions in the interval $(-\infty, \infty)$. Since the value of $x^2 + x + 3$ at 0 is the positive number 3, the value of $x^2 + x + 3$ must be positive for every real number. In other words, every real number is a solution of $x^2 + x + 3 > 0$. In addition, the preceding argument shows that $x^2 + x + 3 < 0$ has no real solutions. ∎

The preceding techniques can also be used to solve inequalities involving rational expressions. To see why, note that the sign of a quotient, such as $\frac{3x+1}{x-1}$, is determined by the signs of its numerator and denominator. For instance, if $3x + 1$ is positive and $x - 1$ is negative, then the quotient is negative. Therefore, even though the quotient $\frac{3x+1}{x-1}$ is *not* equal to the product $(3x + 1)(x - 1)$, the quotient has the same *sign* as this product. Consequently, in an interval in which both numerator and denominator are nonzero,* the value of the quotient $\frac{3x+1}{x-1}$ will be either positive for every number in the interval or negative for every number in the interval.

Example 6 To solve $\frac{x+3}{x-1} > -2$, we first change it to an equivalent inequality with 0 on one side and a single rational expression on the other:†

$$\frac{x+3}{x-1} + 2 > 0$$

$$\frac{x+3}{x-1} + \frac{2(x-1)}{x-1} > 0$$

$$\frac{x+3+2x-2}{x-1} > 0$$

$$\frac{3x+1}{x-1} > 0$$

* That is, where the equation $(3x + 1)(x - 1) = 0$ has no solutions.
† If you are tempted to solve it quickly by multiplying both sides by $x - 1$, see the warning below.

The numbers at which the numerator or denominator is 0 are:

$$3x + 1 = 0 \quad \text{or} \quad x - 1 = 0$$
$$x = -\frac{1}{3} \quad\quad\quad x = 1$$

Consequently, the numerator and denominator of $\dfrac{3x+1}{x-1}$ are nonzero in each of these intervals:

$$x < -\frac{1}{3} \quad -\frac{1}{3} < x < 1 \quad x > 1$$

Now determine where $\dfrac{3x+1}{x-1}$ is positive or negative by evaluating it at one number in each interval:

Interval	$x < -\frac{1}{3}$	$-\frac{1}{3} < x < 1$	$x > 1$
Test number in this interval	-1	0	2
Value of $\dfrac{3x+1}{x-1}$ at test number	1	-1	7
Sign of $\dfrac{3x+1}{x-1}$ at every number in this interval	$+$	$-$	$+$

The last line of the chart shows that the solutions of $\dfrac{3x+1}{x-1} > 0$, and hence of the original inequality, are all numbers x such that $x < -1/3$ or $x > 1$. ∎

WARNING

Don't treat rational inequalities as if they were equations, as in this *incorrect* "solution" of the preceding example:

$$\frac{x+3}{x-1} > -2$$
$$x + 3 > -2(x - 1) \quad \text{[Both sides multiplied by } x - 1\text{]}$$
$$x + 3 > -2x + 2$$
$$3x > -1$$
$$x > -\frac{1}{3}$$

According to this, 0 should be a solution. But putting $x = 0$ in the original inequality produces the false statement $-3 > -2$.*

Example 7 The solutions of $\dfrac{3x + 1}{x - 1} \geq 0$ are the numbers at which the quotient is positive or zero. Example 6 shows that the quotient is positive when $x < -1/3$ or $x > 1$. The quotient $\dfrac{3x + 1}{x - 1}$ is 0 only when the numerator is 0 and the denominator is nonzero.† This occurs when $x = -1/3$. Therefore the solutions of $\dfrac{3x + 1}{x - 1} \geq 0$ are all numbers x such that

$$x \leq -\frac{1}{3} \quad \text{or} \quad x > 1. \blacksquare$$

A slight variation of this technique can be used to solve inequalities involving factored polynomials of high degree.

Example 8 To solve $x^3(2x + 10)^{14}(x - 9)^{27} < 0$, we first solve the equation $x^3(2x + 10)^{14}(x - 9)^{27} = 0$:

$$x^3 = 0 \quad \text{or} \quad (2x + 10)^{14} = 0 \quad \text{or} \quad (x - 9)^{27} = 0$$
$$x = 0 \qquad\qquad 2x + 10 = 0 \qquad\qquad x = 9$$
$$x = -5$$

We then choose a test number in each of the intervals determined by $-5, 0,$ and 9. Instead of finding the value of $x^3(2x + 10)^{14}(x - 9)^{27}$ at each test number, which would involve extensive computation, we need only find its *sign* by finding the sign of each factor. For instance, let -6 be the test number for the interval $(-\infty, -5)$ and observe that

$(-6)^3$ is negative (an odd power of a negative number);

$(2(-6) + 10)^{14}$ is positive (even power of a nonzero number);

$(-6 - 9)^{27}$ is negative (odd power of a negative number).

Therefore $(-6)^3(2(-6) + 10)^{14}(-6 - 9)^{27}$ is the product of a negative, a positive, and a negative factor and hence is positive. Similar arguments lead to the following chart:

---

* The source of the error is multiplying by $x - 1$. This quantity is negative for some values of x and positive for others. To do this calculation correctly, you must consider two separate cases and reverse the direction of the inequality when $x - 1$ is negative.

† When the denominator is 0 (that is, when $x = 1$), the quotient is *not defined*.

Interval	$x<-5$	$-5<x<0$	$0<x<9$	$x>9$
Test number in this interval	-6	-1	1	10
Sign of factors at test number x^3	$-$	$-$	$+$	$+$
$(2x+10)^{14}$	$+$	$+$	$+$	$+$
$(x-9)^{27}$	$-$	$-$	$-$	$+$
Sign of $x^3(2x+10)^{14}(x-6)^{27}$ at test number and hence at every number in this interval	$+$	$+$	$-$	$+$

The last line shows that the solutions of $x^3(2x+10)^{14}(x-9)^{27}<0$ are all numbers x such that $0<x<9$. ■

Absolute Value Inequalities

To solve an inequality such as

$$|x^2-x-4|>3 \quad \text{or} \quad \left|\frac{x+3}{x-1}\right|\le 2$$

translate it into an equivalent inequality that doesn't involve absolute values by using the fact in the box on page 106. For instance, $|x^2-x-4|>3$ is equivalent to

$$x^2-x-4<-3 \quad \text{or} \quad x^2-x-4>3.$$

This can be solved by the techniques described above. Similarly, $\left|\frac{x+3}{x-1}\right|\le 2$ is equivalent to

$$-2\le \frac{x+3}{x-1}\le 2.$$

EXERCISES

In Exercises 1–30; solve the inequality.

1. $x^2-4x+3\le 0$
2. $x^2-7x+10\le 0$
3. $x^2+9x+14\ge 0$
4. $x^2+8x+15\ge 0$
5. $x^2\ge 9$
6. $x^2\ge 16$
7. $6+x-x^2\le 0$
8. $4-3x-x^2\ge 0$
9. $6x^2-x\ge 2$
10. $4x^2+11x+6\ge 0$
11. $(x-\sqrt{2})(x+\sqrt{3})<0$
12. $(x+\sqrt{5})(x-\sqrt{7})>0$

13. $x(4x + 10) > 12 - 10x - 4x^2$

14. $x^2(x - 4) \leq 5x$ 15. $x^3 - x \geq 0$

16. $x^3 + 2x^2 + x > 0$ 17. $x^3 - 2x^2 - 3x < 0$

18. $x^4 - 14x^3 + 48x^2 \geq 0$

19. $x^4 - 5x^2 + 4 < 0$

20. $x^4 - 10x^2 + 9 \leq 0$

21. $\dfrac{3x + 1}{2x - 4} > 0$ 22. $\dfrac{2x - 1}{5x + 3} \geq 0$

23. $\dfrac{x^2 + x - 2}{x^2 - 2x - 3} < 0$ 24. $\dfrac{2x^2 + x - 1}{x^2 - 4x + 4} \geq 0$

25. $\dfrac{x - 2}{x - 1} < 1$ 26. $\dfrac{-x + 5}{2x + 3} \geq 2$

27. $\dfrac{x - 3}{x + 3} \leq 5$ 28. $\dfrac{2x + 1}{x - 4} > 3$

29. $\dfrac{2}{x + 3} \geq \dfrac{1}{x - 1}$ [Hint: Collect all terms on one side and find a common denominator.]

30. $\dfrac{1}{x - 1} < \dfrac{-1}{x + 2}$

In Exercises 31–36, solve the inequality by checking the signs of factors, as in Example 8.

31. $x^4(x + 3)^5(x - 5)^7 > 0$

32. $x^5(x + 2)^6(x + 5)^8 < 0$

33. $(4x + 1)^8(x - 1)^9(x + 4)^{10} \leq 0$

34. $(2x - 1)^{11}(2x + 7)^{12}(3x + 2)^{13} \geq 0$

35. $\dfrac{(x + 3)(x + 2)^5}{x(x - 1)} \geq 0$

36. $\dfrac{(x^2 - 9)^7}{x^3 + 3x - 10x} > 0$

In Exercises 37–42, solve the inequality by using the hint for Exercise 37.

37. $x^2 - 2x - 4 \geq 0$ [Hint: Use the quadratic formula to show that the solutions of $x^2 - 2x - 4 = 0$ are $1 \pm \sqrt{5}$. So the intervals are: $x < 1 - \sqrt{5}$; $1 - \sqrt{5} < x < 1 + \sqrt{5}$; $x > 1 + \sqrt{5}$; choose a test number in each interval and proceed as above.]

38. $x^2 - 2x - 1 \leq 0$ 39. $x^2 + 2x > 2$

40. $x^2 - 4x < 2$

41. $(x^2 - 6x + 7)(x^2 - 10) > 0$

42. $(x^2 - 4x - 3)(x^2 - 3) \leq 0$

In Exercises 43–54, solve the inequality by first transforming it into an equivalent one that doesn't involve absolute values.

43. $\left|\dfrac{x - 1}{x + 2}\right| \leq 3$ 44. $\left|\dfrac{x + 1}{3x + 5}\right| < 2$

45. $\left|\dfrac{2x - 1}{x + 5}\right| > 1$ 46. $\left|\dfrac{x + 1}{x + 2}\right| \geq 2$

47. $|x^2 - 2| < 1$ 48. $|x^2 - 4| \leq 3$

49. $|x^2 - 4| < 5$ 50. $|x^2 - 5| > 4$

51. $|x^2 + x - 1| \geq 1$ 52. $|x^2 + x - 4| \leq 2$

53. $|3x^2 - 8x + 2| < 2$ 54. $|x^2 + 3x - 4| < 6$

55. Find all pairs of numbers that satisfy these two conditions: Their sum is 20 and the sum of their squares is less than 362.

56. The length of a rectangle is 6 inches longer than its width. What are the possible widths if the area of the rectangle is at least 667 square inches?

57. It costs a craftsman $5 in materials to make a medallion. He has found that if he sells the medallions for $50 - x$ dollars each, where x is the number of medallions produced each week, then he can sell all that he makes. His fixed costs are $350 per week. If he wants to sell all he makes and show a profit each week, what are the possible numbers of medallions he should make?

58. A retailer sells file cabinets for $80 - x$ dollars each, where x is the number of cabinets she receives from the supplier each week. She pays $10 for each file cabinet and has fixed costs of $600 per week. How many file cabinets should she order from the supplier each week in order to guarantee that she makes a profit?

59. It costs a company $25 to manufacture one widget. The retail price of a widget is set at $475 - x$ dollars, where x is the number of widgets manufactured. If all the widgets are sold and the fixed costs are $20,000, then what are the smallest number and the largest number of widgets on which the company can make a profit?

60. The Junkfood Company can produce a bag of

Munchies for 50¢. Building, equipment, and other fixed costs run $500 per day. If x is the number of bags produced per day, all of them will be sold at a price of $1.95 - \dfrac{x}{2000}$ dollars per bag. How many bags can be produced each day at a profit?

In Exercises 61–64, you will need the formula for the height h of an object above the ground at time t seconds: $h = -16t^2 + v_0 t + h_0$; this formula was explained on page 86.

61. A toy rocket is fired straight up from ground level with an initial velocity of 80 feet per second. During what time interval will it be at least 64 feet above the ground?

62. A projectile is fired straight up from ground level with an initial velocity of 72 feet per second. During what time interval is it at least 37 feet above the ground?

63. A ball is dropped from the roof of a 120-foot-high building. During what time period will it be strictly between 56 and 39 feet above the ground?

64. A ball is thrown straight up from a 40-foot-high tower with an initial velocity of 56 feet per second.
    (a) During what time interval is the ball at least 8 feet above the ground?
    (b) During what time interval is the ball between 53 feet and 80 feet above the ground?

65. (a) Solve the inequalities $x^2 < x$ and $x^2 > x$.
    (b) Use the results of part (a) to show that for any nonzero real number c with $|c| < 1$, it is always true that $c^2 < |c|$.
    (c) Use the results of part (a) to show that for any nonzero real number c with $|c| > 1$, it is always true that $c^2 > c$.

66. (a) If $0 < a \leq b$, show that $1/a \geq 1/b$.
    (b) If $a \leq b < 0$, show that $1/a \geq 1/b$.
    (c) If $a < 0 < b$, how are $1/a$ and $1/b$ related?

CHAPTER REVIEW

Important Concepts

Section 2.1	Basic principles for solving equations	62
	First-degree equations	62
Section 2.2	Guidelines for solving applied problems	68
Section 2.3	Quadratic equations	76
	Factoring	76
	Completing the square	77
	The quadratic formula	79
	Discriminant	79
	Real solutions	80
Section 2.4	Applications of quadratic equations	82
Section 2.5	Complex number	87
	Imaginary number	88
	Conjugate	89
	Square roots of negative numbers	90
	Solutions of equations	91
Section 2.6	Equations of quadratic type	94
	Radical equations	95
	Absolute value equations	98
Excursion 2.6.A	Rational Solutions Test	100
	Factor Theorem	101

Section 2.7	Basic Principles for solving inequalities	103
	Linear inequalities	103
	Linear absolute value inequalities	105
Section 2.8	Value of an algebraic expression	108
	Sign chart method for solving polynomial inequalities	109
	Rational inequalities	111
	Absolute value inequalities	114

Important Facts and Formulas

- Quadratic Formula: If $a \neq 0$, then the solutions of $ax^2 + bx + c = 0$ are $x = \dfrac{-b \pm \sqrt{b^2 - 4ac}}{2a}$.
- If $a \neq 0$, then the number of real solutions of $ax^2 + bx + c = 0$ is 0, 1, or 2, depending on whether the discriminant $b^2 - 4ac$ is negative, zero, or positive.
- The solutions of $|x| \leq k$ are all numbers x such that $-k \leq x \leq k$.
- The solutions of $|x| \geq k$ are all numbers x such that $x \leq -k$ or $x \geq k$.

Review Questions

1. Solve for x: $2\left(\dfrac{x}{5} + 7\right) - 3x = \dfrac{x+2}{5} - 4$

2. Solve for t: $\dfrac{t+1}{t-1} = \dfrac{t+2}{t-3}$

3. Solve for x: $\dfrac{3}{x} - \dfrac{2}{x-1} = \dfrac{1}{2x}$

4. Solve for z: $2 - \dfrac{1}{z-2} = \dfrac{z-3}{z-2}$

5. Solve for r: $Q = \dfrac{b-a}{2r}$

6. Solve for R: $\dfrac{v}{t} = 2\pi R h$

7. Solve for x in terms of y: $xy + 3 = x - 2y$

8. Bert weighs 10 times as much as the Thanksgiving turkey and Sally weighs 7 times as much as the turkey. If Bert is 48 pounds heavier than Sally, how much does Bert weigh?

9. A jeweler wants to make a 1-ounce ring consisting of gold and silver, using $200 worth of metal. If gold costs $600 per ounce and silver $50 per ounce, how much of each metal should she use?

10. A calculator is on sale for 15% less than the list price. The sale price, plus a 5% shipping charge, totals $210. What is the list price?

11. Jack can do a job in 5 hours and Walter can do the same job in 4 hours. How long will it take them to do the job together?

12. A car leaves the city traveling at 54 mph. One-half hour later, a second car leaves from the same place and travels at 63 mph along the same road. How long will it take for the second car to catch up with the first?

13. A 12-foot long rectangular board is cut in two pieces so that one piece is 4 times as long as the other. How long is the bigger piece?

14. George owns 200 shares of stock, 40% of which are in the computer industry. How many more shares must he buy in order to have 50% of his total shares in computers?

15. Solve for x: $3x^2 - 2x + 5 = 0$

16. Solve for x: $5x^2 + 6x = 7$

17. Solve for y: $3y^2 - 2y = 5$

18. Find the *number* of real solutions of the equation $20x^2 + 12 = 31x$.

19. Find the number of real solutions of the equation $\dfrac{1}{3x+6} = \dfrac{2}{x+2}$.

20. For what value of k does the equation $kt^2 + 5t + 2 = 0$ have exactly one real solution for t?

21. Do there exist two real numbers whose sum is 2 and whose product is 2? Justify your answer.

22. Find two consecutive integers, the sum of whose squares is 481.

23. A square region is changed into a rectangular one by making it 2 feet longer and twice as wide. If the area of the rectangular region is 3 times larger than the area of the original square region, what was the length of a side of the square before it was changed?

24. The radius of a circle is 10 inches. By how many inches should the radius be increased so that the area increases by 5π square inches?

In Questions 25–34, express the given complex number in the form $a + bi$, with a and b real numbers.

25. $(2 + i)(4 - 7i)$

26. $(3 + i)(3 - i)$

27. i^{37}

28. i^{-59}

29. $(2 + \sqrt{-9})(5 - \sqrt{-4})$

30. $(\sqrt{-4} + 1)(\sqrt{-3} - 2)$

31. $\dfrac{1}{3 + i}$

32. $\dfrac{1}{2i}$

33. $\dfrac{2 + 3i}{1 + i}$

34. $\dfrac{3 + i}{1 - i} + \dfrac{2}{3 - 2i}$

35. Express the *conjugate* of $\dfrac{1}{2+3i}$ in the form $a+bi$, with a and b real.

36. Express in terms of i and simplify: $\dfrac{-\sqrt{-16}}{\sqrt{-36}}$.

In Questions 37–44, solve the equation in the complex number system.

37. $x^2 + 36 = 0$

38. $x^2 + 3x + 10 = 0$

39. $x^2 + 2x + 5 = 0$

40. $5x^2 + 2 = 3x$

41. $-3x^2 + 4x - 5 = 0$

42. $3x^4 + x^2 - 2 = 0$

43. $8x^4 + 10x^2 + 3 = 0$

44. $x^3 - 8 = 0$

In Questions 45–52, find all real solutions of the equation.

45. $x^4 - 11x^2 + 18 = 0$

46. $x^4 - x^2 = 6$

47. $x^6 - 4x^3 + 4 = 0$

48. $3y^7 - 3y^5 - 15y^3 = 0$

49. $\sqrt{x-1} = 2 - x$

50. $\sqrt[3]{1 - t^2} = -2$

51. $\sqrt{x+1} + \sqrt{x-1} = 1$

52. $x^{1/4} - 5x^{1/2} + 6 = 0$

53. Solve for s: $t = \sqrt{\dfrac{2s}{g}}$

54. Solve for x: $|3x - 1| = 4$

55. Find the rational solutions of $x^4 - 2x^3 - 4x^2 + 1 = 0$.

56. Consider the equation $2x^3 - 8x^2 + 5x + 3 = 0$:
  (a) List all the rational numbers that could *possibly* be solutions.
  (b) Find one rational solution of the equation.
  (c) Find all the solutions of the equation.

57. Find all rational solutions of $x^3 + 2x^2 - 2x - 2 = 0$.

58. Find all real solutions of $x^3 - 6x^2 + 11x - 6 = 0$.

59. The equation $x^3 - 2x + 1 = 0$ has
  (a) no real solutions;
  (b) only one real solution;
  (c) three rational solutions;
  (d) only one rational solution;
  (e) exactly two real solutions.

In Questions 60–64, solve the inequality.

60. $-3(x - 4) \leq 5 + x$

61. $-4 < 2x + 5 < 9$

62. $2x - 3 \leq 5x + 9 < -3x + 4$; express your answer in interval notation.

63. $|3x + 2| \geq 2$

64. $\left|\dfrac{y + 2}{3}\right| \geq 5$

65. If $0 < r \leq s - t$, then which of these statements is *false*?

   (a) $s \geq r + t$ (d) $\dfrac{s - t}{r} \geq 0$

   (b) $t - s \leq -r$ (e) $s - r \geq t$

   (c) $-r \geq s - t$

In Questions 66–71, solve the inequality.

66. $x^2 + x > 12$

67. $x^2 + x - 20 > 0$

68. $(x - 1)^2(x^2 - 1)x \leq 0$

69. $\dfrac{2x - 1}{3x + 1} < 1$ 70. $\dfrac{1}{x + 1} < x$

71. $\dfrac{x - 2}{x + 4} \leq 3$

72. If $\dfrac{x + 3}{2x - 3} > 1$, then which of these statements is *true*?

   (a) $\dfrac{x - 3}{2x + 3} < 1$ (d) $2x + 3 > x - 3$

   (b) $\dfrac{2x - 3}{x + 3} < -1$ (e) none of these

   (c) $\dfrac{3 - 2x}{x + 3} > 1$

CHAPTER 3

The Coordinate Plane and Lines

One of the great discoveries of 17th-century mathematics was that the techniques of algebra could be applied to geometry. The resulting subject was called analytical geometry. We now present the fundamental concepts of analytic geometry that are needed for the study of functions and later for calculus. Other topics in analytical geometry are presented in a later chapter.*

3.1 THE COORDINATE PLANE

The number line associates points on the line with real numbers. A similar construction allows us to associate points in the plane with ordered *pairs* of real numbers.

Draw two number lines in the plane, one horizontal with the positive direction to the right, and one vertical with the positive direction upward, as shown in Figure 3–1, page 122. The lines are called the **coordinate axes** and their point of intersection the **origin**.† The horizontal axis is usually called the **x-axis**, and the vertical axis is called the **y-axis**.‡ The coordinate axes divide the plane into four regions, called **quadrants**, that are numbered as in Figure 3–1. A plane equipped with coordinate axes is called a **coordinate plane** and is said to have a **rectangular** (or **Cartesian**) **coordinate system**.

Given a point P in the plane, draw a vertical line from P to the x-axis and a horizontal line from P to the y-axis. These lines intersect the axes at some numbers c and d, as shown in Figure 3–2, page 122. P is now associated with the ordered pair of numbers (c, d) and we say that P has **coordinates** (c, d). The number c is called the **x-coordinate** of P, and d is called the **y-coordinate** of P.

---

* The chapter entitled "The Conic Sections" may be covered immediately after this chapter if you wish.

† Any pair of perpendicular lines may be used as coordinate axes. For convenience, however, it is customary to have axes that are vertical and horizontal.

‡ Although x and y are the traditional labels for the coordinate axes, there is nothing sacred about them. Other letters may be used when convenient.

3 THE COORDINATE PLANE AND LINES

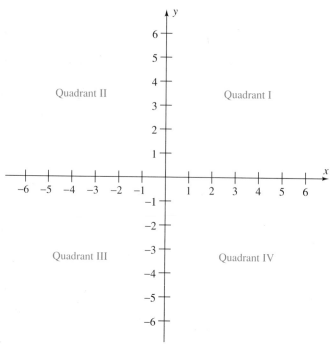

FIGURE 3-1

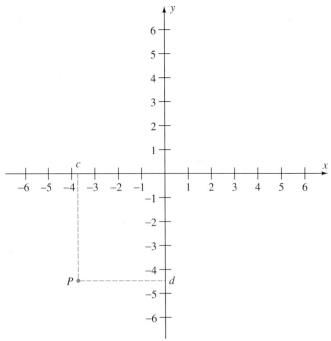

FIGURE 3-2

Here are some examples of points and their coordinates:

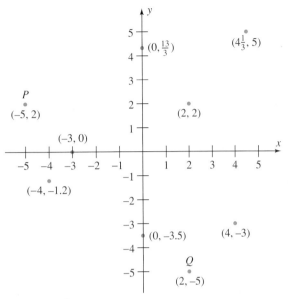

FIGURE 3-3

WARNING

The coordinates of a point are an *ordered* pair. Figure 3-3 shows that the point P with coordinates $(-5, 2)$ is quite different from the point Q with coordinates $(2, -5)$. The same numbers (2 and -5) occur in both cases, but in *different order*.

We shall often identify a point with its coordinates and refer, for example, to the point (2, 3). When dealing with several points simultaneously, it is customary to label the coordinates of the first point (x_1, y_1), the second point (x_2, y_2), the third point (x_3, y_3), and so on.* Once the plane is coordinatized, it's easy to compute the distance between any two points:

THE DISTANCE FORMULA

The distance between points (x_1, y_1) and (x_2, y_2) is
$$\sqrt{(x_1 - x_2)^2 + (y_1 - y_2)^2}$$

Before proving the distance formula, we shall see how it is used.

---

*"x_1" is read "x-one" or "x-sub-one"; it is a *single symbol* denoting the first coordinate of the first point, just as c denotes the first coordinate of (c, d). Analogous remarks apply to y_1, x_2, etc.

Example 1 To find the distance between the points $(-1, -3)$ and $(2, -4)$ in Figure 3-4, substitute $(-1, -3)$ for (x_1, y_1) and $(2, -4)$ for (x_2, y_2) in the distance formula:

$$\sqrt{(x_1 - x_2)^2 + (y_1 - y_2)^2} = \sqrt{(-1 - 2)^2 + (-3 - (-4))^2}$$
$$= \sqrt{9 + (-3 + 4)^2} = \sqrt{9 + 1} = \sqrt{10}$$

The order in which the points are used doesn't make a difference. If we substitute $(2, -4)$ for (x_1, y_1) and $(-1, -3)$ for (x_2, y_2), we get the same answer:

$$\sqrt{(2 - (-1))^2 + (-4 - (-3))^2} = \sqrt{3^2 + (-1)^2} = \sqrt{10}. \ \blacksquare$$

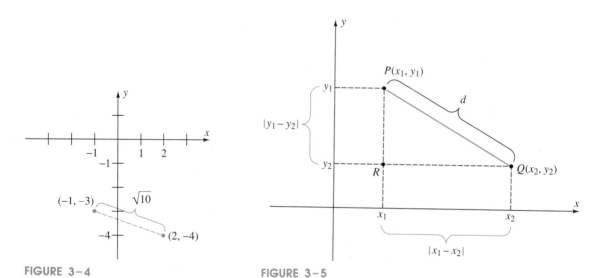

FIGURE 3-4

FIGURE 3-5

Example 2 To find the distance from (a, b) to $(2a, -b)$, where a and b are fixed real numbers, substitute a for x_1, b for y_1, $2a$ for x_2, and $-b$ for y_2 in the distance formula:

$$\sqrt{(x_1 - x_2)^2 + (y_1 - y_2)^2} = \sqrt{(a - 2a)^2 + (b - (-b))^2}$$
$$= \sqrt{(-a)^2 + (b + b)^2} = \sqrt{a^2 + (2b)^2}$$
$$= \sqrt{a^2 + 4b^2}. \ \blacksquare$$

Proof of the Distance Formula Let P and Q be points that don't lie on the same vertical or horizontal line, as illustrated in Figure 3-5. As shown in Figure 3-5, the length of RQ is the same as the distance from x_1 to x_2 on the x-axis (number line), namely, $|x_1 - x_2|$. Similarly, the length of PR is the same as the distance from y_1 to y_2 on the y-axis, namely, $|y_1 - y_2|$. According to the Pythagorean Theorem* the length d of PQ is given by:

$$(\text{length } PQ)^2 = (\text{length } RQ)^2 + (\text{length } PR)^2$$
$$d^2 = |x_1 - x_2|^2 + |y_1 - y_2|^2$$

Since $|c|^2 = c^2$ for any real number c (see page 26), this equation becomes:

$$d^2 = (x_1 - x_2)^2 + (y_1 - y_2)^2$$

Since the length d is nonnegative we must have

$$d = \sqrt{(x_1 - x_2)^2 + (y_1 - y_2)^2}$$

This completes the proof in this case. The proof of the cases when P and Q lie on the same vertical or horizontal line is outlined in Exercise 56. ∎

The distance formula can be used to prove this useful fact:

THE MIDPOINT FORMULA

> The midpoint of the line segment from (x_1, y_1) to (x_2, y_2) is
> $$\left(\frac{x_1 + x_2}{2}, \frac{y_1 + y_2}{2} \right)$$

Example 3 The midpoint of the segment joining $(-4, 5)$ and $(6, 2)$ is $\left(\frac{-4 + 6}{2}, \frac{5 + 2}{2} \right) = \left(1, \frac{7}{2} \right)$, as shown in Figure 3-6. ∎

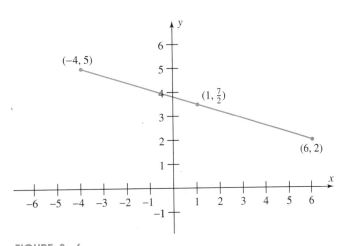

FIGURE 3-6

---

* See the Geometry Review Appendix.

Proof of the Midpoint Formula The situation is shown in Figure 3–7.

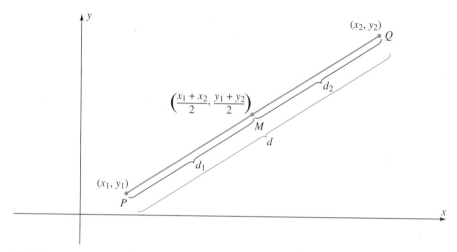

FIGURE 3–7

In order to show that M is the midpoint of segment PQ, we must show that
$$d_1 = d_2 \quad \text{and} \quad d_1 + d_2 = d.$$
Using the distance formula with the points P and M, we have

$$d_1 = \sqrt{\left(\frac{x_1 + x_2}{2} - x_1\right)^2 + \left(\frac{y_1 + y_2}{2} - y_1\right)^2}$$

$$= \sqrt{\left(\frac{x_1 + x_2}{2} - \frac{2x_1}{2}\right)^2 + \left(\frac{y_1 + y_2}{2} - \frac{2y_1}{2}\right)^2}$$

$$= \sqrt{\left(\frac{x_2 - x_1}{2}\right)^2 + \left(\frac{y_2 - y_1}{2}\right)^2} = \sqrt{\frac{(x_2 - x_1)^2}{4} + \frac{(y_2 - y_2)^2}{4}}$$

$$= \sqrt{\frac{1}{4}((x_2 - x_1)^2 + (y_2 - y_1)^2)} = \frac{1}{2}\sqrt{(x_2 - x_1)^2 + (y_2 - y_1)^2} = \frac{1}{2}d$$

A similar calculation for M and Q (Exercise 58) shows that $d_2 = \frac{1}{2}d$. Hence

$$d_1 = \frac{1}{2}d = d_2 \quad \text{and} \quad d_1 + d_2 = \frac{1}{2}d + \frac{1}{2}d = d$$

and the proof is complete. ∎

A **solution** of an equation in two variables, x and y, is a pair of numbers such that the substitution of the first number for x and the second for y produces a true statement. In this case, we say that the pair of numbers **satisfies the equation**. For example, $(3, -2)$ is a solution of $5x + 7y = 1$ because $5 \cdot 3 + 7 \cdot (-2) = 1$, and $(-2, 3)$ is *not* a solution because $5 \cdot (-2) + 7 \cdot 3 \neq 1$.

The **graph of an equation or inequality** in two variables is the set of points whose coordinates are solutions of the given equation or inequality. Thus the graph is a *geometric picture of the solutions*. The distance formula enables us to find the graphs of a few equations quickly and exactly, as we now demonstrate.

Circles

If (c, d) is a point in the plane and r a positive number, then the **circle with center (c, d) and radius r** consists of all points (x, y) that lie r units from (c, d), as shown in Figure 3–8.

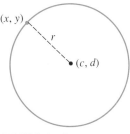

FIGURE 3–8

According to the distance formula, the statement that "the distance from (x, y) to (c, d) is r units" is equivalent to:

$$\sqrt{(x - c)^2 + (y - d)^2} = r$$

Squaring both sides shows that (x, y) satisfies this equation:

$$(x - c)^2 + (y - d)^2 = r^2$$

Reversing the procedure shows that any solution (x, y) of this equation is a point on the circle. Therefore

EQUATION OF THE CIRCLE

> The circle with center (c, d) and radius r is the graph of
> $$(x - c)^2 + (y - d)^2 = r^2$$

We say that $(x - c)^2 + (y - d)^2 = r^2$ is the **equation of the circle** with center (c, d) and radius r. If the center is at the origin, then $(c, d) = (0, 0)$ and the equation has a simpler form:

CIRCLE WITH CENTER AT THE ORIGIN

> The circle with center $(0, 0)$ and radius r is the graph of
> $$x^2 + y^2 = r^2$$

Example 4

(a) Letting $r = 1$ shows that the graph of $x^2 + y^2 = 1$ is the circle of radius 1 centered at the origin, as shown in Figure 3-9. This circle is called the **unit circle**.

(b) The circle with center $(-3, 2)$ and radius 2, shown in Figure 3-10, is the graph of the equation
$$(x - (-3))^2 + (y - 2)^2 = 2^2$$
or equivalently,
$$(x + 3)^2 + (y - 2)^2 = 4. \blacksquare$$

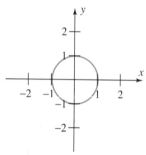

FIGURE 3-9

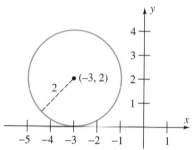

FIGURE 3-10

Example 5

To find the equation of the circle with center $(3, -1)$ that passes through $(2, 4)$, we must first find the radius. Since $(2, 4)$ is on the circle, the radius is the distance from $(2, 4)$ to $(3, -1)$ as shown in Figure 3-11 below, namely,
$$\sqrt{(2 - 3)^2 + (4 - (-1))^2} = \sqrt{1 + 25} = \sqrt{26}.$$

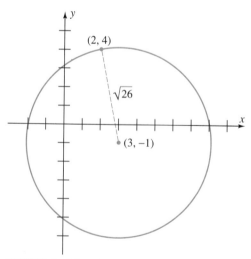

FIGURE 3-11

The equation of the circle with center at $(3, -1)$ and radius $\sqrt{26}$ is

$$(x - 3)^2 + (y - (-1))^2 = (\sqrt{26})^2$$
$$(x - 3)^2 + (y + 1)^2 = 26$$
$$x^2 - 6x + 9 + y^2 + 2y + 1 = 26$$
$$x^2 + y^2 - 6x + 2y - 16 = 0. \blacksquare$$

The equation of any circle can always be written in the form

$$x^2 + y^2 + Bx + Cy + D = 0$$

for some constants B, C, D, as in the last example (where $B = -6$, $C = 2$, $D = -16$). Conversely, the graph of such an equation can always be determined, as in the following example.

Example 6 To find the graph of $3x^2 + 3y^2 - 12x - 30y + 45 = 0$, we divide both sides by 3 and rewrite the equation as

$$(x^2 - 4x) + (y^2 - 10y) = -15$$

Next we complete the square in both expressions in parentheses (see page 77). To complete the square in $x^2 - 4x$ we add 4 (the square of half the coefficient of x) and to complete the square in $y^2 - 10y$ we add 25 (why?). In order to have an equivalent equation we must add these numbers to *both* sides:

$$(x^2 - 4x + 4) + (y^2 - 10y + 25) = -15 + 4 + 25$$
$$(x - 2)^2 + (y - 5)^2 = 14$$

Since $14 = (\sqrt{14})^2$, this is the equation of the circle with center $(2, 5)$ and radius $\sqrt{14}$. $\blacksquare$

EXERCISES

1. Find the coordinates of the points in Figure 3–12.

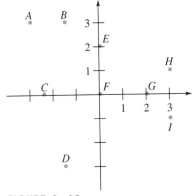

FIGURE 3–12

2. Draw coordinate axes and plot these points: $(0, 0)$, $(-3, 2.1)$, $(2.1, -3)$, $(-4/3, 1)$, $(5, \pi)$, $(2, \sqrt{2})$, $(-3, \pi)$, $(4, 6)$, $(-\sqrt{3}, \sqrt{3})$, $(\sqrt{3}, -\sqrt{3})$, $(5/2, 17/3)$.

In Exercises 3–8, determine whether the point is on the graph of the given equation.

3. $(1, -2)$; $3x - y - 5 = 0$

4. $(2, -1)$; $x^2 + y^2 - 6x + 8y = -15$

5. $(6, 2)$; $3y + x = 12$

6. $(1, -2)$; $3x + y = 12$

7. $(3, 4)$; $(x - 2)^2 + (y + 5)^2 = 4$

8. $(1, -1)$; $\dfrac{x^2}{2} + \dfrac{y^2}{3} = 1$

In Exercises 9–16, find the distance between the two points and the midpoint of the segment joining them.

9. $(-3, 5), (2, -7)$ 10. $(2, 4), (1, 5)$
11. $(1, -5), (2, -1)$ 12. $(-2, 3), (-3, 2)$
13. $(\sqrt{2}, 1), (\sqrt{3}, 2)$ 14. $(-1, \sqrt{5}), (\sqrt{2}, -\sqrt{3})$
15. $(a, b), (b, a)$ 16. $(s, t), (0, 0)$

In Exercises 17–22, describe the set of all points (x, y) that satisfy the given condition.

17. $y = 5$ 18. $x = -5$
19. $xy = 0$ [Hint: When is a product zero?]
20. $xy > 0$ 21. $|x| < 1$
22. $|y| \leq 2$

In Exercises 23–26, find the equation of the circle with given center and radius r.

23. $(-3, 4); r = 2$ 24. $(-2, -1); r = 3$
25. $(0, 0); r = \sqrt{2}$ 26. $(5, -2); r = 1$

In Exercises 27–30, sketch the graph of the equation.

27. $(x - 2)^2 + (y - 4)^2 = 1$
28. $(x + 1)^2 + (y - 3)^2 = 9$
29. $(x - 5)^2 + (y + 2)^2 = 5$
30. $(x + 6)^2 + y^2 = 4$

In Exercises 31–36, find the center and radius of the circle whose equation is given.

31. $x^2 + y^2 + 8x - 6y - 15 = 0$
32. $15x^2 + 15y^2 = 10$
33. $x^2 + y^2 + 6x - 4y - 15 = 0$
34. $x^2 + y^2 + 10x - 75 = 0$
35. $x^2 + y^2 + 25x + 10y = -12$
36. $3x^2 + 3y^2 + 12x + 12 = 18y$

In Exercises 37–39, show that the three points are the vertices of a right triangle and state the length of the hypotenuse. [You may assume that a triangle with sides of lengths a, b, c is a right triangle with hypotenuse c provided that $a^2 + b^2 = c^2$.]

37. $(0, 0), (1, 1), (2, -2)$

38. $\left(\frac{\sqrt{2}}{2}, 0\right), \left(\frac{\sqrt{2}}{2}, \frac{\sqrt{2}}{2}\right), (0, 0)$

39. $(3, -2), (0, 4), (-2, 3)$

40. What is the perimeter of the triangle with vertices $(1, 1), (5, 4),$ and $(-2, 5)$?

In Exercises 41–46, find the equation of the circle.

41. Center $(2, 2)$; passes through the origin.
42. Center $(-1, -3)$; passes through $(-4, -2)$.
43. Center $(1, 2)$; intersects x-axis at -1 and 3.
44. Center $(3, 1)$; diameter 2.
45. Center $(-5, 4)$; tangent (touching at one point) to the x-axis.
46. Center $(2, -6)$; tangent to the y-axis.

47. One diagonal of a square has endpoints $(-3, 1)$ and $(2, -4)$. Find the endpoints of the other diagonal.

48. Find the vertices of all possible squares with this property: Two of the vertices are $(2, 1)$ and $(2, 5)$. [Hint: There are three such squares.]

49. Do Exercise 48 with (c, d) and (c, k) in place of $(2, 1)$ and $(2, 5)$.

50. Find the three points that divide the line segment from $(-4, 7)$ to $(10, -9)$ into four parts of equal length.

51. Find all points P on the x-axis that are 5 units from $(3, 4)$. [Hint: P must have coordinates $(x, 0)$ for some x and the distance from P to $(3, 4)$ is 5.]

52. Find all points on the y-axis that are 8 units from $(-2, 4)$.

53. Find all points with first coordinate 3 that are 6 units from $(-2, -5)$.

54. Find all points with second coordinate -1 that are 4 units from $(2, 3)$.

Unusual Problems

55. For each real number k, the graph of $(x - k)^2 + y^2 = k^2$ is a circle. Describe all possible such circles.

56. (a) Prove that the distance formula is valid for two points on the same vertical line. [Hint: The points must have the same first coordinate (why?), say (x_1, y_1) and (x_1, y_2). The distance between them is $|y_1 - y_2|$ (why?). See the box on page 26.]

(b) Prove that the distance formula is valid for two points on the same horizontal line.

57. Suppose every point in the coordinate plane is moved 5 units straight up.
    (a) To what points do each of these points go: $(0, -5), (2, 2), (5, 0), (5, 5), (4, 1)$?
    (b) Which points go to each of the points in part (a)?
    (c) To what point does (a, b) go?
    (d) To what point does $(a, b - 5)$ go?
    (e) What point goes to $(-4a, b)$?
    (f) What points go to themselves?

58. Complete the proof of the Midpoint Formula by showing that $d_2 = \frac{1}{2}d$.

59. Let M be the midpoint of AC in the right triangle shown in Figure 3-13. Prove that M is equidistant from the vertices of the triangle. [*Hint:* A has coordinates $(0, r)$ for some r and C has coordinates $(s, 0)$ for some s.]

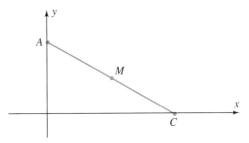

FIGURE 3-13

3.2 SLOPES OF LINES

The question of how steeply a straight line rises as you move from left to right is easily understood *geometrically*. In this section we develop a *numerical* way to measure the steepness of a line.

When you move from a point P to a point Q on a line, there are two numbers involved, as shown in Figure 3-14 below:

(i) The vertical distance you move (the **rise**).

(ii) The horizontal distance you move (the **run**).

The steepness of the line depends on these two numbers, as can be seen in Figure 3-14. In the left-hand line (the least steep), the rise (4 units) is only half the run (8 units), whereas in the right-hand line (the most steep), the rise (4 units) is four times greater than the run (1 unit). We can express the relationship of rise and run by the fraction rise/run, as illustrated in Figure 3-14:

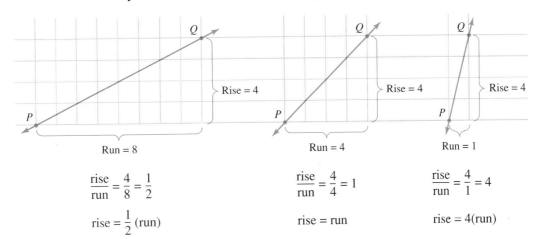

FIGURE 3-14

The numbers $\frac{1}{2}$, 1, 4 are increasing, just as the steepness of the corresponding lines increases. So the fraction rise/run provides a numerical means of describing the steepness of a line.

If P has coordinates (x_1, y_1) and Q has coordinates (x_2, y_2), then Figure 3–15 shows that the vertical rise from P to Q is the number $y_2 - y_1$ and the horizontal run from P to Q is $x_2 - x_1$.

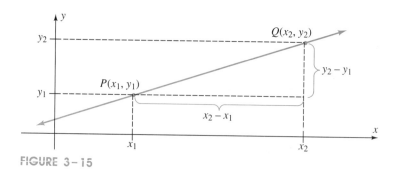

FIGURE 3–15

So $\dfrac{\text{rise}}{\text{run}}$ is the number $\dfrac{y_2 - y_1}{x_2 - x_1}$ and we have this definition:

DEFINITION OF SLOPE

The *slope* of a nonvertical straight line is

$$\frac{y_2 - y_1}{x_2 - x_1}$$

where (x_1, y_1) and (x_2, y_2) are any two distinct points on the line. In short, slope = rise/run.

Example 1 To find the slope of the line through $(0, -1)$ and $(4, 1)$ we apply the formula above with $x_1 = 0$, $y_1 = -1$ and $x_2 = 4$, $y_2 = 1$:

$$\text{slope} = \frac{y_2 - y_1}{x_2 - x_1} = \frac{1 - (-1)}{4 - 0} = \frac{2}{4} = \frac{1}{2}.$$

The order of the points makes no difference; if we use $(4, 1)$ for (x_1, y_1) and $(0, -1)$ for (x_2, y_2) we obtain the same number:

$$\text{slope} = \frac{y_2 - y_1}{x_2 - x_1} = \frac{-1 - 1}{0 - 4} = \frac{-2}{-4} = \frac{1}{2}. \blacksquare$$

Example 1 shows that the slope does not depend on the order of the points. In fact, much more is true:

3.2 SLOPES OF LINES

SLOPE THEOREM

The slope of a nonvertical straight line is independent of the choice of points used to compute it. Any two points may be used in any order to compute the slope.

Proof Let P, Q, R, S be points on the line, with coordinates as shown in Figure 3–16.

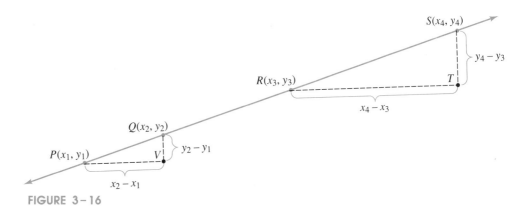

FIGURE 3–16

Then the slope using P and Q is $\dfrac{y_2 - y_1}{x_2 - x_1}$ and the slope using R and S is $\dfrac{y_4 - y_3}{x_4 - x_3}$.

To prove that these two numbers are equal we consider the triangles PQV and RST (as in Figure 3–16). Since angle V and angle T are both right angles, they are equal. Basic facts about parallel lines show that

$$\text{angle } P = \text{angle } R \qquad \text{angle } Q = \text{angle } S$$

Therefore triangles PQV and RST are similar.* Consequently, the ratios of the corresponding sides of the triangles are equal. In particular,

$$\frac{\text{length } QV}{\text{length } PV} = \frac{\text{length } ST}{\text{length } RT}$$

But we can express these lengths in terms of the coordinates of the points, as shown in Figure 3–16:

$$\frac{y_2 - y_1}{x_2 - x_1} = \frac{\text{length } QV}{\text{length } PV} = \frac{\text{length } ST}{\text{length } RT} = \frac{y_4 - y_3}{x_4 - x_3}$$

This proves the claim in the box above. ∎

Example 2 The lines L_1, L_2, L_3, L_4, and L_5 shown in Figure 3–17 on the next page are determined by these points:

---
* Similar triangles are discussed in the Geometry Review Appendix.

L_1: $(-2, -1)$ and $(-1, 6)$ L_2: $(-2, -1)$ and $(1, 8)$
L_3: $(-2, -1)$ and $(2, 3)$ L_4: $(-2, -1)$ and $(1, 0)$
L_5: $(-2, -1)$ and $(5, -1)$

We can compute the slopes of these lines as shown in Figure 3–17.

slope $L_1 = \dfrac{6-(-1)}{-1-(-2)} = \dfrac{7}{1} = 7$

slope $L_2 = \dfrac{8-(-1)}{1-(-2)} = \dfrac{9}{3} = 3$

slope $L_3 = \dfrac{3-(-1)}{2-(-2)} = \dfrac{4}{4} = 1$

slope $L_4 = \dfrac{0-(-1)}{1-(-2)} = \dfrac{1}{3}$

slope $L_5 = \dfrac{-1-(-1)}{5-(-2)} = \dfrac{0}{7} = 0$

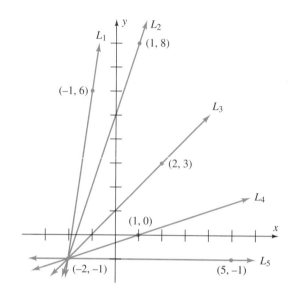

FIGURE 3–17

Observe how the slopes correspond to the steepness of the lines: Horizontal lines have slope 0, and *the greater the slope, the more steeply the lines rises* from left to right. ■

Example 3 The lines L_6, L_7, L_8, and L_9 shown in Figure 3–18 are determined by these points:

L_6: $(-3, 2)$ and $(4, 1)$ L_7: $(-3, 2)$ and $(2, -4)$
L_8: $(-3, 2)$ and $(0, -6)$ L_9: $(-3, 2)$ and $(-2, -3)$

Each of these lines falls from left to right and each has *negative* slope, as computed in Figure 3–18. In this case, too, slopes correspond to the steepness of the lines: The larger the slope in absolute value, the more steeply the line falls from left to right. ■

The slope formula doesn't work for vertical lines. The reason is that any two points on the same vertical line have the same first coordinate, say $x = c$.

slope $L_6 = \dfrac{1-2}{4-(-3)} = \dfrac{-1}{7}$

slope $L_7 = \dfrac{-4-2}{2-(-3)} = \dfrac{-6}{5}$

slope $L_8 = \dfrac{-6-2}{0-(-3)} = \dfrac{-8}{3}$

slope $L_9 = \dfrac{-3-2}{-2-(-3)} = \dfrac{-5}{1} = -5$

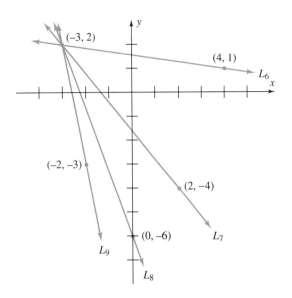

FIGURE 3-18

So the slope formula becomes $\dfrac{y_2 - y_1}{c - c} = \dfrac{y_2 - y_1}{0}$. But this is *not* a real number. Therefore

PROPERTIES OF SLOPE

> The slope of a vertical line is not defined.
>
> The slope of a nonvertical line is a number m that measures how steeply the line rises or falls:
>
> If $m > 0$, the line rises from left to right; the larger m is, the more steeply the line rises.
>
> If $m < 0$, the line falls from left to right; the larger $|m|$ is, the more steeply the line falls.
>
> If $m = 0$, the line is horizontal.

Example 4 If L is a line with slope $\dfrac{1}{3}$ and $(2, -4)$ is on L, find three more points on L. ■

Solution To say that the slope of L is $\dfrac{\text{rise}}{\text{run}} = \dfrac{1}{3}$ means that if you start at $(2, -4)$, run 3 units to the *right*, then *rise* 1 unit, the point where you finish (namely, $(5, -3)$) is also on L, as shown in Figure 3-19.

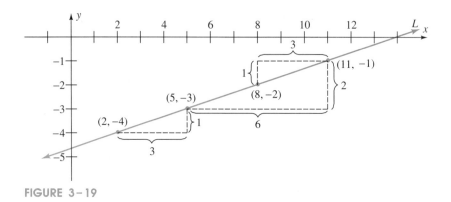

FIGURE 3-19

But rise/run = 1/3 = 2/6, so we can also say that if you start at a point on L, run 6 units to the *right,* then *rise* 2 units, you will end up at a point on L. If you do this, beginning at $(5, -3)$, you end up at the point $(11, -1)$ on L, as in Figure 3-19. Finally, it is also true that rise/run = 1/3 = −1/−3. This means that if you start at a point on L, say, $(11, -1)$, and run 3 units to the *left* (negative horizontal direction), then *fall* 1 unit (negative vertical direction), you will end up at a point on L, namely, $(8, -2)$, as shown in Figure 3-19. The reason these techniques work is that the slope is a *ratio*. Each of the fractions 1/3, 2/6, −1/−3, and so on, expresses the same basic fact: The rise is one third of the run. ■

The slope of a line measures how steeply it rises or falls. Since parallel lines rise or fall equally steeply, the following fact is plausible (see Exercise 34 for a proof):

SLOPES OF PARALLEL LINES

Two nonvertical straight lines are parallel exactly when they have the same slope.

Example 5 If L is the line through $(0, 2)$ and $(1, 5)$ and M is the line through $(2, 1)$ and $(3, 4)$ as shown in Figure 3-20, then

$$\text{slope } L = \frac{5-2}{1-0} = \frac{3}{1} = 3 \quad \text{and} \quad \text{slope } M = \frac{4-1}{3-2} = \frac{3}{1} = 3$$

Therefore L and M are parallel. ■

Two lines that meet at a right angle (90° angle) are said to be **perpendicular**. As you might suspect, there is a close relationship between the slopes of two perpendicular lines.

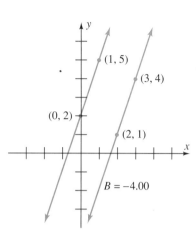

FIGURE 3-20

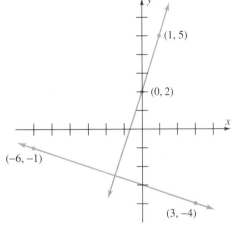

FIGURE 3-21

SLOPES OF PERPENDICULAR LINES

Let L be a line with slope k and M a line with slope m. Then L and M are perpendicular exactly when $km = -1$.

A proof of this fact is outlined in Exercise 44 of Section 3.3.

Example 6 Let L be the line through $(0, 2)$ and $(1, 5)$. Let M be the line through $(-6, -1)$ and $(3, -4)$, as shown in Figure 3-21 above. Then

$$\text{slope } L = \frac{5-2}{1-0} = 3 \quad \text{and} \quad \text{slope } M = \frac{-4-(-1)}{3-(-6)} = \frac{-3}{9} = -\frac{1}{3}$$

Since $3(-\frac{1}{3}) = -1$, the lines L and M are perpendicular. ∎

EXERCISES

In Exercises 1–5, find the slope of the line through the given points.

1. $(1, 2), (3, 7)$
2. $(-1, -2), (2, -1)$
3. $\left(\frac{1}{4}, 0\right), \left(\frac{3}{4}, 2\right)$
4. $(\sqrt{2}, -1), (2, -9)$
5. $(\pi, 1), (-1, \pi)$

6. On one graph, sketch five lines satisfying these conditions: (i) one line has slope 0, two lines have positive slope, and two lines have negative slope; (ii) all five lines meet at a single point.

7. On one graph sketch five line segments, not all meeting at a single point, whose slopes are five different positive numbers. Do this in such a way that the left-hand line has the largest slope, the second line from the left the next largest slope, and so on.

In Exercises 8–12, determine whether the line through P and Q is parallel or perpendicular to the line through R and S, or neither.

8. $P = (1, -4) \qquad Q = (2, -3)$
   $R = (-1, 3) \qquad S = (2, 0)$

9. $P = (2, 5)$ $Q = (-1, -1)$
   $R = (4, 2)$ $S = (6, 1)$

10. $P = \left(0, \dfrac{3}{2}\right)$ $Q = (1, 1)$
    $R = (2, 7)$ $S = (3, 9)$

11. $P = \left(-3, \dfrac{1}{3}\right)$ $Q = (1, -1)$
    $R = (2, 0)$ $S = \left(4, -\dfrac{2}{3}\right)$

12. $P = (3, 3)$ $Q = (-3, -1)$
    $R = (2, -2)$ $S = (4, -5)$

In Exercises 13–14, find five points on the given line in addition to the given points.

13. Line with slope 3/2 through (3, 4).

14. Line with slope 5 through $(1, -2)$. (*Remember:* $5 = 5/1$.)

In Exercises 15–18, graph the given line.

15. Through (1, 2) with slope 1.
16. Through $(-3, 4)$ with slope -1.
17. Through (6, 1) with slope 0.
18. Through $(-2, 3)$ with no slope.

19. The doorsill of a campus building is 5 ft above ground level. To allow wheelchair access the steps in front of the door are to be replaced by a straight ramp with constant slope 1/12, as shown in Figure 3–22. How long must the ramp be?

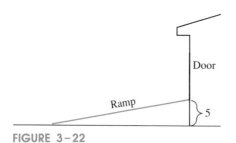

FIGURE 3–22

20. Use slopes to show that the points $(-4, 6), (-1, 12)$, and $(-7, 0)$ all lie on the same straight line.

21. Use slopes to show that the points $(-5, -2)$, $(-3, 1), (3, 0)$, and $(5, 3)$ are the vertices of a parallelogram.

22. Are the points $(-10, 9)$, $\left(-\dfrac{16}{3}, \dfrac{13}{3}\right)$, $(-3, -2)$, and $(4, -9)$ the vertices of a parallelogram?

In Exercises 23–24, use slopes to determine whether the three given points are the vertices of a right triangle.

23. $(9, 6), (-1, 2), (1, -3)$
24. $(1, 6), (-5, -8), (5, -4)$

In Exercises 25–30, find a number t such that the line passing through the two given points has slope -2.

25. $(0, t), (9, 4)$ 26. $(1, t), (-3, 5)$
27. $(-2, t), (-1, -7)$ 28. $(t, t), (5, 9)$
29. $\left(\dfrac{t}{3}, -2\right), \left(4, \dfrac{t}{4}\right)$
30. $(t + 1, 5), (6, -3t + 7)$

31. Let L be the line through $(-4, 5)$ that is perpendicular to the line through $(1, 3)$ and $(-4, 2)$. Find three points on L.

32. Let L be the line through $(4, -3)$ that is parallel to the line through $(-2, 4)$ and $(4, -1)$. Find three points on L.

Unusual Problems

33. Let L be a nonvertical straight line through the origin. L intersects the vertical line through $(1, 0)$ at a point P. Show that the second coordinate of P is the slope of L.

34. Prove that nonvertical parallel lines L and M have the same slope, as follows. Suppose M lies above L and choose two points (x_1, y_1) and (x_2, y_2) on L.
    (a) Let P be the point on M with first coordinate x_1. Let b denote the vertical distance from P to (x_1, y_1). Show that the second coordinate of P is $y_1 + b$.
    (b) Let Q be the point on M with first coordinate x_2. Use the fact that L and M are parallel to show that the second coordinate of Q is $y_2 + b$.
    (c) Compute the slope of L using (x_1, y_1) and (x_2, y_2). Compute the slope of M using the

points P and Q. Verify that the two slopes are the same. For a proof of the fact that nonvertical lines with the same slope are parallel, see Exercise 42 in Section 3.3.

35. For which line segment in Figure 3–23 is the slope the
   (a) largest?
   (b) smallest?
   (c) largest in absolute value?
   (d) closest to zero?

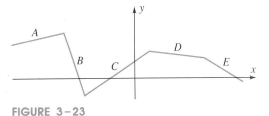

FIGURE 3–23

3.3 EQUATIONS OF LINES

We now show that every straight line is the graph of a first-degree equation, and vice versa.*

First suppose L is a vertical line. Then L crosses the x-axis at some point $(c, 0)$, as shown in Figure 3–24.

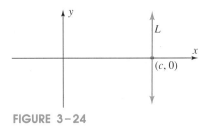

FIGURE 3–24

The points on L are exactly those points with first coordinate c, that is, the points (x, y) with $x = c$. Therefore

VERTICAL LINES

> The graph of the equation $x = c$ is the vertical line through $(c, 0)$.

For example, the vertical line through $(14, 79)$ also goes through $(14, 0)$, so it is the graph of $x = 14$.

For nonvertical lines, the slope is the key:

POINT-SLOPE FORM OF THE EQUATION OF A LINE

> The line L with slope m through the point (x_1, y_1) is the graph of the equation
> $$y - y_1 = m(x - x_1).$$

---

* Throughout this section "equation" means an equation in *two* variables. If only one variable is mentioned, it is understood that the other appears with zero coefficient. For instance, $x = 7$ indicates the equation $x + 0y = 7$ and $y = 4$ the equation $0x + y = 4$.

Proof Let (c, d) be any other point on L. Using (c, d) and (x_1, y_1) to compute the slope of L, we have:

$$\frac{d - y_1}{c - x_1} = \text{slope } L = m, \quad \text{or equivalently,} \quad d - y_1 = m(c - x_1)$$

Therefore (c, d) satisfies the equation $y - y_1 = m(x - x_1)$. Since (x_1, y_1) also satisfies this equation, *every* point on L is on the graph of this equation. The rest of the proof (namely, that every point on the graph of the equation is necessarily on L) will be omitted. ∎

Example 1 The line L with slope 2 through $(1, -6)$ is the graph of $y - y_1 = m(x - x_1)$, with $m = 2$ and $(x_1, y_1) = (1, -6)$; that is,

$$y - (-6) = 2(x - 1)$$

which simplifies to $y = 2x - 8$. We can use the equation to find other points on the line. For instance, if $x = 0$, then $y = 2 \cdot 0 - 8 = -8$, so that $(0, -8)$ is on the graph. Since two points determine a straight line, this is enough information to obtain the entire graph (see Figure 3–25).* ∎

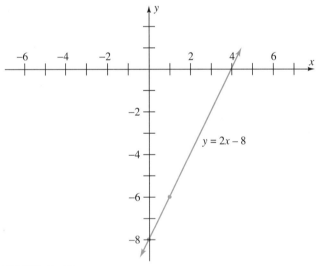

FIGURE 3–25

Example 2 To find the equation of the line M† through $(5, 3)$ and $(-3, 1)$ we must first find the slope of M:

$$\frac{1 - 3}{-3 - 5} = \frac{-2}{-8} = \frac{1}{4}.$$

---

* Since we can only picture a small part of the line, we use arrows to indicate that the line continues infinitely far in each direction.

† That is, the equation whose graph is M.

3.3 EQUATIONS OF LINES

Now we use the slope $\frac{1}{4}$ and the point $(5, 3)$ to find the equation:

$$y - y_1 = m(x - x_1)$$

$$y - 3 = \frac{1}{4}(x - 5)$$

which simplifies to $y = \frac{1}{4}x + \frac{7}{4}$. If you use the point $(-3, 1)$ instead of $(5, 3)$, you get the same equation:

$$y - 1 = \frac{1}{4}(x - (-3)), \quad \text{or equivalently,} \quad y = \frac{1}{4}x + \frac{7}{4}. \blacksquare$$

If a graph intersects the y-axis at $(0, b)$, then the number b is called a **y-intercept** of the graph. If a graph intersects the x-axis at $(a, 0)$, then the number a is an **x-intercept** of the graph. For example, the line in Figure 3–25 on the opposite page has x-intercept 4 and y-intercept -8.

To find the y-intercepts without graphing, just set $x = 0$ in the equation and solve for y: These values are precisely the y-coordinates of points with x-coordinate 0. For instance, the y-intercept of the line $y = \frac{1}{4}x + \frac{7}{4}$ is $\frac{7}{4}$. To find the x-intercepts, set $y = 0$ and solve for x.

If a line has slope m and y-intercept b, then $(0, b)$ is on the line. The graph of the line through $(0, b)$ with slope m is $y - b = m(x - 0)$. By rearranging terms, we have:

SLOPE-INTERCEPT FORM OF THE EQUATION OF A LINE	The line with slope m and y-intercept b is the graph of the equation $$y = mx + b.$$

Example 3 The graph of $x - 2y = 6$ can be easily determined by rewriting the equation:

$$2y = x - 6$$

$$y = \frac{1}{2}x - 3$$

This equation is of the form $y = mx + b$. So the graph is a line with slope $m = 1/2$ and y-intercept $b = -3$. Therefore $(0, -3)$ is on the graph. To sketch this graph, we first use the equation to find another point on it. For example, when $x = 2$, then $y = \frac{1}{2}(2) - 3 = -2$ so that $(2, -2)$ is on the graph. Thus we obtain the graph in Figure 3-26 on the next page. $\blacksquare$

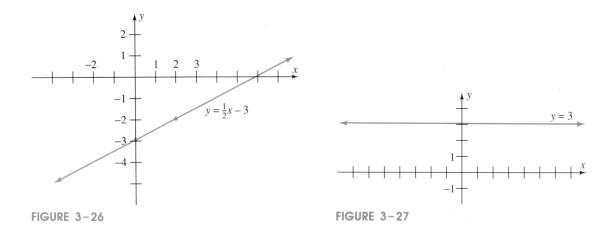

FIGURE 3-26   FIGURE 3-27

Example 4 The equation $y = 3$ can be written as $y = 0x + 3$. So its graph is a line with slope 0 and y-intercept 3, that is, a horizontal line through $(0, 3)$, as shown in Figure 3-27 above. ■

Example 5 To find the equation of the line L through $(2, -1)$ that is parallel to the line M with equation $3x - 2y + 6 = 0$, we first find the slope of M by rewriting its equation in slope-intercept form:

$$2y = 3x + 6, \quad \text{or equivalently,} \quad y = \frac{3}{2}x + 3.$$

Therefore M has slope 3/2. The parallel line L must have the same slope, 3/2. Since $(2, -1)$ is on L, the equation of L is

$$y - (-1) = \frac{3}{2}(x - 2), \quad \text{or equivalently,} \quad y = \frac{3}{2}x - 4. \ \blacksquare$$

Example 6 To find the equation of the line L through $(8, 3)$ that is perpendicular to the line M with equation $4x - y = 5$, we first write the equation of M in slope-intercept form: $y = 4x - 5$. Hence M has slope 4. The perpendicular line L must have slope $-\frac{1}{4}$ (because the product of slopes of perpendicular lines is -1; see page 137). So the equation of L is $y - 3 = -\frac{1}{4}(x - 8)$. ■

The preceding discussion shows that the equation of any straight line is a **first-degree equation,** that is, an equation that can be written in the form $Ax + By + C = 0$ for some constants A, B, C (with at least one of A, B nonzero). Conversely, the graph of the first-degree equation $Ax + By + C = 0$ is a straight line. If $B \neq 0$, this equation can be written

$$By = -Ax - C$$

$$y = \frac{-A}{B}x - \frac{C}{B}$$

3.3 EQUATIONS OF LINES

Therefore the graph is a straight line with slope $-A/B$ and y-intercept $-C/B$. If $A \neq 0$ and $B = 0$, then

$$Ax = -C, \quad \text{or equivalently,} \quad x = -\frac{C}{A}$$

So the graph of the equation is a vertical line with x-intercept $-C/A$. We have proved:

FIRST-DEGREE EQUATIONS

The graph of every first-degree equation $Ax + By + C = 0$ is a straight line, and every straight line is the graph of a first-degree equation.

For this reason, first-degree equations are called **linear equations**.

EXERCISES

In Exercises 1–4, find the equation of the line with given slope m and passing through the given point.

1. $m = 1$; $(3, 5)$
2. $m = 2$; $(-2, 1)$
3. $m = -1$; $(6, 2)$
4. $m = 0$; $(-4, -5)$

In Exercises 5–8, find the equation of the line through the two given points.

5. $(0, -5)$ and $(-3, -2)$
6. $(4, 3)$ and $(2, -1)$
7. $\left(\frac{4}{3}, \frac{2}{3}\right)$ and $\left(\frac{1}{3}, 3\right)$
8. $(6, 7)$ and $(6, 15)$

In Exercises 9–12, find the equation of the line with given slope m and given y-intercept b.

9. $m = 1$; $b = 2$
10. $m = 2$; $b = 5$
11. $m = -4$; $b = 2$
12. $m = -5$; $b = -2$

In Exercises 13–16, find the slope and y-intercept of the line whose equation is given. (*Hint:* Put the equation in slope-intercept form.)

13. $2x - y + 5 = 0$
14. $3x + 4y = 7$
15. $3(x - 2) + y = 7 - 6(y + 4)$
16. $2(y - 3) + (x - 6) = 4(x + 1) - 2$

In Exercises 17–26, find an equation for the line satisfying the given conditions.

17. Through $(-2, 1)$ with slope 3.
18. y-intercept -7 and slope 1.
19. Through $(2, 3)$ and parallel to $3x - 2y = 5$.
20. Through $(1, -2)$ and perpendicular to $y = 2x - 3$.
21. x-intercept 5 and y-intercept -5.
22. Through $(-5, 2)$ and parallel to the line through $(1, 2)$ and $(4, 3)$.
23. Through $(-1, 3)$ and perpendicular to the line through $(0, 1)$ and $(2, 3)$.
24. y-intercept 3 and perpendicular to $2x - y + 6 = 0$.
25. y-intercept 0 and parallel to $x - 3y + 7 = 0$.
26. Parallel to $3x - 3y + 5 = 7$, with the same y-intercept as $2x - 5y + 7 = 4$.

In Exercises 27–30, determine whether the given lines are parallel, perpendicular, or neither. [*Hint:* Put the equations in slope-intercept form and use well-known facts about slopes.]

27. $2x + y - 2 = 0$ and $4x + 2y + 18 = 0$
28. $3x + y - 3 = 0$ and $6x + 2y + 17 = 0$
29. $x + y - 3 = 0$ and $y - x + 2 = 0$

30. $y - ax + 5 = 0$ and $ay + x - 7 = 0$ (where a is a constant)

31. Find a real number k such that $(3, -2)$ is on the line $kx - 2y + 7 = 0$.

32. Find a real number k such that the line $3x - ky + 2 = 0$ has y-intercept -3.

In Exercises 33–38, find the equation of the tangent line to the circle at the given point. [*Note:* If P is a point on a circle with center C, then the tangent line to the circle at P is the straight line through P that is perpendicular to the radius CP.]

33. $x^2 + y^2 = 25$ at $(3, 4)$ [*Hint:* Here C is $(0, 0)$ and P is $(3, 4)$; what is the slope of radius CP?]

34. $x^2 + y^2 = 169$ at $(-5, 12)$

35. $(x - 1)^2 + (y - 3)^2 = 5$ at $(2, 5)$

36. $(x - 3)^2 + (y + 2)^2 = 26$ at $(-2, -3)$

37. $x^2 + y^2 + 2x + 4y = 0$ at $(0, 0)$

38. $x^2 + y^2 + 6x - 8y + 15 = 0$ at $(-2, 1)$

39. Let A, B, C, D be nonzero real numbers. Show that the lines $Ax + By + C = 0$ and $Ax + By + D = 0$ are parallel.

40. Graph the equation $|x| + |y| = 1$. [*Hint:* Look at one quadrant at a time. For instance, all points in the second quadrant have $x < 0$ and $y > 0$, so that $|x| = -x$ and $|y| = y$. Hence in the second quadrant, the graph of $|x| + |y| = 1$ is the same as the graph of $-x + y = 1$.]

41. Let L be a line which is neither vertical nor horizontal and which does *not* pass through the origin. Show that L is the graph of $\dfrac{x}{a} + \dfrac{y}{b} = 1$, where a is the x-intercept and b is the y-intercept of L.

42. Show that two nonvertical lines with the same slope are parallel. [*Hint:* The equations of distinct lines with the same slope must be of the form $y = mx + b$ and $y = mx + c$ with $b \neq c$ (why?). If (x_1, y_1) were a point on both lines, its coordinates would satisfy both equations. Show that this leads to a contradiction and conclude that the lines have no point in common.]

43. A road is being built along the floor of a valley. At the end of the valley it must pass over the hill, as shown in Figure 3–28. Because of soil conditions and other factors, the road is to lie along a straight line of slope $\tfrac{1}{25}$ which passes through point A (as indicated by the colored line below). The top of the hill will be cut away, and the bottom of the hill will be filled in to road level. The point A lies 400 ft above the floor of the valley. The peak of the hill (point B) lies 500 ft above the valley floor. The horizontal distance d_1 is 600 ft, distance d_2 is 300 ft.
(a) Find the point where the road should meet the valley floor by determining the distance from P to Q. [*Hint:* Sketch the picture on a coordinate system, with the valley floor as the x-axis and the vertical line through A as the y-axis.]
(b) Find out how much of the hilltop must be cut off by determining the vertical distance from point B to the road.

44. This exercise provides a proof of the statement in the box on page 137. Let L be a line with slope k and M a line with slope m and assume *both* L and M pass through the origin.

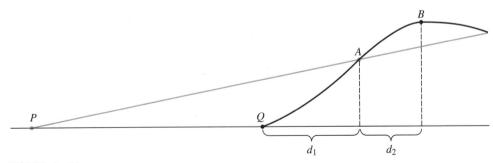

FIGURE 3–28

(a) Show that L passes through $(1, k)$ and M passes through $(1, m)$.
(b) Compute the length of each side of the triangle with vertices $(0, 0)$, $(1, k)$ and $(1, m)$.
(c) Suppose L and M are perpendicular. Then the triangle of part (b) has a right angle at $(0, 0)$. Use part (b) and the Pythagorean Theorem to find an equation involving k, m, and various constants. Simplify this equation to show that $km = -1$.
(d) Suppose instead that $km = -1$ and prove that L and M are perpendicular. [*Hint:* You may assume that a triangle whose sides a, b, c satisfy $a^2 + b^2 = c^2$ is a right triangle with hypotenuse c. Use this fact and $km = -1$ to "reverse" the argument in part (c)].
(e) Finally, assume L and M are any two nonvertical lines (which don't necessarily go through the origin), with slope $L = k$ and slope $M = m$. Use the preceding material to prove that L is perpendicular to M exactly when $km = -1$. [*Hint:* Every line is parallel to a line through the origin and parallel lines have the same slope.]

45. Let (c, d) be a point in the plane with $c \neq 0$. Use parts (a) and (b) to prove that (c, d) and $(-c, -d)$ lie on the same straight line through the origin:
   (a) Find the slope and the equation of the straight line L through (c, d) and $(0, 0)$.
   (b) Show that $x = -c$, $y = -d$ satisfy the equation of L. Hence $(-c, -d)$ lies on the line L.
   (c) Explain why (c, d) and $(-c, -d)$ lie on opposite sides of the origin on the line L.
   (d) Show that (c, d) and $(-c, -d)$ are the same distance from the origin. [*Hint:* What is the midpoint of the line segment from (c, d) to $(-c, -d)$?]

CHAPTER REVIEW

Important Concepts

Section 3.1 Coordinate plane 121
Distance formula 123
Midpoint formula 125
Graphs of equations and inequalities 127
Circle, center, radius 127
Equation of the circle 127
Unit circle 128

Section 3.2 Rise 131
Run 131
Slope 132
Slopes of parallel lines 136
Slopes of perpendicular lines 137

Section 3.3 Equations of vertical lines 139
Point-slope form of the equation of a line 139
Slope-intercept form of the equation of a line 141
Graphs of first-degree equations 143

Important Facts and Formulas

- *Distance Formula:* Distance from (x_1, y_1) to (x_2, y_2) is

$$\sqrt{(x_1 - x_2)^2 + (y_1 - y_2)^2}$$

- *Midpoint Formula:* Midpoint of segment from (x_1, y_1) to (x_2, y_2) is
$$\left(\frac{x_1 + x_2}{2}, \frac{y_1 + y_2}{2}\right)$$
- Equation of the circle with center (c, d) and radius r:
$$(x - c)^2 + (y - d)^2 = r^2$$
- Slope of line through (x_1, y_1) and (x_2, y_2) (where $x_1 \neq x_2$):
$$\frac{y_2 - y_1}{x_2 - x_1}$$
- Nonvertical parallel lines have the same slope.
- Two lines (neither vertical) are perpendicular exactly when the product of their slopes is -1.
- Equation of the line through (x_1, y_1) with slope m:
$$y - y_1 = m(x - x_1)$$
- Equation of the line with slope m and y-intercept b:
$$y = mx + b$$

Review Questions

1. Find the distance from $(1, -2)$ to $(4, 5)$.
2. Find the distance from $(3/2, 4)$ to $(3, 5/2)$.
3. Find the distance from (c, d) to $(c - d, c + d)$.
4. Find the midpoint of the line segment from $(-4, 7)$ to $(9, 5)$.
5. Find the midpoint of the line segment from (c, d) to $(2d - c, c + d)$.
6. Find the equation of the circle with center $(-3, 4)$ that passes through the origin.
7. (a) If $(1, 1)$ is on a circle with center $(2, -3)$, what is the radius of the circle?
   (b) Find the equation of the circle in part (a).
8. Sketch the graph of $3x^2 + 3y^2 = 12$.
9. Sketch the graph of $(x - 5)^2 + y^2 - 9 = 0$.
10. Find the center and radius of the circle whose equation is $x^2 + y^2 - 2x + 6y + 1 = 0$.
11. Which of statements i–iv are descriptions of the circle with center $(0, -2)$ and radius 5?
    i. The set of points (x, y) that satisfy $|x| + |y + 2| = 5$.
    ii. The set of all points whose distance from $(0, -2)$ is 5.
    iii. The set of all points (x, y) such that $x^2 + (y + 2)^2 = 5$.
    iv. The set of all points (x, y) such that $\sqrt{x^2 + (y + 2)^5} = 5$.
    (a) ii and iv.
    (b) i and iv.

(c) i, ii, and iv.
(d) ii and iii.
(e) i, ii, iii, and iv.

12. If the equation of a circle is $3x^2 + 3(y-2)^2 = 12$, which of the following statements is true?
    (a) The circle has diameter 3.
    (b) The center of the circle is (2, 0).
    (c) The point (0, 0) is on the circle.
    (d) The circle has radius $\sqrt{12}$.
    (e) The point (1, 1) is on the circle.

13. The graph of one of the equations below is *not* a circle. Which one?
    (a) $x^2 + (y+5)^2 = \pi$
    (b) $7x^2 + 4y^2 - 14x + 3y^2 - 2 = 0$
    (c) $3x^2 + 6x + 3 = 3y^2 + 15$
    (d) $2(x-1)^2 - 8 = -2(y+3)^2$
    (e) $\dfrac{x^2}{4} + \dfrac{y^2}{4} = 1$

14. Find the equation of the line passing through (1, 3) and (2, 5).
15. Find the equation of the line passing through (2, −1) with slope 3.
16. Find a point on the graph of $y = 3x$ whose distance to the origin is 2.
17. Find the equation of the line that crosses the y-axis at $y = 1$ and is perpendicular to the line $2y - x = 5$.
18. (a) Find the y-intercept of the line $2x + 3y - 4 = 0$.
    (b) Find the equation of the line through (1, 3) which has the same y-intercept as the line in part (a).
19. Find the equation of the line through (−4, 5) that is parallel to the line through (1, 3) and (−4, 2).
20. Graph the line $3x + y - 1 = 0$.
21. As a balloon is launched from the ground, the wind is blowing it due east. The conditions are such that the balloon is ascending along a straight line with slope 1/5. After 1 hour the balloon is 5000 ft vertically above the ground. How far east has the balloon blown?
22. The point (u, v) lies on the line $y = 5x - 10$. What is the slope of the line passing through (u, v) and the point $(0, -10)$?

In Questions 23–29, determine whether the statement is true or false.

23. The graph of $x = 5y + 6$ has y-intercept 6.
24. The graph in Figure 3–29 on the next page is the graph of $x^2 + y^2 = 4$.

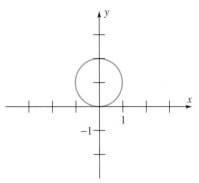

FIGURE 3-29

25. The lines $3x + 4y = 12$ and $4x + 3y = 12$ are perpendicular.
26. Slope is not defined for horizontal lines.
27. The line in Figure 3-30 has positive slope.

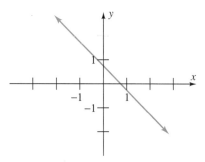

FIGURE 3-30

28. The line in Figure 3-30 does not pass through the third quadrant.
29. The y-intercept of the line in Figure 3-30 is negative.
30. Consider the *slopes* of the lines shown in Figure 3-31.

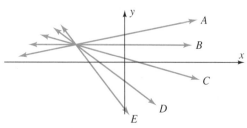

FIGURE 3-31

Which slope has the largest *absolute value*?

31. Which of the following lines rises most steeply from left to right?
    (a) $y = -4x - 10$
    (b) $y = 3x + 4$
    (c) $20x + 2y - 20 = 0$
    (d) $4x = y - 1$
    (e) $4x = 1 - y$

32. Which of the following lines is *not* perpendicular to the line $y = x + 5$?
    (a) $y = 4 - x$
    (b) $y + x = -5$
    (c) $4 - 2x - 2y = 0$
    (d) $x = 1 - y$
    (e) $y - x = \frac{1}{5}$

33. Which of the following lines does *not* pass through the third quadrant?
    (a) $y = x$
    (b) $y = 4x - 7$
    (c) $y = -2x - 5$
    (d) $y = 4x + 7$
    (e) $y = -2x + 5$

34. Let a, b be fixed real numbers. Where do the lines $x = a$ and $y = b$ intersect?
    (a) Only at (b, a).
    (b) Only at (a, b).
    (c) These lines are parallel, so they don't intersect.
    (d) If $a = b$, then these are the same line, so they have infinitely many points of intersection.
    (e) Since these equations are not of the form $y = mx + b$, the graphs are not lines.

35. What is the *y-intercept* of the line $2x - 3y + 5 = 0$?

36. For what values of k will the graphs of $2y + x + 3 = 0$ and $3y + kx + 2 = 0$ be perpendicular lines?

CHAPTER 4

Functions and Graphs

The concept of a function and functional notation are central to modern mathematics. In this chapter you will be introduced to functions and operations on functions, learn how to use functional notation, and develop skill in constructing and interpreting graphs of functions.

4.1 FUNCTIONS

To understand the origin of the concept of function it may help to consider some "real-life" situations in which one numerical quantity depends on, corresponds to, or determines another.

Example 1 The amount of income tax you pay depends on the amount of your income. The way in which the income determines the tax is given by the tax law. ■

Example 2 The weather bureau records the temperature over a 24-hour period in the form of a graph (Figure 4–1). The graph shows the temperature that corresponds to each given time. ■

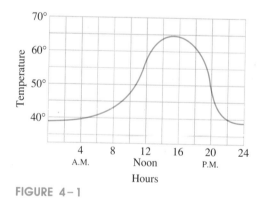

FIGURE 4–1

Example 3 Suppose a rock is dropped straight down from a high place. Physics tells us that the distance traveled by the rock in t seconds is $16t^2$ feet. So the distance depends on the time. ■

The first common feature shared by these examples is that each involves *two sets* of numbers, as summarized in this chart:

	Set 1	Set 2
Example 1	All incomes	All tax amounts
Example 2	Hours since midnight	Temperatures during the day
Example 3	Seconds elapsed after dropping the rock	Distances rock travels

The second common feature is that in each example there is a definite *rule* by which an element of Set 1 determines an element of Set 2. This rule may be given in a variety of ways: the tax law in Example 1, the time/temperature graph in Example 2, and the formula (distance = $16t^2$) in Example 3.

By focusing on these common features in the examples we obtain this definition:

DEFINITION OF FUNCTION

A *function* consists of:

A set of numbers (called the *domain*); and

A *rule* that assigns to each number in the domain one and only one number.

The set of all numbers that are assigned to numbers in the domain by the rule is called the *range* of the function.

The domain corresponds to Set 1 and the range to Set 2 in the examples. Thus in Example 1, the domain consists of all possible income amounts and the range consists of all possible tax amounts.

The domain in Example 2 is the set of hours in the day (that is, all real numbers from 0 to 24). The rule is given by the time/temperature graph, which shows the temperature at each time. The graph also shows that the range (the temperatures that occur during the day) consists of the numbers from 38 to 63. For each time of day (number in the domain) there is one and only one temperature (number in the range). Notice, however, that it is possible to have the same temperature (number in the range) corresponding to two different times (numbers in the domain). In general,

The rule of a function assigns to each number in the domain *one and only one* number in the range. But the rule may possibly assign the *same* number in the range to several different numbers in the domain.

Although real-world situations, such as Examples 1–3, are the *motivation* for studying functions, the emphasis in mathematics courses is on the functions themselves, independent of possible interpretations in specific situations. Here are some examples of functions in a purely mathematical context.

Example 4 The domain of the **absolute value function** is the set of all real numbers. The rule assigns to each real number x its absolute value $|x|$. For instance, $|-6| = 6$, so 6 in the range is assigned to -6 in the domain. Since $|x| \geq 0$ for every x, the range consists of all nonnegative real numbers, that is, the interval $[0, \infty)$. ∎

Example 5 For each real number s that is not an integer, let $[s]$ denote the *integer* that is closest to s on the *left* side of s on the number line; if s is itself an integer, we define $[s] = s$. Here are some examples:

$$[-4.7] = -5 \quad [-3] = -3 \quad [-1.5] = -2 \quad [0] = 0 \quad \left[\frac{5}{3}\right] = 1 \quad [\pi] = 3$$

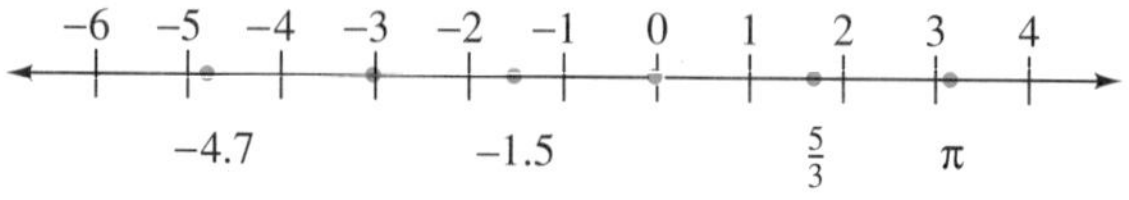

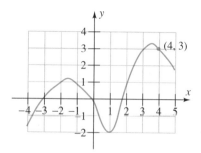

FIGURE 4-2

The function whose domain is the set of all real numbers and whose rule is:

Assign to the real number x the integer $[x]$

is called the **greatest integer function**. The range of the greatest integer function is the set of all integers. ∎

Example 6 The graph in Figure 4-3

FIGURE 4-3

defines a function whose domain is the interval $[-4, 5]$ and whose rule is:

Assign to a number x the number y such that (x, y) lies on the graph.

For instance, the rule assigns to 4 the number 3 since $(4, 3)$ is on the graph. Since all the second coordinates of points on the graph lie between -2 and 3.3,* the range of this function is the interval $[-2, 3.3]$. ∎

WARNING

Not every graph defines a function in this way; see Exercise 28.

EXERCISES

The notation $[r]$ used in Exercises 1–10 is explained in Example 5.

1. $[6.75] = ?$
2. $[.75] = ?$
3. $[-4/3] = ?$
4. $[-10/3] = ?$
5. $[2/3] = ?$
6. $[-2/3] = ?$
7. $[-16\frac{1}{2}] = ?$
8. $[16.0001] = ?$

In Exercises 9 and 10, give examples of two numbers, u and v, for which the given statement is true and two numbers for which it is false.

9. $[u + v] = [u]$
10. $[u] + [v] < [u + v]$

In Exercises 11–14, a functional situation is described that involves at least one function. Verbally describe the domain, range, and rule of each such function, providing there is sufficient information to do so. For example,

Situation: Harry Hamburger owns a professional baseball team. He decides that the only factor to be considered in determining a player's salary for this year is his batting average last season. *Function:* Salaries are a function of batting averages. *Domain:* All team members' batting averages. *Range:* All team members' salaries. *Rule:* There isn't enough information given to determine the rule.

11. The area of a circle depends on its radius.
12. If you drive your car at a constant rate of 55 mph, then your distance from the starting point varies with the time elapsed.
13. When more is spent on advertising at Grump's Department Store, sales increase. If less is spent on advertising, sales drop.
14. A widget manufacturer has a contract to produce widgets at a price of $1 each. His profit per widget is determined by the cost of manufacturing the widget. (Note that if it costs more than $1 to make a widget, he loses money; interpret such a loss as "negative profit.") Find an algebraic formula for the rule of this function.

Exercises 15–18 refer to Example 1 in the text. Assume that the state income tax law reads as follows:

Annual Income	Amount of Tax
Less than $2000	0
$2000–$6000	2% of income over $2000
More than $6000	$80 plus 5% of income over $6000

15. Find the number in the range (tax amount) that is assigned to each of the following numbers in the domain (incomes):

    $500, $1509, $3754,
    $6783, $12,500, $55,342

16. Find four different numbers in the domain of this function that are associated with the same number in the range.

17. Explain why your answer in Exercise 16 does *not* contradict the definition of a function (in the box on page 151).

---

* We are assuming here that we can read the graph with perfect accuracy. In actual practice your results will only be as accurate as your measuring ability. But the basic idea should be clear.

18. Is it possible to do Exercise 16 if all four numbers in the domain are required to be greater than 2000? Why or why not?

19. The amount of postage required to mail a first-class letter is determined by its weight. In this situation, is weight a function of postage? Or vice versa? Or both?

20. Could the following statement ever be the rule of a function?

Assign to a number x in the domain the number in the range whose square is x.

Why or why not? If there is a function with this rule, what is its domain and range?

Use Figure 4–4 for Exercises 21–27. Each of the graphs in the figure defines a function as in Example 6.

21. State the domain and range of the function defined by graph (a).

22. State the number in the range that the function of Exercise 21 assigns to the following numbers in the domain: $-2, -1, 0, 1$.

23. Do Exercise 22 for these numbers in the domain: $1/2, 5/2, -5/2$.

24. State the domain and range of the function defined by graph (b).

25. State the number in the range that the function of Exercise 24 assigns to the following numbers in the domain: $-2, 0, 1, 2.5, -1.5$.

26. State the domain and range of the function defined by graph (c).

27. State the number in the range that the function of Exercise 26 assigns to the following numbers in the domain: $-2, -1, 0, 1/2, 1$.

28. Explain why none of the graphs in Figure 4–5 defines a function according to the procedure in Example 6. What goes wrong?

Unusual Problems

29. Take a calculator equipped with keys labeled "COS" and "ln"—you don't have to know what these labels mean in order to do this exercise. Consider the function whose rule is "enter a number from the domain, then press the ln key, and then press the COS key." Experiment with this function, then answer the following questions. You may not be able to prove your answers—just make the best estimate you can based on the evidence from your experiments.
(a) What is the largest set of real numbers that could be the domain of this function?
(b) Using the domain in part (a), what is the range of the function?

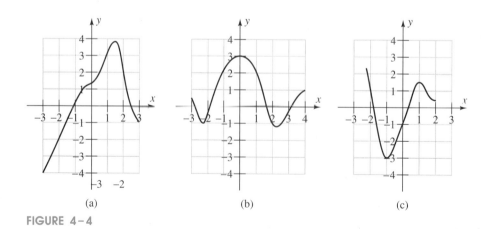

(a) (b) (c)

FIGURE 4–4

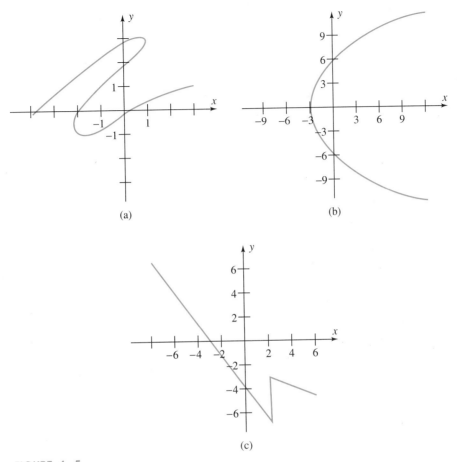

FIGURE 4-5

4.2 FUNCTIONAL NOTATION

Functional notation is a convenient shorthand language which facilitates the analysis of mathematical problems involving functions. Here are some examples from the last section.

Example 1 (Income Tax) One state's income tax law reads:

Income	Amount of Tax
$5000 or less	0
More than $5000	10% of income over $5000

Let I denote income and write $T(I)$ (read "T of I")to denote the amount of tax on income I. In this shorthand language $T(7500)$ denotes "the tax on an income of \$7500." The sentence "The tax on an income of \$7500 is \$250" is abbreviated as $T(7500) = 250$. There is nothing that forces us to use the letters T and I here:

Any choice of letters or symbols will do, provided we make clear what is meant by the letters or symbols chosen. ∎

Example 3 (Falling Rock) Let t denote the time (in seconds) and $d(t)$ "the distance (in feet) that the rock has traveled after t seconds." The fact that "the distance traveled after t seconds is $16t^2$ feet" can then be abbreviated as $d(t) = 16t^2$. For instance,

$$d(1) = 16 \cdot 1^2 = 16$$

means "the distance the rock has traveled after 1 second is 16 feet" and

$$d(4) = 16 \cdot 4^2 = 256$$

means "the distance the rock has traveled after 4 seconds is 256 feet." ∎

WARNING

The parentheses in $d(t)$ do *not* denote multiplication as in the algebraic equation $3(a + b) = 3a + 3b$. The entire symbol $d(t)$ is part of a *shorthand language*. In particular,

$$d(1 + 4) \text{ is } not \text{ equal to } d(1) + d(4).$$

For we saw above that $d(1) = 16$ and $d(4) = 256$, so that $d(1) + d(4) = 16 + 256 = 272$. But $d(1 + 4)$ is "the distance traveled after $1 + 4$ seconds," that is, the distance after 5 seconds, namely, $16 \cdot 5^2 = 400$. In general,

Functional notation is a convenient shorthand for phrases and sentences in the English language. It is *not* the same as ordinary algebraic notation.

Functional notation is easily adapted to the usual mathematical setting where the particulars of time, distance, etc., are eliminated. Suppose a function is given. Denote the function by f and let x denote a number in the domain. Then

$f(x)$ denotes the number in the range that is assigned to the number x by the rule of the function f.

For example, $f(6)$ denotes the number in the range assigned to the number 6 in the domain. The sentence "*y* is the number in the range assigned to the number *x* in the domain by the rule of the function *f*"

is then abbreviated as

$$y = f(x)$$

which is read "*y* equals *f* of *x*."

Each of the following sentences means *exactly the same thing* as $y = f(x)$:

The **value of the function *f*** at *x* is *y*.

The function *f* **maps** *x* to *y*.

y is the **image** of *x* under (the function) *f*.

Similarly, the number $f(x)$ is sometimes called

the **value** of (the function) *f* at *x*, or

the **image** of *x* (under the function *f*).

In actual practice, functions are seldom presented in the style of domain, rule, range, as they have been here. Usually the rule is given by an equation or formula, and the domain is determined by this convention:*

DOMAIN CONVENTION

> **Unless specific information to the contrary is given, the domain of a function *f* is taken to be the set consisting of every real number for which the rule of *f* produces a real number.**

Example 1 The expression $f(x) = \sqrt{x^2 + 1}$ defines the function *f* whose rule is: Assign to *x* the number $\sqrt{x^2 + 1}$. Since $x^2 \geq 0$ always, $\sqrt{x^2 + 1}$ is a real number for every *x*. So the domain of the function *f* consists of all real numbers.

To find $f(3)$, simply replace *x* by 3 in the formula:

$$f(3) = \sqrt{3^2 + 1} = \sqrt{10}$$

Similarly, replacing *x* by -5 and 0 shows that

$$f(-5) = \sqrt{(-5)^2 + 1} = \sqrt{26} \quad \text{and} \quad f(0) = \sqrt{0^2 + 1} = 1$$

For any real numbers, *a*, *b*, *c*, the quantities $f(a)$, $f(a + b)$, and $f(c^4)$ are computed in the same way: Replace *x* in the formula by the expression in parentheses:

$$f(a) = \sqrt{a^2 + 1}$$
$$f(a + b) = \sqrt{(a + b)^2 + 1} = \sqrt{a^2 + 2ab + b^2 + 1}$$
$$f(c^4) = \sqrt{(c^4)^2 + 1} = \sqrt{c^8 + 1}. \blacksquare$$

---

* For most purposes it isn't necessary to specify the range more precisely than saying that it is some set of real numbers.

Example 2 The expression $h(u) = \dfrac{u^2 + 5}{u - 1}$ defines the function h whose rule is: Assign to u the number $\dfrac{u^2 + 5}{u - 1}$. The domain of h consists of all real numbers except 1, because $\dfrac{u^2 + 5}{u - 1}$ is a real number whenever $u \neq 1$.

To find $h(\sqrt{3})$ and $h(-2)$, replace u by $\sqrt{3}$ and -2, respectively, in the rule of h:

$$h(\sqrt{3}) = \frac{(\sqrt{3})^2 + 5}{\sqrt{3} - 1} = \frac{8}{\sqrt{3} - 1} \quad \text{and} \quad h(-2) = \frac{(-2)^2 + 5}{-2 - 1} = -3.$$

Similarly, by substituting the expression in parentheses for u in the formula, we see that

$$h(-a) = \frac{(-a)^2 + 5}{-a - 1} = \frac{a^2 + 5}{-a - 1}$$

$$h(r^2 + 3) = \frac{(r^2 + 3)^2 + 5}{(r^2 + 3) - 1} = \frac{r^4 + 6r^2 + 9 + 5}{r^2 + 2} = \frac{r^4 + 6r^2 + 14}{r^2 + 2}$$

$$h(\sqrt{c + 2}) = \frac{(\sqrt{c + 2})^2 + 5}{\sqrt{c + 2} - 1} = \frac{c + 2 + 5}{\sqrt{c + 2} - 1} = \frac{c + 7}{\sqrt{c + 2} - 1}. \blacksquare$$

Example 3 For each number t, there is exactly *one* number y that satisfies the equation $y = t^3 + 6t^2 - 5$. The equation is understood to define the function whose rule is: assign to t the number $t^3 + 6t^2 - 5$. So we can also describe the rule of this function by $f(t) = t^3 + 6t^2 - 5$. Hence

$$f(0) = 0^3 + 6 \cdot 0^2 - 5 = -5 \quad \text{and} \quad f(2) = 2^3 + 6 \cdot 2^2 - 5 = 27. \blacksquare$$

> **WARNING**
>
> Some equations *don't* define functions. For instance, if $t = 3$ in $y^2 = 4t^2$, then $y^2 = 36$. There are *two* values of y (6 and -6) that satisfy $y^2 = 36$. Hence $y^2 = 4t^2$ is not the rule of a function.

Example 4 More than one formula may be involved in the rule of a function, as here:

$$f(x) = \begin{cases} 2x + 3 & \text{if } x < 4 \\ x^2 - 1 & \text{if } 4 \leq x \leq 10 \end{cases}$$

The rule of f gives no directions for numbers x with $x > 10$. So the domain of f is the interval $(-\infty, 10]$. If c is a real number, you cannot find $f(c)$ unless you know whether $c < 4$ or $4 \leq c \leq 10$.

To find $f(-5)$, note that $-5 < 4$, so the first part of the rule applies: $f(-5) = 2(-5) + 3 = -7$. To find $f(8)$, however, use the second part of the rule (because 8 is between 4 and 10): $f(8) = 8^2 - 1 = 63$. Also use the second part of the rule to find $f(4) = 4^2 - 1 = 15$. ∎

When functional notation is used in expressions such as $f(-x)$ or $f(x + h)$, the same basic rule applies: Replace x in the formula by the *entire* expression in parentheses.

Example 5 If $f(x) = x^2 + x - 2$, then

$$f(-x) = (-x)^2 + (-x) - 2 = x^2 - x - 2$$

Note that in this case $f(-x)$ is *not* the same as $-f(x)$, because $-f(x)$ is the negative of the number $f(x)$, that is,

$$-f(x) = -(x^2 + x - 2) = -x^2 - x + 2. \quad \blacksquare$$

Example 6 If $f(x) = x^2 + 1$ and $h \neq 0$, then

$$\frac{f(x + h) - f(x)}{h} = \frac{[(x + h)^2 + 1] - [x^2 + 1]}{h}$$

$$= \frac{[x^2 + 2xh + h^2 + 1] - x^2 - 1}{h}$$

$$= \frac{2xh + h^2}{h} = \frac{h(2x + h)}{h} = 2x + h. \quad \blacksquare$$

There are a number of common mistakes in using functional notation, most of which arise from treating it as ordinary algebra rather than a specialized shorthand language.

WARNING

COMMON MISTAKES WITH FUNCTIONAL NOTATION

Each of the following statements may be FALSE:
1. $f(a + b) = f(a) + f(b)$
2. $f(a - b) = f(a) - f(b)$
3. $f(ab) = f(a)f(b)$
4. $f(ab) = af(b)$
5. $f(ab) = f(a)b$

Example 7 Here are examples of three of the errors listed above.

1. If $f(x) = x^2$, then $f(3 + 2) = f(5) = 5^2 = 25$. But $f(3) + f(2) = 3^2 + 2^2 = 9 + 4 = 13$. So $f(3 + 2) \neq f(3) + f(2)$.

3. If $f(x) = x + 7$, then $f(3 \cdot 4) = f(12) = 12 + 7 = 19$. But $f(3)f(4) = (3 + 7)(4 + 7) = 10 \cdot 11 = 110$. So $f(3 \cdot 4) \neq f(3)f(4)$.

5. If $f(x) = x^2 + 1$, then $f(2 \cdot 3) = (2 \cdot 3)^2 + 1 = 36 + 1 = 37$. But $f(2) \cdot 3 = (2^2 + 1)3 = 5 \cdot 3 = 15$. So $f(2 \cdot 3) \neq f(2) \cdot 3$. ∎

Applications of functional notation usually require one quantity to be expressed as a function of another.

Example 8 A 20- by 20-inch sheet of cardboard is to be used to make an open-top box by cutting x- by x-inch squares from each corner and bending up the sides, as shown in Figure 4–6. Express the volume of the box as a function of the length x of the side of the cut-out squares.

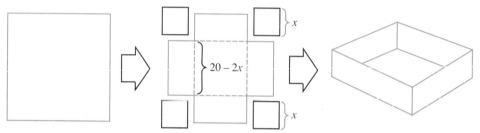

FIGURE 4–6

Solution Since the sheet is 20 by 20, the dotted line segment in Figure 4–6 has length $20 - 2x$ inches. When the sides are folded up, the box will measure $(20 - 2x)$ by $(20 - 2x)$ by x inches high. Hence the volume of the box is given by:

$$\text{length} \cdot \text{width} \cdot \text{height} = (20 - 2x)(20 - 2x)x = x(20 - 2x)^2$$

Therefore the rule of the volume function is $V(x) = x(20 - 2x)^2$. ∎

EXERCISES

Exercises 1–24 refer to these three functions:

$$f(x) = \sqrt{x + 3} - x + 1 \qquad g(t) = t^2 - 1$$

$$h(x) = x^2 + \frac{1}{x} + 2$$

In each case find the indicated value of the function.

1. $f(0)$
2. $f(1)$
3. $f(5/2)$
4. $f(\pi)$
5. $f(\sqrt{2})$
6. $f(\sqrt{2} - 1)$
7. $f(-2)$
8. $f(-3/2)$
9. $h(3)$
10. $h(-4)$
11. $h(3/2)$
12. $h(\pi + 1)$
13. $h(a + k)$
14. $h(-x)$
15. $h(2 - x)$
16. $h(x - 3)$
17. $g(3)$
18. $g(-2)$
19. $g(0)$
20. $g(x)$
21. $g(s + 1)$
22. $g(1 - r)$
23. $g(-t)$
24. $g(t + h)$

4.2 FUNCTIONAL NOTATION

In Exercises 25–32, compute:

(a) $f(r)$ (b) $f(r) - f(x)$ (c) $\dfrac{f(r) - f(x)}{r - x}$

In part (c) assume $r \neq x$ and simplify your answer. Example: If $f(x) = x^2$, then $\dfrac{f(r) - f(x)}{r - x} =$
$\dfrac{r^2 - x^2}{r - x} = \dfrac{(r + x)(r - x)}{r - x} = r + x.$

25. $f(x) = x$
26. $f(x) = -10x$
27. $f(x) = 3x + 7$
28. $f(x) = x^3$
29. $f(x) = x - x^2$
30. $f(x) = x^2 + 1$
31. $f(x) = \sqrt{x}$
32. $f(x) = 1/x$

Exercises 33–36 refer to the graphs in Figure 4-7, each of which defines a function with domain $[-3, 4]$, as in Example 6 on page 152.

33. In graph (a), find $f(-3)$, $f(-3/2)$, $f(0)$, $f(1)$, $f(5/2)$, $f(4)$. Careful approximate answers are acceptable.
34. In graph (a), find $f(0) - f(2)$; $f(5/2) - f(3)$; $f(4) + 3f(-2)$.
35. In graph (b), find $f(-5/2)$, $f(-3/2)$, $f(0)$, $f(3)$, $f(4)$.
36. In graph (c), find $f(-3) - f(3)$; $f(0) + f(1)$; $3f(2) + 2$.
37. In a certain state the sales tax $T(p)$ on an item of price p dollars is 5% of p. Which of the following formulas gives the correct sales tax in all cases?

(i) $T(p) = p + 5$ (ii) $T(p) = 1 + 5p$
(iii) $T(p) = p/20$
(iv) $T(p) = p + (5/100)p = p + .05p$
(v) $T(p) = (5/100)p = .05p$

38. Let T be the sales tax function of Exercise 37 and find $T(3.60)$, $T(4.80)$, $T(.60)$, and $T(0)$.

In Exercises 39–56, determine the domain of the function according to the usual conventions.

39. $f(x) = x^2$
40. $g(x) = (1/x^2) + 2$
41. $h(t) = |t| - 1$
42. $k(u) = \sqrt{u}$
43. $f(x) = [x]^2$ [Hint: See Example 5 in Section 4.1.]
44. $g(t) = |t - 1|$
45. $k(x) = |x| + \sqrt{x} - 1$
46. $h(x) = \sqrt{(x + 1)^2}$
47. $g(u) = \dfrac{|u|}{u}$
48. $h(x) = \dfrac{\sqrt{x - 1}}{x^2 - 1}$
49. $g(t) = \sqrt{t^2}$
50. $y = x^3 + 2$
51. $g(y) = [-y]$
52. $f(t) = \sqrt{-t}$
53. $g(u) = \dfrac{u^2 + 1}{u^2 - u - 6}$
54. $f(t) = \sqrt{4 - t^2}$
55. $y = -\sqrt{9 - (x - 9)^2}$
56. $y = \sqrt{-x} + \dfrac{2}{x + 1}$

57. Give an example of two different functions f and g that have all of the following properties:

$f(-1) = 1 = g(-1)$ and $f(0) = 0 = g(0)$
and $f(1) = 1 = g(1)$.

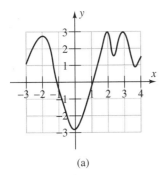

(a)

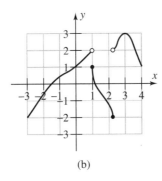

(b)

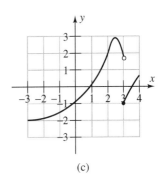

(c)

FIGURE 4-7

58. Give an example of a function h that has the property that $h(u) = h(2u)$ for every real number u.
59. Give an example of a function f that has the property that $f(x) = 2f(x)$ for every real number x.
60. Give an example of a function g with the property that $g(x) = g(-x)$ for every real number x.

In Exercises 61–68, compute the *difference quotient* $\dfrac{f(x+h) - f(x)}{h}$.

61. $f(x) = x + 5$
62. $f(x) = 7x + 2$
63. $f(x) = x^2 + 3$
64. $f(x) = x^2 + 2x$
65. $f(x) = x^2 + 3x - 1$
66. $f(x) = 2x^2 - 5x + 1$
67. $f(x) = 1/x$
68. $f(x) = x^3$

In Exercises 69–72, determine which of the following statements are true for *all* numbers in the domain of the given function. If a statement is not true, give a numerical example to demonstrate this fact.

(i) $f(x^2) = (f(x))^2$
(ii) $f(|x|) = |f(x)|$
(iii) $f(-x) = f(x)$
(iv) $f(3x) = 3f(x)$

69. $f(x) = x$
70. $f(x) = 4x$
71. $f(x) = |x|$
72. $f(x) = x^2$

73. Define a function that expresses the circumference of a circle as a function of the radius.
74. Define a function that expresses the area of a circle as a function of the diameter.
75. Define a function that expresses the area of a square as a function of the length of a side of the square.
76. Define a function that expresses the area of a square as a function of the length of a diagonal of the square.
77. Suppose a car travels at a constant rate of 55 mph for 2 hours and travels at 45 mph thereafter. Show that distance traveled is a function of time and find the rule of the function.
78. A man walks for 45 minutes at a rate of 3 mph, then jogs for 75 minutes at a rate of 5 mph, then sits and rests for 30 minutes, and finally walks for $1\frac{1}{2}$ hours. Find the rule of the function that expresses his distance traveled as a function of time. (*Warning:* Don't mix up the units of time; use either minutes or hours, not both.)
79. The distance between city C and city S is 2000 miles. A plane flying directly to S passes over C at noon. If the plane travels at 475 mph, express the distance of the plane from city S as a function of time.
80. Do Exercise 79 for a plane that travels at 325 mph.
81. The list price of a workbook is $12. But if 10 or more copies are purchased, then the price per copy is reduced by 25¢ for every copy above 10. (That is, $11.75 per copy for 11 copies, $11.50 per copy for 12 copies, and so on.)
   (a) The price per copy is a function of the number of copies purchased. Find the rule of this function.
   (b) The total cost of a quantity purchase is (number of copies) × (price per copy). Show that the total cost is a function of the number of copies and find the rule of the function.
82. A potato chip factory has a daily overhead from salaries and building costs of $1800. The cost of ingredients and packaging to produce a pound of potato chips is 50¢. A pound of potato chips sells for $1.20. Show that the factory's daily profit is a function of the number of pounds of potato chips sold and find the rule of this function. (Assume that the factory sells all the potato chips it produces each day.)
83. A rectangular region of 6000 sq ft is to be fenced in on three sides with fencing costing $3.75 per ft and on the fourth side with fencing costing $2.00 per ft. Express the cost of the fence as a function of the length x of the fourth side.
84. A box with a square base measuring $t \times t$ ft is to be made of three kinds of wood. The cost of the wood for the base is 85¢ per sq ft; the wood for the sides costs 50¢ per sq ft and the wood for the top $1.15 per sq ft. The volume of the box is to be 10 cu ft. Express the total cost of the box as a function of the length t.

4.3 GRAPHS OF FUNCTIONS

The graph of a function f is the graph of the *equation* $y = f(x)$. So the point (x, y) is on the graph of f precisely when x is a number in the domain of f and y is the number $f(x)$, the value of the function at x. In other words,

The graph of the function f consists of all points $(x, f(x))$, where x is any number in the domain of f.

Graphs of functions can be distinguished from other graphs by:

THE VERTICAL LINE TEST

The graph of a function has this property:

No vertical line intersects the graph more than once.

Conversely, any graph with this property is the graph of a function.

To see why this statement is true, note that a point with first coordinate $x = c$ on the graph of a function f must have second coordinate $f(c)$. Since a function cannot have two different values at c, there can't be two different points on the graph of f with first coordinate c. But distinct points on a vertical line necessarily have the same first coordinate and different second coordinates. So the graph of a function cannot contain more than one point from any vertical line. For the converse statement, see Exercise 40.

Sometimes the rule of a function will give you enough algebraic information to determine its graph precisely.

Example 1 To graph $f(x) = \sqrt{4 - x^2}$, we must graph the equation $y = \sqrt{4 - x^2}$. Note that y is always nonnegative. Squaring both sides shows that $y^2 = 4 - x^2$, so that $x^2 + y^2 = 4$; the graph of this equation is a circle with center at $(0, 0)$ and radius 2 (see page 127). So the graph of f will be the upper half of this circle (the points with nonnegative second coordinates), as shown in Figure 4–8. ■

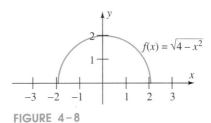

FIGURE 4–8

In most cases there isn't enough information to determine the graph completely. In such cases, the usual procedure is to plot a reasonable number of points and make an "educated guess" about the rest. The point plotting is essentially a mechanical task. In fact, there are now computer programs and hand-held calculators that will graph any function whose rule is given by an algebraic or trigonometric formula. Such calculators plot a huge number of points much more quickly and accurately than is possible by hand.

Even if you use a graphing calculator, however, there is more to graphing functions than simply plotting points. Consequently, the emphasis here and in later sections is on analyzing various algebraic facts to "get a feel" for the graph *before* plotting any points. Skill in doing this makes it much easier to graph functions by hand, and enables you to use a graphing calculator with maximum efficiency. More important, it gives you the deeper understanding of functions that is essential for understanding calculus and higher mathematics.

Example 2 To graph the **square root function** $g(x) = \sqrt{x}$, we note first that $g(x)$ is defined only when $x \geq 0$ and in this case $\sqrt{x} \geq 0$ also. So all the points $(x, \sqrt{x})$ on the graph lie in the first quadrant or on the positive axes. Furthermore, we know that $\sqrt{x}$ gets larger as x takes larger values. This means that as you move to the right on the graph (x takes larger values), the graph rises ($\sqrt{x}$ gets larger). Using a calculator, we see that

x	0	.25	1	2	3	4	6	9	10
$g(x) = \sqrt{x}$	0	.5	1	1.414	1.732	2	2.449	3	3.162

The corresponding points $(x, g(x))$ are plotted in Figure 4–9 below. The information about the rising graph shows that the graph must look like the one in Figure 4–10.* ■

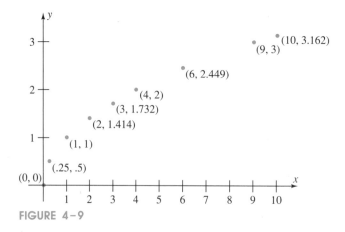

FIGURE 4–9

---

*Here and below, an arrow indicates that the graph continues on indefinitely in the direction indicated.

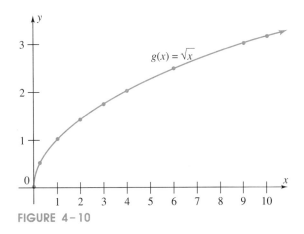
FIGURE 4-10

Example 3 The graph of $f(x) = x^2$ consists of all points (x, x^2). Since $x^2 \geq 0$ for all x, the graph lies on or above the x-axis. We evaluate the function at various numbers and plot the corresponding points (Figure 4-11 below). Since x^2 gets very large as x gets large in absolute value, we know that the graph rises sharply as you move to the right or the left. This fact and the plotted points suggest the graph in Figure 4-12 on the next page.

x	$f(x) = x^2$
-2.5	6.25
-2	4
-1.5	2.25
-1	1
$-.5$	.25
0	0
.5	.25
1	1
1.5	2.25
2	4
2.5	6.25

FIGURE 4-11

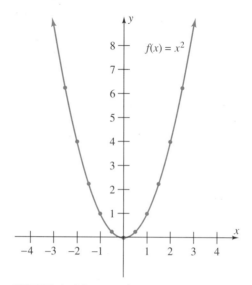

FIGURE 4-12

The graph of $f(x) = x^2$ is an example of a **parabola**. Its lowest point $(0, 0)$ is the **vertex** of the parabola. ∎

Example 4 To graph $f(x) = \dfrac{1}{x^2 + 1}$, we first compute the value of the function at several numbers:

x	-10	-7	-3	-1	$-\frac{1}{2}$	0	$\frac{1}{3}$	1	2	3	7	10
$f(x) = \dfrac{1}{x^2+1}$	$\dfrac{1}{101}$	$\dfrac{1}{50}$	$\dfrac{1}{10}$	$\dfrac{1}{2}$	$\dfrac{4}{5}$	1	$\dfrac{9}{10}$	$\dfrac{1}{2}$	$\dfrac{1}{5}$	$\dfrac{1}{10}$	$\dfrac{1}{50}$	$\dfrac{1}{101}$

As x gets larger and larger in absolute value, $f(x) = \dfrac{1}{x^2 + 1}$ is a positive number that gets closer and closer to zero. In geometric terms this means that as you move to the far left or far right ($|x|$ large), the graph lies above the x-axis [$f(x)$ positive] but gets closer and closer to the x-axis. Furthermore, $x^2 + 1 \geq 1$ for every number x, so that $f(x) = \dfrac{1}{x^2 + 1} \leq 1$.

Using these facts and the points obtained from the table above, we see that the entire graph looks like the one shown in Figure 4-13. ∎

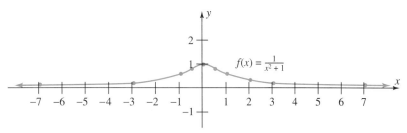

FIGURE 4-13

Example 5 It is easy to graph **linear functions** such as

$$f(x) = 3x - 4, \quad g(t) = -2t + 2, \quad h(x) = -6.$$

For instance, the graph of $f(x) = 3x - 4$ is the graph of the linear equation $y = 3x - 4$ and hence is a straight line (see Section 3.3). You need only plot two points to determine the entire graph. The graphs of g and h may be obtained in the same way, as shown in Figure 4-14. ∎

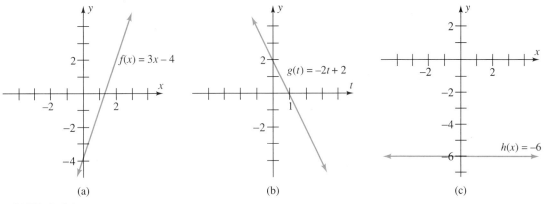

FIGURE 4-14

Example 6 The function $f(x) = |x|$ has a two-part rule:

$$f(x) = \begin{cases} x & \text{if } x \geq 0 \\ -x & \text{if } x < 0 \end{cases}$$

So the right half of the graph of $f(x) = |x|$ (that is, where $x \geq 0$) is the same as the right half of the graph of $g(x) = x$. The graph of $g(x) = x$ is the line plotted in Figure 4-15, page 168. The left half of the graph of $f(x) = |x|$ (that is, where $x < 0$) is the same as the left half of the graph of $h(x) = -x$. The graph of h is the straight line shown in Figure 4-16. Combining these two halves produces the graph of $f(x) = |x|$ in Figure 4-17. ∎

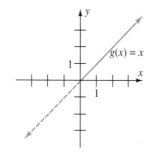

FIGURE 4-15

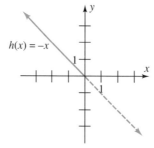

FIGURE 4-16

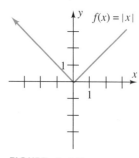

FIGURE 4-17

Example 7 To graph

$$h(x) = \begin{cases} x^2 & \text{if } x < 0 \\ x & \text{if } 0 \leq x < 5 \\ -2x + 11 & \text{if } x \geq 5 \end{cases}$$

we deal separately with each part of the rule of the function:

$x < 0$ For these values of x, the graph of h coincides with the graph of $f(x) = x^2$, which is sketched in Figure 4-12 above.

$0 \leq x < 5$ For these values, the graph of h coincides with the graph of $g(x) = x$, which is a straight line.

$x \geq 5$ For these values of x, the graph of h coincides with the graph of $k(x) = -2x + 11$, which is also a straight line.

Combining these facts and plotting a few points for *each* of the three parts produces the graph of h in Figure 4-18.

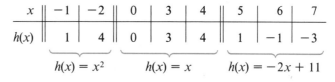

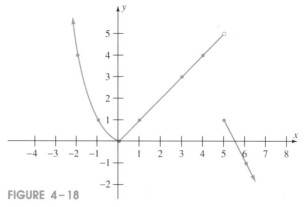

FIGURE 4-18

The open circle at (5, 5) indicates that (5, 5) is *not* on the graph of h (although it is on the graph of $g(x) = x$). The closed circle at (5, 1) indicates that (5, 1) *is* on the graph of h. ∎

Example 8 To graph a function whose rule involves the greatest integer function,* such as $g(x) = x - [x]$, it's best to consider what the graph looks like between each two consecutive integers. For instance,

If	Then [x] =	So That x − [x] =
$-2 \leq x < 1$	-2	$x - (-2) = x + 2$
$-1 \leq x < 0$	-1	$x - (-1) = x + 1$
$0 \leq x < 1$	0	x
$1 \leq x < 2$	1	$x - 1$
$2 \leq x < 3$	2	$x - 2$

Thus the rule g really consists of many parts. The graph of each part is a straight line segment. Plotting some points in each of the intervals $[-2, -1)$, $[-1, 0)$, etc., shows that for $-2 \leq x \leq 3$ the graph is as in Figure 4-19. ∎

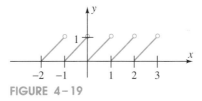

FIGURE 4-19

WARNING

Making an algebraic analysis of the function and plotting a few points does not always produce an accurate graph. Even when plotting points does suggest a pattern, you may not be able to answer questions like these:

(i) When the graph rises from one point to another, which way does it bend?

* See page 152.

(ii) Does the graph wiggle between two points?

(iii) When the graph appears to change from rising to falling between points P and Q, exactly where does it change?

(iv) Is the graph a continuous, unbroken curve between two points? Or are there gaps, jumps, holes, or isolated points?

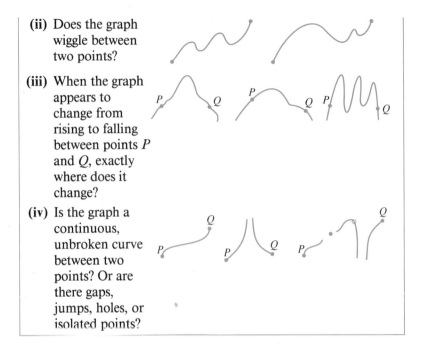

The answer to Question (iv) in the box above is especially important. Unless there was specific information to the contrary in the examples above, we *assumed* that the graph was a continuous, unbroken curve. You should do the same. This assumption is justified in calculus, which also provides rigorous answers to Questions (i)–(iii).

Although a properly used graphing calculator may answer most of the questions above, it has other shortcomings. Because of its limited resolution, for example, straight lines may look "wiggly," sharp corners as in the graph of the absolute value function (page 168) may appear rounded, and graphs involving the greatest integer function may include vertical line segments (contrary to the Vertical Line Test on page 163). If an inappropriate "viewing window" is chosen, the calculator may produce no graph at all.

The moral here is that you must know the fundamental algebraic properties of functions and how they affect the graph. These topics will be explored further in the following sections.

EXERCISES

In Exercises 1–24, sketch the graph of the given function.

1. $f(x) = x + 2$
2. $g(x) = x - 1$
3. $h(x) = 2x - 1$
4. $g(x) = \frac{3}{2}x + \frac{1}{2}$
5. $h(x) = 2 - x^2$
6. $f(x) = 3 - 2x^2$
7. $g(x) = x^3$
8. $f(x) = -x^3 + 1$
9. $h(x) = \sqrt{x - 2}$
10. $g(x) = \sqrt{x + 3}$
11. $f(x) = \sqrt{-x}$
12. $f(x) = \sqrt{3 - x}$

13. $h(x) = \sqrt{x} - 1$ 14. $g(x) = \sqrt{x} + 2$
15. $h(x) = \sqrt[3]{x}$ 16. $g(x) = \sqrt[3]{x + 4}$
17. $h(x) = |x| + 1$ 18. $g(x) = |x| - 4$
19. $f(x) = |x - 5|$ 20. $g(x) = |x + 3|$

21. $f(x) = \begin{cases} x^2 & \text{if } x \geq -1 \\ 2x + 3 & \text{if } x < -1 \end{cases}$

22. $g(x) = \begin{cases} |x| & \text{if } x < 1 \\ -3x + 4 & \text{if } x \geq 1 \end{cases}$

23. $k(u) = \begin{cases} -2u - 2 & \text{if } u < -3 \\ u - [u] & \text{if } -3 \leq u \leq 1 \\ 2u^2 & \text{if } u > 1 \end{cases}$

24. $f(x) = \begin{cases} x^2 & \text{if } x < -2 \\ x & \text{if } -2 \leq x < 4 \\ \sqrt{x} & \text{if } x \geq 4 \end{cases}$

25. The greatest integer function f, whose rule is $f(x) = [x]$, is called a **step function**. To see why, carefully sketch the graph of f. [*Hint:* What does the graph look like on the interval $[0, 1)$? on $[1, 2)$? on $[-1, 0)$? on $[-2, -1)$? and so forth.]

26. Graph $f(x) = -[x]$ for $-3 \leq x \leq 3$. [The hint for Exercise 25 also applies here.]

27. Graph $g(x) = [-x]$ for $-3 \leq x \leq 3$. [*Note:* This is *not* the same as $-[x]$.]

28. Graph $h(x) = [x] + [-x]$ for $-3 \leq x \leq 3$.

29. At this writing first-class postage rates are 29¢ for the first ounce or fraction thereof, plus 23¢ for each additional ounce or fraction thereof. Assume that each first-class letter carries one 29¢ stamp and as many 23¢ stamps as are necessary. Then the *number* of stamps required for a first-class letter is a function of the weight of the letter (in ounces). Call this function the postage stamp function.
    (a) Describe the rule of the postage stamp function algebraically.
    (b) Graph the postage stamp function.
    (c) Graph the function f whose rule is $f(x) = p(x) - [x]$, where p is the postage stamp function.

30. Sketch the graph of $f(x) = 1/x$. [*Hint:* The function is not defined when $x = 0$; what does the graph look like *near* $x = 0$? What does the graph look like when $|x|$ is very large?]

31. Sketch the graph of $f(x) = x^3 - x$. [*Hint:* Determine whether $f(x)$ is always positive or always negative in each of the intervals $(-\infty, -1)$, $(-1, 0)$, $(0, 1)$, $(1, \infty)$. What happens at $x = -1$, $x = 0$, $x = 1$?]

32. When you have done Exercises 7 and 15, answer this question: How is the graph of $g(x) = x^3$ related geometrically to the graph of $h(x) = \sqrt[3]{x}$?

In Exercises 33–39, graph the given function as best you can. Some of these graphs may be harder than the examples or other exercises.

33. $f(x) = \sqrt{1 - x^2}$ 34. $g(t) = -\sqrt{16 - t^2}$

35. $f(x) = 2x^2 - 3x + 2$

36. $g(x) = -3x^2 + 4x - 7$

37. $h(t) = \dfrac{t}{1 - t}$

38. $k(x) = x^3 - 3x + 1$

39. $h(x) = \dfrac{x}{x^2 + 1}$

Unusual Problem

40. Show that a graph which passes the Vertical Line Test is necessarily the graph of a function. [*Hint:* See Example 6 in Section 4.1.]

4.3.A EXCURSION: Graph Reading

Graph reading deals with translating graphical information into equivalent statements in English or functional notation.

Example 1 A recording device in the weather bureau in City F produces the graph of the function that relates time of day to temperature, as shown in Figure

4–20. We shall adopt this functional notation: tem(t) denotes the temperature at time t (measured in hours after midnight). For example, since (4, 39) is on the graph* we know that tem(4) = 39, that is, the temperature at 4 A.M. was 39°.

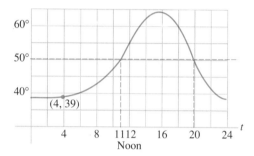

FIGURE 4–20

At what times during the day was the temperature below 50°? To answer this question we translate it into functional and graphical terms. To say that the temperature is below 50° at time t means that the second coordinate of the point (t, tem(t)) is less than 50. The points with this property are the ones that lie *below* the horizontal line through 50, as shown in Figure 4–20. The first coordinates of all such points are the *times* when the temperature was below 50. The graph shows that

$$\text{tem}(t) < 50 \quad \text{whenever} \quad 0 \leq t < 11 \quad \text{or} \quad 20 < t \leq 24.$$

In other words, the temperature was below 50° from midnight to 11 A.M. and again from 8 P.M. ($t = 20$) to midnight. ■

Example 2 A slightly more complicated problem is to determine the time period *before* 4 P.M. during which the temperature was at least 60°. Remembering that 4 P.M. is 16 hours after midnight we make these translations:

Statement	Functional Notation	Graph
The time is before 4 P.M.	$t < 16$	(t, tem(t)) lies to the left of the vertical line through $t = 16$.
The temperature is at least 60°.	tem(t) $\geq$ 60	(t, tem(t)) lies on or above the horizontal line through 60.

* Here and below our results are only as accurate as our measuring ability. But the basic idea should be clear.

Figure 4–21 below shows that the points $(t, \text{tem}(t))$ with $t < 16$ and $\text{tem}(t) \geq 60$ are those with first coordinates between 12.8 and 16. So the temperature was at least $60°$ from 12:48 P.M. ($t = 12.8$) to 4 P.M. ($t = 16$). ■

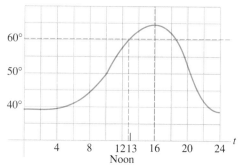

FIGURE 4–21

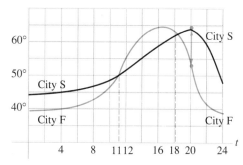

FIGURE 4–22

Example 3 Suppose the temperature graph for City S is superimposed on the one for City F, as in Figure 4–22 above. Denote the temperature at time t in City S by $\text{tem}_S(t)$ and answer these questions:

(i) At what times was it warmer in City F than in City S?

(ii) Was there any time when it was at least $10°$ warmer in City S than in City F?

Once again, it's a matter of three-way translation:

Statement	Functional Notation	Graph
City F is warmer than City S at time t.	$\text{tem}(t) > \text{tem}_S(t)$	The point $(t, \text{tem}(t))$ lies directly above $(t, \text{tem}_S(t))$.
City S is at least $10°$ warmer than City F at time t.	$\text{tem}_S(t) \geq \text{tem}(t) + 10$	The point $(t, \text{tem}_S(t))$ is at least 10 units above $(t, \text{tem}(t))$.

Careful measurement in Figure 4–22 shows that

(i) $\text{tem}(t) > \text{tem}_S(t)$ for all t in the interval $(11, 18)$.

(ii) $\text{tem}_S(t) \geq \text{tem}(t) + 10$ for many values of t, including $t = 20$.

In other words,

(i) It was warmer in City F between 11 A.M. and 6 P.M.

(ii) It was at least $10°$ warmer in City S at 8 P.M. ($t = 20$) and at other times. ■

174 4 FUNCTIONS AND GRAPHS

Example 4 The graphs of the functions g and h in Figure 4–23 below can be used to solve these problems:

(i) Find all numbers in the interval $[-3, 3]$ such that $g(x) = 2$.

(ii) Find the largest interval over which the function g is increasing, *and* the function h is decreasing,* *and* $h(x) \geq g(x)$ for every number x in the interval.

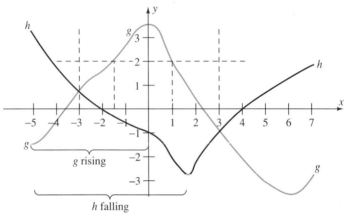

FIGURE 4–23

(i) The graph of g consists of the points $(x, g(x))$. The points on the graph that lie *between* the vertical lines through -3 and 3 and *on* the horizontal line through 2 have first coordinate x in the interval $[-3, 3]$ and second coordinate $g(x) = 2$. Figure 4–23 shows that the only such points are $(-1.5, 2)$ and $(1, 2)$. So the answer to Problem (i) is $x = -1.5$ and $x = 1$.

(ii) Figure 4–23 shows that $[-5, 0]$ is the only interval over which the graph of h is falling *and* the graph of g is rising. Clearly, $h(x) \geq g(x)$ exactly when the point $(x, h(x))$ lies above the point $(x, g(x))$. The only time this occurs in the interval $[-5, 0]$ is when $-5 \leq x \leq -3$. Therefore the answer to Problem (ii) is the interval $[-5, -3]$. ■

EXERCISES

Exercises 1–12 deal with the function g whose entire graph is shown in Figure 4–24.

1. If $t = 1.5$, then $g(2t) = ?$

2. If $t = 1.5$, then $2g(t) = ?$

3. If $y = 2$, then $g(y + 1.5) = ?$

4. If $y = 2$, then $g(y) + g(1.5) = ?$

---

*A function is increasing on an interval if its graph rises from left to right over the interval; it is decreasing if its graph falls from left to right over the interval. See Excursion 4.4.A for more information.

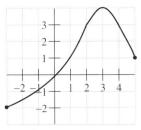
FIGURE 4-24

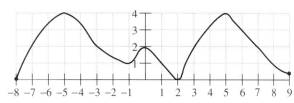
FIGURE 4-25

5. If $y = 2$, then $g(y) + 1.5 = $?
6. $g(0) = $?
7. If $v = 1.5$, then $g(3v - 1.5) = $?
8. If $s = 2$, then $g(-s) = $?
9. For what values of z is $g(z) = 1$?
10. For what values of z is $g(z) = -1$?
11. What is the largest interval over which the graph is rising from left to right?
12. At what number t in the interval $[-1, 2]$ is $g(t)$ largest?
13. Draw the graph of a function f, that satisfies the following four conditions:

   (i) domain $f = [-2, 4]$;
   (ii) range $f = [-5, 6]$;
   (iii) $f(-1) = f(2)$;
   (iv) $f\left(\dfrac{1}{2}\right) = 0$.

14. Draw the graph of a function different from the one in Exercise 13 that also satisfies all the conditions of Exercise 13.

Exercises 15-24 deal with the function f whose entire graph is shown in Figure 4-25.

15. What is the domain of f?
16. What is the range of f?
17. Find all numbers x such that $f(x) = 2$.
18. Find all numbers x such that $f(x) > 2$.
19. Find at least three numbers x such that $f(x) = f(-x)$.
20. Find all numbers x such that $f(x) = f(7)$.
21. Find a number x such that $f(x + 1) = 0$.
22. Find two numbers x such that $f(x - 2) = 4$.
23. Find a number x such that $f(x + 1) = f(x - 2)$.
24. Find a number x such that $f(x) + 1 = f(x - 4)$.

Exercises 25-32 deal with the two functions f and g whose entire graphs are shown in Figure 4-26.

25. What is the domain of f? The domain of g?
26. What is the range of f? The range of g?
27. Find all numbers x in the interval $[-3, 1]$ such that $f(x) = 2$.
28. Find all numbers x in the interval $[-3, 3]$ such that $g(x) \geq 2$.

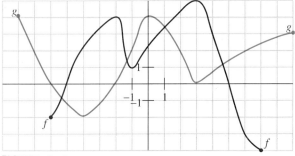
FIGURE 4-26

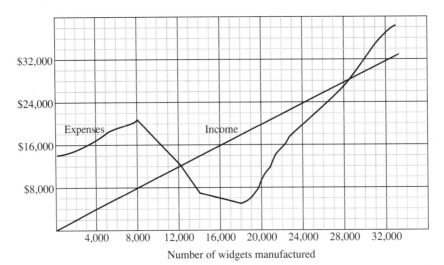

FIGURE 4-27

29. Find the number x for which $f(x) - g(x)$ is largest.
30. For how many values of x is it true that $f(x) = g(x)$?
31. Find all intervals over which both functions are defined, the graph of f is falling, *and* the graph of g is rising (from left to right).
32. Find all intervals over which the graph of $g(x)$ is falling from left to right *and* $0 \leq f(x) \leq 2$.

Exercises 33–38 deal with this situation: The owners of the Rieben & Tabares Deluxe Widget Works have determined that both their weekly manufacturing expenses and their weekly sales income are functions of the number of widgets manufactured each week. Figure 4–27 shows the graphs of these two functions.

33. Use careful measurement on the graph and the fact that profit = income − expenses to determine the weekly profit if 5000 widgets are manufactured.
34. Do the same if 10,000, 14,000, 18,000, or 22,000 widgets are manufactured.
35. What is the smallest number of widgets that can be manufactured each week without losing money?
36. What is the largest number of widgets that can be manufactured without losing money?
37. The owners build a new lounge and swimming pool for their employees. This raises their expenses by approximately $5000 per week. Draw the graph of the new "expense function."
38. Owing to competitive pressure, widget prices cannot be increased and the income function remains the same. Answer Exercises 33–36 with the expense function of Exercise 37 in place of the old one.

Exercises 39–47 deal with the weather bureau graph of the temperature as a function of time (measured in hours after midnight), as shown in Figure 4–28.

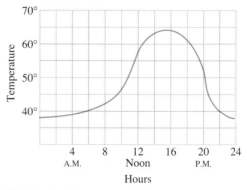

FIGURE 4-28

We write tem(t) for the temperature at t hours after midnight.

39. Find tem(10). Find tem(3 + 12).

40. Is tem(6) bigger than, equal to, or less than tem(18)?

41. At which time is the temperature 50°?

42. Find a 4-hour period for which tem(h) > 40 for all h in this 4-hour period.

43. Find the difference in temperature at 10 and 16 hours. Identify this difference in the graph; that is, express difference in terms of points and their location on the graph.

44. Is it true that tem(6) = tem(8)? Explain in terms of the graph.

45. Is it true that tem(6·2) = 6·tem(2)?

46. The temperature graph above was recorded in city F. City B is 500 miles to the south of city F, and its temperature is 7° higher all day long. Sketch into the above drawing the temperature for city B during the same day.

47. Find an hour h at which the temperature in city B is the same as tem(12) in city F. (See Exercise 46.)

4.4 GRAPHING TECHNIQUES*

A circle can be graphed quickly and accurately because its shape can be determined *algebraically* from its equation, without plotting any points. The same idea is the theme of this section: Use algebra to discover the essential geometric properties of a graph, *before* plotting any points. When these are known, it is often possible to draw the graph precisely, with a minimum of point plotting.

Symmetry

Informally, a graph is **symmetric with respect to a line** if the part of the graph on one side of the line is the mirror image of the part on the other side, with the line being the mirror. More precisely, the graph is symmetric with respect to the line L if for every point P on the graph (and not on L), there is a point Q on the graph such that L is the perpendicular bisector of line segment PQ. P and Q are mirror image points. The most important types of line symmetry are:

symmetry with respect to the y-axis

The part of the graph on the right side of the y-axis is the mirror image of the part on the left side of the y-axis.

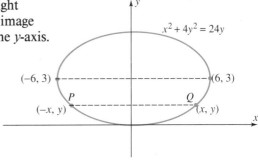

FIGURE 4–29

---

* This section may be omitted or postponed if desired; it is not needed for Section 4.5 or 4.6. Except for the comprehensive examples at the end of the section, the three subsections are essentially independent of each other and may be covered in any order.

symmetry with respect to the *x*-axis

The part of the graph above the *x*-axis is the mirror image of the part below the *x*-axis.

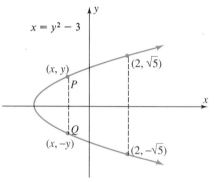

FIGURE 4-30

A graph is **symmetric with respect to a point** *K* if each point *P* on the graph has an "image point" *Q* on the graph such that *K* is the midpoint of the line segment from *P* to *Q*. Here's an example:

symmetry with respect to the origin

A straight line through any point *P* on the graph and the origin also intersects the graph at a point *Q* such that the origin is the midpoint of segment *PQ*.

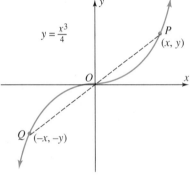

FIGURE 4-31

Knowing that a graph has one of these symmetry properties cuts in half the number of points that must be plotted. For instance, if the graph is known to be symmetric with respect to the *y*-axis, then you only have to plot points on the right side of the *y*-axis; symmetry can be used to get the rest of the graph.

By examining the graphs in the figures above we can translate the definition of symmetry into algebraic terms (coordinates and equations). For example, suppose the graph of an equation is symmetric with respect to the *y*-axis. Figure 4-29 shows that *P* and *Q* are mirror image points precisely when their second coordinates are the same and their first coordinates negatives of each other, that is:

Whenever (x, y) is on the graph, then $(-x, y)$ is also on it.

4.4 GRAPHING TECHNIQUES

So replacing x by $-x$ in the equation leads to the same number y. In other words, replacing x by $-x$ produces an equivalent equation. A similar analysis in the other cases (Exercise 78) leads to these symmetry tests:

SYMMETRY TESTS

Symmetry with Respect to	Coordinate Test for Symmetry	Algebraic Test for Symmetry
y-axis	(x, y) on graph implies $(-x, y)$ on graph.	Replacing x by $-x$ produces an equivalent equation.
x-axis	(x, y) on graph implies $(x, -y)$ on graph.	Replacing y by $-y$ produces an equivalent equation.
origin	(x, y) on graph implies $(-x, -y)$ on graph.	Replacing x by $-x$ and y by $-y$ produces an equivalent equation.

Example 1

(a) Replacing x by $-x$ in the equation $x^2 + 4y^2 = 24y$ produces the same equation because $(-x)^2 = x^2$. Therefore the graph of this equation is symmetric with respect to the y-axis, as shown in Figure 4–29 above.

(b) Replacing y by $-y$ in the equation $x = y^2 - 3$ produces the same equation, so the graph is symmetric with respect to the x-axis, as shown in Figure 4–30.

(c) Replacing x by $-x$ and y by $-y$ in the equation $y = x^3/4$ yields:

$$-y = \frac{(-x)^3}{4}, \quad \text{that is,} \quad -y = -\frac{x^3}{4}$$

This equation is equivalent to $y = x^3/4$ since it can be obtained from it by multiplying by -1. Therefore the graph of $y = x^3/4$ is symmetric with respect to the origin, as shown in Figure 4–31. ∎

For *functions*, the algebraic description of symmetry takes a different form. A function f whose graph is symmetric with respect to the y-axis is called an **even function**. To say that the graph of $y = f(x)$ is symmetric with respect to the y-axis means that replacing x by $-x$ produces the same y-value. In other words, the function takes the same value at both x and $-x$. Therefore

EVEN FUNCTIONS

A function f is even provided that

$f(x) = f(-x)$ for every number x in the domain of f.

For example, $f(x) = x^4 + x^2$ is even because
$$f(-x) = (-x)^4 + (-x)^2 = x^4 + x^2 = f(x).$$

Except for zero functions ($f(x) = 0$ for every x in the domain), *the graph of a function is never symmetric with respect to the x-axis.* The reason is the Vertical Line Test: The graph of a function never contains two points with the same first coordinate. If both (5, 3) and (5, −3), for instance, were on the graph, this would say that $f(5) = 3$ and $f(5) = -3$, which is impossible when f is a function.

A function whose graph is symmetric with respect to the origin is called an **odd function.** If both (x, y) and $(-x, -y)$ are on the graph of such a function f, then we must have both

$$y = f(x) \quad \text{and} \quad -y = f(-x)$$

so that $f(-x) = -y = -f(x)$. Therefore

ODD FUNCTIONS

> **A function f is odd provided that**
>
> $f(-x) = -f(x)$ **for every number x in the domain of f.**

For example, $f(x) = x^3$ is an odd function because
$$f(-x) = (-x)^3 = -x^3 = -f(x).$$

Vertical and Horizontal Shifts

If the rule of a function is changed algebraically to produce a new function, how can the graph of the new function be obtained from the graph of the original one?

Consider, for example, the functions $f(x) = x^2$ and $g(x) = x^2 + 2$. The rule of g is obtained from the rule of f by adding the constant 2. What does this say about the graphs? If (x, x^2) is on the graph of f, then the point on the graph of g with first coordinate x is $(x, x^2 + 2)$. Since these two points have the same first coordinate, they are on the same vertical line. Since the second coordinates differ by 2, $(x, x^2 + 2)$ lies 2 units directly above (x, x^2). Since this is true for every x, the graph of g is just the graph of f *shifted 2 units upward,* as shown in Figure 4–32 on the opposite page.

A similar argument shows that the graph of $h(x) = x^2 - 3$ is just the graph of $f(x) = x^2$ *shifted 3 units downward,* as shown in Figure 4–33 on the next page. This argument can be used with any function f and any constant c:

VERTICAL SHIFTS

> **If $c > 0$, then the graph of $g(x) = f(x) + c$ is the graph of f shifted upward c units.**
>
> **If $c > 0$, then the graph of $h(x) = f(x) - c$ is the graph of f shifted downward c units.**

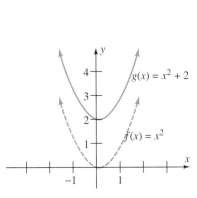

FIGURE 4-32

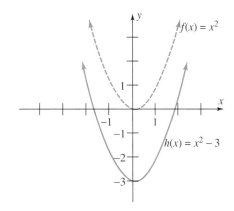

FIGURE 4-33

Starting with the function $f(x) = x^2$, we can form a new function g whose rule is $g(x) = f(x + 5) = (x + 5)^2$. If (x, y) is on the graph of f, then $y = f(x)$. Note that

$$g(x - 5) = f((x - 5) + 5) = f(x) = y$$

so $(x - 5, y)$ is on the graph of g. Since they have the same second coordinate, (x, y) and $(x - 5, y)$ lie on the same horizontal line. Since their first coordinates differ by 5, $(x - 5, y)$ on the graph of g lies 5 units to the left of (x, y) on the graph of f. Since this is true for every x, the graph of g is just the graph of f shifted 5 units to the *left,* as shown in Figure 4-34.

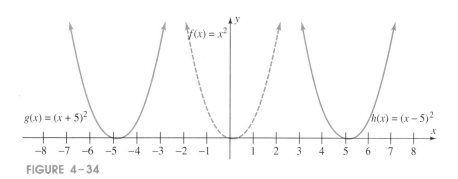

FIGURE 4-34

An analogous argument shows that the graph of $h(x) = f(x - 5) = (x - 5)^2$ (note the minus sign) can be obtained by shifting the graph of $f(x) = x^2$ five units to the *right,* as shown in Figure 4-34. In the general case we have:

HORIZONTAL SHIFTS

> Let f be a function and c a positive constant.
>
> The graph of $g(x) = f(x + c)$ is the graph of f shifted horizontally c units to the left.
>
> The graph of $h(x) = f(x - c)$ is the graph of f shifted horizontally c units to the right.

Expansions, Contractions, and Reflections

How is the graph of $f(x) = x^2 - 3$ related to the graph of $g(x) = 2(x^2 - 3)$? For each x, the point $(x, g(x))$ is on the graph of g. But $(x, g(x))$ is the point $(x, 2f(x))$. So it lies on the same vertical line as $(x, f(x))$, on the same side of the x-axis, but *twice as far* from the x-axis. Consequently, the graph of $g(x) = 2(x^2 - 3) = 2f(x)$ is the graph of $f(x)$ *stretched vertically away from the x-axis* by a factor of 2, as shown in Figure 4–35.

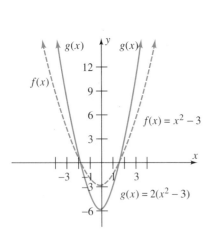

FIGURE 4–35

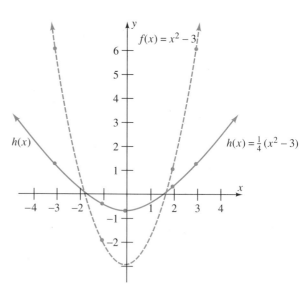

FIGURE 4–36

A similar analysis shows that the graph of $h(x) = \frac{1}{4}(x^2 - 3)$ (that is, $h(x) = \frac{1}{4}f(x)$), is the graph of f *shrunk vertically toward the x-axis* by a factor of $\frac{1}{4}$, as shown in Figure 4–36 above. Analogous arguments work for any function f and any positive constant c:

EXPANSIONS AND CONTRACTIONS

> If $c > 1$, then the graph of $g(x) = cf(x)$ is the graph of f stretched vertically away from the x-axis by a factor of c.
>
> If $0 < c < 1$, then the graph of $h(x) = cf(x)$ is the graph of f shrunk vertically toward the x-axis by a factor of c.

If we multiply the rule of a function f by -1 we obtain the new function $g(x) = -f(x)$. For every point $(x, f(x))$ on the graph of f, the point $(x, -f(x))$ is on the graph of g. Since their first coordinates are equal, these two points are on the same vertical line. Since their second coordinates are negatives of each other, the points lie on *opposite sides* of the x-axis, the *same distance* from the axis. So each point on the graph of f has a mirror image point on the graph of g, with the x-axis being the mirror.

Example 2 If $f(x) = x^2 - 3$, then the graph of $g(x) = -f(x) = -(x^2 - 3)$ is a reflection of the graph of f in the x-axis, as shown in Figure 4-37. ■

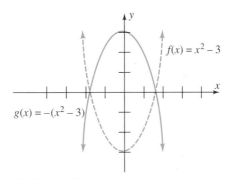

FIGURE 4-37

Similar arguments can be used with any function and lead to this conclusion:

REFLECTIONS

> The graph of $g(x) = -f(x)$ is the graph of f reflected in the x-axis.

Here are some comprehensive examples, employing several of the techniques developed above.

Example 3 To graph $g(x) = 2(x - 3)^2 - 1$, note that the rule of g may be obtained from the rule of $f(x) = x^2$ in three steps:

$$f(x) = x^2 \xrightarrow{\text{Step 1}} (x - 3)^2 \xrightarrow{\text{Step 2}} 2(x - 3)^2 \xrightarrow{\text{Step 3}} 2(x - 3)^2 - 1 = g(x)$$

Step 1 shifts the graph of f horizontally 3 units to the right; step 2 stretches the resulting graph away from the x-axis by a factor of 2; step 3 shifts this graph 1 unit downward, thus producing the graph of g in Figure 4–38. ∎

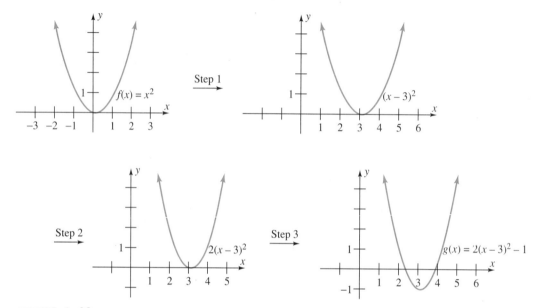

FIGURE 4–38

Example 4 The rule of the function $h(x) = -\frac{1}{2}(x^2 - 3)$ can be obtained in two steps from the rule of $f(x) = x^2 - 3$:

$$f(x) = x^2 - 3 \xrightarrow{\text{Step 1}} \frac{1}{2}(x^2 - 3) \xrightarrow{\text{Step 2}} -\frac{1}{2}(x^2 - 3) = h(x)$$

Step 1 shrinks the graph of f toward the x-axis by a factor of $\frac{1}{2}$, and step 2 reflects the resulting graph in the x-axis, thus producing the graph of h in Figure 4–39. ∎

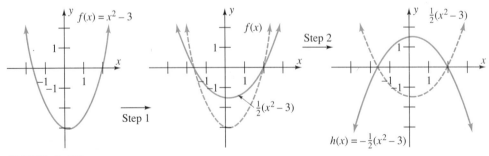

FIGURE 4–39

EXERCISES

In Exercises 1–10, determine whether the given function is even, odd, or neither (that is, whether its graph is symmetric with respect to the y-axis, the origin, or neither).

1. $f(x) = 4x$
2. $k(t) = -5t$
3. $f(x) = x^2 - |x|$
4. $h(u) = |3u|$
5. $k(t) = t^4 - 6t^2 + 5$
6. $f(x) = x(x^4 - x^2) + 4$
7. $f(t) = \sqrt{t^2 - 5}$
8. $h(x) = \sqrt{7 - 2x^2}$
9. $f(x) = \dfrac{x^2 + 2}{x - 7}$
10. $g(x) = \dfrac{x^2 + 1}{x^2 - 1}$

In Exercises 11–14, determine whether or not the graph of the given equation is symmetric with respect to the x-axis.

11. $x^2 - 6x + y^2 + 8 = 0$
12. $x^2 + 8x + y^2 = -15$
13. $x^2 - 2x + y^2 + 2y = 2$
14. $x^2 - x + y^2 - y = 0$

In Exercises 15–20, determine whether the given graph is symmetric with respect to the y-axis, the x-axis, or the origin.

15.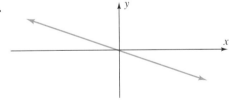

16.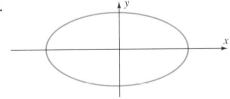

17.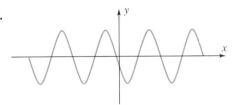

18.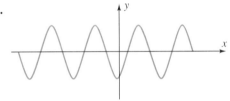

19.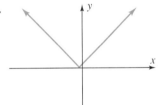

20.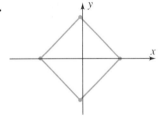

In Exercises 21–24, complete the graph of the given function, assuming that it satisfies the given symmetry condition.

21. even

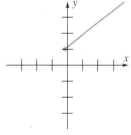

22. even

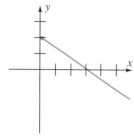

23. odd

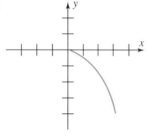

24. odd

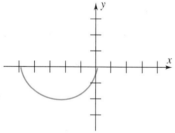

In Exercises 25–28, sketch half the graph of the given equation. Then use symmetry to obtain the rest of the graph.

25. $y = \sqrt{x^2 - 2}$ 26. $y = -2x^3$

27. $|y| + 3 = x$ 28. $y = x^2 - |x|$

In Exercises 29–36, use the graph of $f(x) = x^2$ to graph the given function quickly.

29. $g(x) = x^2 + 1$ 30. $h(x) = x^2 - \dfrac{3}{2}$

31. $g(x) = 4x^2$ 32. $h(x) = \dfrac{1}{2} x^2$

33. $k(x) = -2x^2$ 34. $t(x) = -\dfrac{1}{3} x^2$

35. $g(x) = (x + 2)^2$ 36. $h(x) = (x - 4)^2$

In Exercises 37–50, use the graphs of $g(x) = \sqrt{x}$ and $f(x) = |x|$ (given in Section 4.3) to sketch the graph of the given function.

37. $h(x) = |x| - 5$ 38. $k(x) = |x| + 2$

39. $k(x) = 3|x|$ 40. $r(u) = \dfrac{1}{2} |u|$

41. $h(u) = -4|u|$ 42. $g(x) = -\dfrac{1}{3} |u|$

43. $s(x) = |x - 5|$ 44. $t(x) = |x + 5|$

45. $k(x) = \sqrt{x} + 3$ 46. $r(x) = 5\sqrt{x}$

47. $s(x) = -\sqrt{x}$ 48. $h(x) = -\dfrac{1}{2} \sqrt{x}$

49. $h(x) = \sqrt{x - 2}$ 50. $t(x) = \sqrt{x + 3}$

In Exercises 51–58, use the graph of the function f in Figure 4–40 to sketch the graph of the given function.

51. $g(x) = f(x) + 3$ 52. $h(x) = f(x) - 1$

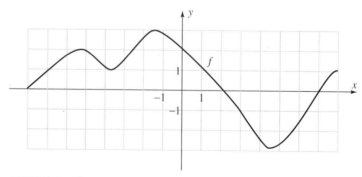

FIGURE 4–40

53. $g(x) = 5f(x)$ 54. $h(x) = \frac{1}{2}f(x)$

55. $k(t) = -f(t)$ 56. $g(u) = -4f(u)$

57. $h(x) = f(x+3)$ 58. $k(x) = f\left(x - \frac{5}{2}\right)$

In many of the following exercises you may need to use several techniques in sequence, as in Examples 3 and 4. In Exercises 59–68, use the graph of $f(x) = x^2$ to sketch the graph of the given function.

59. $g(x) = 3x^2 + 1$ 60. $h(x) = -2x^2 + 4$

61. $g(x) = -\frac{1}{2}x^2 + \frac{5}{2}$ 62. $h(x) = -2x^2 - 3$

63. $g(x) = 2(x+2)^2$

64. $t(x) = -2(x-2)^2 - 2$

65. $g(x) = (x-1)^2 + 3$

66. $h(x) = -(x+2)^2 + 4$

67. $k(x) = 3(x-1)^2 + 2$

68. $t(x) = -3(x+4)^2 - 1$

In Exercises 69–72, use the graphs of $g(x) = \sqrt{x}$ and $f(x) = |x|$ (given in Section 4.3) to graph the given function.

69. $h(x) = 3|x| + 5$ 70. $k(x) = -2|x| - 2$

71. $k(u) = -\sqrt{u+1} + 2$

72. $h(x) = 4\sqrt{x-2} + \frac{3}{2}$

73. (a) Draw some coordinate axes and plot the points $(0, 1)$, $(1, -3)$, $(-5, 2)$, $(-3, 5)$, $(2, 3)$, and $(4, 1)$.

(b) Suppose the points in part (a) lie on the graph of an *even* function f. Plot the points $(0, f(0))$, $(-1, f(-1))$, $(5, f(5))$, $(3, f(3))$, $(-2, f(-2))$, and $(-4, f(-4))$.

74. Draw the graph of an *even* function that includes the points $(0, -3)$, $(-3, 0)$, $(2, 0)$, $(1, -4)$, $(2.5, -1)$, $(-4, 3)$, and $(-5, 3)$.*

75. (a) Plot the points $(0, 0)$; $(2, 3)$; $(3, 4)$; $(5, 0)$; $(7, -3)$; $(-1, -1)$; $(-4, -1)$; $(-6, 1)$.

(b) Suppose the points in part (a) lie on the graph of an *odd* function f. Plot the points $(-2, f(-2))$; $(-3, f(-3))$; $(-5, f(-5))$; $(-7, f(-7))$; $(1, f(1))$; $(4, f(4))$; $(6, f(6))$.

(c) Draw the graph of an odd function f that includes all the points plotted in parts (a) and (b).*

76. Draw the graph of an odd function that includes the points $(-3, 5)$, $(-1, 1)$, $(2, -6)$, $(4, -9)$, and $(5, -5)$.*

77. Show that any graph that has two of the three types of symmetry (x-axis, y-axis, origin) necessarily has the third type also.

78. (a) Use the midpoint formula to show that $(0, 0)$ is the midpoint of the segment joining (x, y) and $(-x, -y)$. Conclude that the coordinate test for symmetry with respect to the origin (page 179) is correct.

(b) Explain why the coordinate test for symmetry with respect to the x-axis (page 179) is correct.

79. Let f be a function whose graph is symmetric with respect to the x-axis. Explain why the rule of f must be $f(x) = 0$ for every x in the domain. [*Hint:* What can be said, for example, if both $(5, 3)$ and $(5, -3)$ are on the graph of f?]

4.4.A EXCURSION: Increasing and Decreasing Functions

A function is said to be **increasing on an interval** if its graph always *rises* as you move from left to right in the interval. A function whose graph rises from left to right throughout its domain is called an **increasing function**. The first two

---

* There are many correct answers.

functions pictured in Figure 4-41 are increasing on the stated intervals. The third is an increasing function.

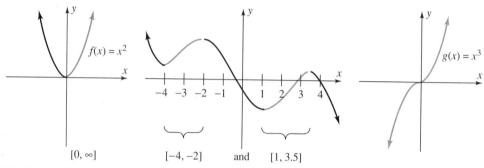

FIGURE 4-41

A function is said to be **decreasing on an interval** if its graph always *falls* as you move from left to right in the interval. For example, the first function in Figure 4-41 above is decreasing on the interval $(-\infty, 0]$. The second function is decreasing on the intervals $(-\infty, -4]$, $[-2, 1]$, and $[3.5, \infty)$. A function whose graph falls from left to right throughout its domain is called a **decreasing function**.

By examining Figure 4-42, we obtain an algebraic description of increasing and decreasing functions:

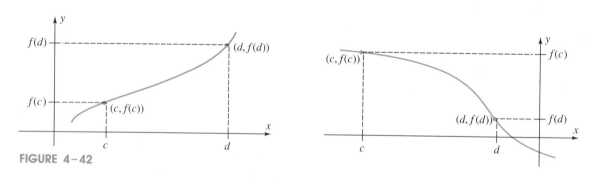

FIGURE 4-42

Suppose $c < d$. If the graph moves upward as you go from $x = c$ to $x = d$, then the second coordinate at d must be larger than the second coordinate at c. Thus

> The function f is **increasing** on an interval provided that for any numbers c, d in the interval,
>
> whenever $c < d$, then $f(c) < f(d)$.

Similarly, if the graph moves downward, the second coordinate at d must be *smaller* than the second coordinate at c. Hence

4.4.A INCREASING AND DECREASING FUNCTIONS

The function f is **decreasing** on an interval provided that for any numbers c and d in the interval,

whenever $c < d$, then $f(c) > f(d)$.

EXERCISES

In Exercises 1–3, state the intervals on which the function whose graph is shown is increasing and the intervals on which it is decreasing.

1.
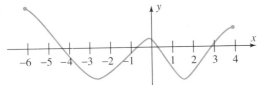

2.
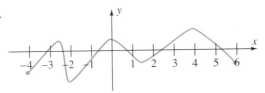

3.
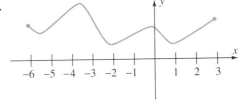

In Exercises 4–6, answer the following questions about the function whose graph is shown.

(i) Is the function even?
(ii) Is the function increasing on [1, 5]?
(iii) Is the function decreasing on [−3, 2]?

4.
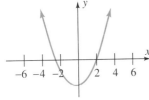

5.
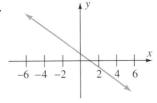

6.
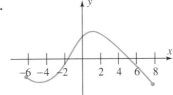

In Exercises 7–10, use algebra to show that the function is increasing on the interval (0, 10]. Some of them may also be increasing on other intervals. You may assume the usual facts about inequalities, including the following:

if $0 \leq c < d$, then $c^2 < d^2$

if $c < d$, then $c^3 < d^3$

7. $f(x) = x^2 + 3$ **8.** $g(x) = x^3 - 10{,}000$

9. $h(t) = t^2 + t + 5$ **10.** $k(u) = u^3 + u^2 + 1$

In Exercises 11–15, find one or more functions (among those whose graphs appear in Figure 4–43) for which the given statement is true.

11. $f(2) < f(1)$.

12. $f(x)$ is negative and increasing from $x = 1$ to $x = 2$.

13. $f(0) < 0$ but $f(2) > 0$.

14. $f(x)$ is negative and decreasing from $x = -4$ to $x = -3$.

15. $f(x) > f(-x)$ for some number x with $|x| \leq 5$.

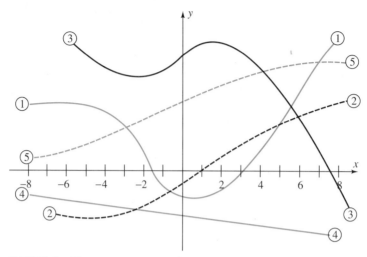

FIGURE 4-43

In Exercises 16-21, find a pair of functions (or several pairs) from among the five whose graphs appear in Figure 4-43 for which the given statement is true.

16. $f(1) - g(1) > 0$.

17. $f(x) < g(3)$ for $0 \leq x \leq 2$.

18. $f(x)g(x) < 0$ for $x > 4$.

19. $\dfrac{f(x)}{g(x)} < 1$ for some $x < 0$.

20. $f(x) = g(x)$ for some x with $|x| \geq 2$.

21. $f(x) \leq g(-x)$ for some x with $|x| < 4$.

22. Sketch the graph of a function f that satisfies these five conditions:

   (i) $f(-1) = 2$

   (ii) $f(x) \geq 2$ when x is in the interval $(-1, \frac{1}{2})$

   (iii) $f(x)$ starts decreasing when $x = 1$

   (iv) $f(3) = 3 = f(0)$

   (v) $f(x)$ starts increasing when $x = 5$

   [*Note:* The function whose graph you sketch need not be given by an algebraic formula.]

4.5 OPERATIONS ON FUNCTIONS

We now examine ways in which two or more given functions can be used to create new functions. If f and g are functions, then their **sum** is the function h defined by the rule

$$h(x) = f(x) + g(x).$$

For example, if $f(x) = 3x^2 + x$ and $g(x) = 4x - 2$, then

$$h(x) = f(x) + g(x) = (3x^2 + x) + (4x - 2) = 3x^2 + 5x - 2.$$

Instead of using a different letter h for the sum function, we shall usually denote it by $f + g$. Thus the sum $f + g$ is defined by the rule:

$$(f + g)(x) = f(x) + g(x).$$

This rule is *not* just a formal manipulation of symbols. If x is a number, then so are $f(x)$ and $g(x)$. The plus sign in $f(x) + g(x)$ is addition of *numbers* and the result is a number. But the plus sign in $f + g$ is addition of *functions* and the result is a new function.

The **difference** $f - g$ and the **product** fg of functions f and g are the functions defined by the rules:

$$(f - g)(x) = f(x) - g(x) \quad \text{and} \quad (fg)(x) = f(x)g(x).$$

The domain of the sum, difference, and product functions is the set of all real numbers that are in both the domain of f and the domain of g. The **quotient** f/g is the function defined by

$$\left(\frac{f}{g}\right)(x) = \frac{f(x)}{g(x)}.$$

Its domain is the set of all real numbers x in both the domain of f and the domain of g such that $g(x) \neq 0$.

Example 1 If $f(x) = \sqrt{3x}$ and $g(x) = x^2 - 1$, then

$$(f + g)(x) = \sqrt{3x} + x^2 - 1$$
$$(f - g)(x) = \sqrt{3x} - (x^2 - 1) = \sqrt{3x} - x^2 + 1$$
$$(fg)(x) = \sqrt{3x}(x^2 - 1) = \sqrt{3x} \cdot x^2 - \sqrt{3x}$$
$$\left(\frac{f}{g}\right)(x) = \frac{\sqrt{3x}}{x^2 - 1}.$$

The domain of $f + g$, $f - g$, and fg consists of all x in both the domain of g (all real numbers) and the domain of f (all nonnegative reals), that is, all $x \geq 0$. The domain of f/g consists of the nonnegative x for which $g(x) \neq 0$, that is, all nonnegative reals *except* $x = 1$. ∎

If c is a real number and f is a function, then the product of f and the constant function $g(x) = c$ is usually denoted cf. For example, if the function $f(x) = x^3 - x + 2$, and $c = 5$, then $5f$ is the function given by

$$(5f)(x) = 5 \cdot f(x) = 5(x^3 - x + 2) = 5x^3 - 5x + 10.$$

Composition of Functions

Another way of combining functions is illustrated by the function $h(x) = \sqrt{x^3}$. To compute $h(4)$, for example, you first find $4^3 = 64$ and then take the square root $\sqrt{64} = 8$. So the rule of h may be rephrased as:

First apply the function $f(x) = x^3$,

Then apply the function $g(t) = \sqrt{t}$ to the result.

The same idea can be expressed in functional notation like this:

$$x \xrightarrow{\text{first apply } f} f(x) \xrightarrow{\text{then apply } g \text{ to the result}} g(f(x))$$

$$x \xrightarrow{\phantom{\text{first apply } f}} x^3 \xrightarrow{\phantom{\text{then apply } g \text{ to the result}}} \sqrt{x^3}$$
$$\underbrace{}_{\text{apply } h}$$

So the rule of h may be written as $h(x) = g(f(x))$, where $f(x) = x^3$ and $g(t) = \sqrt{t}$. We can think of h as being made up of two simpler functions f and g, or we can think of f and g being "composed" to create the function h. Both viewpoints are useful.

Example 2 Suppose $f(x) = 4x^2 + 1$ and $g(t) = \dfrac{1}{t+2}$. Define a new function h whose rule is "first apply f; then apply g to the result." In functional notation

$$x \xrightarrow{\text{first apply } f} f(x) \xrightarrow{\text{then apply } g \text{ to the result}} g(f(x))$$

So the rule of the function h is $h(x) = g(f(x))$. Evaluating $g(f(x))$ means that whenever t appears in the formula for $g(t)$, we must replace it by $f(x) = 4x^2 + 1$:

$$h(x) = g(f(x)) = \frac{1}{f(x) + 2} = \frac{1}{(4x^2 + 1) + 2} = \frac{1}{4x^2 + 3}. \blacksquare$$

The function h is an illustration of the following definition.

COMPOSITE FUNCTIONS

Let f and g be functions. The *composite function* of f and g is the function which assigns to x the number $g(f(x))$. It is denoted $g \circ f$.

The symbol "$g \circ f$" is read "g circle f" or "f followed by g." (Note the order carefully; the functions are applied *right to left*.) So the rule of the composite function is:

$$(g \circ f)(x) = g(f(x))$$

Example 3 If $f(x) = 2x + 5$ and $g(t) = 3t^2 + 2t + 4$, then

$$(f \circ g)(2) = f(g(2)) = f(3 \cdot 2^2 + 2 \cdot 2 + 4) = f(20) = 2 \cdot 20 + 5 = 45.$$

Similarly,

$$(g \circ f)(-1) = g(f(-1)) = g(2(-1) + 5) = g(3) = 3 \cdot 3^2 + 2 \cdot 3 + 4 = 37. \blacksquare$$

The domain of $g \circ f$ is determined by the usual convention:

DOMAINS OF COMPOSITE FUNCTIONS

The domain of the composite function $g \circ f$ is the set of all real numbers x such that x is in the domain of f and $f(x)$ is in the domain of g.

Example 4 If $f(x) = \sqrt{x}$ and $g(t) = t^2 - 5$, then
$$(g \circ f)(x) = g(f(x)) = (f(x))^2 - 5 = (\sqrt{x})^2 - 5 = x - 5.$$
Although $x - 5$ is defined for every real number x, the domain of $g \circ f$ is *not* the set of all real numbers. The domain of g is the set of all real numbers, but the function $f(x) = \sqrt{x}$ is defined only when $x \geq 0$. So the domain of $g \circ f$ is the set of nonnegative real numbers, that is, the interval $[0, \infty)$. ∎

Example 5 If $h(x) = \sqrt{3x^2 + 1}$, then h may be considered as the composite $g \circ f$, where $f(x) = 3x^2 + 1$ and $g(u) = \sqrt{u}$:
$$(g \circ f)(x) = g(f(x)) = g(3x^2 + 1) = \sqrt{3x^2 + 1} = h(x). ∎$$

Example 6 If $k(x) = (x^2 - 2x + \sqrt{x})^3$, then k is $g \circ f$, where $f(x) = x^2 - 2x + \sqrt{x}$ and $g(t) = t^3$:
$$(g \circ f)(x) = g(f(x)) = g(x^2 - 2x + \sqrt{x}) = (x^2 - 2x + \sqrt{x})^3 = k(x). ∎$$

As you may have noticed, there are two possible ways to form a composite function from two given functions. If f and g are functions, we can consider either

$(g \circ f)(x) = g(f(x))$,   [*the composite of f and g*]

$(f \circ g)(x) = f(g(x))$,   [*the composite of g and f*]

The *order is important*, as we shall now see:

$g \circ f$ and $f \circ g$ usually are *not* the same function.

Example 7 If $f(x) = x^2$ and $g(x) = x + 3$,* then
$$(g \circ f)(x) = g(f(x)) = g(x^2) = x^2 + 3$$
but
$$(f \circ g)(x) = f(g(x)) = f(x + 3) = (x + 3)^2 = x^2 + 6x + 9.$$
Obviously, $g \circ f \neq f \circ g$ since, for example, they have different values at $x = 0$. ∎

---

* Now that you have the idea of composite functions, we'll use the same letter for the variable in both functions.

> **WARNING**
>
> Don't confuse the product function fg with the composite function $f \circ g$ (g followed by f). For instance, if $f(x) = 2x^2$ and $g(x) = x - 3$, then the product fg is given by:
>
> $$(fg)(x) = f(x)g(x) = 2x^2(x - 3) = 2x^3 - 6x^2.$$
>
> It is *not* the same as the composite $f \circ g$ because
>
> $$(f \circ g)(x) = f(g(x)) = f(x - 3) = 2(x - 3)^2 = 2x^2 - 12x + 18.$$

By using the operations above, a complicated function may be considered as being built up from simple parts.

Example 8 The function $f(x) = \sqrt{\dfrac{3x^2 - 4x + 5}{x^3 + 1}}$ may be considered as the composite $f = g \circ h$, where

$$h(x) = \frac{3x^2 - 4x + 5}{x^3 + 1} \quad \text{and} \quad g(x) = \sqrt{x}$$

since

$$(g \circ h)(x) = g(h(x)) = g\left(\frac{3x^2 - 4x + 5}{x^3 + 1}\right) = \sqrt{\frac{3x^2 - 4x + 5}{x^3 + 1}} = f(x).$$

The function $h(x) = \dfrac{3x^2 - 4x + 5}{x^3 + 1}$ is the quotient $\dfrac{p}{q}$, where

$$p(x) = 3x^2 - 4x + 5 \quad \text{and} \quad q(x) = x^3 + 1.$$

The function $p(x) = 3x^2 - 4x + 5$ may be written $p = k - s + r$, where

$$k(x) = 3x^2, \quad s(x) = 4x, \quad r(x) = 5.$$

The function k, in turn, can be considered as the product $3I^2$, where I is the *identity function* [whose rule is $I(x) = x$]:

$$(3I^2)(x) = 3(I^2(x)) = 3(I(x)I(x)) = 3 \cdot x \cdot x = 3x^2 = k(x).$$

Similarly, $s(x) = (4I)(x) = 4I(x) = 4x$. The function $q(x) = x^3 + 1$ may be "decomposed" in the same way.

Thus the complicated function f is just the result of performing suitable operations on the identity function I and various constant functions. ∎

EXERCISES

In Exercises 1–4, find the indicated values, where $g(t) = t^2 - t$ and $f(x) = 1 + x$.

1. $g(f(0))$
2. $(f \circ g)(3)$
3. $g(f(2) + 3)$
4. $f(2g(1))$

In Exercises 5–8, find $(g \circ f)(3)$, $(f \circ g)(1)$, and $(f \circ f)(0)$.

5. $f(x) = 3x - 2$, $g(x) = x^2$
6. $f(x) = |x + 2|$, $g(x) = -x^2$
7. $f(x) = x$, $g(x) = -3$
8. $f(x) = x^2 - 1$, $g(x) = \sqrt{x}$

In Exercises 9–14, find the rule of the function $f \circ g$ and the rule of $g \circ f$.

9. $f(x) = x^2$, $g(x) = x + 3$
10. $f(x) = -3x + 2$, $g(x) = x^3$
11. $f(x) = 1/x$, $g(x) = \sqrt{x}$
12. $f(x) = \dfrac{1}{2x + 1}$, $g(x) = x^2 - 1$
13. $f(x) = \sqrt[3]{x}$, $g(x) = x^2 + 1$
14. $f(x) = 2x^2 + 2x - 1$, $g(x) = |x - 1| + 2$

In Exercises 15–18, verify that $f \circ g = I$ and $g \circ f = I$, where I is the identity function whose rule is $I(x) = x$ for every x.

15. $f(x) = 9x + 2$, $g(x) = \dfrac{x - 2}{9}$
16. $f(x) = \sqrt[3]{x - 1}$, $g(x) = x^3 + 1$
17. $f(x) = \sqrt[3]{x} + 2$, $g(x) = (x - 2)^3$
18. $f(x) = 2x^3 - 5$, $g(x) = \sqrt[3]{\dfrac{x + 5}{2}}$

In Exercises 19–22, find $(f + g)(x)$, $(f - g)(x)$, and $(g - f)(x)$.

19. $f(x) = -3x + 2$, $g(x) = x^3$
20. $f(x) = x^2 + 2$, $g(x) = -4x + 7$
21. $f(x) = 1/x$, $g(x) = x^2 + 2x - 5$
22. $f(x) = \sqrt{x}$, $g(x) = x^2 + 1$

In Exercises 23–24, find $(fg)(x)$, $(f/g)(x)$, and $(g/f)(x)$.

23. $f(x) = -3x + 2$, $g(x) = x^3$
24. $f(x) = 4x^2 + x^4$, $g(x) = \sqrt{x^2 + 4}$

Exercises 25 and 26 refer to the function f whose graph is shown in Figure 4–44.

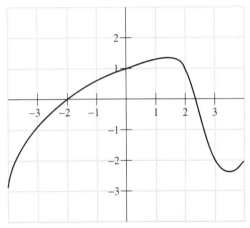

FIGURE 4–44

25. Let g be the composite function $f \circ f$ [that is, $g(x) = (f \circ f)(x) = f(f(x))$]. Use the graph of f to fill in the following table (approximate where necessary).

x	f(x)	g(x) = f(f(x))
-4		
-3		
-2	0	1
-1		
0		
1		
2		
3		
4		

26. Use the information obtained in Exercise 25 to sketch the graph of the function g.

In Exercises 27–30, fill the blanks in the given table. In each case the values of the functions f and g are given by these tables:

x	f(x)
1	3
2	5
3	1
4	2
5	3

t	g(t)
1	5
2	4
3	4
4	3
5	2

27.

x	(g∘f)(x)
1	4
2	
3	5
4	
5	

28.

t	(f∘g)(t)
1	
2	2
3	
4	
5	

29.

x	(f∘f)(x)
1	
2	
3	3
4	
5	

30.

t	(g∘g)(t)
1	
2	
3	
4	4
5	

In Exercises 31–36, write the given function as the composite of two functions, neither of which is the identity function. (There may be more than one way to do this.) Example: $f(x) = \sqrt{x^2 + 1}$ can be written $f = g \circ h$ with $g(x) = \sqrt{x}$ and $h(x) = x^2 + 1$.

31. $f(x) = \sqrt[3]{x^2 + 2}$

32. $g(x) = \sqrt{x + 3} - \sqrt[3]{x + 3}$

33. $h(x) = (7x^3 - 10x + 17)^7$

34. $k(x) = \sqrt[3]{(7x - 3)^2}$

35. $f(x) = \dfrac{1}{3x^2 + 5x - 7}$

36. $g(t) = \dfrac{3}{\sqrt{t - 3}} + 7$

37. If $f(x) = x + 1$ and $g(t) = t^2$, then
$$(g \circ f)(x) = g(f(x)) = g(x + 1) = (x + 1)^2$$
$$= x^2 + 2x + 1$$

Find two other functions $h(x)$ and $k(t)$ such that $(k \circ h)(x) = x^2 + 2x + 1$.

38. If f is any function and I is the identity function, what are $f \circ I$ and $I \circ f$?

In Exercises 39–42, determine whether the functions $f \circ g$ and $g \circ f$ are defined. If a composite function is defined, find its domain.

39. $f(x) = x^3$, $g(x) = \sqrt{x}$

40. $f(x) = x^2 + 1$, $g(x) = \sqrt{x}$

41. $f(x) = \sqrt{x + 10}$, $g(x) = 5x$

42. $f(x) = -x^2$, $g(x) = \sqrt{x}$

43. (a) If $f(x) = 2x^3 + 5x - 1$, find $f(x^2)$.
(b) If $f(x) = 2x^3 + 5x - 1$, find $(f(x))^2$.
(c) Are the answers in parts (a) and (b) the same? What can you conclude about $f(x^2)$ and $(f(x))^2$?

44. Give an example of a function f such that
$$f\left(\dfrac{1}{x}\right) \neq \dfrac{1}{f(x)}.$$

4.6 INVERSE FUNCTIONS*

Suppose that f and g are functions such that

$(f \circ g)(x) = x$ for every number x in the domain of g;
$(g \circ f)(x) = x$ for every number x in the domain of f.

Then we say that f and g are **inverse functions** (or that g is the **inverse** of f or f is the inverse of g). The basic idea is that "f undoes what g does" and "g undoes what f does."

Example 1 Let $f(x) = -2x + 1$ and $g(x) = \dfrac{x-1}{-2}$. Both these functions have the set of all real numbers as domain. For any number x we have:

$(f \circ g)(x) = f(g(x))$ *[definition of $f \circ g$]*

$= f\left(\dfrac{x-1}{-2}\right)$ *[definition of g]*

$= -2\left(\dfrac{x-1}{-2}\right) + 1 = x$ *[definition of f and arithmetic]*

On the other hand,

$(g \circ f)(x) = g(f(x))$ *[definition of $g \circ f$]*

$= g(-2x + 1)$ *[definition of f]*

$= \dfrac{(-2x+1) - 1}{-2} = x$ *[definition of g and arithmetic]*

Therefore f and g are inverse functions. ■

There do exist functions that don't have inverses, but for now we shall deal only with ones that do. When such a function is given, we can often find its inverse by algebraic means. An analogy may be helpful in understanding the method. When getting dressed you first put on your socks, and then your shoes. When undressing, you must reverse both the *order* of the original steps and the steps themselves: First take off your shoes, and then take off your socks.

Example 2 The function $f(x) = 3x - 2$ does this to the number x:

Multiply it by 3, and then subtract 2.

The inverse function g of f "undoes what f does." So it must reverse each step, and do so in the opposite order (remember the shoes and socks):

---

* The material in this section is not needed until Chapter 6.

First add 2 (the reverse of subtracting 2),

Then divide by 3 (the reverse of multiplying by 3).

If we begin with a number y, then the rule of g is:

Add 2 (producing $y + 2$), then divide by 3 $\left(\text{producing } \dfrac{y+2}{3}\right)$,

that is, $g(y) = \dfrac{y+2}{3}$. It's easy to check that g actually *is* the inverse of f:

$$(g \circ f)(x) = g(f(x)) = g(3x - 2) = \frac{(3x-2)+2}{3} = x \quad \text{for every } x$$

$$(f \circ g)(y) = f(g(y)) = f\left(\frac{y+2}{3}\right) = 3\left(\frac{y+2}{3}\right) - 2 = y \quad \text{for every } y. \quad \blacksquare$$

The verbal reasoning method in the preceding example is equivalent to the following algebraic method. The function f is given by the equation $y = f(x)$, that is, $y = 3x - 2$. We solve this equation for x:

Add 2 to both sides: $\quad 3x = y + 2$

Divide both sides by 3: $\quad x = \dfrac{y+2}{3}$

The steps used to solve the equation are precisely those used previously to find the rule of g (add 2, then divide by 3). So it's not surprising that the inverse function g is given by the equation

$$x = \frac{y+2}{3}, \quad \text{that is,} \quad x = g(y)$$

Example 3 The inverse of $f(x) = x^3 + 5$ can be found by solving for x in the equation $y = f(x)$, that is, $y = x^3 + 5$:

Subtract 5 from both sides: $\quad x^3 = y - 5$

Take cube roots on both sides: $\quad x = \sqrt[3]{y-5}$

You can easily check that the function $x = g(y)$, that is, $x = \sqrt[3]{y-5}$ is the inverse of f:

$$(f \circ g)(y) = f(g(y)) = f(\sqrt[3]{y-5}) = (\sqrt[3]{y-5})^3 + 5 = y \quad \text{for every } y$$

$$(g \circ f)(x) = g(f(x)) = g(x^3 + 5) = \sqrt[3]{(x^3+5) - 5} = x \quad \text{for every } x$$

Since the letter used for the variable of a function doesn't matter, the rule of the inverse function g is usually written $g(x) = \sqrt[3]{x-5}$. $\blacksquare$

If g is the inverse function of f, then the graph of g may be obtained by using the graph of f and this fact:

If the point (u, v) is on the graph of f, then the point (v, u) is on the graph of the inverse function g, and vice versa.

To see why this statement is true, let (u, v) be on the graph of f. Then $v = f(u)$. Applying g shows that $g(v) = g(f(u)) = u$ (by the definition of inverse function). Hence $(v, u) = (v, g(v))$ is on the graph of g. A similar argument, beginning with a point on the graph of g, shows that reversing its coordinates produces a point on the graph of f.

If you plot a pair of points whose coordinates are reversed, such as $(2, 7)$ and $(7, 2)$, you find that they are mirror images of each other, with the line $y = x$ being the mirror:*

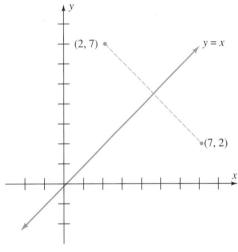

FIGURE 4–45

The same thing is true for any pair of points with reversed coordinates: (u, v) and (v, u) are mirror images of each other, with the line $y = x$ being the mirror (Exercise 36). Consequently,

GRAPHS OF INVERSE FUNCTIONS

If g is the inverse function of f, then the graph of g is the reflection of the graph of f in the line $y = x$.

Example 4 The preceding example shows that the inverse of $f(x) = x^3 + 5$ is the function $g(x) = \sqrt[3]{x - 5}$. Figure 4–46 on the next page shows that the graph of g is the reflection of the graph of f in the line $y = x$. ∎

---

* You can verify this fact algebraically by showing that the line $y = x$ is the perpendicular bisector of the line segment from $(2,7)$ to $(7,2)$ (Exercise 35).

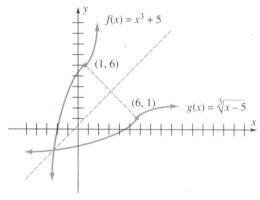

FIGURE 4-46

In order to determine exactly which functions have inverses, we must introduce a new concept. A function f is said to be **one-to-one** if it never takes the same value at two different numbers, that is,

$$\text{if } a \neq b, \text{ then } f(a) \neq f(b)$$

In graphical terms this means that two points on the graph, $(a, f(a))$ and $(b, f(b))$, that have different x-coordinates can't have the same y-coordinate. Hence these points cannot lie on the same horizontal line because all points on a horizontal line have the same y-coordinate. Therefore we have this geometric description of one-to-one functions:

THE HORIZONTAL LINE TEST	If a function f is one-to-one, then no horizontal line intersects the graph of f more than once. Conversely, if the graph of a function has this property, then the function is one-to-one.

Example 5 The graph of $f(x) = x^2$ (Figure 4-47 on the opposite page) fails the Horizontal Line Test because every horizontal line above the x-axis intersects the graph twice. Hence f is not one-to-one. But the graph of $g(x) = x^3$ passes the test: Every horizontal line intersects the graph of g just once. Therefore g is one-to-one. ∎

We can now determine which functions have inverses:

INVERSE FUNCTION THEOREM	If a function f has an inverse function, then f is one-to-one. Conversely, if a function is one-to-one, then it has an inverse function.

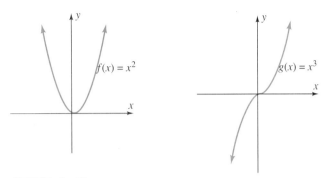

FIGURE 4–47

To see why a function f with an inverse g must be one-to-one, recall that the graphs of f and g are reflections of each other in the line $y = x$. Every horizontal line that intersects the graph of f corresponds under reflection to a vertical line intersecting the graph of g, as shown in Figure 4–48:

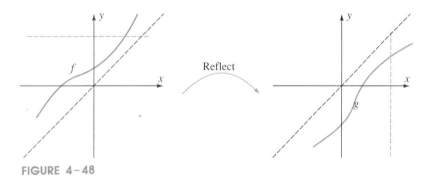

FIGURE 4–48

But no vertical line intersects the graph of the function g more than once (see the *Vertical* Line Test on page 163). Consequently, no horizontal line can intersect the graph of f more than once. The proof that every one-to-one function has an inverse is in Exercise 48.

Example 6 Example 5 shows that $f(x) = x^2$ is not one-to-one. Hence f does not have an inverse function. That example also shows that $g(x) = x^3$ is one-to-one. Therefore g does have an inverse function. By solving $y = x^3$ for x, (namely, $x = \sqrt[3]{y}$), we see that the rule of the inverse function is $h(x) = \sqrt[3]{x}$. ∎

A function that is **increasing** (graph *always* rising from left to right) or **decreasing** (graph *always* falling from left to right) is necessarily one-to-one, because its graph can never touch the same horizontal line twice (it would have to change direction to do so). So every such function has an inverse. Furthermore, it can be shown that

Every increasing or decreasing function has an inverse function with the same property.

NOTE In many texts the inverse function of a function f is denoted by f^{-1}. In this notation, for instance, we would write $f(x) = x^3 + 5$ and $f^{-1}(x) = \sqrt[3]{x - 5}$ and the definition of inverse functions becomes:

$$f(f^{-1}(x)) = x \text{ for every } x \text{ in the domain of } f^{-1}; \text{ and}$$

$$f^{-1}(f(x)) = x \text{ for every } x \text{ in the domain of } f.$$

In this context, f^{-1} does *not* mean $1/f$ (see Exercise 37).

EXERCISES

In Exercises 1–6, verify that the given functions are inverses of each other by calculating $f \circ g$ and $g \circ f$.

1. $f(x) = x + 1$, $g(x) = x - 1$
2. $f(x) = 2x - 6$, $g(x) = \dfrac{x}{2} + 3$
3. $f(x) = \dfrac{1}{x + 1}$, $g(x) = \dfrac{1 - x}{x}$
4. $f(x) = \dfrac{-3}{2x + 5}$, $g(x) = \dfrac{-3 - 5x}{2x}$
5. $f(x) = x^5$, $g(x) = \sqrt[5]{x}$
6. $f(x) = x^3 - 1$, $g(x) = \sqrt[3]{x + 1}$

In Exercises 7–12, use the verbal reasoning method (as in Example 2) to find the inverse function g of the given function.

7. $f(x) = 5x + 1$
8. $f(x) = \dfrac{x}{2} + 2$
9. $f(x) = x^3$
10. $f(x) = (x + 4)^3$
11. $f(x) = 2(x + 4)^3 - 1$
12. $f(x) = \dfrac{2(x + 4)^3 - 1}{5}$

In Exercises 13–26, use algebra to find the inverse function g of the given function.

13. $f(x) = -x$
14. $f(x) = -x + 1$
15. $f(x) = 5x - 4$
16. $f(x) = -3x + 5$
17. $f(x) = 5 - 2x^3$
18. $f(x) = (x^5 + 1)^3$
19. $f(x) = \sqrt{4x - 7}$
20. $f(x) = 5 + \sqrt{3x - 2}$
21. $f(x) = 1/x$
22. $f(x) = 1/\sqrt{x}$
23. $f(x) = \dfrac{1}{2x + 1}$
24. $f(x) = \dfrac{x}{x + 1}$
25. $f(x) = \dfrac{x^3 - 1}{x^3 + 5}$
26. $f(x) = \sqrt[5]{\dfrac{3x - 1}{x - 2}}$

In Exercises 27–32, each given function has an inverse. Graph the function; then graph its inverse without plotting any points (reflect carefully).

27. $f(x) = \dfrac{x}{3} - 2$
28. $f(x) = \dfrac{x - 5}{7}$
29. $f(x) = \sqrt{x + 3}$
30. $f(x) = \sqrt{3x - 2}$
31. $f(x) = x^3 + 1$
32. $f(x) = \sqrt[3]{x + 3}$

33. List three different functions, each of which is its own inverse.

34. Let C be the temperature in degrees Celsius. Then the temperature in degrees Fahrenheit is given by $f(C) = \dfrac{9}{5}C + 32$. Let g be the function that converts degrees Fahrenheit to degrees Celsius. Show that g is the inverse function of f and find the rule of g.

35. Show that the line $y = x$ is the perpendicular bisector of the line segment from $(2, 7)$ to $(7, 2)$ as follows.

(a) Find the slope of the line L through $(2, 7)$ and $(7, 2)$.
(b) Show that L is perpendicular to the line $y = x$.
(c) Show that the midpoint of the line segment from $(2, 7)$ to $(7, 2)$ is on the line $y = x$.

36. If u and v are any real numbers with $u \neq v$, show that the line $y = x$ is the perpendicular bisector of the line segment from (u, v) to (v, u) as follows.
    (a) Find the slope of the line L through (u, v) and (v, u).
    (b) Show that L is perpendicular to the line $y = x$.
    (c) Show that the midpoint of the segment from (u, v) to (v, u) lies on the line $y = x$.

37. (a) Using the f^{-1} notation for inverse functions, find $f^{-1}(x)$ when $f(x) = 3x + 2$.
    (b) Find $f^{-1}(1)$ and $1/f(1)$. Conclude that f^{-1} is not the same function as $1/f$.

38. Show that the inverse function of the function f whose rule is $f(x) = \dfrac{2x + 1}{3x - 2}$ is f itself.

In Exercises 39–46, none of the functions has an inverse. State at least one way of restricting the domain of the function (that is, find a function with the same rule and a smaller domain) so that the restricted function has an inverse. Then find the rule of the inverse function.

Example: $f(x) = x^2$ has no inverse. But the function h with domain all $x \geq 0$ and rule $h(x) = x^2$ is increasing (its graph is the right half of the graph of f—see Figure 4-47 on page 201) and therefore has an inverse.

39. $f(x) = |x|$
40. $f(x) = |x - 3|$
41. $f(x) = -x^2$
42. $f(x) = x^2 + 4$
43. $f(x) = \dfrac{x^2 + 6}{2}$
44. $f(x) = \sqrt{4 - x^2}$
45. $f(x) = \dfrac{1}{x^2 + 1}$
46. $f(x) = 3(x + 5)^2 + 2$

Unusual Problems

47. Let m and b be constants with $m \neq 0$. Show that the function $f(x) = mx + b$ has an inverse function g and find the rule of g.

48. If f is a one-to-one function, prove that f has an inverse function as follows.
    (a) Let v be any number in the range of f. Use the definition of range (page 151) and the definition of one-to-one to show that there is one and only one number u in the domain of f such that $f(u) = v$.
    (b) Define a function g whose domain is the same as the range of f and whose rule is: $g(v) = u$, where u is the unique number such that $f(u) = v$. Show that g is the inverse function of f.

4.7 VARIATION*

If a plane flies at 400 mph for t hours, then it travels a distance of $400t$ miles. So the distance d is related to the time t by the equation $d = 400t$. This is an example of *variation:*

DIRECT VARIATION

The statement *v varies directly as u* or *v is directly proportional to u* means that $v = ku$ for some constant k. The constant k is called the *constant of variation.*

In the example of the plane, where $d = 400t$, d varies directly as t and the constant of variation is 400.

---
* This section is not needed in the sequel.

Example 1 When you swim underwater, the pressure p varies directly with the depth d at which you swim. At a depth of 20 feet the pressure is 8.6 pounds per square inch. What will the pressure be at 65 feet deep?

Solution Since p varies directly with d, we know that $p = kd$ for some constant k. We first use the given information to find k. Since $p = 8.6$ when $d = 20$, the equation $p = kd$ becomes

$$8.6 = k(20), \quad \text{or equivalently,} \quad k = \frac{8.6}{20} = .43$$

Therefore the equation relating pressure and depth is $p = .43d$. To find the pressure at a depth of 65 feet we substitute 65 for d and find that $p = .43(65) = 27.95$ pounds per square inch. ∎

The basic idea in direct variation is that the two quantities grow or shrink together. For instance, when $v = 3u$, then as u gets large, so does v. But two quantities can also be related in such a way that one grows as the other shrinks, or vice versa. For example if $v = 5/u$, then when u is large (say $u = 500$), v is small: $v = 5/500 = .01$. This is an example of *inverse variation*:

INVERSE VARIATION

> The statement v *varies inversely as* u or v *is inversely proportional to* u means that $v = k/u$ for some constant k. The constant k is called the *constant of variation*.

Example 2 According to one of Parkinson's Laws, the amount of time the Math Department spends discussing an item in its budget is inversely proportional to the cost of the item. If the Department spent 40 minutes discussing the purchase of a ditto machine for $450, how much time will it spend on a $100 appropriation for the annual picnic?

Solution If t denotes the discussion time and c the cost of the item, then $t = k/c$ for some constant k. We know that $t = 40$ when $c = 450$. Substituting these numbers in $t = k/c$ shows that

$$40 = \frac{k}{450}, \quad \text{or equivalently,} \quad k = 40 \cdot 450 = 18{,}000.$$

So time and cost are related by the equation $t = 18000/c$. Therefore the time spent discussing the $100 for the picnic is

$$t = \frac{18000}{100} = 180 \text{ minutes } (= 3 \text{ hours!}). \quad \blacksquare$$

The terminology of variation also applies to situations involving powers of quantities or more than two quantities, as illustrated in these examples.

Example 3
(a) The equation $y = 7x^4$ says that **y varies directly as the fourth power of x,** or equivalently, **y is directly proportional to the fourth power of x.**
(b) The equation $y = 10/x^3$ says that **y varies inversely (or is inversely proportional to) the third power of x.** ■

Example 4
(a) The equation $z = 10xy$ says that **z varies jointly as x and y.**
(b) The equation $W = 4xy^3z^2$ says that **W varies jointly as x and the cube of y and the square of z.**
(c) The equation $K = 4t^2/d^4$ says that **K varies directly as the square of t and inversely as the fourth power of d.** ■

Example 5 The electrical resistance R of wire (of uniform material) varies directly as the length L and inversely as the square of the diameter d. A 2-meter-long piece of wire with diameter 4.4 millimeters has a resistance of 500 ohms. What diameter wire should be used if a 10-meter piece is to have a resistance of 1300 ohms?

Solution The given information means that $R = kL/d^2$ for some constant k. Furthermore, $R = 500$ when $L = 2$ and $d = 4.4$. Hence

$$500 = \frac{k \cdot 2}{(4.4)^2}$$

$$k = \frac{500(4.4)^2}{2} = 4840$$

Therefore the formula is $R = 4840L/d^2$. We must find d when $L = 10$ and $R = 1300$:

$$1300 = \frac{4840 \cdot 10}{d^2}$$

$$1300d^2 = 48400$$

$$d^2 = \frac{48400}{1300} \approx 37.2308$$

Since the diameter is positive, we must have $d \approx \sqrt{37.2308} \approx 6.1$ mm. ■

EXERCISES

In Exercises 1–6, express the given statement as an equation using k for the constant of variation.

1. a varies inversely as b.

2. r is proportional to t.

3. z varies jointly as x, y, and w.

4. The weight w of an object varies inversely as the square of the distance d from the object to the center of the earth.

5. The distance d one can see to the horizon varies directly as the square root of the height h above sea level.

6. The pressure p exerted on the floor by a person's shoe heel is directly proportional to the weight w of the person and inversely proportional to the square of the width r of the heel.

In Exercises 7–16, express the given statement as an equation and find the constant of variation.

7. v varies directly as u; $v = 8$ when $u = 2$.
8. v is directly proportional to u; $v = .4$ when $u = .8$.
9. v varies inversely as u; $v = 8$ when $u = 2$.
10. v is inversely proportional to u; $v = .12$ when $u = .1$.
11. t varies jointly as r and s; $t = 24$ when $r = 2$ and $s = 3$.
12. B varies inversely as u and v; $B = 4$ when $u = 1$ and $v = 3$.
13. w varies jointly as x and y^2; $w = 96$ when $x = 3$ and $y = 4$.
14. p varies directly as the square of z and inversely as r; $p = 32/5$ when $z = 4$ and $r = 10$.
15. T varies jointly as p and the cube of v and inversely as the square of u; $T = 24$ when $p = 3$, $v = 2$, and $u = 4$.
16. D varies jointly as the square of r and the square of s and inversely as the cube of t; $D = 18$ when $r = 4$, $s = 3$, and $t = 2$.

17. If r varies directly as t and $r = 6$ when $t = 3$, find r when $t = 2$.
18. If r is directly proportional to t and $r = 4$ when $t = 2$, find t when $r = 2$.
19. If b varies inversely as x and $b = 9$ when $x = 3$, find b when $x = 12$.
20. If b is inversely proportional to x and $b = 10$ when $x = 4$, find x when $b = 12$.
21. Suppose w is directly proportional to the sum of u and the square of v. If $w = 200$ when $u = 1$ and $v = 7$, then find u when $w = 300$ and $v = 5$.
22. Suppose z varies jointly as x and y. If $z = 30$ when $x = 5$ and $y = 2$, then find x when $z = 45$ and $y = 3$.
23. Suppose r varies inversely as s and t. If $r = 12$ when $s = 3$ and $t = 1$, then find r when $s = 6$ and $t = 2$.
24. Suppose u varies jointly as r and s and inversely as t. If $u = 1.5$ when $r = 2$, $s = 3$, and $t = 4$, then find r when $u = 27$, $s = 9$, and $t = 5$.
25. Suppose D varies jointly as b and c, and inversely as the square of d. If $D = 2.4$ when $b = 16$, $c = 3$, and $d = 5$, then find c when $D = 5$, $b = 6$, and $d = 10$.
26. In the State of Confusion your income tax varies directly as your income. If the tax is $200 on an income of $16,000, find the tax on an income of $7,000.
27. By experiment you discover that the amount of water that comes from your garden hose varies directly with the water pressure. A pressure of 10 pounds per square inch is needed to produce a flow of 3 gallons per minute.
   (a) What pressure is needed to produce a flow of 4.2 gallons per minute?
   (b) If the pressure is 5 pounds per square inch, what is the flow rate?
28. At a fixed temperature the pressure of an enclosed gas is inversely proportional to its volume. The pressure is 50 kilograms per square centimeter when the volume is 200 cubic centimeters. If the gas is compressed to 125 cubic centimeters, what is the pressure?
29. The electrical resistance in a piece of wire of a given length and material varies inversely as the square of the diameter of the wire. If a wire of diameter .01 cm has a resistance of .4 ohm, what is the resistance of a wire of the same length and material, but with diameter .025 cm?
30. The weight of a cylindrical can of glop varies jointly as the height and the square of the base radius. The weight is 250 ounces when the height is 20 inches and the base radius is 5 inches. What is the height when the weight is 960 ounces and the base radius is 8 inches?
31. The force needed to keep a car from skidding on a circular curve varies inversely as the radius of the curve and jointly as the weight of the car and the square of the speed. It takes 1500 kilograms of force to keep a 1000-kilogram car from skidding on a curve of radius 200 meters at a speed of 50 kilometers per hour. What force is needed to keep the same car from skidding on a curve of radius 320 meters at 100 kilometers per hour?

32. The period of a pendulum is the time it takes for the pendulum to make one complete swing and return to its starting point. The period varies directly with the *square root* of the length of the pendulum. A pendulum 3 meters long has a period of 4 seconds. If a grandfather clock has a 1.92-meter pendulum, what is its period?

CHAPTER REVIEW

Important Concepts

Section 4.1	Function 151
	Domain 151
	Range 151
	Rule 151
	Greatest integer function 152
Section 4.2	Functional notation 155
	Domain convention for functions 157
	Common mistakes with functional notation 159
Section 4.3	Vertical Line Test 163
	Graphs of functions 163
Excursion 4.3.A	Graph reading 171
	Translating verbal statements into functional notation 172
	Translating functional notation into graphical information 172
Section 4.4	Symmetry with respect to y-axis 177
	Symmetry with respect to x-axis 178
	Symmetry with respect to the origin 178
	Coordinate tests for symmetry 179
	Algebraic tests for symmetry 179
	Even and odd functions 179
	Vertical and horizontal shifting of graphs 180
	Stretching a graph away from, or shrinking it toward, the x-axis 182
	Reflecting a graph in the x-axis 183
Excursion 4.4.A	Increasing and decreasing functions 187
Section 4.5	Sums, differences, products, and quotients of functions 190
	Composition of functions 191
Section 4.6	Inverse functions 197
	Graphs of inverse functions 199
	One-to-one functions 200
	Increasing and decreasing functions 201
Section 4.7	Direct variation 203
	Inverse variation 204
	Joint variation 205

Review Questions

1. Let [x] denote the greatest integer function and evaluate
   (a) [−5/2] = ──── .
   (b) [1755] = ──── .
   (c) [18.7] + [−15.7] = ──── .
   (d) [−7] − [7] = ──── .

2. If $f(x) = x + |x| + [x]$, then find $f(0), f(-1), f(1/2)$, and $f(-3/2)$.

3. Let f be the function given by the rule $f(x) = 7 - 2x$. Complete this table:

x	0	1	2	−4	t	k	b − 1	1 − b	6 − 2u
f(x)	7								

4. What is the domain of the function g given by
$$g(t) = \frac{\sqrt{t-2}}{t-3}?$$

5. In each case give a *specific* example of a function and numbers a, b to show that the given statement may be *false*.
   (a) $f(a + b) = f(a) + f(b)$
   (b) $f(ab) = f(a)f(b)$

6. If $f(x) = |3 - x|\sqrt{x-3} + 7$, then $f(7) - f(4) = $ ──── .

7. What is the domain of the function given by
$$g(r) = \sqrt{r-2} + \sqrt{r-2}?$$

8. What is the domain of the function $f(x) = \sqrt{-x+2}$?

9. If $h(x) = x^2 - 3x$, then $h(t + 2) = $ ──── .

10. Which of the following statements about the greatest integer function $f(x) = [x]$ is true for *every* real number x?
    (a) $x - [x] = 0$
    (b) $x - [x] \leq 0$
    (c) $[x] + [-x] \leq 0$
    (d) $[-x] \geq [x]$
    (e) $3[x] = [3x]$

11. If $f(x) = 2x^3 + x + 1$, then $f(x/2) = $ ──── .

12. If $g(x) = x^2 - 1$, then $g(x - 1) - g(x + 1) = $ ──── .

13. Sketch the graph of the function f given by
$$f(x) = \begin{cases} x^2 & \text{if } x \leq 0 \\ x + 1 & \text{if } 0 < x < 4 \\ \sqrt{x} & \text{if } x \geq 4 \end{cases}$$

14. Sketch the graph of the function $f(x) = |x| + 1$.

15. Sketch the graph of the function f given by:
$$f(x) = \begin{cases} 2x - 3 & \text{if } x < 1 \\ x^2 - x & \text{if } x \geq 1 \end{cases}$$
16. Sketch the graph of $g(x) = x - |x|$.
17. Sketch the graph of $f(x) = \sqrt{4 - x^2}$.
18. Sketch the graph of a function f that satisfies all of these conditions:

   (i) domain of $f = [-3, 4]$

   (ii) range of $f = [-2, 5]$

   (iii) $f(-2) = 0$

   (iv) $f(1) > 2$

   [*Note:* There are many possible correct answers and the function whose graph you sketch need *not* have a simple algebraic rule.]
19. Sketch the graph of $g(x) = |x - 2|$.

Use the graph of the function f in Figure 4-49 to answer Questions 20-23.

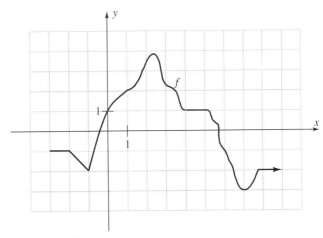

FIGURE 4-49

20. What is the domain of f?   21. What is the range of f?
22. Find all numbers x such that $f(x) = 1$.
23. Find a number x such that $f(x + 1) < f(x)$. (Many correct answers are possible.)

Use the graph of the function f in Figure 4-50 on the next page to answer Questions 24-30.

24. What is the domain of f?   25. $f(-3) = $ _____.

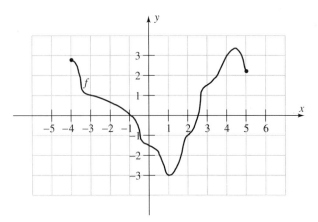

FIGURE 4-50

26. $f(2 + 2) =$ _____. 27. $f(-1) + f(1) =$ _____.
28. True or false: $2f(2) = f(4)$.   29. True or false: $3f(2) = -f(4)$.
30. True or false: $f(x) = 3$ for exactly one number x.

Use the graphs of the functions f and g in Figure 4-51 to answer Questions 31-36.

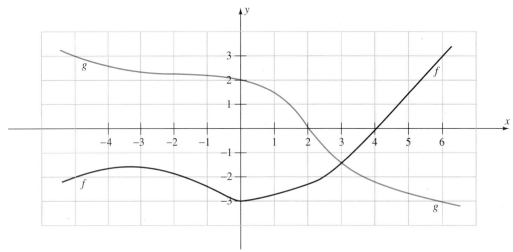

FIGURE 4-51

31. For which values of x is $f(x) = 0$?
32. True or false: if a and b are numbers such that $-5 \le a < b \le 6$, then $g(a) < g(b)$.
33. For which values of x is $g(x) \ge f(x)$?

34. Find $f(0) - g(0)$.
35. For which values of x is $f(x + 1) < 0$?
36. What is the distance from the point $(-5, g(-5))$ to the point $(6, g(6))$?
37. Sketch a graph that is symmetric with respect to both the x-axis and the y-axis. (There are many correct answers and your graph need not be the graph of an equation.)
38. Sketch the graph of a function that is symmetric with respect to the origin. (There are many correct answers and you don't have to state the rule of your function.)

In Questions 39–44, determine whether the graph of the given equation is symmetric with respect to the x-axis, the y-axis, or the origin.

39. $y = (x + 3)^4 - 2$
40. $y = x(x - 1)^2(x + 5)^2$
41. $y = 2x^3(x + 2)$
42. $x^2 = y^2 + 2$
43. $x^2 + y^4 + 5 = 0$
44. $5y = 7x^2 - 2x$

In Questions 45–47, determine whether the given function is even, odd, or neither.

45. $g(x) = 9 - x^2$
46. $f(x) = |x|x + 1$
47. $h(x) = 3x^5 - x(x^4 - x^2)$

48. (a) Draw some coordinate axes and plot the points $(-2, 1), (-1, 3), (0, 1), (3, 2), (4, 1)$.
    (b) Suppose the points plotted in part (a) lie on the graph of an *even* function f. Plot these points: $(2, f(2))$, $(1, f(1))$, $(0, f(0))$, $(-3, f(-3))$, $(-4, f(-4))$.

49. Determine whether the circle with equation $x^2 + y^2 + 6y = -5$ is symmetric with respect to the x-axis, the y-axis, or the origin.
50. On what intervals is the function $f(x) = -x^2 + 1$ *increasing*?
51. On what intervals is the function $f(x) = (x - 3)^2 + 3$ *decreasing*?
52. Sketch the graph of $f(x) = x^2$ and then use it to sketch the graphs of these functions (on the same set of axes):

$$g(x) = x^2 - 4 \quad \text{and} \quad h(x) = (x - 2)^2$$

53. Sketch the graph of $f(x) = |x|$ and use it to sketch the graphs of these functions on the same set of axes:

$$g(x) = \frac{1}{4}|x| \quad \text{and} \quad h(x) = -|x + 4|$$

54. The graph of a function f is shown in Figure 4–52. On the same coordinate plane, carefully draw the graphs of the functions g and h whose rules are:

$$g(x) = -f(x) \quad \text{and} \quad h(x) = 1 - f(x)$$

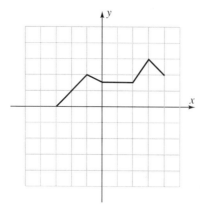

FIGURE 4-52

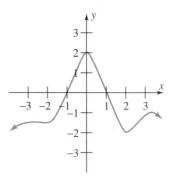

FIGURE 4-53

55. Figure 4–53 shows the graph of a function f. If g is the function given by $g(x) = f(x + 2)$, then which of these statements about the graph of g is true?

   (a) It does not cross the x-axis.
   (b) It does not cross the y-axis.
   (c) It crosses the y-axis at $y = 4$.
   (d) It crosses the y-axis at the origin.
   (e) It crosses the x-axis at $x = -3$.

56. If $f(x) = 3x + 2$ and $g(x) = x^3 + 1$, find:

   (a) $(f + g)(-1)$
   (b) $(f - g)(2)$
   (c) $(fg)(0)$

57. If $f(x) = \dfrac{1}{x - 1}$ and $g(x) = \sqrt{x^2 + 5}$, find:

   (a) $(f/g)(2)$
   (b) $(g/f)(x)$
   (c) $(fg)(c + 1)\ (c \neq 1)$

58. Find two functions f and g such that neither is the identity function and
$$(f \circ g)(x) = (2x + 1)^2$$

59. Use the graph of the function g in Figure 4–54 to fill in the table below, in which h is the composite function $g \circ g$.

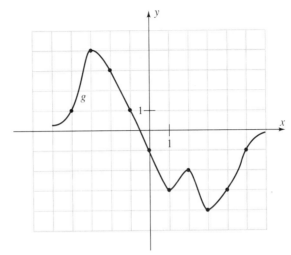

FIGURE 4-54

x	-4	-3	-2	-1	0	1	2	3	4
$g(x)$					-1				
$h(x) = g(g(x))$									

Questions 60–65 refer to the functions $f(x) = \dfrac{1}{x+1}$ **and** $g(t) = t^3 + 3$.

60. $(f \circ g)(1) = $ _____. **61.** $(g \circ f)(2) = $ _____.

62. $g(f(-2)) = $ _____. **63.** $(g \circ f)(x-1) = $ _____.

64. $g(2 + f(0)) = $ _____. **65.** $f(g(1) - 1) = $ _____.

66. Let f and g be the functions given by
$$f(x) = 4x + x^4 \quad \text{and} \quad g(x) = \sqrt{x^2 + 1}$$
(a) $(f \circ g)(x) = $ _____. **(b)** $(g - f)(x) = $ _____.

67. If $f(x) = \dfrac{1}{x}$ and $g(x) = x^2 - 1$, then
$(f \circ g)(x) = $ _____ and $(g \circ f)(x) = $ _____.

68. Let $f(x) = x^2$. Give an example of a function g with domain all real numbers such that $g \circ f \neq f \circ g$.

69. If $f(x) = \dfrac{1}{1-x}$ and $g(x) = \sqrt{x}$, then find the domain of the composite function $f \circ g$.

70. These tables show the values of the functions f and g at certain numbers:

x	-1	0	1	2	3
$f(x)$	1	0	1	3	5

and

t	0	1	2	3	4
$g(t)$	-1	0	1	2	5

Which of the following statements is *true*?

(a) $(g - f)(1) = 1$
(d) $(g \circ f)(2) = 1$
(b) $(f \circ g)(2) = (f - g)(0)$
(e) None of the above are true.
(c) $f(1) + f(2) = f(3)$

71. If $f(x) = 2x^2$ and $g(x) = -1$, then sketch the graph of the composite function $f \circ g$.
72. Find the inverse of the function $f(x) = 2x + 1$.
73. Find the inverse of the function $f(x) = \sqrt{5 - x} + 7$.
74. Find the inverse of the function $f(x) = \sqrt[5]{x^3 + 1}$.
75. The graph of a function f is shown in Figure 4–55. Sketch the graph of the inverse function of f.

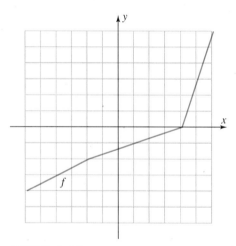

FIGURE 4–55

76. Which of the following functions have inverse functions (give reasons for your answers):

(a) $f(x) = x^3$ (b) $f(x) = 1 - x^2$, $x \le 0$ (c) $f(x) = |x|$

77. If x and y are directly proportional and $x = 12$ when $y = 36$, then what is y when $x = 2$?

78. Driving time varies inversely as speed. If it takes 3 hours to drive to the beach at an average speed of 48 mph, how long will it take to drive home at an average speed of 54 mph?

79. Find the constant of variation when T varies directly as the square of R and inversely as S, provided $T = .6$ when $R = 3$ and $S = 15$.

80. The statement "r varies directly as s and the square root of t and inversely as the cube of x" means that for some constant k:

(a) $r = kstx^3$ (b) $rx^3 = ks\sqrt{t}$

(c) $r = \dfrac{kx^3\sqrt{t}}{s}$ (d) $k = \dfrac{rs\sqrt{t}}{x^3}$

(e) $kx^3 s = k\sqrt{t}$

CHAPTER 5

Polynomial and Rational Functions

Polynomial functions arise naturally in many applications. Since their values can be computed using only simple arithmetic (addition, subtraction, and multiplication) they are ideally suited for high-speed computers. This fact is crucial since many complicated functions that arise in applied mathematics can be approximated by polynomial functions or their quotients (rational functions).

> **ROADMAP:** The sections of this chapter are essentially independent of one another and may be read in any order.

5.1 QUADRATIC FUNCTIONS*

A function f whose rule can be written in the form

$$f(x) = ax^2 + bx + c$$

for some constants a, b, c, with $a \neq 0$, is called a **quadratic function.** Here are some examples of quadratic functions:

$$f(x) = x^2, \quad g(x) = 3x^2 - 2x + 6, \quad h(x) = -x^2 + x$$

The graph of a quadratic function is a curve called a **parabola.** All such parabolas have the same basic shape and open either upward or downward, as shown in Figures 5–1 and 5–2. If a parabola opens upward, its **vertex** is the lowest point on the graph. If it opens downward, its **vertex** is the highest point on the graph. In every case, the graph is symmetric with respect to the vertical line through the vertex; this line is called the **axis** of the parabola.

Example 1 The graph of $f(x) = x^2$ was obtained in Section 4.3 and is shown in Figure 5–1. As we saw in Section 4.4, the graph of $g(x) = -\frac{1}{4}x^2$ shown in Figure 5–2 is the graph of $f(x) = x^2$ shrunk by a factor of $\frac{1}{4}$ and reflected in the

---

* Section 4.4 (Graphing Techniques) is a prerequisite for this section.

x-axis. In both graphs, (0, 0) is the vertex, and the y-axis is the axis of the parabola. ∎

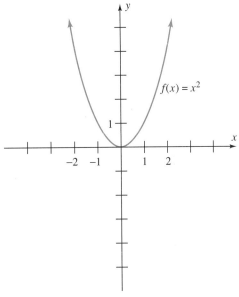

FIGURE 5-1

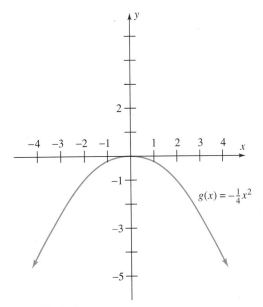

FIGURE 5-2

Example 2 The function $g(x) = 2(x - 3)^2 + 1$ is quadratic because its rule can be written in the required form:

$$g(x) = 2(x - 3)^2 + 1 = 2(x^2 - 6x + 9) + 1 = 2x^2 - 12x + 19$$

As we saw in Section 4.4, the graph of $g(x) = 2(x - 3)^2 + 1$ is the graph $f(x) = x^2$ shifted horizontally 3 units to the right, stretched by a factor of 2, and shifted vertically 1 unit upward. It is a parabola with vertex (3, 1) and axis the vertical line $x = 3$, as shown in Figure 5–3. ∎

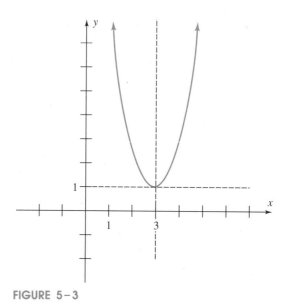

FIGURE 5–3

The graph of the function g in Example 2 was easily determined because the rule of g had a special algebraic form. The graph of any quadratic function can be determined in a similar fashion by first rewriting its rule.

Example 3 The rule of the quadratic function $g(x) = 3x^2 + 30x + 77$ may be rewritten as $g(x) = 3(x^2 + 10x) + 77$. Next we complete the square in the expression in parentheses by adding 25 (the square of half the coefficient of x). In order not to change the rule of the function we must also *subtract* 25:

$$g(x) = 3(x^2 + 10x + 25 - 25) + 77$$

Using the distributive law and factoring, we have:

$$g(x) = 3(x^2 + 10x + 25) - 3 \cdot 25 + 77$$
$$= 3(x + 5)^2 + 2$$

The graph of $g(x) = 3(x + 5)^2 + 2$ is the graph of $f(x) = x^2$ shifted horizontally 5 units to the left, stretched by a factor of 3, then shifted vertically 2 units upward, as shown in Figure 5-4 below. Hence it is a parabola with vertex $(-5, 2)$ and axis the vertical line $x = -5$. Note that the rule of g can be rewritten in terms of the coordinates of the vertex $(-5, 2)$:

$$g(x) = 3(x + 5)^2 + 2 = 3(x - (-5))^2 + 2. \blacksquare$$

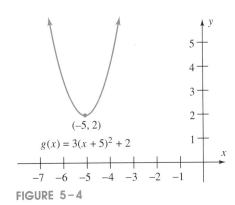

FIGURE 5-4

The techniques in the preceding examples work in the general case and may be summarized as follows.

QUADRATIC FUNCTIONS

> The graph of the quadratic function $f(x) = ax^2 + bx + c$ is a parabola. The rule of f may be rewritten in the form:
>
> $$f(x) = a(x - h)^2 + k$$
>
> The parabola opens upward if $a > 0$ and downward if $a < 0$. Its vertex is (h, k), and its axis is the vertical line $x = h$.

Example 4 To graph $f(x) = -4x^2 + 12x - 8$ we first write $f(x) = -4(x^2 - 3x) - 8$. Then complete the square in the expression in parentheses by adding $\frac{9}{4}$ (square of half the coefficient of x). In order to leave the rule of f unchanged we also subtract $\frac{9}{4}$:

$$f(x) = -4\left(x^2 - 3x + \frac{9}{4} - \frac{9}{4}\right) - 8$$

$$= -4\left(x^2 - 3x + \frac{9}{4}\right) - 4\left(-\frac{9}{4}\right) - 8$$

$$= -4\left(x - \frac{3}{2}\right)^2 + 1$$

The statement in the box above (with $a = -4$, $h = \frac{3}{2}$, $k = 1$) shows that the graph of f is a downward-opening parabola with vertex $(\frac{3}{2}, 1)$ and axis the line $x = \frac{3}{2}$. Its graph in Figure 5-5 on the next page is the graph of $h(x) = x^2$ shifted horizontally $\frac{3}{2}$ units to the right, stretched by a factor of 4, reflected in the x-axis, and shifted vertically 1 unit upward. ∎

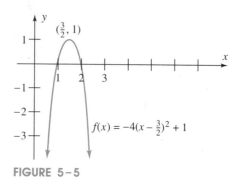

FIGURE 5-5

Example 5 Find the rule of a quadratic function f whose graph is a parabola with vertex $(-2, 5)$ that contains the point $(-3, 8)$.

Solution We know the rule of f can be written in the form $f(x) = a(x - h)^2 + k$, with (h, k) being the vertex. Thus $(h, k) = (-2, 5)$, so that $h = -2$ and $k = 5$. Hence $f(x) = a(x - (-2))^2 + 5 = a(x + 2)^2 + 5$. Since $(-3, 8)$ is on the graph we must have $f(-3) = 8$, that is,

$$a(-3 + 2)^2 + 5 = 8$$

Solving this equation for a shows that $a = 3$. Therefore the rule of f is $f(x) = 3(x + 2)^2 + 5 = 3x^2 + 12x + 17$. ∎

If $f(x) = ax^2 + bx + c$ is any quadratic function, then we can complete the square as in the examples above. The details are omitted, but you can easily check the result by multiplying out the right side here:

$$f(x) = ax^2 + bx + c = a\left(x - \left(\frac{-b}{2a}\right)\right)^2 + \left(c - \frac{b^2}{4a}\right).$$

Therefore

VERTEX OF A PARABOLA

> The graph of the quadratic function $f(x) = ax^2 + bx + c$ is a parabola whose vertex has x-coordinate $-b/2a$.

It isn't necessary to memorize the y-coordinate—just compute $f(-b/2a)$ to find it.

Example 6 The graph of $f(x) = -3x^2 + 12x - 18$ is a downward-opening parabola whose vertex has x-coordinate

$$-\frac{b}{2a} = -\frac{12}{2(-3)} = -\frac{12}{-6} = 2$$

and y-coordinate $f(2) = -3 \cdot 2^2 + 12 \cdot 2 - 18 = -6$. ∎

Applications

Suppose $(r, f(r))$ is the vertex of the graph of a quadratic function f. Depending on whether the parabola opens downward or upward, one of these cases must hold:

On the Graph	In Functional Notation
1. $(r, f(r))$ is the highest point	$f(x) \leq f(r)$ for every x
2. $(r, f(r))$ is the lowest point	$f(x) \geq f(r)$ for every x

In the first case we say that $f(r)$ is the **maximum value** of the function and in the second that $f(r)$ is the **minimum value**. The solution of many applied problems depends on finding the maximum or minimum value of a quadratic function.

Example 7 Find the area and dimensions of the largest rectangular field that can be enclosed with 3000 feet of fence.

Solution Let x denote the length and y the width of the field, as shown in Figure 5-6.

perimeter $= x + y + x + y = 2x + 2y$
area $= xy$

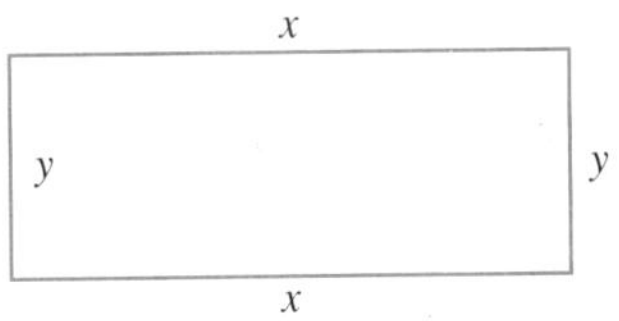

FIGURE 5-6

Since the perimeter is the length of the fence, $2x + 2y = 3000$. Hence $2y = 3000 - 2x$ and $y = 1500 - x$. Consequently, the area is

$$A = xy = x(1500 - x) = 1500x - x^2 = -x^2 + 1500x.$$

The largest possible area is just the maximum value of the quadratic function $A(x) = -x^2 + 1500x$. This maximum occurs at the vertex of the graph of $A(x)$ (which is a downward-opening parabola because the coefficient of x^2 is negative). The vertex may be found by completing the square, or by using the fact in the box opposite (with $a = -1$ and $b = 1500$):

$$\text{The } x\text{-coordinate of the vertex is } -\frac{1500}{2(-1)} = 750 \text{ feet.}$$

Hence the y-coordinate of the vertex, the maximum value of $A(x)$, is

$$A(750) = -750^2 + 1500 \cdot 750 = 562{,}500 \text{ sq ft}$$

It occurs when the length is $x = 750$. In this case the width is $y = 1500 - x = 1500 - 750 = 750$. ∎

Example 8 Find real numbers c and d whose difference is 5 and whose product is as small as possible.

Solution Since $c - d = 5$, we have $c = d + 5$. We want the product
$$cd = (5 + d)d = d^2 + 5d$$
to be a minimum. Since the graph of $f(d) = d^2 + 5d$ is an upward-opening parabola, the minimum value of $f(d)$ occurs at the vertex of the parabola. We can find the vertex by completing the square or using the fact in the box above (with $a = 1$ and $b = 5$):

The first coordinate of the vertex is $d = \dfrac{-5}{2 \cdot 1} = -\dfrac{5}{2}$.

Hence the smallest product occurs when $d = -\frac{5}{2}$ and $c = 5 + d = 5 - \frac{5}{2} = \frac{5}{2}$. The smallest value of cd is $(\frac{5}{2})(-\frac{5}{2}) = -\frac{25}{4}$. ∎

Example 9 When the fare on an airport bus is \$3, there are, on the average, 20 passengers per trip. With each fare increase of 25¢ the average number of passengers decreases by 1. What fare should be charged to have the largest possible revenue per trip?

Solution For each *dollar* the fare is raised, there are 4 fewer passengers per trip. So if the fare is increased by x dollars to $(3 + x)$ dollars, then there will be $20 - 4x$ passengers per trip. Then
$$\text{revenue} = (\text{fare}) \cdot (\text{number of passengers})$$
$$= (3 + x)(20 - 4x)$$
$$= -4x^2 + 8x + 60$$
We must find the value of x at which the quadratic function $r(x) = -4x^2 + 8x + 60$ takes its maximum value. The graph of $r(x)$ is a downward-opening parabola (why?), and the maximum value occurs at the vertex, whose x-coordinate is $x = \dfrac{-b}{2a} = \dfrac{-8}{2(-4)} = 1$. Therefore the fare should be $3 + 1 = \$4$. ∎

EXERCISES

In Exercises 1–12, *without graphing*, determine the vertex of the given parabola and state whether it opens upward or downward.

1. $f(x) = 3(x - 5)^2 + 2$
2. $g(x) = -6(x - 2)^2 - 5$
3. $y = -(x - 1)^2 + 2$
4. $h(x) = -x^2 + 1$
5. $y = -\dfrac{3}{2}\left(x + \dfrac{3}{2}\right)^2 + \dfrac{3}{2}$

6. $v = 656(t - 590)^2 + 7284$
7. $f(x) = -3x^2 + 4x + 5$
8. $g(x) = 2x^2 - x - 1$
9. $y = -x^2 + x$
10. $h(t) = \frac{1}{2}t^2 - \frac{3}{2}t + \frac{5}{2}$
11. $y + (x - 3)^2 = 1$
12. $y = -2x^2 + 2x - 1$

In Exercises 13–20, sketch the graph of the given quadratic function and label its vertex.

13. $h(x) = 4(x - 1)^2 + 2$
14. $k(x) = 3(x - 2)^2 - 3$
15. $p(x) = x^2 - 4x - 1$
16. $q(x) = x^2 + 8x + 6$
17. $f(x) = x^2 - 10x + 20$
18. $h(x) = x^2 + 3x + 1$
19. $f(x) = 2x^2 - 4x + 1$
20. $r(x) = -3x^2 + 9x - 5$

In Exercises 21–28, find the rule of the quadratic function whose graph is a parabola satisfying the given conditions.

21. vertex $(0, 0)$; $(2, 12)$ on graph
22. vertex $(0, 1)$; $(2, -7)$ on graph
23. vertex $(1, 0)$; $(3, 8)$ on graph
24. vertex $(-2, 0)$; $(-4, 16)$ on graph
25. vertex $(-3, 3)$; $(-2, 2)$ on graph
26. vertex $(2, -2)$; $(1, 1)$ on graph
27. vertex $\left(\frac{1}{2}, 2\right)$; $(3, 3)$ on graph
28. vertex $\left(-\frac{1}{4}, 3\right)$; $\left(\frac{3}{2}, 10\right)$ on graph

29. Find the number b such that the vertex of the parabola $y = x^2 + bx + c$ lies on the y-axis.
30. Find the number c such that the vertex of the parabola $y = x^2 + 8x + c$ lies on the x-axis.

31. What is the minimum product of two numbers whose difference is 4? What are the numbers?
32. Find numbers c and d whose sum is -18 and whose product is as large as possible.
33. Find two numbers such that the sum of one and twice the other is 36 and the product of the two numbers is as large as possible.
34. Find two numbers such that the difference of one and three times the other is 48 and the product of the two numbers is as small as possible.
35. The sum of the height h and the base b of a triangle is 30. What height and base will produce a triangle of maximum area?
36. A trough is to be made by bending a long, flat piece of tin 10 inches wide into a rectangular shape. What depth should the trough be in order to have the maximum possible cross sectional area?
37. A field bounded on one side by a river is to be fenced on three sides so as to form a rectangular enclosure. If the total length of fence to be used is 200 feet, what dimensions will yield an enclosure of the largest possible area?
38. A rectangular box (with top) has a square base. The sum of the lengths of its 12 edges is 8 feet. What dimensions should the box have in order that its surface area be as large as possible?
39. A salesperson finds that her sales average 40 cases per store when she visits 20 stores a week. Each time she visits an additional store per week, the average sales per store decrease by 1 case. How many stores should she visit each week if she wants to maximize her sales?
40. A miniature golf course averages 200 patrons per evening when it charges $2 per person. For each 5¢ increase in the admission price, the average attendance drops by 2 people. What admission price will produce the largest ticket revenue?
41. A potter can sell 120 bowls per week at $4 per bowl. For each 50¢ decrease in price 20 more bowls are sold. What price should be charged in order to maximize sales income?
42. A vendor can sell 200 souvenirs per day at a price of $2 each. Each 10¢ price increase decreases the number of sales by 25 per day. Souvenirs cost the vendor $1.50 each. What price should be charged in order to maximize the profit?
43. When a basketball team charges $4 per ticket, average attendance is 500 people. For each 20¢

decrease in ticket price, average attendance increases by 30 people. What should the ticket price be to insure maximum income?

44. A ballpark concessions manager finds that each salesperson sells an average of 40 boxes of popcorn per game when there are 20 salespeople working. When an additional salesperson is employed, each salesperson averages 1 less box per game. How many salespeople should be hired to insure maximum income?

In Exercises 45–48, use the formula for the height h of an object (that is traveling vertically subject only to gravity) at time t: $h = -16t^2 + v_0 t + h_0$, where h_0 is the initial height and v_0 the initial velocity.

45. A ball is thrown upward from the top of a 96-foot-high tower with an initial velocity of 80 feet per second. When does the ball reach its maximum height and how high is it at that time?

46. A rocket is fired upward from ground level with an initial velocity of 1600 feet per second. When does it attain its maximum height and what is that height?

47. A ball is thrown upward from a height of 6 feet with an initial velocity of 32 feet per second. Find its maximum height.

48. A bullet is fired upward from ground level with an initial velocity of 1500 feet per second. How high does it go?

49. A projectile is fired at an angle of 45° upward. Exactly t sec after firing, its vertical height above the ground is $500t - 16t^2$. What is the greatest height the projectile reaches and at what time does this occur?

5.2 POLYNOMIAL FUNCTIONS

A **polynomial function** is a function whose rule is given by a polynomial, such as

$$f(x) = x^3 + 1 \quad \text{or} \quad g(x) = 3x^4 - 6x^3 + 14x^2 - 6x + 7.$$

The rule of a polynomial function of degree n is of the form

$$f(x) = a_n x^n + a_{n-1} x^{n-1} + \cdots + a_2 x^2 + a_1 x + a_0$$

where $a_n, \ldots, a_0$ are constants and $a_n \neq 0$. The graphs of first-degree polynomial functions are straight lines, and the graphs of second-degree polynomial functions (quadratic functions) are parabolas.* The emphasis here will be on higher degree polynomial functions.

Example 1 The graphs of $g(x) = -x^3$ and $h(x) = 2x^3$ can be obtained by plotting points, as shown in Figure 5–7 on the next page. The graphs of $s(x) = ax^5$ and other *odd* powers of x have the same general shape: They move upward from left to right if $a > 0$, downward if $a < 0$, and are symmetric with respect to the origin (as defined in Section 4.4). ■

---

* See Sections 3.3 and 5.1 for details. It is not necessary to have read those sections in order to understand the material here.

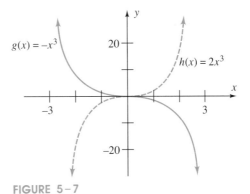

FIGURE 5-7

Example 2 The graphs of $f(x) = x^4$ and $k(x) = -\frac{1}{2}x^4$ shown in Figure 5-8 below can be obtained similarly. The graphs of $t(x) = ax^6$ and other *even* powers of x have the same general shape: They open upward if $a > 0$, downward if $a < 0$, and are symmetric with respect to the y-axis (as defined in Section 4.4). ∎

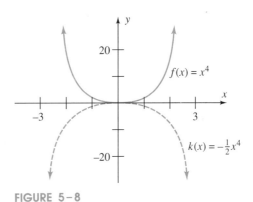

FIGURE 5-8

Here is a list of properties of graphs of polynomial functions. A complete justification of some of these statements requires calculus, so we shall simply illustrate them here.

Extent The domain of every polynomial function is the set of all real numbers, so the graph extends infinitely to the left and to the right. When $|x|$ is large (that is, when x is far from the origin), the graph looks approximately like the graph of the "highest degree term." For instance, the graph of $f(x) = 2x^3 - 6x^2 + 2x - 5$ looks like the graph of $y = 2x^3$ when $|x|$ is very large.* Therefore

---

* See Excursion 5.2.A on page 232 for the reason.

BEHAVIOR FOR LARGE x

> The graph of a polynomial function is far from the x-axis when $|x|$ is large.

x-Intercepts The graph meets the x-axis at the x-intercepts—the numbers c such that $f(c) = 0$. The solutions of $f(x) = 0$ are called the **roots** (or **zeros**)* of the polynomial $f(x)$. So *the x-intercepts occur at the roots of $f(x)$.* We shall see in Section 5.4 that a polynomial of degree n has, at most, n roots. Hence

x-INTERCEPTS

> The graph of a polynomial of degree n meets the x-axis, at most, n times.

Maxima and Minima Although some polynomial graphs are always increasing or always decreasing (such as $g(x) = -x^3$), the typical polynomial graph has "peaks" and "valleys":

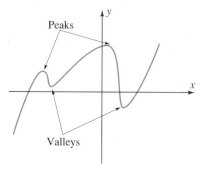

FIGURE 5-9

The technical term for a peak is **relative maximum** (plural maxima) and for a valley **relative minimum** (plural minima). Relative maxima and minima are often called **turning points** of the graph. In calculus it is proved that

TURNING POINTS

> The total number of relative maxima and minima on the graph of a polynomial function of degree n is, at most, $n - 1$.

Continuity The graph of a polynomial function is a *continuous* curve, which means that it is unbroken, with no jumps, gaps, or holes. In addition, polynomial graphs have no sharp corners. Thus none of the graphs in Figure 5-10 is the graph of a polynomial function.

* "Root" is the term used in advanced algebra texts. Mathematicians who work in analysis often prefer the term "zero."

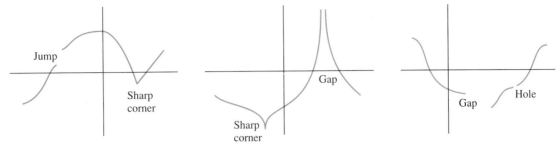

FIGURE 5-10

Since a polynomial graph is continuous, you can draw the graph between any two points *without lifting your pencil* from the paper (which can't always be done for the graphs in Figure 5-10 above). This fact has a very useful consequence. Suppose that f is a polynomial function and that c and d are numbers such that $f(c)$ is negative and $f(d)$ is positive, as shown in Figure 5-11.

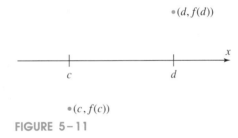

FIGURE 5-11

There is no way to draw the graph of a *function* from $(c, f(c))$ to $(d, f(d))$ without lifting your pencil *unless* the graph crosses the x-axis between c and d (try it!). But the x-intercepts of the graph occur at the roots of $f(x)$. Therefore

INTERMEDIATE VALUE THEOREM

> If $f(c)$ and $f(d)$ have opposite signs, then the polynomial $f(x)$ has a root between c and d.
>
> Consequently, if $f(x)$ has *no* roots in an interval, then
>
> either $f(x)$ is positive for every x in the interval
>
> or $f(x)$ is negative for every x in the interval.

The most efficient way to graph polynomial functions is to use a computer or graphing calculator, together with techniques from calculus, to find the exact location of the relative maxima and minima and the x-intercepts, as well as the manner in which the curve bends (as discussed on page 169).

When these tools are not available, the facts above may be used to make quick sketches of the graphs of factored polynomial functions. These sketches indicate the correct general shape of the graph and can be made reasonably accurate by plotting additional points.

Example 3 To graph $f(x) = 2x^3 + x^2 - 6x$, we begin by factoring:

$$f(x) = x(2x^2 + x - 6) = x(x + 2)(2x - 3).$$

Its *only* roots are $x = 0$, $x = -2$, and $x = \frac{3}{2}$. So $f(x)$ has no roots in any of these intervals:

$$x < -2, \quad -2, < x < 0, \quad 0 < x < \frac{3}{2}, \quad x > \frac{3}{2}.$$

Choose a number in the second interval, say, -1, and note that

$$f(-1) = 2 \cdot (-1)^3 + (-1)^2 - 6 \cdot (-1) = 5.$$

Since $f(x)$ has no roots between -2 and 0 and $f(-1)$ is positive, the Intermediate Value Theorem above implies that $f(x)$ is positive for *every* x between -2 and 0. Hence the graph is above the x-axis between -2 and 0. By choosing a "test number" in each of the other intervals, we can determine the position of the graph relative to the x-axis:

Interval	$x < -2$	$-2 < x < 0$	$0 < x < \frac{3}{2}$	$x > \frac{3}{2}$
Test number in this interval	-3	-1	1	2
Value of $f(x)$ at test number	$f(-3) = -27$	$f(-1) = 5$	$f(1) = -3$	$f(2) = 8$
Sign of $f(x)$ in interval	$-$	$+$	$-$	$+$
Graph of f	Below x-axis	Above x-axis	Below x-axis	Above x-axis

The information in the chart suggests the rough sketch in Figure 5–12.

5.2 POLYNOMIAL FUNCTIONS

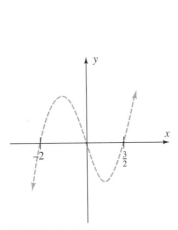

FIGURE 5-12

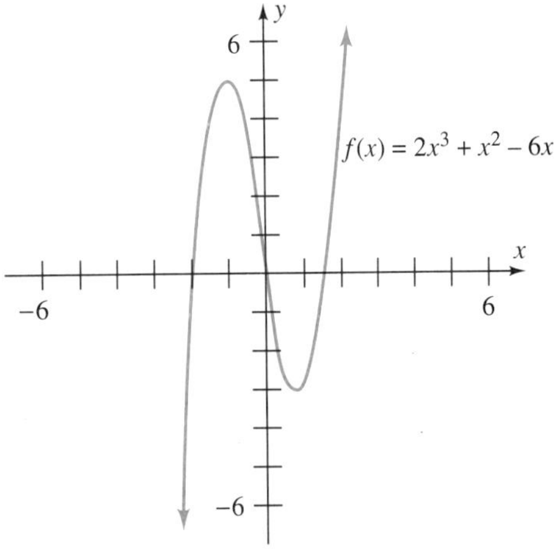

FIGURE 5-13

We claim that this sketch gives the correct basic shape of the graph. The graph of $f(x)$ looks like the graph of $y = 2x^3$ when $|x|$ is large. So the graph must drop sharply below the x-axis on the far left and rise sharply above it on the far right, as shown above. Since the graph touches the x-axis at -2 and 0, lies above the axis in between, and is continuous, it must turn at least once. Similarly, it lies below the x-axis between 0 and $\frac{3}{2}$ and hence must turn at least once there. This accounts for at least two turning points. But $f(x) = 2x^3 + x^2 - 6x$ has degree 3 and therefore has *at most* a total of 2 relative maxima and minima. So the two turning points shown above are the only ones.

Hence the sketch is essentially correct. By plotting additional points, we obtain the more accurate finished graph in Figure 5-13 above. The exact location of the turning points, however, requires calculus. ∎

Example 4 To graph $g(x) = \frac{1}{2}(x + 3)^2(x^2 - 1)$, we first factor:

$$g(x) = \frac{1}{2}(x + 3)^2(x + 1)(x - 1).$$

The roots of $g(x)$ are $x = -3$, $x = -1$, and $x = 1$, so $g(x)$ has no roots in any of these intervals:

$$x < -3, \quad -3 < x < -1, \quad -1 < x < 1, \quad x > 1.$$

By choosing a test number in each interval, we obtain:

Interval	$x < -3$	$-3 < x < -1$	$-1 < x < 1$	$x > 1$
Test number in this interval	-4	-2	0	3
Value of $g(x)$ at test number	$g(-4) = \frac{15}{2}$	$g(-2) = \frac{3}{2}$	$g(0) = -\frac{9}{2}$	$g(3) = 144$
Sign of $g(x)$ in interval	$+$	$+$	$-$	$+$
Graph of g	Above x-axis	Above x-axis	Below x-axis	Above x-axis

When multiplied out, $g(x)$ has leading term $\frac{1}{2}x^4$. So its graph has at most three turning points and looks like the graph of $\frac{1}{2}x^4$ when $|x|$ is large. These facts and the information in the chart suggest the rough sketch in Figure 5–14. Although calculus is needed to determine the exact location of the turning points, plotting more points leads to a reasonably accurate graph in Figure 5–15. ∎

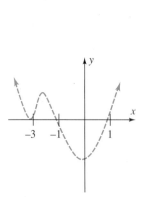

FIGURE 5–14

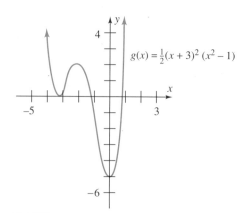

FIGURE 5–15

EXERCISES

1. Sketch an accurate graph of $f(x) = x^6$ for $-3 \leq x \leq 3$.

2. Sketch an accurate graph of $f(x) = x^5$ for $-3 \leq x \leq 3$.

In Exercises 3–8, decide whether the given graph could *possibly* be the graph of a polynomial function.

3.
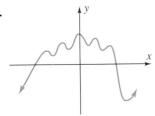

4.
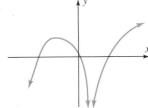

5.

6.
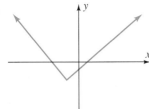

7.
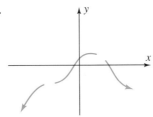

8.
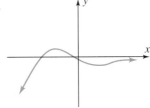

In Exercises 9–14, determine whether the given graph could possibly be the graph of a polynomial function of degree 3, of degree 4, or of degree 5.

9.
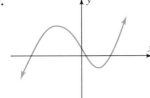

10.
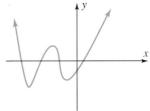

11.
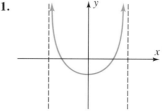

232 5 POLYNOMIAL AND RATIONAL FUNCTIONS

12.

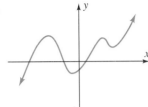

13.

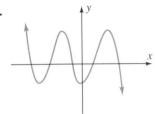

14.

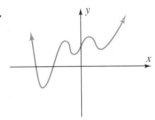

In Exercises 15–20, use the techniques of Section 4.4 and the graphs of $f(x) = x^4$ and $h(x) = 2x^3$ on page 225 to sketch the graph of the given function.

15. $g(x) = 2x^3 + 5$
16. $k(x) = x^4 + 1$
17. $g(x) = -2x^3$
18. $k(x) = 2x^4$
19. $g(x) = (x - 2)^4$
20. $k(x) = 2(x + 3)^4 - 3$

In Exercises 21–32, factor the given function (if necessary) and use the method presented in the text to sketch its graph.

21. $f(x) = (x - 1)(x + 5)(x + 2)$
22. $g(x) = (x - 2)^2(x - 4)$
23. $h(x) = -x(x - 3)^2(2x + 3)$
24. $k(x) = x^2(x + 1)(3x + 7)$
25. $f(x) = \frac{1}{6}(x - 1)(x - 2)(x - 3)(x - 4)$
26. $g(x) = -x(x - 1)(x + 2)^2$
27. $h(x) = x^3 - 4x$
28. $k(x) = x^4 - 5x^2 + 5$
29. $f(x) = x^4 - 6x^2 + 8$
30. $g(x) = x^4 - x^2$
31. $f(x) = x^3 - 3x^2 - 4x$
32. $g(x) = x^3 - 3x^2 - 4x + 2$ [*Hint:* Use Exercise 31.]

In Exercises 33–36, sketch the graph of the function. You may need the Rational Solutions Test (page 100) and the Factor Theorem (page 101) to factor the polynomial.

33. $f(x) = x^3 - 3x^2 - x + 3$
34. $g(x) = x^3 - 3x^2 + x + 1$
35. $f(x) = x^3 - 4x^2 + 2x + 3$
36. $h(x) = x^4 - 4x^3 + 4x - 1$

5.2.A EXCURSION: What Happens for Large x?

Figure 5–16 shows a combined graph of

$$p(x) = x, \quad q(x) = x^2, \quad r(x) = x^3, \quad s(x) = x^4$$

for positive values of x. Figure 5–17 is a combined graph of these same functions from 0 to 100, where the scale has been adjusted to accommodate x^4.

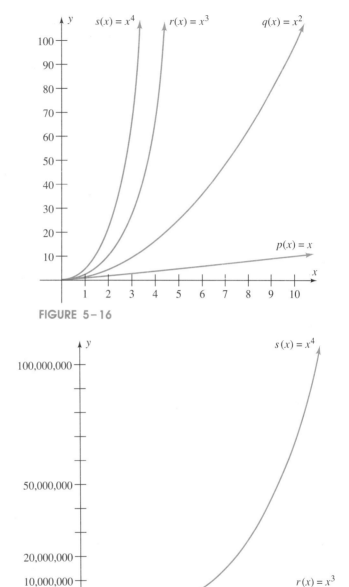

FIGURE 5-16

FIGURE 5-17

In Figure 5-17 the values of x^3, x^2, and x are so small *in comparison to x^4* that their graphs hardly even show up!* For every n, the same thing is true about x^n

* Analogous remarks hold when x is negative. In that case, when $|x|$ is large, then $|x^4|$ is very much larger than $|x^3|$, which is in turn much larger than $|x^2|$.

in comparison to lower powers of x. This is why the graph of a polynomial function looks like the graph of its highest degree term when x is large in absolute value.

Example 1 When x is small, the graph of $f(x) = 2x^3 + x^2 - 6x$ (Figure 5-13 on page 229) looks quite different from the graph of $h(x) = 2x^3$ (Figure 5-7 on page 225). But when x is large in absolute value, we have this table:

x	-100	-50	70	100
$-6x$	600	300	-420	-600
x^2	10,000	2,500	4,900	10,000
$h(x) = 2x^3$	$-2,000,000$	$-250,000$	686,000	2,000,000
$f(x) = 2x^3 + x^2 - 6x$	$-1,989,400$	$-247,200$	690,480	2,009,400

The table shows that when $|x|$ is large, the terms x^2 and $-6x$ are insignificant compared with $2x^3$, and consequently, the values of $f(x)$ and $h(x)$ are relatively close.

Here's what that means in graphical terms. If a scale of 1 inch = 32 units on the x-axis and 1 inch = 450,000 units on the y-axis is used, the graph of $f(x) = 2x^3 + x^2 - 6x$ from $x = -100$ to $x = 100$ will fill this entire page. Even so, it will be barely distinguishable from the graph of $h(x) = 2x^3$. At $x = 100$ the two graphs will be only $\frac{1}{50}$ of an inch apart. If the scale is chosen so as to reduce the overall size of the graphs (to the screen size of a graphing calculator, for instance), then the graphs will look virtually identical. ∎

5.3 RATIONAL FUNCTIONS

A **rational function** is a function whose rule is the quotient of two polynomials, such as

$$f(x) = \frac{1}{x}, \quad t(x) = \frac{4x - 3}{2x + 1}, \quad k(x) = \frac{2x^2 + 5x + 2}{x^3 - 7x + 6}.$$

A rational function $f(x) = g(x)/h(x)$ is defined whenever the denominator is nonzero. So the *domain* of f is the set of all real numbers that are *not* roots of the denominator $h(x)$.

> **WARNING**
>
> You have often cancelled factors, as here:
>
> $$\frac{x^2 - 1}{x - 1} = \frac{(x + 1)(x - 1)}{x - 1} = x + 1$$

> But the *function* with rule $p(x) = x + 1$ is *not* the same as the *function* with the rule $q(x) = \dfrac{x^2 - 1}{x - 1}$. For when $x = 1$,
>
> $$p(1) = 1 + 1 = 2 \quad \text{but} \quad q(1) = \dfrac{1^2 - 1}{1 - 1} = \dfrac{0}{0}.$$
>
> So $q(1)$ is *not defined*, whereas $p(1)$ *is* defined. For any number *except* 1, the two functions have the same value. But the difference for $x = 1$ makes them *different functions*.

A computer or graphing calculator is the most efficient way to graph rational functions. But you won't be able to use one effectively or to appreciate the key features of the graph it produces unless you understand the behavior of rational functions. The key to this behavior is a simple fact from arithmetic:

THE BIG-LITTLE PRINCIPLE

> The farther a number c is from 0, the closer the number $1/c$ is to 0. Conversely, the closer c is to 0, the farther $1/c$ is from 0. In less precise, but more suggestive, terms:
>
> $$\dfrac{1}{\text{big}} = \text{little} \quad \text{and} \quad \dfrac{1}{\text{little}} = \text{big}$$

For example, 5000 is bigger (farther from 0) than 100, but 1/5000 is smaller (closer to 0) than 1/100. Similarly, $-1/1000$ is very close to 0, but $1/(-1/1000) = -1000$ is far from 0.

Example 1 To graph $f(x) = \dfrac{3x + 5}{2x - 4}$ we first note that 2 is the root of the denominator, so f is not defined at $x = 2$. Next we consider the behavior of f near the bad point $x = 2$. When $x > 2$ and x is *very close* to 2, then $f(x)$ is a large positive number. For instance,

$$f(2.0005) = \dfrac{3(2.0005) + 5}{2(2.0005) - 4} = \dfrac{11.0015}{.001} = 11001.5$$

When $x < 2$ and x is very close to 2, $f(x)$ is a negative number, far from 0. For instance,

$$f(1.999) = \dfrac{3(1.999) + 5}{2(1.999) - 4} = \dfrac{10.997}{-.002} = -5498.5$$

The reason why this happens is that when x is very close to 2, the numerator $3x + 5$ is very near $3 \cdot 2 + 5 = 11$ and the denominator $2x - 4$ is very near $2 \cdot 2 - 4 = 0$. So the quotient $f(x)$ is very close to $11 \cdot \dfrac{1}{\text{little}}$, and hence is very far from 0 by the Big-Little Principle. Therefore the graph must move farther and farther from the *x*-axis as *x* gets closer to 2, as shown in Figure 5–18. The graph gets very close to the vertical line $x = 2$, but never touches it. The line is called a **vertical asymptote** of the graph.

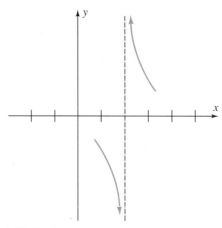

FIGURE 5–18

To determine the shape of the graph when *x* is large in absolute value (far from 0), note that for any nonzero *x*, we can rewrite the rule of *f*:

$$f(x) = \frac{3x + 5}{2x - 4} = \frac{\frac{3x + 5}{x}}{\frac{2x - 4}{x}} = \frac{3 + \frac{5}{x}}{2 - \frac{4}{x}}$$

As *x* gets larger in absolute value (far from 0), both $5/x$ and $4/x$ get very close to 0 by the Big-Little Principle. Consequently, $f(x) = \frac{3 + (5/x)}{2 - (4/x)}$ gets very close to $\frac{3 + 0}{2 - 0} = \frac{3}{2}$. So the graph gets closer and closer to the horizontal line $y = 3/2$, but never touches it, as shown in Figure 5–19. This line is called a **horizontal asymptote** of the graph.

The *x*-intercepts of the graph occur where the numerator is 0 and the denominator nonzero. Setting $3x + 5 = 0$ we see that the only *x*-intercept is $x = -5/3$. The *y*-intercept occurs at $f(0) = -5/4$. This information, along with plotting a few points, produces the graph in Figure 5–19. ∎

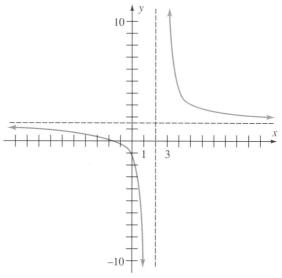

FIGURE 5-19

The same analysis works in the general case (just replace 3, 5, 2, −4 by a, b, c, d):

LINEAR RATIONAL FUNCTIONS

> The graph of $f(x) = \dfrac{ax+b}{cx+d}$ (with $c \neq 0$ and $ad \neq bc$) has a
>
> vertical asymptote at $x = -d/c$ (the root of the denominator);
> horizontal asymptote at $y = a/c$;
> x-intercept at $-b/a$ (the root of the numerator);
> y-intercept at b/d.

The actual graph may be obtained by using a graphing calculator or by plotting some points by hand. Even if you use a graphing calculator, you should know the facts listed in the box since the calculator may not clearly indicate the asymptotes or intercepts.

If you sketch the graph by hand, plot a point or two on *both* sides of the vertical asymptote, to determine whether the graph rises or falls. You may also have to plot points on the far left and far right to see if the graph is above or below the horizontal asymptote. Some examples are shown in Figure 5-20 on the next page.

$f(x) = \dfrac{-5x + 12}{2x - 4}$

Vertical asymptote $x = 2$

Horizontal asymptote $y = -\dfrac{5}{2}$

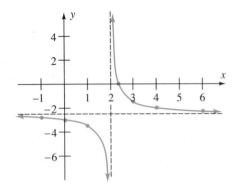

$k(x) = \dfrac{3x + 6}{x} = \dfrac{3x + 6}{1x + 0}$

Vertical asymptote $x = 0$

Horizontal asymptote $y = \dfrac{3}{1} = 3$

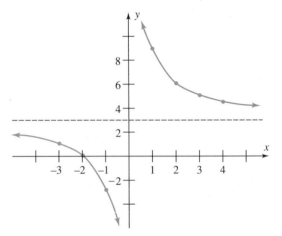

$k(x) = \dfrac{-3}{2x + 5} = \dfrac{0x - 3}{2x + 5}$

Vertical asymptote $x = -\dfrac{5}{2}$

Horizontal asymptote $y = \dfrac{0}{2} = 0$

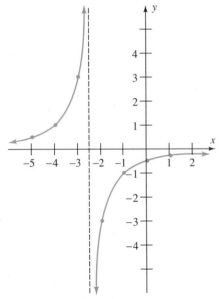

$f(x) = \dfrac{1}{x} = \dfrac{0x + 1}{1x + 0}$

Vertical asymptote $x = 0$

Horizontal asymptote $y = \dfrac{0}{1} = 0$

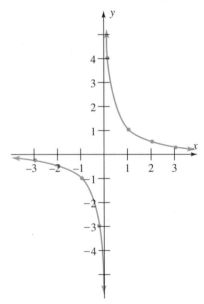

FIGURE 5-20

Other Rational Functions

Here are some important facts about the graphs of more complicated rational functions.

Continuity There will be breaks in the graph (holes or vertical asymptotes) wherever the function is not defined. Except for breaks at these undefined points, the graph is a continuous unbroken curve. In addition, the graph has no sharp corners.

Turning Points The graph may have some relative maxima and minima (peaks and valleys). Without calculus, they cannot be located exactly.

Intercepts The x-intercepts of $f(x) = g(x)/h(x)$, if any, occur at each number c that is a root of the numerator $g(x)$, but *not* of the denominator $h(x)$ (so that $f(c) = g(c)/h(c) = 0/h(c) = 0$). As with any function, the y-intercept occurs at $f(0)$.

Signs If $g(x)$ and $h(x)$ are polynomials, then for any number x the sign ($\pm$) of the product $g(x)h(x)$ is the same as the sign of the quotient $g(x)/h(x)$. So the facts developed for polynomial functions on page 227 can be applied here. In particular, the Intermediate Value Theorem leads to this conclusion:

SIGNS

> On an interval in which neither $g(x)$ nor $h(x)$ has a root, the quotient $g(x)/h(x)$ is
>
> either positive for every x in the interval
>
> or negative for every x in the interval.

Vertical Asymptotes The function $f(x) = g(x)/h(x)$ has a vertical asymptote at every number that is a root of the denominator $h(x)$, but *not* of the numerator $g(x)$. If $x = d$ is a root of both $g(x)$ and $h(x)$, then the graph of $f(x)$ has either a vertical asymptote or a hole at $x = d$; see Figure 5–21 and Exercise 19.

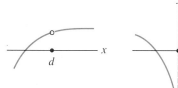

Hole at $x = d$

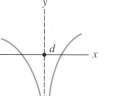

Vertical asymptotes at $x = d$

FIGURE 5–21

Horizontal Asymptotes Consider the function $f(x) = \dfrac{3x^4 - 2x}{2x^4 + 5}$, in which the numerator and denominator are polynomials of the *same* degree. When $x \neq 0$, we can divide both the numerator and denominator by the highest power of x that occurs, namely, x^4, without affecting the value of $f(x)$:

$$f(x) = \frac{3x^4 - 2x}{2x^4 + 5} = \frac{\frac{3x^4 - 2x}{x^4}}{\frac{2x^4 + 5}{x^4}} = \frac{3 - \frac{2}{x^3}}{2 + \frac{5}{x^4}}$$

As x gets farther from 0, the terms $2/x^3$ and $5/x^4$ get closer and closer to 0 by the Big-Little Principle. So $f(x)$ get closer and closer to $\dfrac{3-0}{2+0} = \dfrac{3}{2}$. Thus as x gets farther from 0, the graph of f gets closer and closer to the horizontal line $y = 3/2$. Hence this line is a horizontal asymptote of the graph. Similar arguments work in the general case:

HORIZONTAL ASYMPTOTES

Let $f(x) = \dfrac{ax^n + \cdots}{cx^k + \cdots}$ be a rational function whose numerator has degree n and whose denominator has degree k.

If $n = k$, then the line $y = a/c$ is a horizontal asymptote.
If $n < k$, then the x-axis is a horizontal asymptote.
If $n > k$, then there is no horizontal asymptote.*

WARNING

The functions in the preceding examples were all of the form $f(x) = \dfrac{ax + b}{cx + d}$, with $c \neq 0$ and $ad \neq bc$. The graphs of such functions never cross their horizontal asymptotes. But this may not be the case when higher degree polynomials are involved. The graph of an arbitrary rational function gets very close to, but does not touch, its horizontal asymptote *when x is very large in absolute value*. But for some smaller values of the x, the graph *may* cross the asymptote, as the examples below illustrate.

The facts about asymptotes and intercepts presented above may be used in conjunction with a graphing calculator to determine accurate graphs of rational functions. They may also be used, together with the information on

---

* But there may be other asymptotes, as we shall see below.

5.3 RATIONAL FUNCTIONS

signs, to obtain reasonably accurate graphs by hand, with a minimum of point plotting.

Example 2 To graph $f(x) = \dfrac{x-1}{x^2-x-6}$, we begin by factoring:

$$f(x) = \frac{x-1}{x^2-x-6} = \frac{x-1}{(x+2)(x-3)}.$$

The factored form allows us to read off the necessary information:

Vertical Asymptotes: $x = -2$ and $x = 3$ (roots of the denominator but not of the numerator).

Horizontal Asymptote: x-axis (because denominator has larger degree than the numerator).

Intercepts: y-intercept at $f(0) = \dfrac{0-1}{0^2-0-6} = \dfrac{1}{6}$; x-intercept at $x = 1$ (root of the numerator but not of the denominator).

Signs: The roots of the numerator and denominator ($1, -2$, and 3) determine four intervals in which neither numerator nor denominator has a root:

$$x < -2, \quad -2 < x < 1, \quad 1 < x < 3, \quad x > 3.$$

On each of these intervals $f(x)$ is either always positive or always negative (see the box labeled "Signs" above). On any one of these intervals, the sign of $f(x)$ is the same for every number in the interval. So we can determine that sign by evaluating $f(x)$ at just one number in each interval, as summarized here:

Interval	$x < -2$	$-2 < x < 1$	$1 < x < 3$	$x > 3$
Test number in this interval	-3	0	2	4
Value of $f(x)$ at test number	$f(-3) = -\dfrac{2}{3}$	$f(0) = \dfrac{1}{6}$	$f(2) = -\dfrac{1}{4}$	$f(4) = \dfrac{1}{2}$
Sign of $f(x)$ in interval	$-$	$+$	$-$	$+$
Graph of f	Below x-axis	Above x-axis	Below x-axis	Above x-axis

Now we sketch the asymptotes, as in Figure 5–22 on the next page. The last line of the chart enables us to determine how the graph behaves near each vertical asymptote. For instance, since the graph is always above the x-axis between -2 and 1, it must rise sharply just to the right of -2. Similarly, since the graph is below the x-axis when $x < -2$, it must fall sharply just to the left of -2. Finally, we plot the intercepts, the points corresponding to the test numbers, and a few additional points, to obtain the graph in Figure 5–22. ∎

5 POLYNOMIAL AND RATIONAL FUNCTIONS

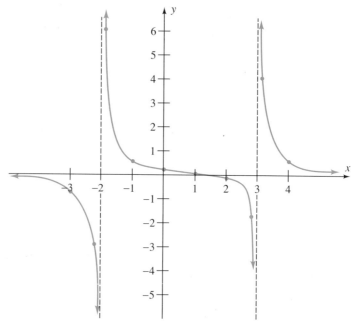

FIGURE 5-22

Example 3 To graph $f(x) = \dfrac{2x^2}{x^2 + x - 2}$ we factor and then read off the necessary information:

$$f(x) = \frac{2x^2}{x^2 + x - 2} = \frac{2x^2}{(x + 2)(x - 1)}.$$

Vertical Asymptotes: $x = -2$ and $x = 1$ (roots of denominator).

Horizontal Asymptote: $y = 2/1 = 2$ (because numerator and denominator have the same degree; see the box on page 240).

Intercepts: x-intercept at $x = 0$ (root of numerator); y-intercept at $f(0) = 0$.

Signs: The roots of numerator and denominator $(0, -2, 1)$ determine the intervals to test:

Interval	$x < -2$	$-2 < x < 0$	$0 < x < 1$	$x > 1$
Test number in this interval	-3	-1	$\frac{1}{2}$	3
Value of $f(x)$ at test number	$f(-3) = \frac{9}{2}$	$f(-1) = -1$	$f\left(\frac{1}{2}\right) = -\frac{2}{5}$	$f(3) = \frac{9}{5}$
Sign of $f(x)$ in interval	$+$	$-$	$-$	$+$
Graph of f	Above x-axis	Below x-axis	Below x-axis	Above x-axis

We sketch the asymptotes and use the last line of the chart to determine how the graph looks on each side of each vertical asymptote. Plotting some points leads to the sketch in Figure 5–23.

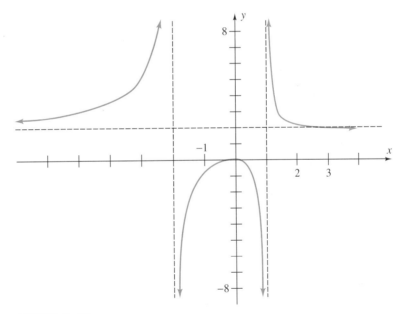

FIGURE 5–23

Since $f(0) = 0$ and $f(x) < 0$ for other x between -2 and 1, we know that the turning point shown at $x = 0$ is accurate.* The only problem with Figure 5–23 is that you can't really see what's going on when $x > 2$. By using the magnifying feature of a graphing calculator or by greatly expanding the scale and plotting many points, we see that this portion of the graph looks like Figure 5–24:

---

* Calculus is usually needed to determine the exact location of the turning points.

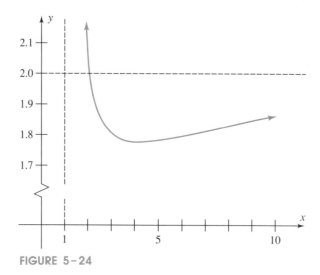

FIGURE 5-24

So the graph crosses the horizontal asymptote at $x = 2$, has a turning point near $x = 4$, and then stays below the asymptote, moving closer and closer as x takes larger values. ∎

When the degree of the denominator of a rational function is smaller than the degree of the numerator, then the graph has no horizontal asymptotes. But it may have an **oblique asymptote** (a straight line that is neither horizontal nor vertical). Or it may have curves other than straight lines as asymptotes (Exercise 25).

Example 4 Although the numerator of $p(x) = \dfrac{x^2 + 1}{x - 1}$ can't be factored, we can still determine the salient features of its graph.

Vertical Asymptote: $x = 1$ (root of the denominator).

Intercepts: There is no x-intercept because the numerator has no real roots. The y-intercept is $p(0) = -1$.

Signs: Since $x = 1$ is the only real root of the numerator or denominator, there are just two intervals to consider: $x < 1$ and $x > 1$. You can make a sign chart, as in the preceding example, or simply observe that $x^2 + 1$ is always positive. So $p(x) = \dfrac{x^2 + 1}{x - 1}$ is positive when $x - 1$ is positive (namely, when $x > 1$) and negative when $x - 1$ is negative ($x < 1$).

Other Asymptotes: There is no horizontal asymptote because the degree of the numerator is larger than the degree of the denominator. Divide the numerator by the denominator (by synthetic or long division) and apply the Division Algorithm:

$$\text{dividend} = (\text{divisor})(\text{quotient}) + \text{remainder}$$
$$x^2 + 1 = (x - 1)(x + 1) + 2$$

Dividing both sides by $x - 1$ shows that

$$p(x) = \frac{x^2 + 1}{x - 1} = (x + 1) + \frac{2}{x - 1}$$

Now when x is large, so is $x - 1$. Hence $2/(x - 1)$ must be very close to 0 by the Big-Little Principle. In this case $p(x)$ is very close to $(x + 1) + 0$. Therefore, as x gets larger in absolute value, the graph of $p(x)$ gets closer and closer to the line $y = x + 1$, as shown in Figure 5-25. This line is an oblique asymptote of the graph.

Finally, we plot some more points and obtain Figure 5-25. ∎

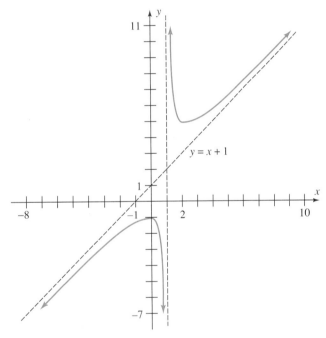

FIGURE 5-25

EXERCISES

In Exercises 1-8, graph the function. (See the box on page 237.)

1. $f(x) = \dfrac{1}{x + 5}$

2. $q(x) = \dfrac{-7}{x - 6}$

3. $k(x) = \dfrac{-3}{2x + 5}$

4. $g(x) = \dfrac{-4}{2 - x}$

5. $f(x) = \dfrac{3x}{x - 1}$

6. $p(x) = \dfrac{x - 2}{x}$

7. $f(x) = \dfrac{2 - x}{x - 3}$

8. $g(x) = \dfrac{3x - 2}{x + 3}$

In Exercises 9-18, graph the function.

9. $f(x) = \dfrac{1}{x(x + 1)^2}$

10. $g(x) = \dfrac{x}{2x^2 - 5x - 3}$

11. $f(x) = \dfrac{x-3}{x^2+x-2}$ 12. $g(x) = \dfrac{x+2}{x^2-1}$

13. $h(x) = \dfrac{(x^2+6x+5)(x+5)}{(x+5)^3(x-1)}$

14. $f(x) = \dfrac{x^2-1}{x^3-2x^2+x}$ 15. $f(x) = \dfrac{-4x^2+1}{x^2}$

16. $k(x) = \dfrac{x^2+1}{x^2-1}$ 17. $q(x) = \dfrac{x^2+2x}{x^2-4x-5}$

18. $F(x) = \dfrac{x^2+x}{x^2-2x+4}$

19. Graph $p(x) = \dfrac{(x+3)(x-3)}{(x+3)(x-5)(x+4)}$. [*Hint:* For all $x \neq -3$, $p(x)$ is the same as $q(x) = \dfrac{x-3}{(x-5)(x+4)}$, so the graph of $p(x)$ is just the graph of $q(x)$ with a *hole* at $x = -3$.]

20. Graph $p(x) = \dfrac{x^3+3x^2}{x^4-4x^2}$. (See the hint for Exercise 19.)

In Exercises 21–24, graph the function. Each graph has an oblique asymptote. (See Example 4.)

21. $f(x) = \dfrac{x^2-x-6}{x-2}$ 22. $k(x) = \dfrac{x^2+x-2}{x}$

23. $Q(x) = \dfrac{4x^2+4x-3}{2x-5}$

24. $K(x) = \dfrac{3x^2-12x+15}{3x+6}$

25. Graph the function $f(x) = \dfrac{x^3-2}{x-1}$ as follows:
   (a) Determine the x-intercepts, vertical asymptotes, signs, and sign changes as usual.
   (b) Divide $x^3 - 2$ by $x - 1$ and use the Division Algorithm (as in Example 4) to show that
   $f(x) = (x^2 + x + 1) + (-1)/(x - 1)$.
   When $|x|$ is large, $-1/(x-1)$ is very close to 0 (why?), so that $f(x)$ is very close to $x^2 + x + 1$. Thus the curve $y = x^2 + x + 1$ is an asymptote of the graph of f. [*Note:* $y = x^2 + x + 1$ is a parabola—see box on page 219.]
   (c) Plot some points and use parts (a) and (b) to sketch the graph of f.

In Exercises 26–29, graph the function. Each graph has a parabolic asymptote. (See Exercise 25.)

26. $p(x) = \dfrac{x^3+8}{x+1}$ 27. $q(x) = \dfrac{x^3-1}{x-2}$

28. $f(x) = \dfrac{x^4-1}{x^2}$

29. $k(x) = \dfrac{(x+1)(x-1)(x+3)}{x+2}$

30. Find the rule of a rational function f that has these properties:
   (i) The curve $y = x^3 - 8$ is an asymptote of the graph of f.
   (ii) $f(2) = 1$.
   (iii) $x = 1$ is a vertical asymptote of the graph.

Unusual Problems

31. The formula for the gravitational acceleration of an object (relative to the earth) is $g(r) = (4 \cdot 10^{14})/r^2$, where r is the distance of the object from the center of the earth (in meters).
   (a) What is the gravitational acceleration at the earth's surface? (The radius of the earth is approximately $(6.4)(10^6)$ m.)
   (b) Graph the function g.
   (c) Does the function g have any roots? Does the gravitational acceleration ever vanish for any value of r? Can you ever "escape the pull of gravity"?

In Exercises 32–34, graph the given function. (*Note:* These graphs are somewhat trickier than those discussed in the text. Some involve peaks and valleys whose existence may not be evident unless you plot many points (or use calculus). Furthermore, you may not be able to find all of the roots of these functions exactly.)

32. $k(x) = \dfrac{2x^3+1}{x^2-1}$

33. $p(x) = \dfrac{(x-2)(x^2+1)}{x^2-1}$

34. $f(x) = \dfrac{3x^3-11x-1}{x^2-4}$

5.4 THEORY OF EQUATIONS*

Most of Chapter 2 was spent solving polynomial equations, that is, finding roots of polynomials. In this section we take a deeper look at this problem and present several important results that describe the number and nature of the roots of various kinds of polynomials.

Long division of polynomials (Section 1.5) led to the Division Algorithm, which we restate here in functional notation:

THE DIVISION ALGORITHM

> If a polynomial $f(x)$ is divided by a nonzero polynomial $h(x)$, then there is a quotient polynomial $q(x)$ and a remainder polynomial $r(x)$ such that
> $$f(x) = h(x)q(x) + r(x)$$
> where either $r(x) = 0$ or $r(x)$ has degree less than the degree of the divisor $h(x)$.

When the divisor is a *first*-degree polynomial of the form $h(x) = x - c$, then

(∗) $$f(x) = (x - c)q(x) + r(x)$$

where $r(x) = 0$ or $r(x)$ has smaller degree than $h(x) = x - c$. But $h(x)$ has degree 1, so the only polynomials of smaller degree are the nonzero constant polynomials (each of which has degree 0). In each case, then, the remainder $r(x)$ is just a real number d (possibly 0), and (∗) becomes

$$f(x) = (x - c)q(x) + d$$

If we evaluate the function f at $x = c$, we obtain

$$f(c) = (c - c)q(c) + d = 0 \cdot q(c) + d = d.$$

So the remainder d is just the value of $f(x)$ at $x = c$. We have proved:

THE REMAINDER THEOREM

> If a polynomial $f(x)$ is divided by $x - c$, then the remainder is the number $f(c)$.

Example 1 To find the remainder when $f(x) = x^{79} + 3x^{24} + 5$ is divided by $x - 1$, we apply the Remainder Theorem with $c = 1$. The remainder is

$$f(1) = 1^{79} + 3 \cdot 1^{24} + 5 = 1 + 3 + 5 = 9. \blacksquare$$

---

* Section 2.5 (Complex Numbers) is a prerequisite for this section.

Example 2 To find the remainder when $f(x) = 3x^4 - 8x^2 + 11x + 1$ is divided by $x + 2$, we must apply the Remainder Theorem *carefully*. The divisor in the theorem is $x - c$, not $x + c$. So we rewrite $x + 2$ as $x - (-2)$ and apply the theorem with $c = -2$. The remainder is

$$f(-2) = 3(-2)^4 - 8(-2)^2 + 11(-2) + 1 = 48 - 32 - 22 + 1 = -5. \blacksquare$$

Example 3 Show that $x - 3$ is a factor of $f(x) = x^3 - 4x^2 + 2x + 3$.

Solution The remainder when $f(x) = x^3 - 4x^2 + 2x + 3$ is divided by $x - 3$ is

$$f(3) = 3^3 - 4 \cdot 3^2 + 2 \cdot 3 + 3 = 27 - 36 + 6 + 3 = 0.$$

Since division by $x - 3$ leaves remainder 0, $x - 3$ is a *factor* of $f(x)$, as explained on page 40. Furthermore 3 is a *root* of $f(x)$ because $f(3) = 0$. $\blacksquare$

Example 3 is an illustration of:

THE FACTOR THEOREM | **The number c is a root of the polynomial $f(x)$ exactly when $x - c$ is a factor of $f(x)$.**

Examples of the use of the Factor Theorem were given on pages 101–102.

Proof of the Factor Theorem. Divide $f(x)$ by $x - c$. The Remainder Theorem shows that the remainder is $f(c)$. Hence by the Division Algorithm,

$$f(x) = (x - c)q(x) + f(c)$$

If c is a root of $f(x)$, then $f(c) = 0$ and we have $f(x) = (x - c)q(x)$. Hence $x - c$ is a factor. Conversely, if $x - c$ is a factor, then the remainder $f(c)$ must be 0, so c is a root. $\blacksquare$

The Fundamental Theorem of Algebra

The preceding discussion can easily be extended to include polynomials with complex number coefficients. *All of the results above are valid for polynomials with complex coefficients.* For example, you can check that i is a root of $f(x) = x^2 + (i - 1)x + (2 + i)$ and that $x - i$ is a factor:

$$f(x) = x^2 + (i - 1)x + (2 + i) = (x - i)(x - (1 - 2i)).$$

Since every real number is also a complex number, polynomials with real coefficients are just special cases of polynomials with complex coefficients. So in the rest of this section, "polynomial" means "polynomial with complex (possibly real) coefficients," unless specified otherwise.

The complex numbers were constructed in order to obtain a solution for the equation $x^2 = -1$, that is, a root for the polynomial $x^2 + 1$. In Section 2.5 we saw that much more is true: *Every* quadratic polynomial with real coefficients has roots in the complex number system. But what about cubics and higher degree polynomials? Do we have to enlarge the complex number system (perhaps many times) to find roots for them? Surprisingly, the answer is no:

THE FUNDAMENTAL THEOREM OF ALGEBRA

> **Every nonconstant polynomial has a root in the complex number system.**

Although this is obviously a powerful result, neither the Fundamental Theorem nor its proof provides a practical method for *finding* a root of a given polynomial.* The proof of the Fundamental Theorem is beyond the scope of this book, but we shall explore some of the useful implications of the theorem, such as this one:

FACTORIZATION OVER THE COMPLEX NUMBERS

> **Let $f(x)$ be a polynomial of degree $n > 0$ with leading coefficient d. Then there are (not necessarily distinct) complex numbers c_1, $c_2, \ldots, c_n$ such that**
> $$f(x) = d(x - c_1)(x - c_2)(x - c_3) \cdots (x - c_n)$$
> **Furthermore, $c_1, c_2, \ldots, c_n$ are the only roots of $f(x)$.**

Proof. By the Fundamental Theorem, $f(x)$ has a complex root c_1. The Factor Theorem shows that $x - c_1$ must be a factor of $f(x)$, say,

$$f(x) = (x - c_1)g(x)$$

where $g(x)$ has degree $n - 1$.† If $g(x)$ is nonconstant, then it has a complex root c_2 by the Fundamental Theorem. Hence $x - c_2$ is a factor of $g(x)$, so that

$$f(x) = (x - c_1)(x - c_2)h(x)$$

for some $h(x)$ of degree $n - 2$ (1 less than the degree of $g(x)$). If $h(x)$ is nonconstant, then it has a complex root c_3 and the argument can be repeated. Continuing in this way, with the degree of the last factor going down by 1 at each step, we reach a factorization in which the last factor is a constant (degree 0 polynomial):

---

* It may seem strange that you can prove that a root exists without actually exhibiting one. But such "existence theorems" are quite common. A rough analogy is the situation that occurs when someone is killed by a sniper's bullet. The police know that there *is* a killer, but *finding* the killer may be impossible.

† The degree of $g(x)$ is 1 less than the degree n of $f(x)$ because $f(x)$ is the product of $g(x)$ and $x - c_1$ (which has degree 1).

(∗) $$f(x) = (x - c_1)(x - c_2)(x - c_3) \cdots (x - c_n)d.$$

If the right side were multiplied out, it would look like

$$dx^n + \text{lower degree terms.}$$

So the constant factor d is the leading coefficient of $f(x)$.

It is easy to see from the factored form (∗) that the numbers $c_1, c_2, \ldots, c_n$ are roots of $f(x)$. If k is *any* root of $f(x)$, then

$$0 = f(k) = d(k - c_1)(k - c_2)(k - c_3) \cdots (k - c_n)$$

The product on the right is 0 only when one of the factors is 0. Since the leading coefficient d is nonzero, we must have

$$k - c_1 = 0 \quad \text{or} \quad k - c_2 = 0 \quad \text{or} \quad \cdots \quad k - c_n = 0$$
$$k = c_1 \quad \text{or} \quad k = c_2 \quad \text{or} \quad \cdots \quad k = c_n$$

Therefore k is one of the c's, and $c_1, \ldots, c_n$ are the only roots of $f(x)$. This completes the proof. ∎

Since the n roots $c_1, \ldots, c_n$ of $f(x)$ may not all be distinct, we see that:

NUMBER OF ROOTS

Every polynomial of degree $n > 0$ has at most n different roots in the complex number system.

Suppose c is a root of a polynomial $f(x)$ of degree n. If $f(x)$ is written in factored form (∗) as above, then c appears one or more times on the list of roots $c_1, c_2, \ldots, c_n$. If c appears *exactly k* times, we say that c is a root of **multiplicity k**. In this case $f(x)$ has exactly k factors of the form $x - c$. Hence $(x - c)^k$ is a factor of $f(x)$, and no higher power of $x - c$ is a factor.

Example 4 The polynomial $f(x) = (x - 1)^2(x + 5)^7(x - \frac{1}{4})$ can be written as

$$f(x) = (x - 1)^2(x - (-5))^7(x - \tfrac{1}{4})^1$$

So 1 is a root of multiplicity 2; -5 is a root of multiplicity 7; and $\frac{1}{4}$ is a root of multiplicity 1. ∎

The results proved above show that when each root is counted as many times as its multiplicity,

A polynomial of degree n has exactly n roots.

Example 5 Find a polynomial $f(x)$ of degree 5 such that 1, -2, and 5 are roots, 1 is a root of multiplicity 3, and $f(2) = -24$.

Solution Since 1 is a root of multiplicity 3, $(x - 1)^3$ must be a factor of $f(x)$. There are at least two other factors corresponding to the roots -2 and 5: $x - (-2) = x + 2$ and $x - 5$. The product of these factors $(x - 1)^3(x + 2)(x - 5)$ has degree 5, as does $f(x)$, so $f(x)$ must look like this:

$$f(x) = d(x - 1)^3(x + 2)(x - 5)$$

where d is the leading coefficient. Since $f(2) = -24$ we have:

$$d(2 - 1)^3(2 + 2)(2 - 5) = f(2) = -24$$

which reduces to $-12d = -24$. Therefore $d = (-24)/(-12) = 2$ and

$$f(x) = 2(x - 1)^3(x + 2)(x - 5)$$
$$= 2x^5 - 12x^4 + 4x^3 + 40x^2 - 54x + 20. \blacksquare$$

Polynomials with Real Coefficients

Recall that the **conjugate** of the complex number $a + bi$ is the number $a - bi$ (see page 89). We usually write a complex number as a single letter, say z, and indicate its conjugate by $\bar{z}$. For instance, if $z = 3 + 7i$, then $\bar{z} = 3 - 7i$. Conjugates play a role whenever a quadratic polynomial with real coefficients has complex roots.

Example 6 The quadratic formula shows that $x^2 - 6x + 13$ has two complex roots:

$$\frac{-(-6) \pm \sqrt{(-6)^2 - 4 \cdot 1 \cdot 13}}{2 \cdot 1} = \frac{6 \pm \sqrt{-16}}{2} = \frac{6 \pm 4i}{2} = 3 \pm 2i$$

The complex roots are $z = 3 + 2i$ and its conjugate $\bar{z} = 3 - 2i$. $\blacksquare$

The preceding example is a special case of a more general theorem, whose proof is outlined in Exercises 59–60:

CONJUGATE ROOTS THEOREM — Let $f(x)$ be a polynomial with *real* coefficients. If the complex number z is a root of $f(x)$, then its conjugate $\bar{z}$ is also a root of $f(x)$.

Example 7 Find a polynomial with real coefficients whose roots include the numbers 2 and $3 + i$.

Solution Since $3 + i$ is a root, its conjugate $3 - i$ must also be a root. Consider the polynomial

$$f(x) = (x - 2)(x - (3 + i))(x - (3 - i)).$$

Obviously 2, $3 + i$, and $3 - i$ are roots of $f(x)$. Multiplying out this factored form shows that $f(x)$ *does* have real coefficients:

$$f(x) = (x - 2)(x^2 - (3 - i)x - (3 + i)x + (3 + i)(3 - i))$$
$$= (x - 2)(x^2 - 3x + ix - 3x - ix + 9 - i^2)$$
$$= (x - 2)(x^2 - 6x + 10)$$
$$= x^3 - 8x^2 + 22x - 20.$$

The next-to-last line of this calculation also shows that $f(x)$ can be factored as a product of a linear and a quadratic polynomial, each with *real* coefficients. ∎

The technique in Example 7 works because the polynomial $(x - (3 + i))(x - (3 - i))$ turns out to have real coefficients. The proof of the following result shows why this must always be the case:

FACTORIZATION OVER THE REAL NUMBERS

> **Every nonconstant polynomial with real coefficients can be factored as a product of linear and quadratic polynomials with real coefficients in such a way that the quadratic factors, if any, have no real roots.***

Proof. The second box on page 249 shows that

$$f(x) = d(x - c_1)(x - c_2) \cdots (x - c_n)$$

where $c_1, \ldots, c_n$ are the roots of $f(x)$. If some c_i is a real number, then the factor $x - c_i$ is a linear polynomial with real coefficients.† If some c_j is a nonreal complex root, then its conjugate must also be a root. Thus some c_k is the conjugate of c_j, say, $c_j = a + bi$ (with a, b real) and $c_k = a - bi$.‡ In this case,

$$(x - c_j)(x - c_k) = (x - (a + bi))(x - (a - bi))$$
$$= x^2 - (a - bi)x - (a + bi)x + (a + bi)(a - bi)$$
$$= x^2 - ax + bix - ax - bix + a^2 - (bi)^2$$
$$= x^2 - 2ax + (a^2 + b^2)$$

Therefore the factor $(x - c_j)(x - c_k)$ of $f(x)$ is a quadratic with real coefficients (because a and b are real numbers). Its roots (c_j and c_k) are nonreal. By taking the real roots of $f(x)$ one at a time and the nonreal ones in conjugate pairs in this fashion, we obtain the desired factorization of $f(x)$. ∎

---

* A "real root" is a root that is a real number.

† In Example 7, for instance, 2 is a real root and $x - 2$ a linear factor.

‡ In Example 7, for instance, $c_j = 3 + i$ and $c_k = 3 - i$ are conjugate roots.

Example 8 Given that $1 + i$ is a root of $f(x) = x^4 - 2x^3 - x^2 + 6x - 6$, factor $f(x)$ completely over the real numbers.

Solution Since $1 + i$ is a root of $f(x)$, so is its conjugate $1 - i$, and hence $f(x)$ has this quadratic factor:
$$(x - (1 + i))(x - (1 - i)) = x^2 - 2x + 2.$$
Dividing $f(x)$ by $x^2 - 2x + 2$ shows that the other factor is $x^2 - 3$, which factors as $(x + \sqrt{3})(x - \sqrt{3})$. Therefore
$$f(x) = (x + \sqrt{3})(x - \sqrt{3})(x^2 - 2x + 2). \blacksquare$$

Other facts about the roots of polynomials with real coefficients are discussed in Excursion 5.4.A.

EXERCISES

In Exercises 1–6, find the remainder when $f(x)$ is divided by $g(x)$ without using synthetic or long division.

1. $f(x) = x^{10} + x^8$; $g(x) = x - 1$
2. $f(x) = x^6 - 10$; $g(x) = x - 2$
3. $f(x) = 3x^4 - 6x^3 + 2x - 1$; $g(x) = x + 1$
4. $f(x) = x^5 - 3x^2 + 2x - 1$; $g(x) = x - 2$
5. $f(x) = x^3 - 2x^2 + 5x - 4$; $g(x) = x + 2$
6. $f(x) = 10x^{75} - 8x^{65} + 6x^{45} + 4x^{32} - 2x^{15} + 5$; $g(x) = x - 1$

In Exercises 7–10, list the roots of the polynomial and state the multiplicity of each root.

7. $f(x) = x^{54}\left(x + \dfrac{4}{5}\right)$
8. $g(x) = 3\left(x + \dfrac{1}{6}\right)\left(x - \dfrac{1}{5}\right)\left(x + \dfrac{1}{4}\right)$
9. $h(x) = 2x^{15}(x - \pi)^{14}(x - (\pi + 1))^{13}$
10. $k(x) = (x - \sqrt{7})^7(x - \sqrt{5})^5(2x + 1)$

In Exercises 11–22, use the techniques of Sections 2.3 and 2.5 to find all the roots of $f(x)$ in the complex number system; then write $f(x)$ as a product of linear factors.

11. $f(x) = x^2 - 2x + 5$
12. $f(x) = x^2 - 4x + 13$
13. $f(x) = 3x^2 + 2x + 7$
14. $f(x) = 3x^2 - 5x + 2$
15. $f(x) = x^3 - 27$ [Hint: Factor first.]
16. $f(x) = x^3 + 125$ 17. $f(x) = x^3 + 8$
18. $f(x) = x^6 - 64$ [Hint: Let $u = x^3$ and factor $u^2 - 64$ first.]
19. $f(x) = x^4 - 1$ 20. $f(x) = x^4 - x^2 - 6$
21. $f(x) = x^4 - 3x^2 - 10$
22. $f(x) = 2x^4 - 7x^2 - 4$

In Exercises 23–44, find a polynomial $f(x)$ with real coefficients that satisfies the given conditions. Some of these problems have many correct answers.

23. Degree 3; only roots are $1, 7, -4$.
24. Degree 3; only roots are 1 and -1.
25. Degree 6; only roots are $1, 2, \pi$.
26. Degree 5; only root is 2.
27. Degree 3; roots $0, 5, 8$; $f(10) = 17$.
28. Degree 3; roots $-1, \tfrac{1}{2}, 2$; $f(0) = 2$.
29. Roots include $2 + i$ and $2 - i$.
30. Roots include $1 + 3i$; and $1 - 3i$.
31. Roots include 2 and $2 + i$.
32. Roots include 3 and $4i - 1$.
33. Roots include $-3, 1 - i, 1 + 2i$.

34. Roots include 1, $2 + i$, $3i - 1$.
35. Degree 2; roots $1 + 2i$ and $1 - 2i$.
36. Degree 4; roots $3i$ and $-3i$, each of multiplicity 2.
37. Degree 4; only roots are 4, $3 + i$, and $3 - i$.
38. Degree 5; roots 2 (of multiplicity 3), i, and $-i$.
39. Degree 6; roots 0 (of multiplicity 3) and 3, $1 + i$, $1 - i$ each of multiplicity 1.
40. Degree 6; roots include i (of multiplicity 2) and 3.
41. Degree 2; roots include $1 + i$; $f(0) = 6$.
42. Degree 2; roots include $3 + i$; $f(2) = 3$.
43. Degree 3; roots include i and 1; $f(-1) = 8$.
44. Degree 3; roots include $2 + 3i$ and -2; $f(2) = -3$.

In Exercises 45–48, find a polynomial with complex coefficients that satisfies the given conditions.

45. Degree 2; roots i and $1 - 2i$.
46. Degree 2; roots $2i$ and $1 + i$.
47. Degree 3; roots 3, i, and $2 - i$.
48. Degree 4; roots $\sqrt{2}$, $-\sqrt{2}$, $1 + i$, and $1 - i$.

In Exercises 49–56, one root of the polynomial is given; find all the roots.

49. $x^3 - 2x^2 - 2x - 3$; root 3.
50. $x^3 + x^2 + x + 1$; root i.
51. $x^4 + 3x^3 + 3x^2 + 3x + 2$; root i.
52. $x^4 - x^3 - 5x^2 - x - 6$; root i.
53. $x^4 - 2x^3 + 5x^2 - 8x + 4$; root 1 of multiplicity 2.
54. $x^4 - 6x^3 + 29x^2 - 76x + 68$; root 2 of multiplicity 2.
55. $x^4 - 4x^3 + 6x^2 - 4x + 5$; root $2 - i$.
56. $x^4 - 5x^3 + 10x^2 - 20x + 24$; root $2i$.
57. Let $z = a + bi$ and $w = c + di$ be complex numbers (a, b, c, d are real numbers). Prove the given equality by computing each side and comparing the results:

    (a) $\overline{z + w} = \bar{z} + \bar{w}$ (The left side says: First find $z + w$ and then take the conjugate. The right side says: First take the conjugates of z and w and then add.)

    (b) $\overline{z \cdot w} = \bar{z} \cdot \bar{w}$

58. Let $g(x)$ and $h(x)$ be polynomials of degree n and assume that there are $n + 1$ numbers c_1, $c_2, \ldots, c_n, c_{n+1}$ such that
$$g(c_i) = h(c_i) \text{ for every } i.$$
Prove that $g(x) = h(x)$. [*Hint:* Show that each c_i is a root of $f(x) = g(x) - h(x)$. If $f(x)$ is nonzero, what is its largest possible degree? To avoid a contradiction, conclude that $f(x) = 0$.]

59. Suppose $f(x) = ax^3 + bx^2 + cx + d$ has real coefficients and z is a complex root of $f(x)$.

    (a) Use Exercise 57 and the fact that $\bar{r} = r$, when r is a real number, to show that
    $$\overline{f(z)} = \overline{az^3 + bz^2 + cz + d}$$
    $$= a\bar{z}^3 + b\bar{z}^2 + c\bar{z} + d = f(\bar{z}).$$

    (b) Conclude that $\bar{z}$ is also a root of $f(x)$. [*Note:* $f(\bar{z}) = \overline{f(z)} = \bar{0} = 0$.]

60. Let $f(x)$ be a polynomial with real coefficients and z a complex root of $f(x)$. Prove that the conjugate $\bar{z}$ is also a root of $f(x)$. [*Hint:* Exercise 59 is the case when $f(x)$ has degree 3; the proof in the general case is similar.]

61. Use the statement in the box on page 252 to show that every polynomial with real coefficients and *odd* degree must have at least one real root.

62. Give an example of a polynomial $f(x)$ with complex, nonreal coefficients and a complex number z such that z is a root of $f(x)$, but its conjugate is not. Hence the conclusion of the Conjugate Roots Theorem (page 251) may be *false* if $f(x)$ doesn't have real coefficients.

5.4.A EXCURSION: Descartes' Rule of Signs

Descartes' Rule of Signs is a method for determining the possible number of positive and negative real roots of a polynomial with real coefficients. It can be applied to any polynomial with a nonzero constant term whose terms are arranged in order of decreasing powers of x (with terms having a zero coefficient deleted). It is based on counting the *variations in sign* of the polynomial.

A polynomial is said to have a **variation in sign** whenever two consecutive coefficients have opposite signs. For instance,

$$5x^3 - 2x - 1$$

has a variation in sign since the first two coefficients have opposite signs. Similarly, the following polynomial (whose leading coefficient is $+1$) has *two* variations in sign:

$$f(x) = x^5 - 2x^4 - x^3 + 3x + 1$$

$$\underbrace{}_{variation} \quad \underbrace{}_{variation}$$

The polynomial $f(-x)$ is obtained from $f(x)$ by replacing x with $-x$ throughout:

$$f(-x) = (-x)^5 - 2(-x)^4 - (-x)^3 + 3(-x) + 1$$
$$= -x^5 - 2x^4 + x^3 - 3x + 1.$$

Note that $f(-x)$ has *three* variations in sign.

The relationship between the number of variations in sign of $f(x)$ and $f(-x)$ and the roots of $f(x)$ is given by the following theorem whose proof is omitted:

DESCARTES' RULE OF SIGNS

> Let $f(x)$ be polynomial with real coefficients and a nonzero constant term, whose terms are arranged in order of decreasing powers of x (with terms having a zero coefficient deleted).
>
> The number of *positive real roots* of $f(x)$ is either equal to the number of variations in sign of $f(x)$ or is less than that number by an even integer.
>
> The number of *negative real roots* of $f(x)$ is either equal to the number of variations in sign of $f(-x)$ or is less than that number by an even integer.

When applying Descartes' rule, each root is counted the same number of times as its multiplicity. Thus a root of multiplicity 2 is counted as two roots, a root of multiplicity 3 as three roots, and so on.

Example 1 We saw above that $f(x) = x^5 - 2x^4 - x^3 + 3x + 1$ has two variations in sign. Hence it has either 2 or 0 positive real roots. We also saw that $f(-x) = -x^5 - 2x^4 + x^3 - 3x + 1$ has three variations in sign. Therefore $f(x)$ has either 3 or 1 negative real roots. Since a polynomial of degree 5 has exactly 5 roots (counting multiplicities), these are the only possibilities:

Number of Positive Real Roots	Number of Negative Real Roots	Number of Complex Roots
0	1	4
0	3	2
2	1	2
2	3	0

In the next section we shall see that $f(x)$ actually has 1 negative and 2 positive real roots. ∎

A useful consequence of Descartes' Rule is this fact:

If a polynomial $f(x)$ has just one variation in sign, then $f(x)$ has exactly one positive real root.

This is true because the number of positive real roots is either 1 or less than 1 by an even integer (that is, $1, -1, -3$, etc.). Since the number of positive roots can't be less than 0, there must be exactly one positive real root. An analogous conclusion for negative roots holds when $f(-x)$ has just one variation in sign.

Example 2 Determine the nature of the roots of

$$f(x) = 4x^8 + x^6 + 2x^4 + x^2 - 6x.$$

Solution Since Descartes' Rule applies only to polynomials with a nonzero constant term, we first factor:

$$f(x) = x(4x^7 + x^5 + 2x^3 + x - 6).$$

Thus $x = 0$ is one root of $f(x)$ and the others are the roots of

$$g(x) = 4x^7 + x^5 + 2x^3 + x - 6.$$

Applying Descartes' Rule to $g(x)$, we see that $g(x)$ has one variation in sign and, hence, exactly one positive real root. Since

$$g(-x) = 4(-x)^7 + (-x)^5 + 2(-x)^3 + (-x) - 6$$
$$= -4x^7 - x^5 - 2x^3 - x - 6$$

$g(-x)$ has no variations in sign. Hence $g(x)$ has no negative real roots. Therefore the original polynomial $f(x)$ (of degree 8) has 2 real roots (0 and a positive one) and 6 complex roots. ∎

EXERCISES

In Exercises 1-10, use Descartes' Rule of Signs to determine the possible number of positive and negative real roots of the polynomial.

1. $f(x) = 4x^5 + 2x^3 + 5x - 8$
2. $f(x) = x^3 - x + 1$
3. $g(x) = x^4 + 7x^3 - 4x^2 - 39x - 20$
4. $g(x) = 2x^5 - 9x^4 + 6x^2 + x - 4$
5. $h(x) = 4x^4 - 6x^3 + x^2 - 3x$
6. $h(x) = 3x^4 + 2x^3 + 5x^2 + 4x$
7. $f(x) = 5x^5 - 6x^3 - 4x^2$
8. $f(x) = 5x^7 + 10x^3 - x^2$
9. $g(x) = x^4 - 2x^3 + 2x^2 - 2x + 1$
10. $g(x) = x^4 - 6x^3 + 13x^2 - 12x + 4$

5.5 APPROXIMATION TECHNIQUES*

Despite the powerful theoretical results of Section 5.4, it may be very difficult to find the irrational roots of a specific polynomial of high degree.† It is often necessary to *approximate* these roots to some stated degree of accuracy. Calculus provides efficient ways of doing this, such as Newton's Method. Even without calculus, however, there are accurate methods of approximating roots. One of these is based on the Intermediate Value Theorem (page 227):

> Let $f(x)$ be a polynomial. If $f(c)$ and $f(d)$ have opposite signs, then $f(x)$ has a root between c and d.

If $f(x)$ has a root between c and d, then the midpoint k of the interval (c, d) will be an approximation to that root. The maximum distance from k to any point in the interval is half the length of the interval, namely, $\frac{1}{2}|c - d|$. Therefore the maximum error in using k as an approximation of the root is $\frac{1}{2}|c - d|$. By choosing c and d close enough together, you can make this maximum error as small as you want. The **bisection method** of approximating roots, which is illustrated in the next example, is a systematic way of doing this.

Example 1 To find the roots of $f(x) = x^5 - 2x^4 - x^3 + 3x + 1$, we first evaluate $f(x)$ at consecutive integers:

x	0	1	2
$f(x) = x^5 - 2x^4 - x^3 + 3x + 1$	1	2	−1

There is a root between 1 and 2 because $f(1)$ is positive and $f(2)$ negative. Our first approximation is $x = 1.5$, the midpoint of the interval $(1, 2)$. Using a calculator, we find that

$$f(1) = 2 \quad \text{and} \quad f(1.5) = -.40625$$

---

* Throughout this section all polynomials have real coefficients, and "root" means a root that is a real number.

† The rational roots of a polynomial $h(x)$ can always be found by using the Rational Solutions Test (page 100).

Since $f(1)$ and $f(1.5)$ have opposite signs, the root is actually between 1 and 1.5. So our second approximation is $x = 1.25$, the midpoint of the interval $(1, 1.5)$. Since

$$f(1.25) \approx .96582 \quad \text{and} \quad f(1.5) = -.40625$$

the root is between 1.25 and 1.5. Our third approximation is $x = 1.375$, the midpoint of $(1.25, 1.5)$. The maximum error in this approximation is half the length of the interval $(1.25, 1.5)$, namely, $\frac{1}{2}|1.25 - 1.5| = .125$. Continuing the process, we have*

Step	Interval (c, d)	f(c)	f(d)	Mid-point k	f(k)	Maximum Error
3	(1.25, 1.5)	.9658	−.4063	1.3750	.2914	.1250
4	(1.375, 1.5)	.2914	−.4063	1.4375	−.0599	.0625
5	(1.375, 1.4375)	.2914	−.0599	1.4063	.1159	.0313
6	(1.4063, 1.4375)	.1159	−.0599	1.4219	.0278	.0156
7	(1.4219, 1.4375)	.0278	−.0599	1.4297	−.0160	.0078
8	(1.4219, 1.4297)	.0278	−.0160	1.4258	.0059	.0039
9	(1.4258, 1.4297)	.0059	−.0160	1.4278	−.0051	.0020
10	(1.4258, 1.4278)	.0059	−.0051	1.4268	.0003	.0010

Our tenth approximation $x = 1.4268$ is within .0010 of the root. If more accuracy is needed, the process can be carried further.

To find another root of $f(x)$, evaluate $f(x)$ at consecutive integers (starting either at -1 and working down, or at 3 and working up) until you find two at which the function values have opposite signs. Then repeat the procedure above. ∎

Graphing Calculator Method (Optional)

A graphing calculator can also be used to approximate the roots of a polynomial $f(x)$ because the graph of $y = f(x)$ crosses the x-axis at these roots. The basic idea is to magnify the portion of the graph at which the graph appears to cross the x-axis.† Then display the x-coordinate of the point where the graph appears to cross the axis;† it is an approximation of the root.

---

* The midpoint formula (page 125) is used to calculate k. All numbers in the chart were rounded to four decimal places after the calculations were made.

† The specific procedures needed to do this vary with the calculator, but can be done with relatively few keystrokes.

Example 2 The graph $f(x) = x^5 - 2x^4 - x^3 + 3x + 1$, given by a graphing calculator, is shown in Figure 5–26. It shows that $f(x)$ has a root between -1 and 0.

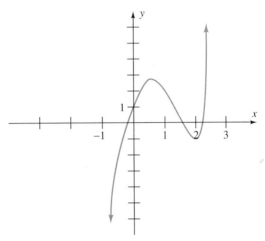

FIGURE 5.26

Figure 5–27 shows successive magnifications of the graph near this root, each by a factor of 100. In each case, the x-coordinate of the point where the graph appears to cross the x-axis is displayed.*

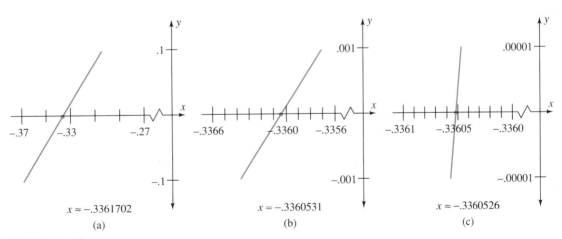

FIGURE 5–27

The graph in Figure 5–27(c) shows that $x = -.3360526$ is an approximation of the root.

*With a large degree of magnification, a tiny portion of any graph looks like a straight line. The value of the calculator is not the picture, but its ability to display the coordinates of the points it plots.

To determine the accuracy of this approximation, we display the coordinates of nearby points on the graph and find that

$$x = -.3360527 \qquad x = -.3360526$$
$$f(x) \approx -.00000025 \qquad f(x) \approx +.0000000687$$

Since $f(x)$ takes opposite signs, the actual root is between $-.3360527$ and our approximation, $-.3360526$. The maximum possible error in our approximation is the distance between these two x-values, namely, $.0000001$. ∎

Bounds*

Since a polynomial of degree n has, at most, n distinct roots, it has a smallest and a largest root. Hence there are **lower** and **upper bounds** for the roots, that is, numbers r and s such that $r < s$ and every root lies between r and s.† Here is one technique for finding such bounds.

Example 3 The graph of $f(x) = x^5 - 2x^4 - x^3 + 3x + 1$ in Figure 5-26 above suggests that 3 is an upper bound for the roots. To prove this, we first use synthetic division‡ to divide $f(x)$ by $x - 3$:

```
3 | 1  -2  -1   0   3   1
  |      3   3   6  18  63
    1   1   2   6  21 |64
```

Read off the quotient and remainder and apply the Division Algorithm:

$$f(x) = (x - 3)(x^4 + x^3 + 2x^2 + 6x + 21) + 64.$$

When $x > 3$, then $x - 3$ *is positive* and the quotient $x^4 + x^3 + 2x^2 + 6x + 21$ is also positive (because all the coefficients are). The remainder is also positive. Therefore, $f(x) > 0$ whenever $x > 3$. In particular, there are no roots greater than 3. So 3 *is* an upper bound for the roots of $f(x)$.

We can check -1 as a possible lower bound for the roots by dividing $f(x)$ by $x - (-1) = x + 1$:

```
-1 | 1  -2  -1   0   3   1
   |    -1   3  -2   2  -5
     1  -3   2  -2   5 |-4
```

The Division Algorithm shows that:

$$f(x) = (x + 1)(x^4 - 3x^3 + 2x^2 - 2x + 5) - 4.$$

When $x < -1$, then $x + 1$ *is negative*. When x is negative, its odd powers are

---

* Synthetic division (Excursion 1.5.A) is a prerequisite for the remainder of this section.

† The bounds are not unique. Any number smaller than r is a lower bound, and any number larger than s is an upper bound.

‡ See page 42.

negative and its even powers are positive. Consequently, $x^4 - 3x^3 + 2x^2 - 2x + 5$ is positive (the negative powers of x are multiplied by negative coefficients). The product of this positive factor with the negative $x + 1$ is negative. The remainder is also negative. Hence $f(x) < 0$ whenever $x < -1$. So there are no roots less than -1, and -1 is a lower bound for the roots. ∎

The technique in the example works because the coefficients of the quotient and the remainder (the last row in the synthetic division process) have the right signs (all nonnegative for the upper bound, and alternating for the lower bound). The same is true in the general case:

BOUNDS ON THE ROOTS OF A POLYNOMIAL

Let $f(x)$ be a polynomial with positive leading coefficient.

If $d > 0$ and every number in the last row in the synthetic division of $f(x)$ by $x - d$ is *nonnegative,* then d is an upper bound for the roots of $f(x)$.

If $c < 0$ and the numbers in the last row of the synthetic division of $f(x)$ by $x - c$ are alternately positive and negative (with 0 considered as either), then c is a lower bound for the roots of $f(x)$.

EXERCISES

In Exercises 1–6,

(a) Locate one *irrational* root of the polynomial between two integers.
(b) Use the bisection method to approximate this root with a maximum error of .01.

1. $12x^3 - 28x^2 - 7x - 10$
2. $x^5 + x^2 - 7$
3. $x^3 + 4x^2 + 10x + 15$
4. $x^3 - 3x^2 - 6x + 9$
5. $t^4 + 2t^3 - 10t^2 - 6t + 1$
6. $u^4 + 8u^3 + 17u^2 + 4u - 2$

In Exercises 7–10, find lower and upper bounds for the roots of the polynomial.

7. $2x^3 + 5x^2 - 8x - 7$
8. $10x^3 - 15x^2 - 16x + 12$
9. $x^3 + 2x^2 - 7x + 1$
10. $x^5 - x^4 - 5x^3 + 5x^2 + 6x - 6$
11. Two roots of $f(x) = x^5 - 2x^4 - x^3 + 3x + 1$ were approximated in the examples in the text. Use a graphing calculator to approximate the third root, with a maximum error of .0001.
12. (a) Use a graphing calculator to determine the number of real roots of the polynomial in Exercise 10.
    (b) Approximate each of the roots in part (a) with a maximum error of .001.

5 POLYNOMIAL AND RATIONAL FUNCTIONS

CHAPTER REVIEW

Important Concepts

Section 5.1	Quadratic function 216
	Parabola 216
	Vertex 216
	Maximum and minimum values 221
	Applications of quadratic functions 221
Section 5.2	Polynomial function 224
	Graph of $f(x) = x^n$ 225
	Roots of polynomials 226
	Roots and x-intercepts 226
	Relative maxima and minima 226
	Continuity 226
	Technique for graphing polynomial functions 228
Excursion 5.2.A	Growth of $f(x) = x^n$ 232
Section 5.3	Rational function 234
	Big-Little Principle 235
	Vertical asymptote 236
	Graph of $f(x) = \dfrac{ax + b}{cx + d}$ 237
	Holes 239
	Horizontal asymptote 240
	Technique for graphing rational functions 241
	Oblique asymptotes 244
Section 5.4	Division Algorithm 247
	Remainder Theorem 247
	Factor Theorem 248
	Fundamental Theorem of Algebra 249
	Multiplicity of roots 250
	Conjugate pairs of roots 251
	Factoring over the real numbers 252
Excursion 5.4.A	Descartes' Rule of Signs 255
Section 5.5	Bisection method for approximating roots 257
	Upper and lower bounds for roots 260

Important Facts and Formulas

- The graph of $f(x) = ax^2 + bx + c$ is a parabola whose vertex has x-coordinate $-b/2a$.

Review Questions

In Questions 1–4, sketch the graph of the given function or equation.

1. $f(x) = (x - 2)^2 + 3$
2. $g(x) = 2(x + 1)^2 - 1$
3. $y = x^2 - 8x + 12$
4. $y = x^2 - 7x + 6$

Use the graphs in Figure 5–28 to answer Questions 5–10.

5. Which graph could *possibly* be the graph of $y = x^2$?
6. Which graph could *possibly* be the graph of $y = x^2 + 2$?
7. Which graph could *possibly* be the graph of $y = x^2 - 2$?
8. Which graph could *possibly* be the graph of $y = -x^2 - 2$?
9. Which graph could *possibly* be the graph of $y = -x^2 + 2$?
10. Which graph could *possibly* be the graph of $y = -2x^2$?

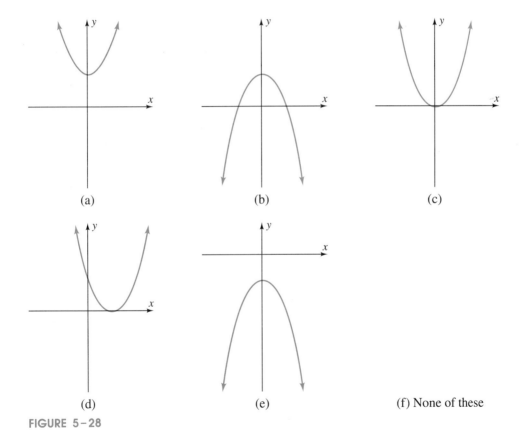

FIGURE 5–28

11. Which of the following statements about the functions

$$f(x) = 3x^2 + 2 \quad \text{and} \quad g(x) = -3x^2 + 2$$

is *false*?
(a) The graphs of f and g are parabolas.
(b) The graphs of f and g have the same vertex.
(c) The graphs of f and g open in opposite directions.
(d) The graph of f is the graph of $y = 3x^2$ shifted 2 units to the right.

12. A model rocket is launched straight up from a platform at time $t = 0$ (where t is time measured in seconds). The altitude $h(t)$ of the rocket above the ground at time t is given by $h(t) = 10 + 112t - 16t^2$ (where $h(t)$ is measured in feet).
(a) What is the altitude of the rocket the instant it is launched?
(b) What is the altitude of the rocket 2 seconds after launching?
(c) What is the maximum *altitude* attained by the rocket?
(d) At what *time* does the rocket return to the altitude at which it was launched?

13. A rectangular garden next to a building is to be fenced with 120 feet of fencing. The side against the building will not be fenced. What should the lengths of the other three sides be in order to assure the largest possible area?

14. A factory offers 100 calculators to a retailer at a price of $20 each. The price per calculator on the entire order will be reduced 5¢ for each additional calculator over 100. What number of calculators will produce the largest possible income for the factory?

15. Which of the following are polynomial functions?
(a) $f(x) = 2^3 + x^2$ (b) $g(x) = x + 1/x$
(c) $h(x) = x^3 - 1/\sqrt{2}$ (d) $f(x) = \sqrt[3]{x^4}$
(e) $g(x) = \pi^3 - x$ (f) $h(x) = \sqrt{2} + 2x^2$
(g) $f(x) = \sqrt{x} + 2x^2$ (h) $g(x) = |x|$

16. What is the remainder when $x^4 + 3x^3 + 1$ is divided by $x^2 + 1$?

17. What is the remainder when $x^{112} - 2x^8 + 9x^5 - 4x^4 + x - 5$ is divided by $x - 1$?

18. Is $x - 1$ a factor of $f(x) = 14x^{87} - 65x^{56} + 51$? Justify your answer.

19. Use synthetic division to show that $x - 2$ is a factor of $x^6 - 5x^5 + 8x^4 + x^3 - 17x^2 + 16x - 4$ and find the other factor.

20. List the roots of this polynomial and the multiplicity of each root:

$$f(x) = 5(x - 4)^3(x - 2)(x + 17)^3(x^2 - 4)$$

21. Find a polynomial f of degree 3 such that $f(1) = 0$, $f(-1) = 0$, and $f(0) = 5$.

22. The graph of one of these functions does not cross the x-axis. Which one?

(a) $f(x) = x^2 - 1$ (b) $g(x) = x^2 - 2x - 1$
(c) $h(x) = x^4 + x^2 + 1$ (d) $f(x) = x^3 + 2x^2 - 3x - 6$
(e) $g(x) = x^3 - 4x + 3$

23. At *how many* places does the graph of $f(x) = x^3 + x$ cross the *x*-axis?

24. At *how many* places does the graph of $f(x) = \dfrac{2x-1}{x-3}$ cross the *x*-axis?

25. Which of the statements that follow about the graph of $f(x) = (x^2 - 1)(x + 1)(x - 5)^3$ is *true*?
    (a) The graph lies above the *x*-axis when $x > 5$.
    (b) The graph lies below the *x*-axis when $x < -1$.
    (c) The graph crosses the *x*-axis at $x = -1$.
    (d) The graph has a vertical asymptote at $x = -1$.
    (e) The graph has a total of at most 2 relative maxima or minima.

26. Sketch the graph of $f(x) = x^3 - 9x$.

27. Draw the graph of a function that could not possibly be the graph of a polynomial function and explain why.

28. Draw a graph that could be the graph of a polynomial function of degree 5. (You need not list a specific polynomial, nor do any computation.)

29. Sketch the graph of $f(x) = (x - 1)^2(x + 2)$.

30. (a) $x = -3$ and $x = 2$ are roots of the polynomial function $f(x) = (x^3 + 2x^2 - 5x - 6)(x - 1)$, as you can easily verify. Find *all* roots of $f(x)$.
    (b) Sketch the graph of the function f.

31. Which of the statements below is *not* true about the polynomial function f whose graph is shown in Figure 5–29?
    (a) f has three roots between -2 and 3.
    (b) $f(x)$ could possibly be a fifth-degree polynomial.
    (c) $(f \circ f)(0) > 0$.
    (d) $f(2) - f(-1) < 3$.
    (e) $f(x)$ is positive for all x in the interval $[-1, 0]$.

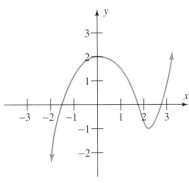

FIGURE 5–29

32. Which of the statements (i)–(v) below about the polynomial function f whose graph is shown in Figure 5–30 are *false*?

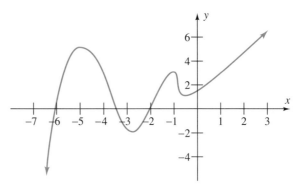

FIGURE 5–30

(i) f has 2 roots in the interval $[-6, -3)$.
(ii) $f(-3) - f(-6) < 0$.
(iii) $f(0) < f(1)$.
(iv) $f(2) - 2 = 0$.
(v) f has degree ≤ 4.

(a) (i) and (iii) only
(b) (ii) and (iv) only
(c) (ii) and (v) only
(d) (iii) and (v) only
(e) (iv) and (v) only

33. Which *one* of the graphs in Figure 5–31 could possibly be the graph of $f(x) = (x + 2)(x^2 + 3)$?

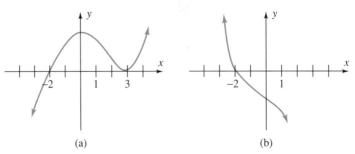

(a)   (b)

FIGURE 5–31 *(continues)*

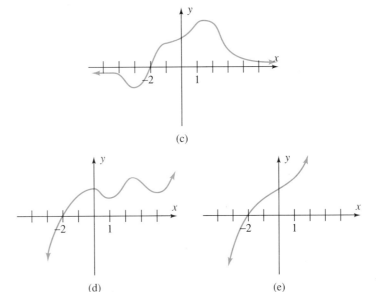

(c)

(d)   (e)

FIGURE 5-31 *(continued)*

34. Sketch the graph of the function $f(x) = \dfrac{1}{x-5}$.

35. Sketch the graph of the function $g(x) = \dfrac{-2}{x+4}$.

36. Sketch the graph of the function $h(x) = \dfrac{3-x}{x-2}$.

37. Sketch the graph of the function $k(x) = \dfrac{4x+10}{3x-9}$.

38. Sketch the graph of $f(x) = \dfrac{x+1}{x^2-1}$.

39. Sketch the graph of $f(x) = \dfrac{(x+3)(x-1)}{(x+1)(x-3)^2}$.

40. Match each of the following rational functions with its graph in Figure 5-32, on the next page.

$$f(x) = \dfrac{x-1}{(x+2)^2(x-3)} \qquad g(x) = \dfrac{x-1}{(x+2)(x-3)}$$

$$h(x) = \dfrac{(x-3)(x+2)}{(x-1)^3} \qquad k(x) = \dfrac{x-3}{(x-1)(x+2)}$$

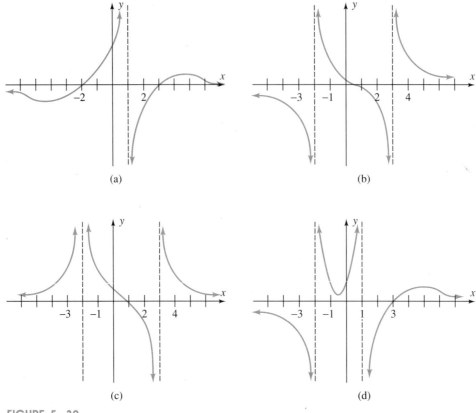

FIGURE 5-32

41. Which of the graphs in Figure 5-33 is the graph of

$$g(x) = \frac{(x-2)^2(x+2)}{x^2(x+4)(2x-6)}?$$

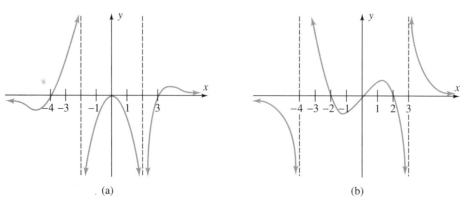

FIGURE 5-33 *(continues)*

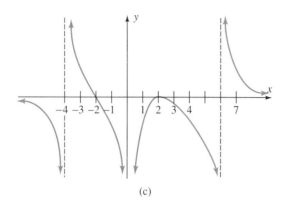

(c)

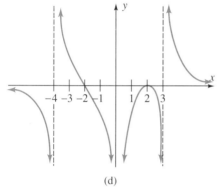

(d)

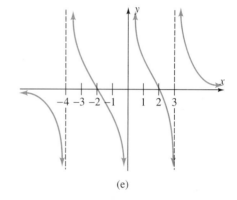

(e)

FIGURE 5-33 *(continued)*

42. Which of the rational functions (a)-(d) listed below could possibly have the graph shown in Figure 5-34?

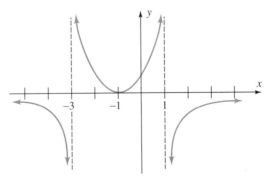

FIGURE 5-34

(a) $f(x) = \dfrac{1 - x^2}{(1 + x)(x - 3)^2}$ (b) $g(x) = \dfrac{(2 - x)^2}{(1 - x)^2(x + 3)}$

(c) $h(x) = \dfrac{(1+x)^2}{(1-x)(x^2+3)}$ (d) $k(x) = \dfrac{(1+x)^2}{(1-x)(3+x)^2}$

(e) none of the functions above

43. Which of these statements about the graph of $f(x) = \dfrac{(x-1)(x+3)}{(x^2+1)(x^2-1)}$ is true?

(a) The graph has two vertical asymptotes.
(b) The graph touches the x-axis at $x = 3$.
(c) The graph lies above the x-axis when $x < -1$.
(d) The graph has a hole at $x = 1$.
(e) The graph has no horizontal asymptotes.

In Questions 44–52, solve the given equation in the complex number system.

44. $x^2 + 36 = 0$ 45. $x^2 + 3x + 10 = 0$
46. $x^2 + 2x + 5 = 0$ 47. $5x^2 + 2 = 3x$
48. $-3x^2 + 4x - 5 = 0$ 49. $3x^4 + x^2 - 2 = 0$
50. $8x^4 + 10x^2 + 3 = 0$ 51. $x^3 + 8 = 0$
52. $x^3 - 27 = 0$

53. One root of $x^4 - x^3 - x^2 - x - 2$ is i. Find all the roots.
54. One root of $x^4 + x^3 - 5x^2 + x - 6$ is i. Find all the roots.
55. Give an example of a fourth-degree polynomial with real coefficients whose roots include 0 and $1 + i$.
56. Find a fourth-degree polynomial f whose only roots are $2 + i$ and $2 - i$ and satisfying $f(-1) = 50$.

What are the possible numbers of positive and negative real roots of

57. $f(x) = x^4 + x^2 - 1$ 58. $g(x) = x^6 - 2x^4 + x^2 + 4x$

59. The polynomial $x^4 - 2x^3 - 2x + 2$ has a root between
(a) -2 and -1 (b) -1 and 0
(c) 1 and 2 (d) 2 and 3
(e) none of these

60. Show that -1 is a lower bound for the roots of $x^4 - 4x^3 + 15$.
61. Show that 5 is an upper bound for the roots of $x^4 - 4x^3 + 16x - 16$.

CHAPTER 6

Exponential and Logarithmic Functions

Exponential and logarithmic functions are essential for the mathematical description of a variety of phenomena in physical science, engineering, and economics. The values of these functions usually cannot be computed by hand. So *a scientific calculator should be used whenever appropriate.*[*]

WARNING

1. A calculator is not a substitute for learning the "theoretical" properties of these functions. You won't be able to use your calculator efficiently or to interpret its answers unless you understand these properties.
2. When computations *can* readily be done by hand, you will be expected to do them without a calculator.

6.1 EXPONENTIAL FUNCTIONS

Rational exponents were discussed in Section 1.3. Before introducing exponential functions we must define a^t for an *irrational* exponent t. An example will illustrate the idea.[†] To compute $10^{\sqrt{2}}$ we use the infinite decimal expansion $\sqrt{2} \approx 1.4142135623 \cdots$ (see Excursion 1.1.A). Each of

$$1.4, \quad 1.41, \quad 1.414, \quad 1.4142, \quad 1.41421, \ldots$$

is a rational number approximation of $\sqrt{2}$, and each is a more accurate approximation than the preceding one. We know how to raise 10 to each of these rational numbers (for instance, $10^{1.4} = 10^{14/10}$):

$$10^{1.4} \approx 25.1189 \qquad 10^{1.4142} \approx 25.9537$$
$$10^{1.41} \approx 25.7040 \qquad 10^{1.41421} \approx 25.9543$$
$$10^{1.414} \approx 25.9418 \qquad 10^{1.414213} \approx 25.9545$$

---

[*] The alternative is to use the tables at the end of the book, but this is slower and less accurate.

[†] This example is not a proof, but should make the idea plausible. Calculus is required for a rigorous proof.

6 EXPONENTIAL AND LOGARITHMIC FUNCTIONS

It appears that as the exponent r gets closer and closer to $\sqrt{2}$, 10^r gets closer and closer to a real number whose decimal expansion begins 25.954 . . . . We define $10^{\sqrt{2}}$ to be this number.

Similarly, for any $a > 0$,

a^t is a well-defined *positive* number for each real exponent t.

We shall also assume this fact:

The exponent laws (page 14) are valid for *all* real exponents.

Consequently, for each positive real number a there is a function (called the **exponential function with base a**) whose domain is all real numbers and whose rule is $f(x) = a^x$. For example,

$$f(x) = 10^x, \quad g(x) = 2^x, \quad h(x) = \left(\frac{1}{2}\right)^x, \quad k(x) = \left(\frac{3}{2}\right)^x$$

When $a > 1$, the graphs of the various exponential functions all have a similar shape.

Example 1 To graph each of the functions

$$f(x) = 10^x, \quad g(x) = \left(\frac{3}{2}\right)^x, \quad h(x) = 2^x, \quad k(x) = 3^x$$

use a graphing calculator or the $\boxed{y^x}$ key on an ordinary calculator (evaluate the function at several numbers,* then plot the corresponding points). In each case, the x-axis is a horizontal asymptote for the graph.

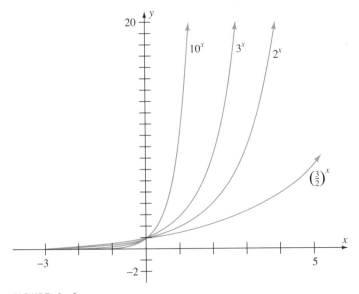

FIGURE 6-1

* To evaluate $3^{2.7}$, for example, press $\boxed{3}\,\boxed{y^x}\,\boxed{2.7}\,\boxed{=}$.

All of these graphs illustrate **exponential growth,** which is much more explosive than polynomial growth. For example, the graph of the polynomial function $s(x) = x^4$ on page 233 is approximately 2.5 inches above the x-axis at $x = 100$. On the same scale, the graph of the exponential function $h(x) = 2^x$ at $x = 100$ would be more than *500 quadrillion miles* high (that's 500 followed by 15 zeros)! ■

Example 2 (Compound Interest) If $5000 is deposited in an account that pays 9.5% interest, compounded annually, then the balance of the account at the end of year t is given by the function $A(t) = 5000(1.095)^t$.* The graph of this function is similar in shape to the graph of $h(x) = 2^x$. In practical terms, this means that if you leave your money in the bank long enough, it will grow to a huge amount. For instance, after 10 years you will have

$$A(10) = 5000(1.095)^{10} = \$12{,}391.14.$$

After 59 years, you will be a millionaire because

$$A(59) = 5000(1.095)^{59} = \$1{,}057{,}798.72.\ ■$$

When $0 < a < 1$, the graph of $f(x) = a^x$ has a different shape which is sometimes described by the term **exponential decay.**

Example 3 A calculator can be used to graph $r(x) = \left(\dfrac{1}{2}\right)^x$ in Figure 6–2.

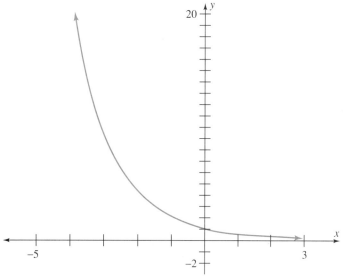

FIGURE 6–2

* For the reason why, see Excursion 6.1.A.

Once again the x-axis is a horizontal asymptote, but the graph of $r(x) = \left(\frac{1}{2}\right)^x$ *falls* from left to right without ever reaching the x-axis. ∎

Example 4 (Radioactive Decay) The **half-life** of a radioactive element is the time it takes a given quantity to decay to one half of its original mass. It is shown in calculus that the mass $M(x)$ at time x is given by $M(x) = c2^{-x/h} = c\left(\frac{1}{2}\right)^{x/h}$, where h is the half-life and c is the original mass of the element.

Plutonium (^{239}Pu) has a half-life of 24,360 years. So the amount remaining from 1 kilogram after x years is given by the function $M(x) = 1\left(\frac{1}{2}\right)^{x/24360}$.

The graph of M is shaped like the graph of $r(x) = \left(\frac{1}{2}\right)^x$, but moves toward the x-axis *very* slowly as x gets larger. In other words, even after an extremely long time a substantial amount will remain. Most of the original kilogram is still there after *ten thousand* years because

$$M(10{,}000) = \left(\frac{1}{2}\right)^{10000/24360} \approx \left(\frac{1}{2}\right)^{.4105} \approx .7524 \text{ kg}$$

This is the reason that nuclear waste disposal is such a serious problem. ∎

The preceding examples illustrate these facts:

GRAPHS OF EXPONENTIAL FUNCTIONS

> If $a > 1$, then $f(x) = a^x$ is a positive, increasing function:
> Its graph is always above the x-axis and rises from left to right.
> If $0 < a < 1$, then $f(x) = a^x$ is a positive, decreasing function:
> Its graph is always above the x-axis and falls from left to right.

You have often worked with the irrational number π, which is the ratio of the circumference of a circle to its diameter. The number π is built into reality. You can't have a circle without π any more than you can repeal the law of gravity.

There is another irrational number, denoted e, which arises naturally in a variety of phenomena* and plays a central role in the mathematical description of the physical universe. The infinite decimal expansion of e begins

$$e = 2.718281828459045 \cdots$$

---

* See Excursion 6.1.A for one example.

The *most important* of the exponential functions is $f(x) = e^x$. Virtually every scientific calculator has an $\boxed{e^x}$ key for evaluating this function.*

Example 5 Since $2 < e < 3$, the graph of $f(x) = e^x$ lies between the graphs of $h(x) = 2^x$ and $k(x) = 3^x$. It is easily obtained by using a calculator, as shown in Figure 6–3. ∎

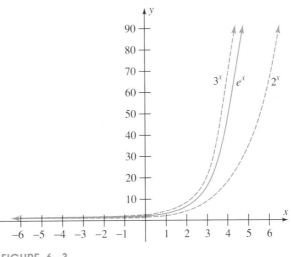

FIGURE 6–3

Example 6 (Population Growth) If the world population continues growing at the present rate, then the population in *billions* in year t will be given by the function $S(t) = 4.2e^{.0198(t-1980)}$. To estimate the world population at the turn of the century, we use a calculator to evaluate this function at $t = 2000$:

$$S(2000) = 4.2e^{.0198(2000-1980)} = 4.2e^{.0198(20)}$$
$$= 4.2e^{.396} \approx 4.2(1.485869)$$
$$\approx 6.24 \text{ billion people.} \blacksquare$$

Example 7 (Inhibited Population Growth) There is an upper limit on the fish population in a certain lake due to the oxygen supply, available food, etc. The population of fish in this lake at time t months is given by the function

$$p(t) = \frac{20{,}000}{1 + 24e^{-t/4}} \quad [t \geq 0]$$

Its graph in Figure 6–4 (on page 276), which can be obtained by plotting a large number of points, shows that the fish population never goes above 20,000. ∎

---

* On calculators without an $\boxed{e^x}$ key, try the two-key combination $\boxed{\text{INV}}\boxed{\text{LN}}$. If 1 $\boxed{\text{INV}}$ $\boxed{\text{LN}}$ $=$ produces 2.718281828, then $\boxed{\text{INV}}$ $\boxed{\text{LN}}$ is the same as $\boxed{e^x}$.

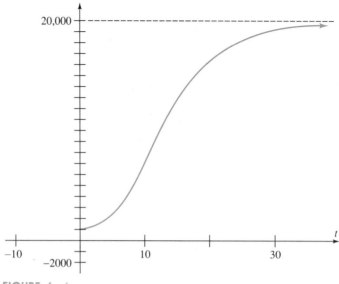

FIGURE 6-4

WARNING

The preceding examples are mathematical models of reality that sometimes ignore important factors in order to keep the functions involved relatively simple. Such simplicity may lead to poor predictions. For instance, the human population growth model in Example 6 above does not take into account factors that may limit growth in the future (major wars, unusual diseases, changing ethical standards, etc.). Hence this model is unlikely to be accurate for long-term predictions over several centuries.

EXERCISES

In Exercises 1–14, graph the function.

1. $f(x) = 4^x$
2. $g(x) = 5^x$
3. $f(x) = (1.2)^x$
4. $g(x) = (.8)^x$
5. $f(x) = 4^{-x}$ $\left[\text{Note: } 4^{-x} = \left(\frac{1}{4}\right)^x.\right]$
6. $f(x) = (5/2)^{-x}$
7. $f(x) = 2^{3x}$
8. $g(x) = 3^{x/2}$
9. $h(x) = 2^{x/3}$
10. $k(x) = 3^{-2x}$
11. $f(x) = e^{2x}$
12. $g(x) = e^{x/2}$
13. $f(x) = 2^{3-x^2}$
14. $h(x) = 3^x - 2^x$

In Exercises 15–22, use the graphs of $h(x) = 2^x$ on page 272 and $f(x) = e^x$ on page 275 and the techniques of Section 4.4 to graph the function.

15. $f(x) = 2^x - 5$
16. $g(x) = -(2^x)$
17. $k(x) = 3(2^x)$
18. $g(x) = 2^{x-1}$
19. $g(x) = e^x + 2$
20. $k(x) = -e^x$

21. $g(x) = e^{x+2}$ 22. $k(x) = -4e^{x-4}$

In Exercises 23–26, find the difference quotient
$$\frac{f(x+h) - f(x)}{h}.$$

23. $f(x) = 10^x$ 24. $f(x) = 5^{x^2}$
25. $f(x) = 2^x + 2^{-x}$ 26. $f(x) = e^x - e^{-x}$

In Exercises 27–31, determine whether the given function is even, odd, or neither (see pages 179–180).

27. $f(x) = 10^x$ 28. $f(x) = 2^x - x$
29. $f(x) = \dfrac{e^x + e^{-x}}{2}$ 30. $h(x) = \dfrac{e^x - e^{-x}}{2}$
31. $f(x) = e^{-x^2}$

32. Use the graph of $f(x) = e^x$ (not a calculator) to explain why $e^x + e^{-x}$ is approximately equal to e^x when x is large.

In Exercises 33–40, graph the function. Exercises 27–32 may be helpful.

33. $k(x) = e^{-x}$ 34. $f(x) = e^{-x^2}$
35. $f(x) = \dfrac{e^x + e^{-x}}{2}$ 36. $h(x) = \dfrac{e^x - e^{-x}}{2}$
37. $g(x) = 2^x - x$ 38. $k(x) = \dfrac{2}{e^x + e^{-x}}$
39. $f(x) = \dfrac{5}{1 + e^{-x}}$ 40. $g(x) = \dfrac{10}{1 + 9e^{-x/2}}$

41. If you deposit $750 at 8.2% interest, compounded annually and paid from the day of deposit to the day of withdrawal, your balance at time t is given by $B(t) = 750(1.082)^t$. How much will you have after 2 years? After 3 years and 9 months?

42. If your interest in Exercise 41 were compounded every second, then your balance at time t would be given by $C(t) = 750e^{.082t}$. How much would you have after 2 years? After 5 years and 4 months?

43. The half-life of carbon-14 is 5730 years. So the amount left from an initial mass of 5 grams in year t is given by $M(t) = 5 \cdot 2^{-t/5730}$, as explained in Example 4. How much would be left after 4000 years? After 8000 years?

44. (a) The half-life of radium is 1620 years. If you start with 100 mg of radium, what is the function that gives the amount remaining after t years? (See Exercise 43 and Example 4.)
(b) How much radium is left after 800 years? After 1600 years? After 3200 years?

45. The population of a colony of fruit flies t days from now is given by the function $p(t) = 100 \cdot 3^{t/10}$. What will the population be in 15 days? In 25 days?

46. A certain type of bacteria grows according to the function $f(t) = 5000e^{.4055t}$, where the time t is measured in hours. What will the population be in 8 hours? In 2 days?

47. If the U.S. population continues to grow as in the past decade, then the population in *millions* in year t will be approximately $P(t) = 250(1.009^{t-1990})$. Estimate the U.S. population in the year 2000 and in the year you reach 65 years old.

48. (a) There were fewer than a billion people in the world when Thomas Jefferson died. Use the world population function in Example 6 to estimate the world population in *billions* in 2010 and 2050.
(b) By trying various values, estimate how many years it will take for the world population in 1995 to double. How many people will there be then?

49. Water and salt are continuously added to a tank in such a way that the number of kilograms of salt in the tank at time t minutes is $g(t) = 200 - 100e^{-t/20}$. How much salt is in the tank at the beginning ($t = 0$)? After 10 minutes? After 20 minutes? After 40 minutes?

50. In a room where the temperature is 70°, a certain hot object cools according to the function $T(x) = 70 + 100e^{-.04x}$, where $T(x)$ is the temperature of the object at time x minutes. What is the temperature after 40 minutes? After an hour and a half?

51. The estimated number of units that will be sold by a certain company t months from now is given by $N(t) = 100{,}000e^{-.09t}$.
(a) What are current sales ($t = 0$)? What will sales be in 2 months? In 6 months?
(b) Will sales ever start to increase again? (What does the graph of $N(t)$ look like?)

52. (a) If inflation runs at a steady 4% per year, then the amount one dollar is worth x years from now is given by the function $f(x) = .96^x$. How much will it be worth in 5 years? In 10 years?

(b) By trying various values, estimate how many years it will be before today's dollar is worth only a dime.

53. (a) The beaver population near a certain lake in year t is approximately $p(t) = \dfrac{2000}{1 + 199e^{-.5544t}}$. What is the population now ($t = 0$) and what will it be in 5 years?
(b) Approximately when will there be 1000 beavers? [Use a calculator to test various possibilities.]

54. (a) The function $g(t) = 1 - e^{-.0479t}$ gives the percentage of the population (expressed as a decimal) that has seen a new TV show t weeks after it went on the air. What percentage of people have seen the show after 24 weeks?
(b) Approximately when will 90% of the people have seen it?

Unusual Problems

55. Find a function $f(x)$ with the property $f(r + s) = f(r)f(s)$ for all real numbers r and s. [Hint: Think exponential.]

56. Find a function $g(x)$ with the property $g(2x) = (g(x))^2$ for every real number x.

57. An eccentric billionaire offers you a job for the month of September. She says that she will pay you 2¢ on the first day, 4¢ on the second day, 8¢ on the third day, and so on, doubling your pay on each successive day.
(a) Let $P(x)$ denote your salary in *dollars* on day x. Find the rule of the function P.
(b) Would you be better off financially if instead you were paid $10,000 per day? [Hint: Consider $P(30)$.]

58. Look back at Section 5.2, where the basic properties of graphs of polynomial functions were listed. Then review the basic properties of the graph of $f(x) = a^x$ discussed in this section. Using these various properties, give an argument to show that for any fixed positive number $a (\neq 1)$, it is *not* possible to find a polynomial function $g(x) = c_n x^n + \cdots + c_1 x + c_0$ such that $a^x = g(x)$ for *all* numbers x. In other words, *no exponential function is a polynomial function.*

6.1.A EXCURSION: Compound Interest and the Number e

If you deposit P dollars at an interest rate of $r\%$, then at the end of one time period you will have P dollars, plus the interest rP,* for a total of $P + rP = P(1 + r)$ dollars. Hence

The amount at the end of the period is the beginning principal multiplied by $(1 + r)$.

If you leave your money in the bank, your principal at the beginning of the second period is $P(1 + r)$. At the end of the second period it has been multiplied by a factor of $(1 + r)$, so you have a total of

$$[P(1 + r)](1 + r) = P(1 + r)^2 \text{ dollars}$$

Your money continues to increase by a factor of $1 + r$ each time period, leading to this conclusion:

---

*Here the rate is written as a decimal; for instance, 8% of P is $.08P$.

6.1.A COMPOUND INTEREST AND THE NUMBER e

COMPOUND INTEREST FORMULA

> If P dollars are invested at interest rate r per time period (expressed as a decimal), then the amount A after t periods is
> $$A = P(1 + r)^t$$

Interest is often paid from day of deposit to day of withdrawal, regardless of the period used for compounding the interest. So the formula is used even when t is not an integer.

Example 1 (Compound Interest) If $7500 is invested at 12% interest compounded yearly, how much is in the account **(a)** after 5 years? **(b)** after 9 years and 3 months?

Solution **(a)** Apply the compound interest formula with $P = 7500$, $r = .12$, and $t = 5$. Then $A = 7500(1.12)^5 = \$13{,}217.56$.

(b) Since 9 years and 3 months is 9.25 years, the amount in the account then is $A = 7500(1.12)^{9.25} = \$21{,}395.77$. ∎

Banks often state an annual interest rate, but compound it more than once a year—for instance, 8% compounded quarterly. If the interest rate i is compounded n times a year, then the interest rate is i/n per period and there are n periods in 1 year.

Example 2 (Compound Interest) If $9000 is invested at 8% annual interest, compounded monthly, how much will the investment be worth after 6 years?

Solution Interest is compounded 12 times per year, so the time period is 1/12 of a year and the interest rate per period is $r = .08/12$. The number of periods in 6 years is $t = 6 \cdot 12 = 72$. Using these numbers in the compound interest formula shows that

$$A = 9000(1 + .08/12)^{72} \approx 9000(1.0067)^{72} = \$14{,}521.52.$$

This is more money than there would be if the interest were compounded annually ($r = .08$ and $t = 6$ in the formula):

$$A = 9000(1 + .08)^6 = 9000(1.08)^6 = \$14{,}281.87. \quad \blacksquare$$

As a general rule, the more often your interest is compounded, the better off you are. But there is, alas, a limit.

Example 3 (The Number e) You have $1 to invest for 1 year. The Exponential Bank offers to pay 100% annual interest, compounded n times per year and rounded to the nearest penny. You can pick any value you want for n. Can you choose n so large that your $1 will grow to some huge amount?

Solution Since the interest rate 100% (= 1.00) is compounded n times per year, the interest rate per period is $r = 1/n$ and the number of periods in 1 year is n. According to the formula, the amount at the end of the year will be $A = \left(1 + \frac{1}{n}\right)^n$. Here's what happens for various values of n:*

Interest is Compounded	$n =$	$\left(1 + \frac{1}{n}\right)^n =$
Annually	1	$(1 + \frac{1}{1})^1 = 2$
Semiannually	2	$(1 + \frac{1}{2})^2 = 2.25$
Quarterly	4	$(1 + \frac{1}{4})^4 \approx 2.4414$
Monthly	12	$(1 + \frac{1}{12})^{12} \approx 2.6130$
Daily	365	$(1 + \frac{1}{365})^{365} \approx 2.71457$
Hourly	8760	$(1 + \frac{1}{8760})^{8760} \approx 2.718127$
Every minute	525,600	$(1 + \frac{1}{525,600})^{525,600} \approx 2.7182792$
Every second	31,536,000	$(1 + \frac{1}{31,536,000})^{31,536,000} \approx 2.7182818$

Since interest is rounded to the nearest penny, your dollar will grow no larger than $2.72, no matter how big n is.

The last entry in the table, 2.7182818, is the number e to seven decimal places. This is just one example of how e arises naturally in real-world situations. In calculus it is shown that e is the *limit* of $\left(1 + \frac{1}{n}\right)^n$, meaning that as n gets larger and larger, $\left(1 + \frac{1}{n}\right)^n$ gets closer and closer to e. ■

EXERCISES

1. If $1000 is invested at 8%, find the value of the investment after 5 years if interest is compounded
   (a) annually   (b) quarterly
   (c) monthly   (d) weekly

2. If $2500 is invested at 11.5%, what is the value of the investment after 10 years if interest is compounded
   (a) annually   (b) monthly
   (c) daily

In Exercises 3–8, determine how much money will be in a savings account if the initial deposit was $500 and the interest rate is:

3. 5% compounded annually for 8 years.
4. 5% compounded annually for 10 years.
5. 5% compounded quarterly for 10 years.
6. 9.3% compounded monthly for 9 years.
7. 8.9% compounded daily for 8.5 years.

---

* The calculations in the table were done on a large computer, using double precision. The interest rate 100% was chosen for computational convenience. Essentially the same point can be made with a more realistic rate.

8. 12.5% compounded weekly for 7 years and 7 months.

9. A typical credit card company charges 18% annual interest, compounded monthly, on the unpaid balance. If your current balance is $520 and you don't make any payments for 6 months, how much will you owe (assuming they don't sue you in the meantime)?

10. By trying various values, estimate how long it will take to double an investment of $100 if the interest rate is 8%, compounded annually. Do the same for a $500 investment.

In Exercises 11–14, use the fact that the compound interest formula applies not just to dollars and interest, but to any items that increase by a specific percentage each time period.

11. If the number of dandelions in your lawn increases by 5% a week and there are 75 dandelions now, how many will there be in 16 weeks?

12. If there are now 3.2 million people who play bridge and the number increases by 3.5% a year, how many people will be playing bridge in 15 years?

13. The population of a certain city is now 840,000 and is increasing at .2% per month. What will the population be 5 years from now?

14. If the population of India is now 750 million people and it continues to increase by 2% a year, what will the population be in 10 years? In 20 years?

6.2 THE NATURAL LOGARITHMIC FUNCTION

From their invention in the 17th century until the development of computers and electronic calculators, logarithms were the only effective tool for numerical computations in astronomy, chemistry, physics, and engineering. Although they are no longer needed for computation, logarithmic functions still play an important role in the sciences and engineering.

> ROADMAP: We begin with natural logarithms because they are the most important ones—practically the only logarithms used in calculus. However, to accommodate those who prefer to begin with logarithms to the base b (Section 6.3), the text is written so that Section 6.3 may be covered before Section 6.2 if desired.

Logarithms are simply *a new language for old ideas*—essentially a special case of exponents. So most of this section is simply a restatement of known facts about exponents in the language and symbolism of logarithms.

The definition depends on the number e (introduced on page 274), whose decimal expansion begins 2.71828. The graph of $f(x) = e^x$ on page 275 shows that for each positive v, there is exactly one number u such that $e^u = v$. For $v > 0$, the term "ln v" is read "the **natural logarithm** of v" * and is defined by:

---

* We shall omit the word "natural" except when it is necessary to distinguish these logarithms from other types that are introduced in Section 6.3.

DEFINITION OF NATURAL LOGARITHM

$$\ln v = u \quad \text{means} \quad e^u = v.$$

In other words, *ln v is the answer to the question:*

To what power must e be raised to produce v?

To find ln 7, for example, raise e to various powers until you find one that works. A little experiment with a calculator shows that $e^{1.946} \approx 7$. Therefore ln 7 ≈ 1.946.

Although such experimentation is useful to give you a feel for logarithms, it's easier to enter the number v in your calculator and press the $\boxed{\text{LN}}$ key. The calculator will produce the exponent (logarithm). For instance,

Logarithmic Statement*	*Equivalent Exponential Statement*
ln 14 = 2.6391	$e^{2.6391} = 14$
ln 65 = 4.1744	$e^{4.1744} = 65$
ln 158 = 5.0626	$e^{5.0626} = 158$

You *don't* need a calculator to understand the essential properties of logarithms. You need only translate logarithmic statements into exponential ones (or vice versa).

Example 1 What is ln(−10)? *Translation:* To what power must e be raised to produce −10? The graph of $f(x) = e^x$ on page 275 shows that *every* power of e is *positive*. So e^u can *never* be −10, or any negative number, or zero. Hence

ln v is defined only when $v > 0$. ■

Example 2 What is ln 1? *Translation:* To what power must e be raised to produce 1? The answer, of course, is $e^0 = 1$. So ln 1 = 0. Similarly, ln e = 1 because 1 is the answer to "what power of e equals e?" Hence

ln 1 = 0 and **ln e = 1.** ■

Example 3 What is ln e^9? *Translation:* To what power must e be raised to produce e^9? Obviously, the answer is 9. So ln $e^9 = 9$, and in general,

ln $e^k = k$ for every real number k. ■

---

* Here and below, all logarithms are rounded to four decimal places. So strictly speaking, the equal sign should be replaced by an "approximately equal" sign (≈).

Example 4 What is $e^{\ln 678}$? Well, $\ln 678$ is the power to which e must be raised to produce 678, that is, $e^{\ln 678} = 678$. Similarly,

$$e^{\ln v} = v \quad \text{for every } v > 0. \blacksquare$$

Here is a summary of the facts introduced in the preceding examples.

PROPERTIES OF NATURAL LOGARITHMS

1. $\ln v$ is defined only when $v > 0$.
2. $\ln 1 = 0$ and $\ln e = 1$.
3. $\ln e^k = k$ for every real number k.
4. $e^{\ln v} = v$ for every $v > 0$.

Example 5 Applying Property 3 with $k = 2x^2 + 7x + 9$ shows that

$$\ln e^{2x^2+7x+9} = 2x^2 + 7x + 9. \blacksquare$$

Example 6 Solve the equation $\ln(x + 1) = 2$.

Solution Since $\ln(x + 1) = 2$, we have:

$$e^{\ln(x+1)} = e^2.$$

Applying Property 4 with $v = x + 1$ shows that

$$x + 1 = e^{\ln(x+1)} = e^2$$
$$x = e^2 - 1 \approx 6.3891. \blacksquare$$

Every exponential function can be expressed in terms of e because for any $a > 0$,

EXPONENTIAL FUNCTIONS

$$a^x = (e^{\ln a})^x = e^{(\ln a)x}$$

The first law of exponents states that $e^m e^n = e^{m+n}$ or, in words,

The exponent of a product is the sum of the exponents of the factors.

Since logarithms are just particular kinds of exponents, this statement translates as:

The logarithm of a product is the sum of the logarithms of the factors.

Here is the same statement in symbolic language:

PRODUCT LAW FOR LOGARITHMS

$$\ln(vw) = \ln v + \ln w \quad \text{for all } v, w > 0.$$

Proof According to Property 4 of logarithms,

$$e^{\ln v} = v \quad \text{and} \quad e^{\ln w} = w.$$

Therefore by the first law of exponents (with $m = \ln v$ and $n = \ln w$):

$$vw = e^{\ln v} e^{\ln w} = e^{\ln v + \ln w}.$$

So raising e to the exponent ($\ln v + \ln w$) produces vw. But the definition of logarithm says that $\ln vw$ is the exponent to which e must be raised to produce vw. Therefore we must have $\ln vw = \ln v + \ln w$. This completes the proof. ∎

Example 7 A calculator shows that $\ln 7 = 1.9459$ and $\ln 9 = 2.1972$. Therefore

$$\ln 63 = \ln 7 \cdot 9 = \ln 7 + \ln 9 = 1.9459 + 2.1972 = 4.1431 \blacksquare$$

WARNING

A common error in applying the Product Law for Logarithms is to write $\ln 7 + \ln 9 = \ln(7 + 9) = \ln 16$ instead of the correct statement $\ln 7 + \ln 9 = \ln 7 \cdot 9 = \ln 63$.

The second law of exponents, namely, $e^m/e^n = e^{m-n}$, may be roughly stated in words as

The exponent of the quotient is the difference of the exponents.

When the exponents are logarithms, this says

The logarithm of a quotient is the difference of the logarithms.

In other words,

QUOTIENT LAW FOR LOGARITHMS

$$\ln\left(\frac{v}{w}\right) = \ln v - \ln w \quad \text{for all } v, w > 0.$$

The proof of the Quotient Law is very similar to the proof of the Product Law (see Exercise 58).

Example 8 $\ln\left(\dfrac{17}{44}\right) = \ln 17 - \ln 44$. ∎

Example 9 For any $w > 0$,

$$\ln\left(\frac{1}{w}\right) = \ln 1 - \ln w = 0 - \ln w = -\ln w. \blacksquare$$

6.2 THE NATURAL LOGARITHMIC FUNCTION

> **WARNING**
>
> Do not confuse $\ln\left(\dfrac{v}{w}\right)$ with the quotient $\dfrac{\ln v}{\ln w}$. They are *different* numbers. For example,
>
> $\ln\left(\dfrac{36}{3}\right) = \ln(12) = 2.4849$, but $\dfrac{\ln 36}{\ln 3} = \dfrac{3.5835}{1.0986} = 3.2619$.

The third law of exponents, namely, $(e^m)^k = e^{mk}$, can also be translated into logarithmic language:

POWER LAW FOR LOGARITHMS

> $\ln(v^k) = k(\ln v)$ for all k and all $v > 0$.

Proof Since $v = e^{\ln v}$, the third law of exponents (with $m = \ln v$) shows that

$$v^k = (e^{\ln v})^k = e^{(\ln v)k} = e^{k(\ln v)}.$$

So raising e to the exponent $k(\ln v)$ produces v^k. But the exponent to which e must be raised to produce v^k is $\ln v^k$. Therefore $\ln v^k = k(\ln v)$, and the proof is complete. ∎

Example 10 $\ln \sqrt{19} = \ln 19^{1/2} = \tfrac{1}{2}(\ln 19)$. ∎

The logarithm laws can be used to simplify various expressions.

Example 11 $\ln 3x + 4 \cdot \ln x - \ln 3xy$

$\qquad\qquad = \ln 3x + \ln x^4 - \ln 3xy \qquad$ *[Power Law]*

$\qquad\qquad = \ln(3x \cdot x^4) - \ln 3xy \qquad$ *[Product Law]*

$\qquad\qquad = \ln \dfrac{3x^5}{3xy} \qquad$ *[Quotient Law]*

$\qquad\qquad = \ln \dfrac{x^4}{y} \qquad$ *[Cancel 3x]* ∎

6 EXPONENTIAL AND LOGARITHMIC FUNCTIONS

Example 12 To simplify $\ln(\sqrt{x}/x) + \ln \sqrt[4]{ex^2}$, we begin by changing to exponential notation:

$$\ln\left(\frac{x^{1/2}}{x}\right) + \ln(ex^2)^{1/4} = \ln(x^{-1/2}) + \ln(ex^2)^{1/4}$$

$$= -\frac{1}{2}\cdot\ln x + \frac{1}{4}\cdot\ln ex^2 \qquad [\text{Power Law}]$$

$$= -\frac{1}{2}\cdot\ln x + \frac{1}{4}(\ln e + \ln x^2) \qquad [\text{Product Law}]$$

$$= -\frac{1}{2}\cdot\ln x + \frac{1}{4}(\ln e + 2\cdot\ln x) \qquad [\text{Power Law}]$$

$$= -\frac{1}{2}\cdot\ln x + \frac{1}{4}\cdot\ln e + \frac{1}{2}\cdot\ln x$$

$$= \frac{1}{4}\cdot\ln e = \frac{1}{4} \qquad [\ln e = 1] \blacksquare$$

The Natural Logarithmic Function

Now that you are familiar with the *language* of logarithms, we consider the **natural logarithmic function** g whose rule is: $g(x) = \ln x$. The domain of g is the set of positive real numbers (the numbers x for which $\ln x$ is defined). The properties of logarithms developed above show that the functions $f(x) = e^x$ and $g(x) = \ln x$ have these properties:

$$(f \circ g)(x) = f(g(x)) = f(\ln x) = e^{\ln x} = x \quad \text{for all } x > 0;$$

$$(g \circ f)(x) = g(f(x)) = g(e^x) = \ln e^x = x \quad \text{for all } x*$$

Therefore, as explained in Section 4.6,

INVERSE FUNCTIONS

$g(x) = \ln x$ is the inverse function of $f(x) = e^x$.

---

* A calculator provides a visual demonstration of these facts; for example,

Enter the number . . . . Press $\boxed{\text{LN}}$ key . . . . Press $\boxed{e^x}$ key

$\qquad\qquad x \qquad\qquad\qquad\qquad g(x) \qquad\qquad\qquad f(g(x)) = (f \circ g)(x)$

The final display is the number x with which you started (with the possible exception of a slight rounding error), thus showing that $(f \circ g)(x) = x$.

6.2 THE NATURAL LOGARITHMIC FUNCTION

Consequently, the graph of $g(x) = \ln x$ is the graph of $f(x) = e^x$ reflected in the line $y = x$ (see page 199).*

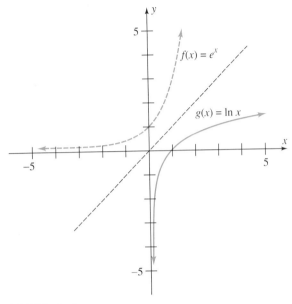

FIGURE 6-5

The key features of the graph of $g(x) = \ln x$ are that the graph is always rising from left to right (that is, $g(x) = \ln x$ is an increasing function) and that the y-axis is a vertical asymptote for the graph (meaning that the graph gets closer and closer to the y-axis, but never touches it).

The graph of $g(x) = \ln x$ illustrates **logarithmic growth,** which is as slow as exponential growth is fast. The graph of $f(x) = e^x$ is in the stratosphere when $x = 150$, whereas the graph of $g(x) = \ln x$, although always rising, is barely more than 5 units high at $x = 150$ and doesn't rise to 6 units high until $x \approx 403.4$.

The graphing techniques of Section 4.4 apply to logarithmic functions.

Example 13 (a) The graph of $h(x) = \ln(x - 2)$ is the graph of $g(x) = \ln x$ shifted horizontally 2 units to the right, as shown in Figure 6-6 on the next page.

(b) The graph of $k(x) = 3 - \ln x$ is the graph of $g(x) = \ln x$ reflected in the x-axis and shifted vertically 3 units upward, as shown in Figure 6-7 on the next page. ∎

---

* The graph of g can also be obtained directly, either by means of a graphing calculator or by using an ordinary calculator and plotting points.

6 EXPONENTIAL AND LOGARITHMIC FUNCTIONS

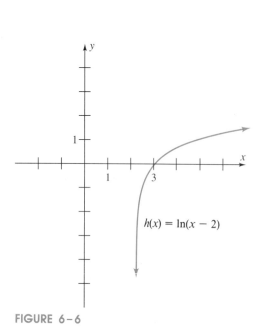

FIGURE 6-6

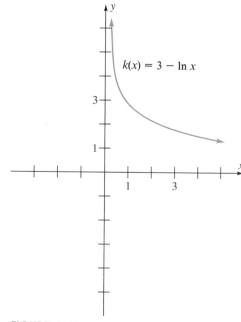

FIGURE 6-7

EXERCISES

Note: Unless stated otherwise, all letters represent positive numbers.

In Exercises 1–6, translate the given logarithmic statement into an equivalent exponential statement.

1. $\ln 3 = 1.0986$
2. $\ln 10 = 2.3026$
3. $\ln 1000 = 6.9078$
4. $\ln \frac{1}{4} = -1.3863$
5. $\ln .01 = -4.6052$
6. $\ln s = r$

In Exercises 7–12, translate the given exponential statement into an equivalent logarithmic one.

7. $e^{3.25} = 25.79$
8. $e^{-4} = .0183$
9. $e^{12/7} = 5.5527$
10. $e^k = t$
11. $e^{2/r} = w$
12. $e^{4uv} = m$

In Exercises 13–24, evaluate the given expression without using a calculator.

13. $\ln e^{15}$
14. $\ln e^{3.78}$
15. $\ln \sqrt{e}$
16. $\ln \sqrt[5]{e}$
17. $e^{\ln 931}$
18. $e^{\ln 34.17}$
19. $e^{\ln \sqrt{37}}$
20. $e^{\ln 7/5}$
21. $\ln e^{x+y}$
22. $\ln e^{x^2+2y}$
23. $e^{\ln x^2}$
24. $e^{\ln \sqrt{x+3}}$

In Exercises 25–30, write the given expression as a single logarithm, as in Example 11.

25. $\ln x^2 + 3 \ln y$
26. $\ln 2x + 2(\ln x) - \ln 3y$
27. $\ln(x^2 - 9) - \ln(x + 3)$
28. $\ln 3x - 2(\ln x - \ln(2 + y))$
29. $2(\ln x) - 3(\ln x^2 + \ln x)$
30. $\ln(e/\sqrt{x}) - \ln \sqrt{ex}$

In Exercises 31–36, let $u = \ln x$ and $v = \ln y$. Write the given expression in terms of u and v. For example, $\ln x^3y = \ln x^3 + \ln y = 3(\ln x) + \ln y = 3u + v$.

31. $\ln(x^2y^5)$
32. $\ln(x^3/y^2)$
33. $\ln(\sqrt{x} \cdot y^2)$
34. $\ln(\sqrt{x}/y)$
35. $\ln(\sqrt[3]{x^2\sqrt{y}})$
36. $\ln(\sqrt{x^2y}/\sqrt[3]{y})$

In Exercises 37–42, find the domain of the given function (that is, the largest set of real numbers for which the rule produces well-defined real numbers).

37. $f(x) = \ln(x + 1)$
38. $g(x) = \ln(x + 2)$
39. $h(x) = \ln(-x)$
40. $k(x) = \ln(2 - x)$
41. $f(x) = \ln(x^2 - x - 2)$
42. $r(x) = \ln(-x - 3) - 10$

In Exercises 43–48, use the graph of $g(x) = \ln x$ on page 287, and the techniques of Section 4.4 to graph the given function.

43. $f(x) = 2 \cdot \ln x$
44. $f(x) = \ln x - 7$
45. $h(x) = \ln(x - 4)$
46. $k(x) = \ln(x + 2)$
47. $h(x) = \ln(x + 3) - 4$
48. $k(x) = \ln(x - 2) + 2$

In Exercises 49–54, graph the function.

49. $f(x) = \ln|x|$
50. $h(x) = |\ln x|$
51. $k(x) = \ln(1/x)$
52. $f(x) = 1/\ln x$
53. $h(x) = \ln x^3$
54. $k(x) = \ln \sqrt{x}$

55. Suppose $f(x) = A \ln x + B$, where A and B are constants. If $f(1) = 10$ and $f(e) = 1$, what are A and B?

56. If $f(x) = A \ln x + B$ and $f(e) = 5$ and $f(e^2) = 8$, what are A and B?

57. Show that $g(x) = \ln\left(\dfrac{x}{1-x}\right)$ is the inverse function of $f(x) = \dfrac{1}{1 + e^{-x}}$. (See Section 4.6.)

58. Prove the Quotient Law for Logarithms: for $v, w > 0$, $\ln\left(\dfrac{v}{w}\right) = \ln v - \ln w$. (Use properties of exponents and the fact that $v = e^{\ln v}$ and $w = e^{\ln w}$.)

59. Show that $f(x) = \ln x^2$ is *not* the same function as $g(x) = 2 \cdot \ln x$. [*Hint:* The domain of each function is the largest set of real numbers for which the rule produces well-defined real numbers.]

60. The doubling function $D(x) = \dfrac{\ln 2}{\ln(1 + x)}$ gives the years required to double your money when it is invested at interest rate x (expressed as a decimal), compounded annually.
    (a) Find the time it takes to double your money at each of these interest rates: 4%, 6%, 8%, 12%, 18%, 24%, 36%.
    (b) Round the answers in part (a) to the nearest year and compare them with these numbers: 72/4, 72/6, 72/8, 72/12, 72/18, 72/24, 72/36. Use this evidence to state a "rule of thumb" for determining approximate doubling time, without using the function D. This rule of thumb, which has long been used by bankers, is called the **rule of 72**.

61. The height h above sea level (in meters) is related to air temperature t (in degrees Celsius), the atmospheric pressure p (in centimeters of mercury at height h), and the atmospheric pressure c at sea level by $h = (30t + 8000)\ln(c/p)$. If the pressure at the top of Mount Rainier is 44 cm on a day when the sea level pressure is 75.126 cm and the temperature is 7°, what is the height of Mount Rainier?

62. If the atmospheric pressure at the top of Mount Everest is 22.13 cm on a day when the sea level pressure is 75 cm, and the temperature is $-25°$, how high is Mount Everest? (See Exercise 61.)

63. A class in elementary Sanskrit is tested at the end of the semester and weekly thereafter on the same material. The average score on the exam taken after t weeks is given by the "forgetting function," $g(t) = 77 - 10 \cdot \ln(t + 1)$. What was the average score on the original exam? On the exam after 2 weeks? After 5 weeks? After 10 weeks?

64. Students in a precalculus class were given a final exam. Each month thereafter they took an equivalent exam. The class average on the exam taken after t months is given by

$$F(t) = 82 - 8 \cdot \ln(x + 1).$$

What was the class average after 6 months? After 9 months? After a year and a half?

65. One person with a flu virus visited the campus. The number T of days it took for the virus to infect n people was given by

$$T = (-16.3)\ln\left(\frac{5450}{n} - \frac{4}{11}\right).$$

How many days will it take for 6000 people to become infected?

66. The perceived loudness L of a sound of intensity I is given by $L = k \cdot \ln I$, where k is a certain constant. By how much must the intensity be increased to double the loudness? (That is, what must be done to I to produce $2L$?)

67. A bicycle store finds that the number N of bikes sold is related to the number d of dollars spent on advertising by $N = 51 + 100 \cdot \ln(d/100 + 2)$.

(a) How many bikes will be sold if nothing is spent on advertising? If $1000 is spent? If $10,000 is spent?
(b) If the average profit is $25 per bike, is it worthwhile to spend $1000 on advertising? What about $10,000?
(c) What are the answers in part (b) if the average profit per bike is $35?

68. The number N of days of training needed for a factory worker to produce x widgets per day is given by $N = -25 \cdot \ln\left(1 - \dfrac{x}{60}\right)$.

(a) How many training days are needed for the worker to be able to produce 40 widgets a day?
(b) It costs $135 to train 1 worker for 1 day. If the profit on 1 widget is $1.85, how many work days does it take before the factory breaks even on the training costs for a worker who can produce 40 widgets a day?

6.3 LOGARITHMIC FUNCTIONS TO OTHER BASES*

From their invention in the 17th century until the development of computers and electronic calculators, logarithms were the only effective tool for numerical computations in astronomy, chemistry, physics, and engineering. Although they are no longer needed for computation, logarithmic functions still play an important role in the sciences and engineering.

Logarithms are simply *a new language for old ideas* — essentially a special case of exponents. So most of this section is simply a restatement of known facts about exponents in the language and symbolism of logarithms.

Throughout this section b is a fixed positive number with $b \neq 1$.

The graph of $h(x) = b^x$ (see Figures 6–1 through 6–3) shows that, for each positive v, there is exactly one number u such that $b^u = v$. For each $v > 0$, the term "$\log_b v$" is read **"the logarithm of v to the base b,"** and is defined by:

DEFINITION OF LOGARITHM

$\log_b v = u$ means $b^u = v.$

So $\log_b v$ is the answer to the question

To what power must b be raised to produce v?

---

* This section is not needed in the sequel. For the most part, it repeats the discussion of Section 6.2 in a more general context, with the number e replaced by any positive number $b \neq 1$.

6.3 LOGARITHMIC FUNCTIONS TO OTHER BASES

Example 1 To find $\log_2 16$, ask yourself "what power of 2 equals 16?" Since $2^4 = 16$, we see that $\log_2 16 = 4$. Similarly, $\log_2(1/8) = -3$ because $2^{-3} = 1/8$. ∎

Example 2 Since logarithms are just exponents, every logarithmic statement can be translated into exponential language:

Logarithmic Statement	*Equivalent Exponential Statement*
$\log_3 81 = 4$	$3^4 = 81$
$\log_4 64 = 3$	$4^3 = 64$
$\log_{125} 5 = \dfrac{1}{3}$	$125^{1/3} = 5$*
$\log_8 \left(\dfrac{1}{4}\right) = -\dfrac{2}{3}$	$8^{-2/3} = \dfrac{1}{4}$ (verify!) ∎

Example 3 The equation $\log_5 x = 3$ is equivalent to the exponential statement $5^3 = x$, so the solution is $x = 125$. ∎

Example 4 Logarithms to the base 10 are called **common logarithms**. It is customary to write log v instead of $\log_{10} v$. Then

$\log 100 = 2 \quad$ because $\quad 10^2 = 100;$

$\log .001 = -3 \quad$ because $\quad 10^{-3} = \dfrac{1}{10^3} = \dfrac{1}{1000} = .001.$

Scientific calculators have a $\boxed{\text{LOG}}$ key for evaluating common logarithms. For instance,†

$\log .4 = -0.3979, \quad \log 45.3 = 1.6561, \quad \log 685 = 2.8357.$ ∎

Example 5 The most frequently used base for logarithms in modern applications is the number $e \ (\approx 2.71828 \ldots)$. Logarithms to the base e are called **natural logarithms** and use a different notation: we write ln v instead of $\log_e v$. Scientific calculators also have an $\boxed{\text{LN}}$ key for evaluating natural logarithms. For example,

$\ln .5 = -0.6931, \quad \ln 65 = 4.1744, \quad \ln 158 = 5.0626.$ ∎

You *don't* need a calculator to understand the essential properties of logarithms. You need only translate logarithmic statements into exponential ones (or vice versa).

---

* because $125^{1/3} = \sqrt[3]{125} = 5$.

† Here and below, all logarithms are rounded to four decimal places. So strictly speaking, the equal sign should be replaced by an "approximately equal" sign ($\approx$).

Example 6 What is $\log(-25)$? *Translation:* To what power must 10 be raised to produce -25? The graph of $f(x) = 10^x$ on page 272 shows that *every* power of 10 is *positive*. So 10^u can *never* be -25, or any negative number, or zero. The same argument works for any base b:

$$\log_b v \text{ is defined only when } v > 0. \blacksquare$$

Example 7 What is $\log_5 1$? *Translation:* To what power must 5 be raised to produce 1? The answer, of course, is $5^0 = 1$. So $\log_5 1 = 0$. Similarly, $\log_5 5 = 1$ because 1 is the answer to "what power of 5 equals 5?" In general,

$$\log_b 1 = 0 \quad \text{and} \quad \log_b b = 1. \blacksquare$$

Example 8 What is $\log_2 2^9$? *Translation:* To what power must 2 be raised to produce 2^9? Obviously, the answer is 9. So $\log_2 2^9 = 9$ and, in general,

$$\log_b b^k = k \quad \text{for every real number } k.$$

This property holds even when k is a complicated expression. For instance, if x and y are positive, then

$$\log_6 6^{\sqrt{3x+y}} = \sqrt{3x+y} \quad (\text{here } k = \sqrt{3x+y}). \blacksquare$$

Example 9 What is $10^{\log 439}$? Well, $\log 439$ is the power to which 10 must be raised to produce 439, that is, $10^{\log 439} = 439$. Similarly,

$$b^{\log_b v} = v \quad \text{for every } v > 0. \blacksquare$$

Here is a summary of the facts illustrated in the preceding examples.

PROPERTIES OF LOGARITHMS

1. $\log_b v$ is defined only when $v > 0$.
2. $\log_b 1 = 0$ and $\log_b b = 1$.
3. $\log_b(b^k) = k$ for every real number k.
4. $b^{\log_b v} = v$ for every $v > 0$.

The first law of exponents states that $b^m b^n = b^{m+n}$, or in words,

> The exponent of a product is the sum of the exponents of the factors.

Since logarithms are just particular kinds of exponents, this statement translates as:

> The logarithm of a product is the sum of the logarithms of the factors.

The second and third laws of exponents, namely $b^m/b^n = b^{m-n}$ and $(b^m)^k = b^{mk}$, can also be translated into logarithmic language:

LOGARITHM LAWS

> Let b, v, w, k be real numbers, with b, v, w positive and $b \neq 1$.
>
> **Product Law:** $\log_b(vw) = \log_b v + \log_b w$.
>
> **Quotient Law:** $\log_b\left(\dfrac{v}{w}\right) = \log_b v - \log_b w$.
>
> **Power Law:** $\log_b(v^k) = k(\log_b v)$.

Proof According to Property 4 in the box after Example 9;

$$b^{\log_b v} = v \quad \text{and} \quad b^{\log_b w} = w.$$

Therefore by the first law of exponents (with $m = \log_b v$ and $n = \log_b w$):

$$vw = b^{\log_b v} b^{\log_b w} = b^{\log_b v + \log_b w}.$$

So raising b to the exponent $(\log_b v + \log_b w)$ produces vw. But the definition of logarithm says that $\log_b vw$ is the exponent to which b must be raised to produce vw. Therefore we must have $\log_b vw = \log_b v + \log_b w$. This completes the proof of the Product Law.

Similarly, by the second law of exponents (with $m = \log_b v$ and $n = \log_b w$) we have:

$$\frac{v}{w} = \frac{b^{\log_b v}}{b^{\log_b w}} = b^{\log_b v - \log_b w}.$$

Since $\log_b(v/w)$ is the exponent to which b must be raised to produce v/w, we must have $\log_b(v/w) = \log_b v - \log_b w$. This proves the Quotient Law. The Power Law is proved in a similar fashion. ■

Example 10 Given that $\log_7 2 = .3562$, $\log_7 3 = .5646$, and $\log_7 5 = .8271$, find: **(a)** $\log_7 10$; **(b)** $\log_7 2.5$; **(c)** $\log_7 48$.

Solution

(a) By the Product Law,

$$\log_7 10 = \log_7(2 \cdot 5) = \log_7 2 + \log_7 5 = .3562 + .8271 = 1.1833.$$

(b) By the Quotient Law,

$$\log_7 2.5 = \log_7\left(\frac{5}{2}\right) = \log_7 5 - \log_7 2 = .8271 - .3562 = .4709.$$

(c) By the Product and Power Laws,

$$\log_7 48 = \log_7(3 \cdot 16) = \log_7 3 + \log_7 16 = \log_7 3 + \log_7 2^4$$
$$= \log_7 3 + 4 \cdot \log_7 2 = .5646 + 4(.3562) = 1.9894. \blacksquare$$

> **WARNING**
>
> 1. A common error in using the Product Law is to write something like $\log 6 + \log 7 = \log(6 + 7) = \log 13$ instead of the correct statement $\log 6 + \log 7 = \log 6 \cdot 7 = \log 42$.
>
> 2. Do not confuse $\log_b\left(\dfrac{v}{w}\right)$ with the quotient $\dfrac{\log_b v}{\log_b w}$. They are *different* numbers. For example, when $b = 10$
>
> $$\log\left(\frac{48}{4}\right) = \log 12 = 1.0792, \text{ but } \frac{\log 48}{\log 4} = \frac{1.6812}{0.6021} = 2.7922.$$

Example 10 worked because we were *given* several logarithms to base 7. But there's no $\boxed{\log_7}$ key on the calculator, so how do you find logarithms to base 7, or to any base other than e or 10? *Answer:* Use the $\boxed{\text{LN}}$ key on the calculator and the following formula:

CHANGE OF BASE FORMULA

> For any positive number v,
>
> $$\log_b v = \frac{\ln v}{\ln b} = \frac{1}{\ln b} \cdot \ln v$$

Proof By Property 4 in the box after Example 9, $b^{\log_b v} = v$. Take the natural logarithm of each side of this equation:

$$\ln(b^{\log_b v}) = \ln v.$$

Apply the Power Law for natural logarithms on the left side:

$$(\log_b v)(\ln b) = \ln v.$$

Dividing both sides by $\ln b$ finishes the proof:

$$\log_b v = \frac{\ln v}{\ln b}. \blacksquare$$

Example 11 To find $\log_7 3$, apply the change of base formula with $b = 7$:

$$\log_7 3 = \frac{\ln 3}{\ln 7} = \frac{1.0986}{1.9459} = .5646. \blacksquare$$

Logarithmic Functions

For each positive b except 1, there is a **logarithmic function** $h(x) = \log_b x$ whose domain is all positive real numbers. The properties of logarithms developed above show that the functions $k(x) = b^x$ and $h(x) = \log_b x$ have these properties:

6.3 LOGARITHMIC FUNCTIONS TO OTHER BASES

$$(k \circ h)(x) = k(h(x)) = k(\log_b x) = b^{\log_b x} = x \quad \text{for all } x > 0;$$
$$(h \circ k)(x) = h(k(x)) = h(b^x) = \log_b b^x = x \quad \text{for all } x.$$

Therefore, as explained in Section 4.6,

INVERSE FUNCTIONS

$h(x) = \log_b x$ **is the inverse function of** $k(x) = b^x$.

Consequently, the graph of $h(x) = \log_b x$ is the graph of $k(x) = b^x$ reflected in the line $y = x$ (see page 199).

Example 12 The graph of $h(x) = \log x$ can be obtained either by reflecting the graph of $k(x) = 10^x$ in the line $y = x$ (as on the left in Figure 6–8) or by using the graph of $g(x) = \ln x$ and the change of base formula:

$$h(x) = \log x = \left(\frac{1}{\ln 10}\right)(\ln x)$$
$$\approx \left(\frac{1}{2.3026}\right)(\ln x) \approx .4343(\ln x) = .4343 g(x).$$

Since $h(x)$ is a constant multiple of $g(x)$, the graph of h is the graph of g shrunk toward the x-axis by a factor of $\frac{1}{\ln 10} \approx .4343$, as explained on page 183 and shown on the right in Figure 6–8. ∎

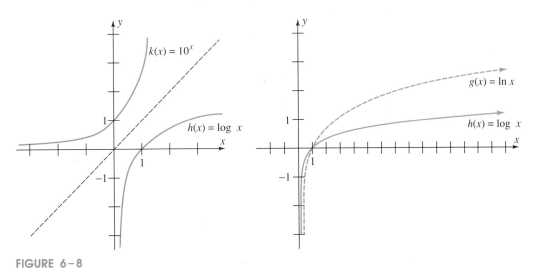

FIGURE 6–8

Because logarithmic growth is so slow, measurements on a logarithmic scale (that is, on a scale determined by a logarithmic function) can sometimes be deceptive.

Example 13 (Earthquakes) The magnitude $R(i)$ of an earthquake on the Richter scale is given by $R(i) = \log(i/i_0)$, where i is the amplitude of the ground motion of the earthquake and i_0 is the amplitude of the ground motion of the so-called zero earthquake.* A moderate earthquake might have 1000 times the ground motion of the zero earthquake (that is, $i = 1000 i_0$). So its magnitude would be

$$\log(1000 i_0 / i_0) = \log 1000 = \log 10^3 = 3.$$

An earthquake with 10 times this ground motion (that is, $i = 10 \cdot 1000 i_0 = 10000 i_0$) would have a magnitude of

$$\log(10000 i_0 / i_0) = \log 10000 = \log 10^4 = 4.$$

So a *tenfold* increase in ground motion produces only a 1-point change on the Richter scale. In general,

Increasing the ground motion by a factor of 10^k increases the Richter magnitude by k units.†

For instance, the 1989 World Series earthquake in San Francisco measured 7.0 on the Richter scale, and the great earthquake of 1906 measured 8.3. The difference of 1.3 points means that the 1906 quake was $10^{1.3} \approx 20$ times more intense than the 1989 one in terms of ground motion. ∎

EXERCISES

Note: Unless stated otherwise, all letters represent positive numbers and $b \neq 1$.

In Exercises 1–8, find the logarithm.

1. $\log 10{,}000$
2. $\log .001$
3. $\log \sqrt[3]{.01}$
4. $\log \sqrt{10}/1000$
5. $\log_2 16$
6. $\log_2 \sqrt[3]{256}$
7. $\log_2 1/(2\sqrt{2})$
8. $\log_2 1/(\sqrt[3]{2})^5$

In Exercises 9–18, translate the given exponential statement into an equivalent logarithmic one.

9. $10^{-2} = .01$
10. $10^3 = 1000$
11. $\sqrt[3]{10} = 10^{1/3}$
12. $10^{.4771} \approx 3$
13. $10^{7k} = r$
14. $10^{(a+b)} = c$
15. $7^8 = 5{,}764{,}801$
16. $2^{-3} = 1/8$
17. $3^{-2} = 1/9$
18. $b^{14} = 3379$

In Exercises 19–28, translate the given logarithmic statement into an equivalent exponential one.

19. $\log 10{,}000 = 4$
20. $\log .001 = -3$
21. $\log 750 \approx 2.88$
22. $\log(.8) = -.097$
23. $\log_5 125 = 3$
24. $\log_8(\tfrac{1}{4}) = -2/3$

---

* It has ground motion amplitude of less than 1 micron on a standard seismograph 100 kilometers from the epicenter.

† *Proof:* If one quake has ground motion amplitude i and the other $10^k i$, then

$$R(10^k i) = \log 10^k i / i_0 = \log 10^k + \log(i/i_0)$$
$$= k + \log(i/i_0) = k + R(i).$$

25. $\log_2(1/4) = -2$
26. $\log_2\sqrt{2} = 1/2$
27. $\log(x^2 + 2y) = z + w$
28. $\log(a + c) = d$

In Exercises 29–36, evaluate the given expression.

29. $\log 10^{\sqrt{43}}$
30. $\log_{17}(17^{17})$
31. $\log 10^{\sqrt{x^2+y^2}}$
32. $\log_{3.5}(3.5^{(x^2-1)})$
33. $\log_{16} 4$
34. $\log_2 64$
35. $\log_{\sqrt{3}}(27)$
36. $\log_{\sqrt{3}}(1/9)$

In Exercises 37–40, the given point is on the graph of $f(x) = \log_b x$. Find b.

37. $(100, 2)$
38. $(8, 3)$
39. $(\sqrt{8}, 3/2)$
40. $(\sqrt[3]{(125)^2}, 2)$

In Exercises 41–46, find x.

41. $\log_3 243 = x$
42. $\log_{81} 27 = x$
43. $\log_{27} x = 1/3$
44. $\log_5 x = -4$
45. $\log_x 64 = 3$
46. $\log_x(1/9) = -2/3$

In Exercises 47–54, b is a positive number such that $\log_b 2 = .13$, $\log_b 3 = .2$, and $\log_b 5 = .3$. Find

47. $\log_b 10$
48. $\log_b 15$
49. $\log_b 4$
50. $\log_b 27$
51. $\log_b(5/3)$
52. $\log_b 48$
53. $\log_b 45$
54. $\log_b 36$

In Exercises 55–62, use a calculator and the change of base formula to find the logarithm.

55. $\log_2 10$
56. $\log_2 22$
57. $\log_7 5$
58. $\log_5 7$
59. $\log_{500} 1000$
60. $\log_{500} 250$
61. $\log_{12} 56$
62. $\log_{12} 725$

In Exercises 63–66, state the magnitude on the Richter scale of an earthquake that satisfies the given condition.

63. 100 times stronger than the zero quake.
64. $10^{4.7}$ times stronger than the zero quake.
65. 350 times stronger than the zero quake.
66. 2500 times stronger than the zero quake.

Exercises 67–70 deal with the energy intensity i of a sound, which is related to the loudness of the sound by the function $L(i) = 10 \log(i/i_0)$, where i_0 is the minimum intensity detectable by the human ear and $L(i)$ is measured in decibels. Find the decibel measure of the sound.

67. Ticking watch (intensity is 100 times i_0).
68. Soft music (intensity is 10,000 times i_0).
69. Loud conversation (intensity is 4 million times i_0).
70. Victoria Falls in Africa (intensity is 10 billion times i_0).

In Exercises 71–76, answer *true* or *false* and give reasons for your answer.

71. $\log_b(r/5) = \log_b r - \log_b 5$
72. $\dfrac{\log_b a}{\log_b c} = \log_b\left(\dfrac{a}{c}\right)$
73. $(\log_b r)/t = \log_b(r^{1/t})$
74. $\log_b(cd) = \log_b c + \log_b d$
75. $\log_5(5x) = 5(\log_5 x)$
76. $\log_b(ab)^t = t(\log_b a) + t$

77. Which is larger: 97^{98} or 98^{97}? [*Hint:* $\log 97 \approx 1.9868$ and $\log 98 \approx 1.9912$ and $f(x) = 10^x$ is an increasing function.]

78. If $\log_b 9.21 = 7.4$ and $\log_b 359.62 = 19.61$, then what is $\log_b 359.62/\log_b 9.21$?

In Exercises 79–82, assume that a and b are positive with $a \neq 1$ and $b \neq 1$.

79. Express $\log_b u$ in terms of logarithms to the base a.
80. Show that $\log_b a = 1/\log_a b$.
81. How are $\log_{10} u$ and $\log_{100} u$ related?
82. Show that $a^{\log b} = b^{\log a}$.

83. If $\log_b x = \dfrac{1}{2}\log_b v + 3$, show that $x = (b^3)\sqrt{v}$.

6.4 EXPONENTIAL AND LOGARITHMIC EQUATIONS

The easiest exponential equations to solve are those in which both sides are powers of the same base.

Example 1 The equation $8^x = 2^{x+1}$ can be rewritten $(2^3)^x = 2^{x+1}$ or, equivalently, $2^{3x} = 2^{x+1}$. Since the powers of 2 are the same, the exponents must be equal, that is,

$$3x = x + 1$$

$$x = \frac{1}{2}. \blacksquare$$

When different bases are involved in an exponential equation, the solution technique is to take the logarithm of each side and then apply the Power Law for Logarithms.*

Example 2 To solve $5^x = 2$,

Take logarithms on each side: $\quad \ln 5^x = \ln 2$

Use the Power Law: $\quad x(\ln 5) = \ln 2$

Divide both sides by $\ln 5$: $\quad x = \dfrac{\ln 2}{\ln 5} \approx \dfrac{.6931}{1.6094} \approx .4307$

Remember: $\dfrac{\ln 2}{\ln 5}$ is *not* $\ln \dfrac{2}{5}$ or $\ln 2 - \ln 5$. $\blacksquare$

Example 3 To solve $2^{4x-1} = 3^{1-x}$,

Take logarithms on each side: $\quad \ln 2^{4x-1} = \ln 3^{1-x}$

Use the Power Law: $\quad (4x - 1)(\ln 2) = (1 - x)(\ln 3)$

Multiply out both sides: $\quad 4x(\ln 2) - \ln 2 = \ln 3 - x(\ln 3)$

Rearrange terms: $\quad 4x(\ln 2) + x(\ln 3) = \ln 2 + \ln 3$

Factor left side: $\quad (4 \cdot \ln 2 + \ln 3)x = \ln 2 + \ln 3$

Divide both sides by $(4 \cdot \ln 2 + \ln 3)$: $\quad x = \dfrac{\ln 2 + \ln 3}{4 \cdot \ln 2 + \ln 3} \approx .4628. \blacksquare$

---

*We shall use natural logarithms, but the same techniques are valid for logarithms to other bases (Exercise 34).

Example 4 (Radiocarbon Dating) When a living organism dies, its carbon-14 decays. The half-life of carbon-14 is 5730 years. As explained in Example 4 on page 274, the amount left at time t is given by $M(t) = c2^{-t/5730}$, where c is the mass of carbon-14 that was present initially.

The skeleton of a mastodon has lost 58% of its original carbon-14.* When did the mastodon die?

Solution Time is measured from the death of the mastodon. The present mass of carbon-14 is $.58c$ less than the original mass c. So the present value of $M(t)$ is $c - .58c = .42c$ and we have:

$$M(t) = c2^{-t/5730}$$
$$.42c = c2^{-t/5730}$$
$$.42 = 2^{-t/5730}$$

The solution of this equation is the time elapsed from the mastodon's death to the present. It can be solved as above:

$$\ln .42 = \ln 2^{-t/5730}$$
$$\ln .42 = \left(\frac{-t}{5730}\right)(\ln 2)$$
$$5730(\ln .42) = -t(\ln 2)$$
$$t = \frac{5730(\ln .42)}{-(\ln 2)} \approx 7171.32$$

Therefore the mastodon died approximately 7200 years ago. ■

Example 5 (Compound Interest)† $3000 is to be invested at 8% per year, compounded quarterly. In how many years will the investment be worth $10,680?

Solution The interest rate per quarter is $.08/4 = .02$. The compound interest formula (page 279) shows that the value of the investment after t quarters is $A = 3000(1 + .02)^t = 3000(1.02)^t$. So we must solve the equation

$$3000(1.02)^t = 10{,}680$$
$$1.02^t = \frac{10{,}680}{3000} = 3.56$$
$$\ln 1.02^t = \ln 3.56$$

---

* Archeologists can determine how much carbon-14 has been lost by a technique that involves measuring the ratio of carbon-14 to carbon-12 in the skeleton.

† Skip this example if you haven't read Excursion 6.1.A.

$$t(\ln 1.02) = \ln 3.56$$

$$t = \frac{\ln 3.56}{\ln 1.02} \approx 64.12 \text{ quarters}$$

Therefore it will take $\frac{64.12}{4} = 16.03$ years. ∎

Example 6 (Population Growth) If there are no inhibiting or stimulating factors, a population of organisms (such as bacteria, animals, or humans) normally grows at a rate proportional to its size. It is shown in calculus that the population $S(t)$ at time t is given by $S(t) = ce^{kt}$, where c is the original population and k is the constant ratio of growth rate to size of population.

A biologist observes that a culture contains 1000 of a certain type of bacteria. Seven hours later there are 5000 bacteria. When will the bacteria population reach 1 billion?

Solution The original population is $c = 1000$, so the population function is $S(t) = 1000e^{kt}$. To determine k we use the fact that $S(7) = 5000$, that is,

$$1000e^{k7} = 5000$$

This equation can be solved for k as above; first, divide both sides by 1000:

$$e^{7k} = 5$$

$$\ln e^{7k} = \ln 5$$

$$7k \cdot \ln e = \ln 5$$

$$k = \frac{\ln 5}{7 \cdot \ln e} = \frac{\ln 5}{7 \cdot 1} \approx .2299 \quad [\ln e = 1]$$

Therefore the population function is $S(t) = 1000e^{.2299t}$. To find the value of t for which $S(t)$ is 1 billion, we must solve:

$$1000e^{.2299t} = 1{,}000{,}000{,}000$$

$$e^{.2299t} = 1{,}000{,}000$$

$$\ln e^{.2299t} = \ln 1{,}000{,}000$$

$$.2299t \cdot \ln e = \ln 1{,}000{,}000$$

$$t = \frac{\ln 1{,}000{,}000}{.2299 \cdot \ln e} = \frac{\ln 1{,}000{,}000}{.2299} \approx 60.0936 \text{ hours.} \blacksquare$$

Example 7 (Inhibited Population Growth) The population of fish in a lake at time t months is given by the function

$$p(t) = \frac{20{,}000}{1 + 24e^{-t/4}}$$

6.4 EXPONENTIAL AND LOGARITHMIC EQUATIONS

How long will it take for the population to reach 15,000?

Solution We must solve this equation for t:

$$15{,}000 = \frac{20{,}000}{1 + 24e^{-t/4}}$$

$$15000(1 + 24e^{-t/4}) = 20{,}000$$

$$1 + 24e^{-t/4} = \frac{20{,}000}{15{,}000} = \frac{4}{3}$$

$$24e^{-t/4} = \frac{1}{3}$$

$$e^{-t/4} = \frac{1}{3} \cdot \frac{1}{24} = \frac{1}{72}$$

$$\ln e^{-t/4} = \ln\left(\frac{1}{72}\right)$$

$$\left(-\frac{t}{4}\right)(\ln e) = \ln 1 - \ln 72$$

$$-\frac{t}{4} = -\ln 72 \qquad [\ln e = 1 \text{ and } \ln 1 = 0]$$

$$t = 4(\ln 72) \approx 17.1067$$

So the population reaches 15,000 in a little over 17 months. ∎

Example 8 To solve $e^x - e^{-x} = 4$, we first multiply both sides by e^x. Since $e^x > 0$ for every x, we get an equivalent equation:

$$e^x e^x - e^{-x} e^x = 4e^x$$

$$(e^x)^2 - 4e^x - 1 = 0$$

Let $u = e^x$ so that the equation becomes

$$u^2 - 4u - 1 = 0$$

The solutions are given by the quadratic formula:

$$u = \frac{-(-4) \pm \sqrt{(-4)^2 - 4 \cdot 1 \cdot (-1)}}{2 \cdot 1} = \frac{4 \pm \sqrt{20}}{2} = \frac{4 \pm 2\sqrt{5}}{2} = 2 \pm \sqrt{5}.$$

Since $u = e^x$, we have

$$e^x = 2 - \sqrt{5} \qquad \text{or} \qquad e^x = 2 + \sqrt{5}$$

But e^x is always positive and $2 - \sqrt{5} < 0$, so the first equation has no solutions. The second can be solved as above:

$$\ln e^x = \ln(2 + \sqrt{5})$$
$$x(\ln e) = \ln(2 + \sqrt{5})$$
$$x(1) = \ln(2 + \sqrt{5}) \qquad [\ln e = 1]$$
$$x = \ln(2 + \sqrt{5}) \approx 1.4436. \quad \blacksquare$$

Logarithmic Equations

Whenever $\ln u = \ln v$, the properties of logarithms show that $u = e^{\ln u} = e^{\ln v} = v$. This fact is useful for solving certain logarithmic equations.

Example 9 To solve $\ln(x - 3) + \ln(2x + 1) = 2 \cdot \ln x$, use the Product Law on the left side and the Power Law on the right side:

$$\ln[(x - 3)(2x + 1)] = \ln x^2$$
$$\ln(2x^2 - 5x - 3) = \ln x^2$$

Since the logarithms are equal, we must have:

$$2x^2 - 5x - 3 = x^2$$
$$x^2 - 5x - 3 = 0$$
$$x = \frac{-(-5) \pm \sqrt{(-5)^2 - 4 \cdot 1 \cdot (-3)}}{2 \cdot 1} = \frac{5 \pm \sqrt{37}}{2}$$

But $x - 3$ is negative when $x = (5 - \sqrt{37})/2$, so $\ln(x - 3)$ is not defined in that case. Therefore the only solution of the original equation is $x = (5 + \sqrt{37})/2 \approx 5.5414.$ $\blacksquare$

A different technique is needed when some terms of the equation involve logarithms and others don't.

Example 10 To solve $\ln(3x + 2) - \ln x = 2$, apply the Quotient Law on the left side:

$$\ln \frac{3x + 2}{x} = 2$$

Now use the fact that $\ln v = u$ means $e^u = v$ (with $v = \dfrac{3x + 2}{x}$ and $u = 2$) to rewrite the equation in exponential form:

$$e^2 = \frac{3x + 2}{x}$$
$$e^2 x = 3x + 2$$
$$e^2 x - 3x = 2$$

6.4 EXPONENTIAL AND LOGARITHMIC EQUATIONS

$$(e^2 - 3)x = 2$$

$$x = \frac{2}{e^2 - 3} \approx .4557. \blacksquare$$

Example 11 To solve $\log_3(x + 3) + \log_3(x - 5) = 2$, we apply the Product Law to obtain: $\log_3(x + 3)(x - 5) = 2$. Now rewrite this equation in exponential form ($\log_3 v = u$ means $3^u = v$):

$$3^2 = (x + 3)(x - 5)$$
$$9 = x^2 - 2x - 15$$
$$0 = x^2 - 2x - 24 = (x - 6)(x + 4)$$

So the possible solutions are $x = 6$ and $x = -4$. But $x - 5$ is negative when $x = -4$, so $\log_3(x - 5)$ is not defined then. Therefore $x = 6$ is the only solution of the original equation. $\blacksquare$

EXERCISES

In Exercises 1–8, solve the equation without using logarithms.

1. $3^x = 81$
2. $3^x + 3 = 30$
3. $3^{x+1} = 9^{5x}$
4. $4^{5x} = 16^{2x-1}$
5. $3^{5x} 9^{x^2} = 27$
6. $2^{x^2+5x} = 1/16$
7. $9^{x^2} = 3^{-5x-2}$
8. $4^{x^2-1} = 8^x$

In Exercises 9–30, solve the equation. Express your answer in terms of natural logarithms (for instance, $x = (2 + \ln 5)/(\ln 3)$). Then use a calculator to find an approximation for the answer.

9. $3^x = 5$
10. $5^x = 4$
11. $2^x = 3^{x-1}$
12. $4^{x+2} = 2^{x-1}$
13. $3^{1-2x} = 5^{x+5}$
14. $4^{3x-1} = 3^{x-2}$
15. $2^{1-3x} = 3^{x+1}$
16. $3^{z+3} = 2^z$
17. $e^{2x} = 5$
18. $e^{-3x} = 2$
19. $6e^{-1.4x} = 21$
20. $3.4e^{-x/3} = 5.6$
21. $2.1e^{(x/2)\ln 3} = 5$
22. $7.8e^{(x/3)\ln 5} = 14$
23. $9^x - 4 \cdot 3^x + 3 = 0$ [*Hint:* Note that $9^x = (3^x)^2$; let $u = 3^x$.]
24. $4^x - 6 \cdot 2^x = -8$
25. $e^{2x} - 5e^x + 6 = 0$ [*Hint:* Let $u = e^x$.]
26. $2e^{2x} - 9e^x + 4 = 0$
27. $6e^{2x} - 16e^x = 6$
28. $8e^{2x} + 8e^x = 6$
29. $4^x + 6 \cdot 4^{-x} = 5$
30. $5^x + 3 = 10 \cdot 5^{-x}$

In Exercises 31–33, solve the equation for x.

31. $\dfrac{e^x - e^{-x}}{2} = t$
32. $\dfrac{e^x + e^{-x}}{e^x - e^{-x}} = t$
33. $\dfrac{e^x - e^{-x}}{e^x + e^{-x}} = t$

34. (a) Solve $7^x = 3$, using natural logarithms. Leave your answer in logarithmic form; don't approximate with a calculator.
    (b) Solve $7^x = 3$, using common (base 10) logarithms. Leave your answer in logarithmic form.
    (c) Use the change of base formula in Section 6.3 to show that your answers in parts (a) and (b) are the same.

In Exercises 35–44, solve the equation as in Example 9.

35. $\ln(3x - 5) = \ln 11 + \ln 2$

36. $\log(4x - 1) = \log(x + 1) + \log 2$

37. $\log(3x - 1) + \log 2 = \log 4 + \log(x + 2)$

38. $\ln(x + 6) - \ln 10 = \ln(x - 1) - \ln 2$

39. $2 \ln x = \ln 36$

40. $2 \log x = 3 \log 4$

41. $\ln x + \ln(x + 1) = \ln 3 + \ln 4$

42. $\ln(6x - 1) + \ln x = \frac{1}{2}\ln 4$

43. $\ln x = \ln 3 - \ln(x + 5)$

44. $\ln(2x + 3) + \ln x = \ln e$

In Exercises 45–54, solve the equation.

45. $\ln(x + 9) - \ln x = 1$

46. $\ln(2x + 1) - 1 = \ln(x - 2)$

47. $\log x + \log(x - 3) = 1$ [Remember: $\log v = u$ means $10^u = v$.]

48. $\log(x - 1) + \log(x + 2) = 1$

49. $\log_5(x + 3) = 1 - \log_5(x - 1)$ [Remember: $\log_5 v = u$ means $5^u = v$.]

50. $\log_4(x - 5) = 2 - \log_4(x + 1)$

51. $\log\sqrt{x^2 - 1} = 2$

52. $\log\sqrt[3]{x^2 + 21x} = 2/3$

53. $\ln(x^2 + 1) - \ln(x - 1) = 1 + \ln(x + 1)$

54. $\dfrac{\ln(x + 1)}{\ln(x - 1)} = 2$

Use the information at the beginning of Example 6 to work Exercises 55–56.

55. One hour after an experiment begins, the number of bacteria in a culture is 100. An hour later there are 500.
    (a) Find the number of bacteria at the beginning of the experiment and the number 3 hours later.
    (b) How long does it take the number of bacteria at any given time to double?

56. A colony of 1000 weevils grows exponentially to 1750 in 1 week. How many weeks does it take for the weevil population to triple to 3000?

In Exercises 57–66, use the function $M(x) = c2^{-x/h}$, which was explained in Example 4 on page 274.

57. How old is a piece of ivory that has lost 36% of its carbon-14? (See Example 4 in this section.)

58. How old is a mummy that has lost 49% of its carbon-14?

59. An American Indian mummy was found recently. If it has lost 26.4% of its carbon-14, approximately how long ago did the Indian die?

60. How old is a wooden statue that has only one third of its original carbon-14?

61. A quantity of uranium decays to two thirds of its original mass in .26 billion years. Find the half-life of uranium.

62. A certain radioactive substance loses one third of its original mass in 5 days. Find its half-life.

63. Krypton-85 loses 6.44% of its mass each year. What is its half-life?

64. Strontium-90 loses 2.5% of its mass each year. What is its half-life?

65. The half-life of a certain substance is 3.6 days. How long will it take for 20 grams to decay to 3 grams?

66. The half-life of cobalt-60 is 4.945 years. How long will it take for 25 grams to decay to 15 grams?

In Exercises 67–73, use the compound interest formula $A = P(1 + r)^t$, which was discussed on page 279.

67. At what annual rate of interest should $1000 be invested so that it will double in 10 years, if interest is compounded quarterly?

68. How long does it take $500 to triple if it is invested at 6% compounded: (a) annually, (b) quarterly, (c) daily?

69. (a) How long will it take to triple your money if you invest $500 at a rate of 5% per year compounded annually?
    (b) How long will it take at 5% compounded quarterly?

70. At what rate of interest (compounded annually) should you invest $500 if you want to have $1500 in 12 years?

71. How much money should be invested at 5% inter-

est, compounded quarterly, so that 9 years later the investment will be worth $5000? This amount is called the **present value** of $5000 at 5% interest.

72. If the inflation rate is 6% per year, how long will it take until the purchasing power of today's dollar is reduced by half?

73. Find a formula that gives the time needed for an investment of P dollars to double, if the interest rate is r% compounded annually. [*Hint:* Solve the compound interest formula for t, when $A = 2P$.]

74. If the population at time t is given by $S(t) = ce^{kt}$, find a formula that gives the time it takes for the population to double.

75. The spread of a flu virus in a community of 45,000 people is given by the function $f(t) = \dfrac{45{,}000}{1 + 224e^{-.899t}}$, where $f(t)$ is the number of people infected in week t.
    (a) How many people had the flu at the outbreak of the epidemic? After 3 weeks?
    (b) When will half the town be infected?

76. The beaver population near a certain lake in year t is approximately $p(t) = \dfrac{2000}{1 + 199e^{-.5544t}}$.
    (a) When will the beaver population reach 1000?
    (b) Will the population ever reach 2000? Why?

77. The probability P percent of having an accident while driving a car is related to the alcohol level of the driver's blood by the formula $P = e^{kt}$, where k is a constant. Accident statistics show that the probability of an accident is 25% when the blood alcohol level is $t = .15$.
    (a) Find k. [Use $P = 25$, not .25.)
    (b) At what blood alcohol level is the probability of having an accident 50%?

78. Under normal conditions, the atmospheric pressure (in millibars) at height h feet above sea level is given by $P(h) = 1015e^{-kh}$, where k is a positive constant.
    (a) If the pressure at 18,000 feet is half the pressure at sea level, find k.
    (b) Using the information from part (a), find the atmospheric pressure at 1000 feet, 5000 feet, and 15,000 feet.

Unusual Problem

79. According to one theory of learning, the number of words per minute N that a person can type after t weeks of practice is given by $N = c(1 - e^{-kt})$, where c is an upper limit that N cannot exceed and k is a constant that must be determined experimentally for each person.
    (a) If a person can type 50 wpm (words per minute) after 4 weeks of practice and 70 wpm after 8 weeks, find the value of k and c for this person. According to the theory, this person will never type faster than c wpm.
    (b) Another person can type 50 wpm after 4 weeks of practice and 90 wpm after 8 weeks. How many weeks must this person practice to be able to type 125 wpm?

CHAPTER REVIEW

Important Concepts

Section 6.1 Irrational exponents 271
Exponential functions 272
Exponential growth and decay 273
The number e 274

Excursion 6.1.A Compound interest formula 279

Section 6.2 Natural logarithm 282
Properties of Natural Logarithms 283
Product Law for Logarithms 283
Quotient Law for Logarithms 284

Section 6.3	Power Law for Logarithms 285 Natural logarithmic function 286 Logarithms to base b 290 Common logarithms 291 Properties of Logarithms 292 Logarithm laws 293 Change of base formula 294 Logarithmic functions and graphs 295 Richter scale 296
Section 6.4	Exponential equations 298 Logarithmic equations 302

- $g(x) = \ln x$ is the inverse function of $f(x) = e^x$:
  $$e^{\ln v} = v \text{ for all } v > 0 \quad \text{and} \quad \ln(e^u) = u \text{ for all } u$$
- $h(x) = \log_b x$ is the inverse function of $k(x) = b^x$:
  $$b^{\log_b v} = v \text{ for all } v > 0 \quad \text{and} \quad \log_b(b^u) = u \text{ for all } u$$
- Logarithm Laws: For all $v, w > 0$ and any k:
  $$\ln(vw) = \ln v + \ln w \qquad \log_b(vw) = \log_b v + \log_b w$$
  $$\ln\left(\frac{v}{w}\right) = \ln v - \ln w \qquad \log_b\left(\frac{v}{w}\right) = \log_b v - \log_b w$$
  $$\ln(v^k) = k(\ln v) \qquad \log_b(v^k) = k(\log_b v)$$
- Change of Base Formula: $\log_b v = \dfrac{\ln v}{\ln b}$
- Compound Interest Formula: $A = P(1 + r)^t$

Review Questions

In Questions 1–4, sketch the graph of the function.

1. $g(x) = 2^x - 1$
2. $f(x) = 2^{x-1}$
3. $h(x) = e^x - 2$
3. $k(x) = e^{-x}$

Questions 5–8 refer to the graphs in Figure 6–9.

5. The graph of $f(x) = |2^x|$ could possibly be:
   (a) I (b) II (c) III (d) IV (e) none of these
6. The graph of $g(x) = -(2^x)$ could possibly be:
   (a) I (b) II (c) III (d) IV (e) none of these
7. The graph of $h(x) = |\ln x|$ could possibly be:
   (a) VII (b) IV (c) V (d) VI (e) none of these

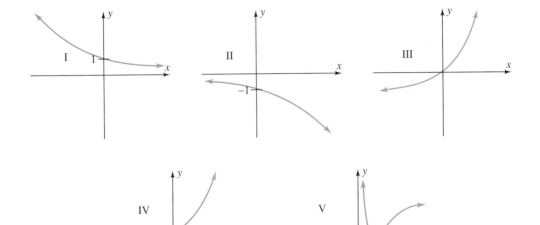

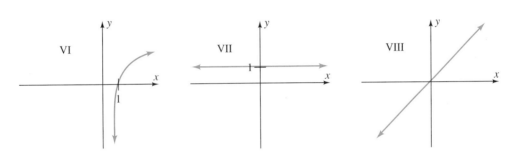

FIGURE 6-9

8. If $f(x) = e^x$ and $g(x) = \ln x$, then the graph of the composite function $g \circ f$ could possibly be:
   (a) III   (b) VI   (c) VII   (d) VIII   (e) none of these

In Questions 9–14, translate the given exponential statement into an equivalent logarithmic one.

9. $e^{6.628} = 756$ 10. $e^{5.8972} = 364$

11. $e^{r^2-1} = u + v$ 12. $e^{a-b} = c$

13. $10^{2.8785} = 756$ 14. $10^{c+d} = t$

In Questions 15–20, translate the given logarithmic statement into an equivalent exponential one.

15. $\ln 1234 = 7.118$ 16. $\ln(ax + b) = y$

17. $\ln(rs) = t$
18. $\log 1234 = 3.0913$
19. $\log_5(cd - k) = u$
20. $\log_d(uv) = w$

In Questions 21–24, evaluate the given expression without using a calculator:

21. $\ln e^3$
22. $\ln \sqrt[3]{e}$
23. $e^{\ln 3/4}$
24. $e^{\ln(x+2y)}$
25. Simplify: $3 \ln \sqrt{x} + (1/2)\ln x$
26. Simplify: $\ln(e^{4e})^{-1} + 4e$
27. Write as a single logarithm: $\ln 3x - 3 \ln x + \ln 3y$
28. Write as a single logarithm: $4 \ln x - 2(\ln x^3 + 4 \ln x)$
29. $\log(-.01) = ?$
30. $\log_{20} 400 = ?$
31. Assume $\ln 5 = 1.6$ and $\ln 11 = 2.4$. Find $\dfrac{\ln 11}{\ln 5}$.

In Questions 32–36, assume $\log_b 2 = .30$ and $\log_b 3 = .48$ and find the logarithm.

32. $\log_b 6$
33. $\log_b(3/2)$
34. $\log_b 27$
35. $\log_b(b/2)$
36. $\dfrac{\log_b 3}{\log_b 4}$

37. Which of the following statements are *true*?
    (a) $\ln 10 = (\ln 2)(\ln 5)$
    (b) $\ln(e/6) = \ln e + \ln 6$
    (c) $\ln(1/7) + \ln 7 = 0$
    (d) $\ln(-e) = -1$
    (e) none of them

38. Which of the following statements are *false*?
    (a) $10(\log 5) = \log 50$
    (b) $\log 100 + 3 = \log 10^5$
    (c) $\log 1 = \ln 1$
    (d) $\log 6/\log 3 = \log 2$
    (e) all of them

Use the graphs in Figure 6–10 for Questions 39–40.

39. The graph of $f(x) = -\ln x$ could possibly be:
    (a) I   (b) IV   (c) V   (d) VI   (e) none of these
40. The graph of $g(x) = 2^x + 1$ could possibly be:
    (a) II   (b) III   (c) IV   (d) VI   (e) none of these

CHAPTER 6 REVIEW 309

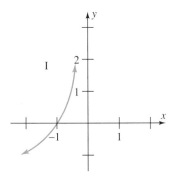
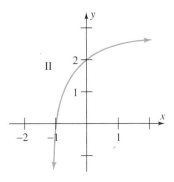
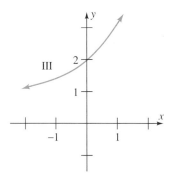
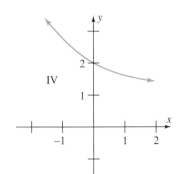
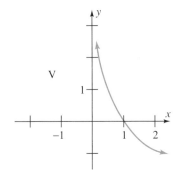
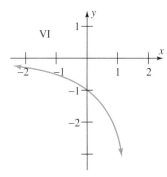

FIGURE 6-10

41. If $\log_3 9^{x^2} = 4$, what is x?
42. What is the domain of the function $f(x) = \ln\left(\dfrac{x}{x-1}\right)$?
43. Solve: $8^x = 4^{x^2-3}$
44. Solve: $e^{3x} = 4$
45. Solve: $2 \cdot 4^x - 5 = -4$
46. Solve: $725e^{-4x} = 1500$
47. Solve for x: $u = c + d \ln x$
48. Solve: $2^x = 3^{x+3}$
49. Solve: $\ln x + \ln(3x - 5) = \ln 2$
50. Solve: $\ln(x + 8) - \ln x = 1$
51. Solve: $\log(x^2 - 1) = 2 + \log(x + 1)$
52. Solve: $\log_4 \sqrt{x^2 + 1} = 1$
53. The half-life of polonium (^{210}Po) is 140 days. If you start with 10 mg, how much will be left at the end of a year?
54. An insect colony grows exponentially from 200 to 2000 in 3 months' time. How long will it take for the insect population to reach 50,000?

55. Hydrogen-3 decays at a rate of 5.59% per year. Find its half-life.
56. The half-life of radium-88 is 1590 years. How long will it take for 10 grams to decay to 1 gram?
57. How much money should be invested at 8% per year, compounded quarterly, in order to have $1000 in 10 years?
58. At what annual interest rate should you invest your money if you want to double it in 6 years?
59. One earthquake measures 4.6 on the Richter scale. A second earthquake is 1000 times more intense than the first. What does it measure on the Richter scale?

CHAPTER 7

The Conic Sections

> **ROADMAP:** Chapters 7–9 are independent of each other and may be read in any order. Chapter 7 may be read at any time after Chapter 3. Its sections are independent of each other and may be read in any order.

When a right circular cone is cut by a plane, the intersection is a curve called a **conic section,** as shown in Figure 7–1:*

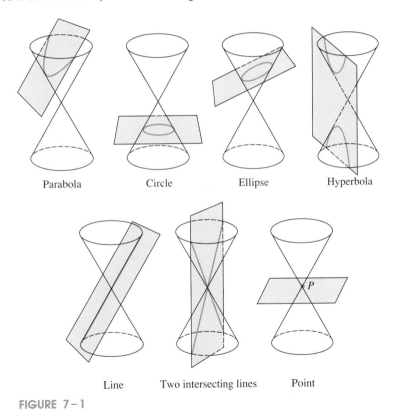

FIGURE 7–1

---

* A point, a line, or two intersecting lines are sometimes called **degenerate conic sections.**

Conic sections were studied by the ancient Greeks and are still of interest. For instance, planets travel in elliptical orbits, parabolic mirrors are used in telescopes, and certain atomic particles follow hyperbolic paths.

In Section 3.1 we saw that a circle can be described solely in terms of the coordinate plane and distance, or as the graph of a particular type of equation. In this chapter, we do the same for parabolas, ellipses, and hyperbolas.

As we did with circles in Section 3.1, we shall show that every nondegenerate conic section is the graph of a second-degree equation of a particular form. Once you have the equation in that form, you can immediately identify the shape of the graph and sketch it easily. The following example illustrates a key technique for carrying out this program.

Example The graph of $x^2 + y^2 = 4$ is a circle with center $(0, 0)$ and radius 2 (see Section 3.1). If we replace x by $x - 5$ and y by $y - 3$ in this equation, we obtain a second equation,

$$(x - 5)^2 + (y - 3)^2 = 4$$

whose graph is a circle of radius 2 with center at $(5, 3)$:

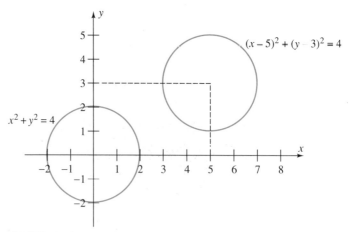

FIGURE 7-2

Thus the graph of $(x - 5)^2 + (y - 3)^2 = 4$ is just the graph of $x^2 + y^2 = 4$ shifted 5 units horizontally to the right and 3 units vertically upward.

The center $(0, 0)$ of $x^2 + y^2 = 4$ moves to $(5, 3)$, the center of $(x - 5)^2 + (y - 3)^2 = 4$. Similarly, a point (r, s) on the first circle moves to the point $(r + 5, s + 3)$ on the second. [*Reason:* if (r, s) is on the first circle, then $r^2 + s^2 = 4$. This can be rewritten as $((r + 5) - 5)^2 + ((s + 3) - 3)^2 = 4$, which says that $(r + 5, s + 3)$ is on the second circle.] ∎

A similar argument works with an arbitrary equation:

VERTICAL AND HORIZONTAL SHIFTS

> Let h and k be constants. Replacing x by $x - h$ and y by $y - k$ in an equation shifts the graph of the equation h units horizontally and k units vertically.* The point (r, s) moves to the point $(r + h, s + k)$.

7.1 ELLIPSES

Let P and Q be points in the plane and r a number greater than the distance from P to Q. The **ellipse** with **foci**† P and Q is the set of all points X such that

(distance from X to P) + (distance from X to Q) = r.

To draw this ellipse, take a piece of string of length r and pin its ends on P and Q. Put your pencil point against the string and move it, keeping the string taut. You will trace out the ellipse, as shown in Figure 7–3.‡

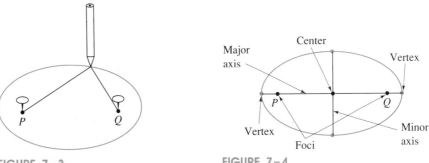

FIGURE 7–3   FIGURE 7–4

The midpoint of the line segment from P to Q is the **center** of the ellipse. The points where the straight line through the foci intersects the ellipse are its **vertices**. The **major axis** of the ellipse is the line segment joining the vertices; its **minor axis** is the line segment through the center, perpendicular to the major axis, as shown in Figure 7–4 above.

The distance formula may be used to find the equation of an ellipse with foci on the x-axis and center at the origin as follows. Suppose the foci are $P = (-c, 0)$ and $Q = (c, 0)$ for some $c > 0$ and that r is a number greater than the distance between the foci. Let $a = r/2$, so that $r = 2a$. Then (x, y) is on the ellipse exactly when

[distance from (x, y) to P] + [distance from (x, y) to Q] = r
$$\sqrt{(x + c)^2 + (y - 0)^2} + \sqrt{(x - c)^2 + (y - 0)^2} = 2a$$
$$\sqrt{(x + c)^2 + y^2} = 2a - \sqrt{(x - c)^2 + y^2}$$

---

* Horizontal shifts are to the right if h is positive and to the left if h is negative. Vertical shifts are upward for positive k and downward for negative k.

† "Foci" is the plural of "focus."

‡ If $P = Q$, you will trace out a circle of radius $r/2$. So a circle is just a special case of an ellipse.

Squaring both sides and simplifying (Exercise 44) we obtain
$$a\sqrt{(x-c)^2 + y^2} = a^2 - cx.$$
Again squaring both sides and simplifying, we have
$$(a^2 - c^2)x^2 + a^2 y^2 = a^2(a^2 - c^2).$$
To simplify the form of this equation, let $b = \sqrt{a^2 - c^2}$* so that $b^2 = a^2 - c^2$ and the equation becomes
$$b^2 x^2 + a^2 y^2 = a^2 b^2.$$
Dividing both sides by $a^2 b^2$ shows that the coordinates of every point on the ellipse satisfy the equation
$$\frac{x^2}{a^2} + \frac{y^2}{b^2} = 1.$$
Conversely, it can be shown that every point whose coordinates satisfy this equation is on the ellipse.

The equation of the ellipse provides additional information. By setting $y = 0$ we see that the x-intercepts are $\pm a$. Similarly, the y-intercepts are $\pm b$. Therefore

ELLIPSE CENTERED AT THE ORIGIN

> Let a and b be real numbers with $a > b > 0$. The graph of
> $$\frac{x^2}{a^2} + \frac{y^2}{b^2} = 1$$
> is an ellipse with center at the origin, x-intercepts $\pm a$, and y-intercepts $\pm b$.
>
> Its vertices are $(-a, 0)$ and $(a, 0)$. The major axis has length $2a$ and the minor axis length $2b$. The foci are $(-c, 0)$ and $(c, 0)$, where $c^2 = a^2 - b^2$.

Example 1 To graph $4x^2 + 9y^2 = 36$, we first divide both sides by 36 and rewrite the equation:
$$\frac{4x^2}{36} + \frac{9y^2}{36} = \frac{36}{36}$$
$$\frac{x^2}{9} + \frac{y^2}{4} = 1$$
$$\frac{x^2}{3^2} + \frac{y^2}{2^2} = 1$$

---

* The distance between the foci is $2c$. Since $r = 2a$ and $r > 2c$ by definition, we have $2a > 2c$ and hence $a > c$. Therefore $a^2 - c^2$ is a positive number and has a real square root.

This equation has the same form as the one in the box above (with $a = 3$ and $b = 2$). Therefore the graph is an ellipse with x-intercepts ± 3, y-intercepts ± 2, as shown in Figure 7–5 below. Since $3 > 2$, the major axis (the longer one) lies on the x-axis and the vertices are $(-3, 0)$ and $(3, 0)$. Since $a^2 - b^2 = 3^2 - 2^2 = 5$, the foci are $(-\sqrt{5}, 0)$ and $(\sqrt{5}, 0)$. ∎

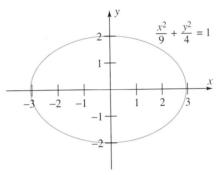

FIGURE 7–5

An argument similar to the one above leads to this description of an ellipse with center at the origin and foci on the y-axis:

ELLIPSE CENTERED AT THE ORIGIN

> Let a and b be real numbers with $b > a > 0$. The graph of
> $$\frac{x^2}{a^2} + \frac{y^2}{b^2} = 1$$
> is an ellipse with center at the origin, x-intercepts $\pm a$, and y-intercepts $\pm b$.
> Its vertices are $(0, -b)$ and $(0, b)$. The major axis has length $2b$ and the minor axis length $2a$. The foci are $(0, -c)$ and $(0, c)$, where $c^2 = b^2 - a^2$.

Example 2 Find the equation and sketch the graph of the ellipse with vertices $(0, \pm 6)$ and foci $(0, \pm 2\sqrt{6})$.

Solution The foci are $(0, -2\sqrt{6})$ and $(0, 2\sqrt{6})$. So the center is $(0, 0)$, and the ellipse has an equation of the form
$$\frac{x^2}{a^2} + \frac{y^2}{b^2} = 1.$$
Since the vertices are $(0, -6)$ and $(0, 6)$, we have $b = 6$ and $c^2 = b^2 - a^2$ with $c = 2\sqrt{6}$. Therefore
$$(2\sqrt{6})^2 = 6^2 - a^2 \quad \text{so that} \quad a^2 = 6^2 - (2\sqrt{6})^2 = 36 - 24 = 12.$$

Hence $a = \sqrt{12}$, and the equation of the ellipse is

$$\frac{x^2}{(\sqrt{12})^2} + \frac{y^2}{6^2} = 1.$$

Therefore the graph is an ellipse with x-intercepts $\pm\sqrt{12} = \pm 2\sqrt{3}$ and y-intercepts ± 6, as shown in Figure 7-6. ∎

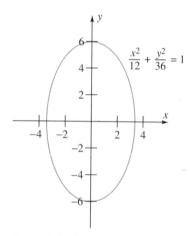

FIGURE 7-6

Let h and k be constants. Replacing x by $x - h$ and y by $y - k$ in an equation shifts the graph horizontally and vertically, with the point (r, s) moving to $(h + r, k + s)$. Therefore

STANDARD EQUATION OF THE ELLIPSE

Let a, b, h, k be real numbers, with a and b positive. The graph of

$$\frac{(x - h)^2}{a^2} + \frac{(y - k)^2}{b^2} = 1$$

is an ellipse with center (h, k).

If $a > b$, the major axis is on the horizontal line $y = k$ and has length $2a$; the minor axis is on the vertical line $x = h$ and has length $2b$. The foci are $(h - c, k)$ and $(h + c, k)$, where $c^2 = a^2 - b^2$.

If $b > a$, the major axis is on the vertical line $x = h$ and has length $2b$; the minor axis is on the horizontal line $y = k$ and has length $2a$. The foci are $(h, k - c)$ and $(h, k + c)$, where $c^2 = b^2 - a^2$.

Example 3 The equation $\dfrac{(x-3)^2}{12} + \dfrac{(y+4)^2}{36} = 1$ can be written as

$$\dfrac{(x-3)^2}{(\sqrt{12})^2} + \dfrac{(y-(-4))^2}{6^2} = 1$$

This is the same form as in the box above (with $a = \sqrt{12}$, $b = 6$, $h = 3$, and $k = -4$). Since $b > a$, the graph is an ellipse with center $(3, -4)$, major axis of length $2 \cdot 6 = 12$ on the vertical line $x = 3$, and minor axis of length $2\sqrt{12}$ on the horizontal line $y = -4$, as shown in Figure 7-7 below. To find the foci, note that $c^2 = b^2 - a^2 = 36 - 12 = 24$ so that $c = \sqrt{24} = 2\sqrt{6}$. Therefore the foci are $(3, 4 - 2\sqrt{6})$ and $(3, 4 + 2\sqrt{6})$. ∎

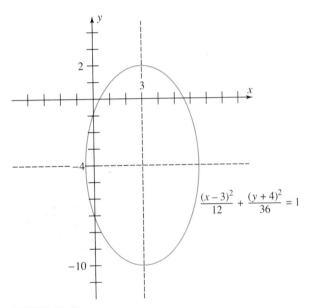

FIGURE 7-7

Example 4 To graph $4x^2 + 9y^2 - 32x - 90y + 253 = 0$, we first rewrite the equation:

$$(4x^2 - 32x) + (9y^2 - 90y) = -253$$
$$4(x^2 - 8x) + 9(y^2 - 10y) = -253$$

Now complete the square in $x^2 - 8x$ and $y^2 - 10y$:

$$4(x^2 - 8x + 16) + 9(y^2 - 10y + 25) = -253 + ? + ?$$

Be careful here: On the left side we haven't just added 16 and 25. When the left side is multiplied out we have actually added in $4 \cdot 16 = 64$ and $9 \cdot 25 = 225$.

Therefore to leave the original equation unchanged, we must add these numbers on the right:

$$4(x^2 - 8x + 16) + 9(y^2 - 10y + 25) = -253 + 64 + 225$$

$$4(x - 4)^2 + 9(y - 5)^2 = 36$$

$$\frac{4(x - 4)^2}{36} + \frac{9(y - 5)^2}{36} = \frac{36}{36}$$

$$\frac{(x - 4)^2}{9} + \frac{(y - 5)^2}{4} = 1$$

$$\frac{(x - 4)^2}{3^2} + \frac{(y - 5)^2}{2^2} = 1$$

Therefore the graph is an ellipse with center at (4, 5), major axis on the horizontal line $y = 5$, and minor axis on the vertical line $x = 4$, as shown in Figure 7-8. Since we have $a = 3$ and $b = 2$, the major axis has length $2a = 6$ and the minor axis has length $2b = 4$. ∎

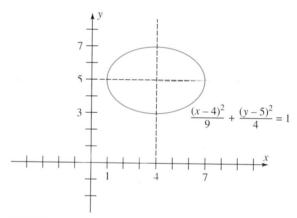

FIGURE 7-8

Eccentricity

The **eccentricity** of an ellipse is denoted e and is defined to be the ratio

$$e = \frac{\text{distance between the foci}}{\text{length of the major axis}}$$

This number is between 0 and 1 because the numerator is less than the denominator (since the foci lie between the ends of the major axis, as shown in Figure 7-9 below). The eccentricity measures the "roundness" of the ellipse, as shown here:

7.1 ELLIPSES

Foci far from vertices	Foci close to vertices
Distance between foci is small in comparison with length of major axis, so e is close to 0 and ellipse is almost circular.	Distance between foci is almost equal to length of major axis, so e is close to 1 and ellipse is elongated.

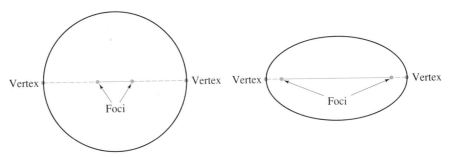

FIGURE 7-9

As we saw in the box on page 316, an ellipse with standard equation

$$\frac{(x-h)^2}{a^2} + \frac{(y-k)^2}{b^2} = 1, \text{ where } a > b \text{ and } c^2 = a^2 - b^2$$

has major axis of length $2a$, and its foci are $(h - c, k)$ and $(h + c, k)$. Since the foci lie on a horizontal line, the distance between them is the difference of their first coordinates, namely, $(h + c) - (h - c) = 2c$. Therefore

$$e = \frac{\text{distance between foci}}{\text{length of major axis}} = \frac{2c}{2a} = \frac{c}{a} = \frac{\sqrt{a^2 - b^2}}{a} \quad (a > b).$$

A similar argument shows that

$$e = \frac{c}{b} = \frac{\sqrt{b^2 - a^2}}{b} \quad (b > a).$$

Example 5 Find the eccentricity of the ellipse with equation:

(a) $\dfrac{x^2}{5} + \dfrac{y^2}{9} = 1$ (b) $\dfrac{(x-8)^2}{16} + \dfrac{(y-3)^2}{2} = 1.$

Solution

(a) We have $a^2 = 5$ and $b^2 = 9$ so that $b > a$. Hence $b = 3$ and

$$\text{eccentricity} = \frac{\sqrt{b^2 - a^2}}{b} = \frac{\sqrt{9-5}}{3} = \frac{\sqrt{4}}{3} = \frac{2}{3}.$$

(b) Here $a^2 = 16$, $a = 4$, $b^2 = 2$, and $a > b$, so that

$$\text{eccentricity} = \frac{\sqrt{a^2 - b^2}}{a} = \frac{\sqrt{16 - 2}}{4} = \frac{\sqrt{14}}{4} \approx .9354. \blacksquare$$

Applications

It can be shown that if a sound or light ray passes through one focus of an ellipse and reflects off the ellipse, then the ray necessarily passes through the other focus of the ellipse, as shown in Figure 7-10.

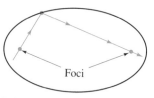

FIGURE 7-10

Exactly this situation occurs under the elliptical dome of the United States Capitol. A person who stands at one focus and whispers toward the ceiling can be clearly heard by anyone at the other focus. Before this fact was widely known, when congress used to sit under the dome, a number of political secrets were inadvertently revealed by congressmen to members of the other party.

The planets and many comets have elliptical orbits with the sun as one focus. The moon travels in an elliptical orbit with the earth as one focus. Satellites are usually put into elliptical orbits around the earth.

Example 6 The earth's orbit around the sun is an ellipse with eccentricity .0167. The sun is one focus and the length of the major axis is approximately 186,000,000 miles. What are the minimum and maximum distances from the earth to the sun?*

Solution The orbit is shown in Figure 7-11. If we use a coordinate system with the major axis on the x-axis and the sun having coordinates $(c, 0)$, then we obtain Figure 7-12. The length of the major axis is $2a = 186,000,000$, so that $a = 93,000,000$. As shown above, the equation of the orbit is:

$$\frac{x^2}{a^2} + \frac{y^2}{b^2} = 1, \quad \text{where } c^2 = a^2 - b^2.$$

Figure 7-12 suggests a fact that can also be proven algebraically: the minimum and maximum distances from a point on the ellipse to the focus $(c, 0)$ occur at the endpoints of the major axis:

$$\text{minimum distance} = a - c \quad \text{and} \quad \text{maximum distance} = a + c.$$

*Distances are measured from the center of each body.

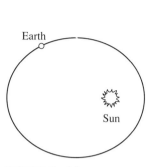

FIGURE 7-11

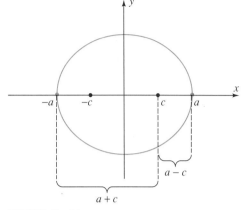

FIGURE 7-12

The definition of eccentricity shows that

$$.0167 = \text{eccentricity} = \frac{c}{a} = \frac{c}{93{,}000{,}000}.$$

Therefore $c = .0167(93{,}000{,}000) = 1{,}553{,}100$, so that

$$\text{minimum distance} = a - c = 93{,}000{,}000 - 1{,}553{,}100$$
$$= 91{,}446{,}900 \text{ miles}$$
$$\text{maximum distance} = a + c = 93{,}000{,}000 + 1{,}553{,}100$$
$$= 94{,}553{,}100 \text{ miles.} \blacksquare$$

EXERCISES

In Exercises 1–8, sketch the graph of the ellipse.

1. $\dfrac{x^2}{25} + \dfrac{y^2}{4} = 1$
2. $\dfrac{x^2}{6} + \dfrac{y^2}{16} = 1$
3. $\dfrac{x^2}{10} - 1 = \dfrac{-y^2}{36}$
4. $\dfrac{y^2}{49} + \dfrac{x^2}{81} = 1$
5. $\dfrac{(x-1)^2}{4} + \dfrac{(y-5)^2}{9} = 1$
6. $\dfrac{(x-2)^2}{16} + \dfrac{(y+3)^2}{12} = 1$
7. $\dfrac{(x+1)^2}{16} + \dfrac{(y-4)^2}{8} = 1$
8. $\dfrac{(x+5)^2}{4} + \dfrac{(y+2)^2}{12} = 1$

Calculus can be used to show that the area of the ellipse with equation $\dfrac{x^2}{a^2} + \dfrac{y^2}{b^2} = 1$ is πab. Use this fact to find the area of each ellipse in Exercises 9–14.

9. $\dfrac{x^2}{16} + \dfrac{y^2}{4} = 1$
10. $\dfrac{x^2}{9} + \dfrac{y^2}{5} = 1$
11. $3x^2 + 4y^2 = 12$
12. $7x^2 + 5y^2 = 35$
13. $6x^2 + 2y^2 = 14$
14. $5x^2 + y^2 = 5$

In Exercises 15–28, find the equation of the ellipse that satisfies the given conditions.

15. Center at origin; foci on x-axis; x-intercepts ± 7; y-intercepts ± 2.

16. Center at origin; foci on y-axis; x-intercepts ± 1; y-intercepts ± 8.
17. Center at origin; foci on x-axis; major axis of length 12; minor axis of length 8.
18. Center at origin; foci on y-axis; major axis of length 20; minor axis of length 18.
19. Center at origin; endpoints of major and minor axes: (−3, 0), (3, 0), (0, −7), (0, 7).
20. Center at origin; vertices (8, 0), (−8, 0); minor axis of length 8.
21. Center at (2, 3); endpoints of major and minor axes: (2, −1), (0, 3), (2, 7), (4, 3).
22. Center at (−5, 2); endpoints of major and minor axes: (0, 2), (−5, 17), (−10, 2), (−5, −13).
23. Center at (7, −4); foci on the line $x = 7$; major axis of length 12; minor axis of length 5.
24. Center at (−3, −9); foci on the line $y = -9$; major axis of length 15; minor axis of length 7.
25. Center at origin; vertices (4, 0), (−4, 0); passing through ($\sqrt{8}$, −2).
26. Center at origin; vertices (0, $\sqrt{6}$), (0, −$\sqrt{6}$); passing through ($2\sqrt{5}$, $\sqrt{3}$).
27. Center at (3, −2); passing through (3, −6) and (9, −2).
28. Center at (2, 5); passing through (2, 4) and (−3, 5).

In Exercises 29–40, sketch the graph of the equation.

29. $4x^2 + 3y^2 = 12$
30. $\dfrac{x^2}{4} + \dfrac{y^2}{9} = 2$
31. $4x^2 + 4y^2 = 1$
32. $x^2 + 4y^2 = 1$
33. $3x^2 + 2y^2 = 6$
34. $(y - 3)^2 - 10 = -x^2 + 2x - 1$
35. $2(x - 3)^2 + 3(y + 3)^2 = 12$
36. $3(x + 4)^2 + 5(y - 5)^2 = 30$
37. $9x^2 + 54x + y^2 - 4y = -76$
38. $4x^2 + y^2 + 8x - 4y = 4$
39. $9x^2 + 4y^2 + 54x - 8y + 49 = 0$
40. $4x^2 + 5y^2 - 8x + 30y + 29 = 0$

In Exercises 41–42, find the equations of two distinct ellipses satisfying the given conditions.

41. Center at (−5, 3); major axis of length 14; minor axis of length 8.
42. Center at (2, −6); major axis of length 15; minor axis of length 6.
43. Consider the ellipse whose equation is $\dfrac{x^2}{a^2} + \dfrac{y^2}{b^2} = 1$. Show that if $a = b$, then the graph is actually a circle.
44. Complete the derivation of the equation of the ellipse on page 313 as follows.
    (a) By squaring both sides, show that the equation
    $$\sqrt{(x + c)^2 + y^2} = 2a - \sqrt{(x - c)^2 + y^2}$$
    may be simplified as
    $$a\sqrt{(x - c)^2 + y^2} = a^2 - cx$$
    (b) Show that the last equation in part (a) may be further simplified as
    $$(a^2 - c^2)x^2 + a^2y^2 = a^2(a^2 - c^2)$$

In Exercises 45–48, find the eccentricity of the ellipse with the given equation.

45. $\dfrac{x^2}{100} + \dfrac{y^2}{99} = 1$
46. $\dfrac{x^2}{18} + \dfrac{y^2}{25} = 1$
47. $\dfrac{(x - 3)^2}{10} + \dfrac{(y - 9)^2}{40} = 1$
48. $\dfrac{(x + 5)^2}{12} + \dfrac{(y - 4)^2}{8} = 1$

49. A satellite is to be placed in an elliptical orbit around the earth in such a way that its maximum distance from the earth is 22,380 km and its minimum distance is 6540 km. Find the eccentricity of the orbit.
50. The first step in landing Apollo 11 on the moon was to place the spacecraft in an elliptical orbit such that the minimum distance from the *surface* of the moon to the spacecraft was 110 km and the maximum distance was 314 km. If the radius of the moon is 1740 km, find the eccentricity of the Apollo 11 orbit.
51. The orbit of the moon around the earth is an ellipse with the earth as one focus. If the length of the major axis of the orbit is 477,736 miles and the length of the minor axis is 477,078 miles, find the

minimum and maximum distances from earth to the moon.

52. Halley's comet has an elliptical orbit with eccentricity .967, and the closest the comet comes to the sun is 54,004,000 miles. What is the maximum distance from the comet to the sun?

Unusual Problems

53. The punch bowl and a table holding the punch cups are placed 50 ft apart at a yard party. A portable fence is then set up so that any guest inside the fence can walk straight to the table, then to the punch bowl, and then return to his or her starting point without traveling more than 150 ft. Describe the longest possible such fence that encloses the largest possible area.

54. An arched footbridge over a 100-ft-wide river is shaped like half an ellipse. The maximum height of the bridge over the river is 20 ft. Find the height of the bridge over a point in the river, exactly 25 ft from the center of the river.

7.2 HYPERBOLAS

Let P and Q be two points in the plane and r a positive real number. The set of all points X such that

$$|(\text{distance from } P \text{ to } X) - (\text{distance from } Q \text{ to } X)| = r$$

is the **hyperbola** with **foci** P and Q; r will be called the **distance difference**. Every hyperbola has the general shape shown by the solid blue lines in Figure 7–13.

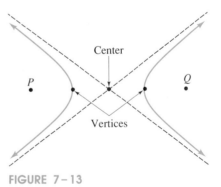

FIGURE 7–13

The dotted straight lines are the **asymptotes** of the hyperbola; it gets closer and closer to the asymptotes, but never touches them. The asymptotes intersect at the midpoint of the line segment from P to Q; this point is called the **center** of the hyperbola. The **vertices** of the hyperbola are the points where it intersects the line segment from P to Q. The line segment joining the vertices is called the **transverse axis** of the hyperbola.

Another complicated exercise in the use of the distance formula, which will be omitted here, leads to the following algebraic description of hyperbolas with center at the origin:

HYPERBOLAS CENTERED AT THE ORIGIN

Let a and b be positive numbers. The graph of

$$\frac{x^2}{a^2} - \frac{y^2}{b^2} = 1$$

is a hyperbola with center at the origin and vertices $(-a, 0)$ and $(a, 0)$, as shown in the figure below. The foci are $(-c, 0)$ and $(c, 0)$, where $c^2 = a^2 + b^2$. The asymptotes are the lines $y = \frac{b}{a}x$ and $y = -\frac{b}{a}x$.

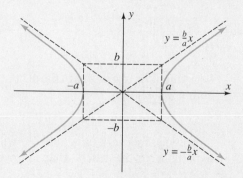

The graph of $\frac{y^2}{b^2} - \frac{x^2}{a^2} = 1$ is a hyperbola with center at the origin and vertices $(0, -b)$ and $(0, b)$, as shown in the figure below. The foci are $(0, -c)$ and $(0, c)$, where $c^2 = a^2 + b^2$. The asymptotes are the lines $y = \frac{b}{a}x$ and $y = -\frac{b}{a}x$.

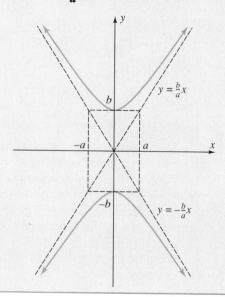

NOTE In the first hyperbola in the box, the distance from vertex $(a, 0)$ to focus $(c, 0)$ is $c - a$, and the distance from $(a, 0)$ to $(-c, a)$ is $a - (-c) = c + a$. Consequently, the distance difference r in the definition of a hyperbola is

$$r = |(c - a) - (c + a)| = 2a.$$

A similar argument shows that the second hyperbola also has $r = 2a$.

As shown in the pictures in the box, the asymptotes pass through the origin and the corners of the rectangle determined by the vertical lines through $\pm a$ and the horizontal lines through $\pm b$. The easiest way to graph a hyperbola is to sketch this rectangle first and use its corners to sketch the asymptotes. This determines the hyperbola's general shape, and only a few points need be plotted to obtain a reasonable graph.

Example 1 To graph $9x^2 - 4y^2 = 36$ we first rewrite the equation:

$$\frac{9x^2}{36} - \frac{4y^2}{36} = \frac{36}{36}$$

$$\frac{x^2}{4} - \frac{y^2}{9} = 1$$

$$\frac{x^2}{2^2} - \frac{y^2}{3^2} = 1$$

Applying the fact in the box above with $a = 2$ and $b = 3$ shows that the graph is a hyperbola with vertices $(2, 0)$ and $(-2, 0)$ and asymptotes $y = \frac{3}{2}x$ and $y = -\frac{3}{2}x$. We first plot the vertices and sketch the rectangle determined by the vertical lines $x = \pm 2$ and the horizontal lines $y = \pm 3$. The asymptotes go through the origin and the corners of this rectangle, as shown on the left in Figure 7–14. It is then easy to sketch the hyperbola. ∎

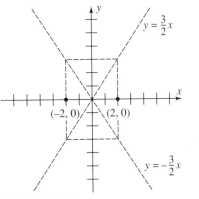
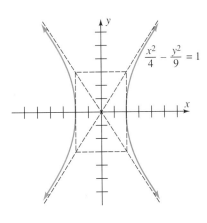

FIGURE 7–14

Example 2 Find the equation and sketch the graph of the hyperbola with vertices $(0, -1)$, $(0, 1)$ that passes through the point $(3, \sqrt{2})$.

Solution Since the vertices are on the y-axis, the equation is of the form

$$\frac{y^2}{b^2} - \frac{x^2}{a^2} = 1$$

and has vertices $(0, \pm b)$. Hence $b = 1$ and the equation is

$$y^2 - \frac{x^2}{a^2} = 1$$

Since $(3, \sqrt{2})$ is on the graph, we must have

$$(\sqrt{2})^2 - \frac{3^2}{a^2} = 1$$

$$(\sqrt{2})^2 a^2 - 3^2 = a^2$$

$$2a^2 - 9 = a^2$$

$$a^2 = 9$$

Therefore $a = 3$, so that the equation of the hyperbola is

$$\frac{y^2}{1^2} - \frac{x^2}{3^2} = 1, \quad \text{or equivalently,} \quad y^2 - \frac{x^2}{9} = 1.$$

The asymptotes are the lines $y = \pm \frac{b}{a} x = \pm \frac{1}{3} x$, and the graph is shown in Figure 7–15. ∎

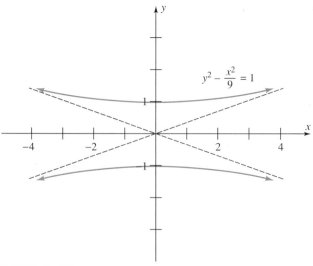

FIGURE 7–15

Let h and k be constants. Replacing x by $x - h$ and y by $y - k$ in an equation shifts the graph horizontally and vertically, with the point (r, s) moving to $(h + r, k + s)$. Therefore

STANDARD EQUATION OF THE HYPERBOLA

> Let a, b, h, k be real numbers with a and b positive.
> The graph of the equation
>
> $$\frac{(x-h)^2}{a^2} - \frac{(y-k)^2}{b^2} = 1$$
>
> is a hyperbola with center (h, k) and vertices $(h - a, k)$ and $(h + a, k)$. The foci are $(h - c, k)$ and $(h + c, k)$, where $c^2 = a^2 + b^2$. The asymptotes are the lines
>
> $$y = \frac{b}{a}(x - h) + k \quad \text{and} \quad y = -\frac{b}{a}(x - h) + k.$$
>
> The graph of the equation
>
> $$\frac{(y-k)^2}{b^2} - \frac{(x-h)^2}{a^2} = 1$$
>
> is a hyperbola with center (h, k) and vertices $(h, k - b)$ and $(h, k + b)$. The foci are $(h, k - c)$ and $(h, k + c)$, where $c^2 = a^2 + b^2$. The asymptotes are the lines
>
> $$y = \frac{b}{a}(x - h) + k \quad \text{and} \quad y = -\frac{b}{a}(x - h) + k.$$

Example 3 The equation $2y^2 - x^2 + 4y + 6x - 11 = 0$ can be written

$$2(y^2 + 2y) - (x^2 - 6x) = 11$$

Now complete the square in $y^2 + 2y$ and in $x^2 - 6x$, by adding 1 and 9, respectively, to these expressions. Since there is a factor of 2 in front of the y expression and a factor of -1 in front of the x expression; we must add $2 \cdot 1$ and -9 to the right side in order to keep the equation unchanged:

$$2(y^2 + 2y + 1) - (x^2 - 6x + 9) = 11 + 2 - 9$$

$$2(y + 1)^2 - (x - 3)^2 = 4$$

$$\frac{(y+1)^2}{2} - \frac{(x-3)^2}{4} = 1$$

$$\frac{(y-(-1))^2}{(\sqrt{2})^2} - \frac{(x-3)^2}{2^2} = 1$$

Therefore the graph is a hyperbola with center $(3, -1)$, as shown in Figure 7–16 on the next page. ■

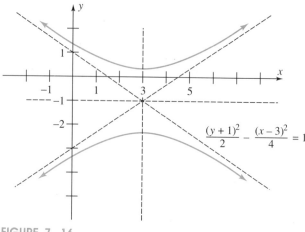

FIGURE 7-16

Eccentricity

The **eccentricity** of a hyperbola is defined in the same way it was for an ellipse,* as the ratio

$$e = \frac{\text{distance between the foci}}{\text{distance between the vertices}}.$$

With hyperbolas, however, the eccentricity is a number greater than 1 because the vertices are closer together than the foci, as shown in Figure 7–17 opposite. The eccentricity measures the "flatness" of the branches of the hyperbola, as shown in Figure 7-17.

As we saw in the box on page 327, a hyperbola with standard equation

$$\frac{(x-h)^2}{a^2} - \frac{(y-k)^2}{b^2} = 1, \quad \text{where } c^2 = a^2 + b^2$$

has vertices $(h - a, k)$ and $(h + a, k)$. Since the vertices lie on the same horizontal line, the distance between them is the difference of their first coordinates, namely $2a$. The foci are $(h - c, k)$ and $(h + c, k)$; the distance between them is $2c$. Therefore

$$e = \frac{\text{distance between foci}}{\text{distance between the vertices}} = \frac{2c}{2a} = \frac{c}{a} = \frac{\sqrt{a^2 + b^2}}{a}.$$

Similarly, the eccentricity of the ellipse with equation

$$\frac{(y-k)^2}{b^2} - \frac{(x-h)^2}{a^2} = 1, \quad \text{where } c^2 = a^2 + b^2$$

is $e = \dfrac{c}{b} = \dfrac{\sqrt{a^2 + b^2}}{b}$.

---

* Keeping in mind that the distance between the vertices of an ellipse is the length of its major axis.

7.2 HYPERBOLAS

Foci close to vertices

Distance between foci is almost the same as the distance between vertices, so e is close to 1 and the branches are pointed.

Foci far from vertices

Distance between foci is much greater than the distance between vertices, so e is large and the branches are nearly flat.

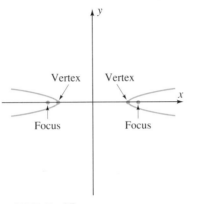
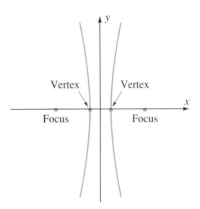

FIGURE 7–17

Applications

It can be shown that if a light ray passes through one focus of a hyperbola and reflects off the hyperbola at a point P, then the reflected ray moves along the straight line determined by P and the other focus of the hyperbola, as shown in Figure 7–18. This property is used in the design of the Cassegrain telescope.

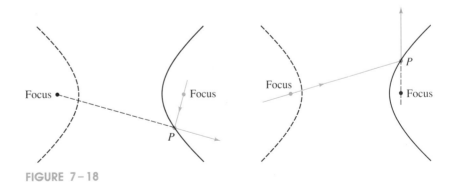

FIGURE 7–18

The long-range navigation system (LORAN) enables a ship to determine its exact location by radio, as follows. As shown in Figure 7–19 on the next page there are three LORAN transmission stations (P, Q, R), each of which continuously transmits signals at regular intervals.

330 7 THE CONIC SECTIONS

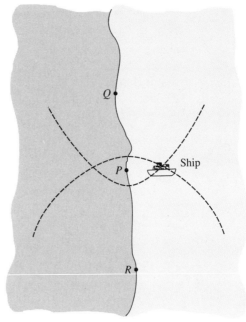

FIGURE 7–19

The ship has equipment which measures the delay between receptions of signals from P and Q. The time difference allows the ship to compute the *difference* in its distances from P and from Q. Since hyperbolas are defined in terms of the difference in distance from two fixed points (the foci), the ship lies on a hyperbola with foci P and Q. Similarly, the delay between receptions of signals from P and R determines a second hyperbola with foci P and R on which the ship lies. The ship's location is found by calculating the equations of the hyperbolas and their intersection point of the appropriate branches.

Example 4 LORAN transmitters Q, P, and R are located 200 miles apart along a straight line and transmit signals that travel at a speed of 980 feet per microsecond. A ship receives a signal from Q 528 microseconds after the signal from P. It receives a signal from R 305 microseconds after the signal from P. Locate the position of the ship.

Solution If the line through the LORAN stations is taken as the x-axis, with the origin located midway between P and Q, and S denotes the ship, then the situation looks like this:

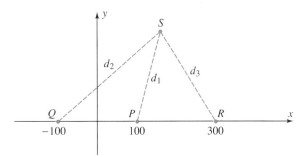

FIGURE 7-20

If the signal takes t microseconds to go from P to S, then

$$d_1 = 980t \quad \text{and} \quad d_2 = 980(t + 528)$$

so that

$$|d_1 - d_2| = |980t - 980(t + 528)| = 980 \cdot 528 = 517{,}440 \text{ feet.}$$

Since one mile is 5280 feet, this means that

$$|d_1 - d_2| = 517{,}440/5{,}280 \text{ miles} = 98 \text{ miles.}$$

As explained on pages 324–325, the hyperbola with foci $Q = (-100, 0)$ and $P = (100, 0)$ and distance difference $r = 98$ has an equation of the form

$$\frac{x^2}{a^2} - \frac{y^2}{b^2} = 1$$

where $c = 100$, $a = r/2 = 98/2 = 49$ and $b = \sqrt{c^2 - a^2} = \sqrt{100^2 - 49^2} \approx 87.2$. So the ship lies on the hyperbola

$$\frac{x^2}{49^2} - \frac{y^2}{(87.2)^2} = 1.$$

A similar argument using P and R as foci shows that the ship lies on the hyperbola with foci $P = (100, 0)$ and $R = (300, 0)$ and center $(200, 0)$ whose distance difference is

$$|d_1 - d_3| = 980 \cdot 305 = 298{,}900 \text{ feet} \approx 56.6 \text{ miles.}$$

In this case, $a = 56.6/2 = 28.3$. The standard equation of the hyperbola shows that the center is $(200, 0)$ and the foci are $(200 - c, k) = (100, 0)$ and $(200 + c, k) = (300, 0)$, which implies that $c = 100$. Thus $b = \sqrt{c^2 - a^2} = \sqrt{100^2 - (28.3)^2} \approx 95.9$, and the ship also lies on the hyperbola with equation

$$\frac{(x - 200)^2}{(28.3)^2} - \frac{y^2}{(95.9)^2} = 1.$$

By solving each of the hyperbola equations for y^2, setting the results equal and solving for x, we find that the point of intersection where the ship lies is at approximately $(130.5, 215.2)$ (coordinates in miles from the origin). ∎

EXERCISES

In Exercises 1–8, sketch the graph of the hyperbola.

1. $\dfrac{x^2}{6} - \dfrac{y^2}{16} = 1$
2. $\dfrac{x^2}{4} - y^2 = 1$
3. $\dfrac{x^2}{10} - \dfrac{y^2}{36} = 1$
4. $\dfrac{y^2}{9} - \dfrac{x^2}{16} = 1$
5. $\dfrac{(y+3)^2}{25} - \dfrac{(x+1)^2}{16} = 1$
6. $\dfrac{(y+1)^2}{9} - \dfrac{(x-1)^2}{25} = 1$
7. $\dfrac{(x+3)^2}{1} - \dfrac{(y-2)^2}{4} = 1$
8. $\dfrac{(y+5)^2}{9} - \dfrac{(x-2)^2}{1} = 1$

In Exercises 9–16, find the equation of the hyperbola that satisfies the given conditions.

9. Center at origin; x-intercepts ± 3; asymptote $y = 2x$.
10. Center at origin; y-intercepts ± 12; asymptote $y = \tfrac{3}{2}x$.
11. Center at origin; vertex $(2, 0)$; passing through $(4, \sqrt{3})$.
12. Center at origin; vertex $(0, \sqrt{12})$; passing through $(2\sqrt{3}, 6)$.
13. Center at $(-2, 3)$; vertex $(-2, 1)$; passing through $(-2 + 3\sqrt{10}, 11)$.
14. Center at $(-5, 1)$; vertex $(-3, 1)$; passing through $(-1, 1 - 4\sqrt{3})$.
15. Center at $(4, 2)$; vertex $(7, 2)$; asymptote $3y = 4x - 10$.
16. Center at $(-3, -5)$; vertex $(-3, 0)$; asymptote $6y = 5x - 15$.
17. Show that the asymptotes of the hyperbola $\dfrac{x^2}{a^2} - \dfrac{y^2}{a^2} = 1$ are perpendicular to each other.
18. Find a number k such that $(-2, 1)$ is on the graph of $3x^2 + ky^2 = 4$. Then graph the equation.

In Exercises 19–26, sketch the graph of the equation.

19. $4x^2 - y^2 = 16$
20. $3y^2 - 5x^2 = 15$
21. $x^2 - 4y^2 = 1$
22. $2x^2 - y^2 = 4$
23. $18y^2 - 8x^2 - 2 = 0$
24. $x^2 - 2y^2 = -1$
25. $(2x - y)(x + 4y) - 7xy = 8$
26. $25y^2 - 4x^2 + 150y + 8x + 121 = 0$

In Exercises 27–32, graph the conic section whose equation is given.

27. $4x^2 + 9y^2 - 8x - 54y + 49 = 0$
28. $x^2 - 16y^2 = 0$
29. $4y^2 - x^2 - 24y + 6x + 11 = 0$
30. $x^2 + 4y^2 + 6x - 8y + 9 = 0$
31. $2x^2 + 2y^2 - 8x - 2y - 4 = 0$
32. $9y^2 - 4x^2 - 18y - 24x - 63 = 0$

33. Two listening stations 1 mile apart record an explosion. One microphone receives the sound 2 seconds after the other does. Use the line through the microphones as the x-axis, with the origin midway between the microphones, and the fact that sound travels at 1100 feet per second to find the equation of the hyperbola on which the explosion is located. Can you determine the exact location of the explosion?

34. Two transmission stations P and Q are located 200 miles apart on a straight shore line. A ship 50 miles from shore is moving parallel to the shore line. A signal from Q reaches the ship 400 microseconds after a signal from P. If the signals travel at 980 feet per microsecond, find the location of the ship (in terms of miles) in the coordinate system with x-axis through P and Q, and origin midway between them.

7.3 PARABOLAS

In Section 5.1 parabolas were defined as the graphs of quadratic functions.*
Parabolas of this kind are a special case of the more general definition, which we now present.

Let L be a line in the plane and P a point not on L. If X is any point not on L, the distance from X to L is defined to be the length of the perpendicular line segment from X to L. The **parabola** with **focus** P and **directrix** L is the set of all points X such that:

distance from X to P = distance from X to L

as shown in blue in Figure 7–21.

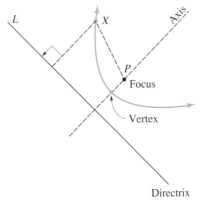

FIGURE 7–21

The line through P perpendicular to L is called the **axis**. The intersection of the axis with the parabola (the midpoint of the segment of the axis from P to L) is the **vertex** of the parabola, as illustrated in Figure 7–21. The parabola is symmetric with respect to its axis.

Suppose that the focus of a parabola is the point $(0, p)$, where p is a nonzero constant, and that the directrix is the horizontal line $y = -p$. If (x, y) is any point on the parabola, then the distance from (x, y) to the horizontal line $y = -p$ is the length of the vertical line segment from (x, y) to $(x, -p)$, as shown in Figure 7–22 on the next page.

---

* Section 5.1 is *not* a prerequisite for this section.

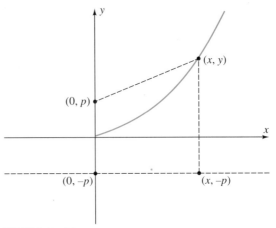

FIGURE 7-22

By the definition of the parabola,

distance from (x, y) to $(0, p)$ = distance from (x, y) to $y = -p$

distance from (x, y) to $(0, p)$ = distance from (x, y) to $(x, -p)$

$$\sqrt{(x-0)^2 + (y-p)^2} = \sqrt{(x-x)^2 + (y-(-p))^2}$$

Squaring both sides and simplifying, we have

$$(x - 0)^2 + (y - p)^2 = (x - x)^2 + (y + p)^2$$
$$x^2 + y^2 - 2py + p^2 = 0^2 + y^2 + 2py + p^2$$
$$x^2 = 4py$$

Conversely, it can be shown that every point whose coordinates satisfy this equation is on the parabola. Therefore

PARABOLA WITH VERTEX AT THE ORIGIN

> If p is a nonzero constant, then the graph of $x^2 = 4py$ is a parabola with
>
> vertex: $(0, 0)$
> focus: $(0, p)$
> directrix: the horizontal line $y = -p$
> axis: the y-axis
>
> The parabola opens upward if $p > 0$ and downward if $p < 0$.

Example 1 Find the focus and directrix and sketch the graph of:

(a) $y = 2x^2$ (b) $y = -x^2/8$.

Solution

(a) $y = 2x^2$ can be rewritten as $x^2 = \frac{1}{2}y$, which is of the form $x^2 = 4py$ with $4p = 1/2$. Therefore $p = 1/8$, and the graph is an upward-opening parabola with focus $(0, p) = (0, 1/8)$ and directrix $y = -p = -1/8$, as shown in Figure 7–23.

(b) $y = -x^2/8$ can be rewritten as $x^2 = -8y$. So $4p = -8$ and $p = -2$. Hence the graph is a downward-opening parabola with focus $(0, p) = (0, -2)$ and directrix $y = -p = -(-2) = 2$, as shown in Figure 7–24. ∎

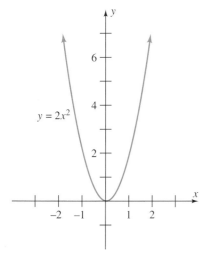

FIGURE 7–23

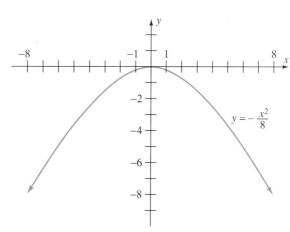

FIGURE 7–24

Parabolas that open to the left or right can be handled by interchanging the roles of x and y in the preceding discussion. In this interchange vertical becomes horizontal, and vice versa. Similarly, upward and downward (positive and negative directions on the y-axis) become right and left (positive and negative directions on the x-axis), and vice versa. Hence

PARABOLA WITH VERTEX AT THE ORIGIN

If p is a nonzero constant, then the graph of $y^2 = 4px$ is a parabola with

vertex: $(0, 0)$
focus: $(p, 0)$
directrix: the vertical line $x = -p$
axis: the x-axis

The parabola opens to the right if $p > 0$ and to the left if $p < 0$.

Example 2 Find the focus, directrix, and equation of the parabola with vertex $(0, 0)$ and focus on the x-axis that passes through the point $(8, 2)$. Sketch the graph.

Solution The equation is of the form $y^2 = 4px$. Since $(8, 2)$ is on the graph, we have $2^2 = 4p \cdot 8$, so that $p = 1/8$. So the focus is $(\frac{1}{8}, 0)$, and the directrix is the vertical line $x = -1/8$. The equation is $y^2 = 4(\frac{1}{8})x = \frac{1}{2}x$, or equivalently, $x = 2y^2$, and the graph is in Figure 7-25. ∎

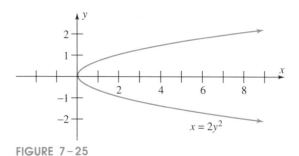

FIGURE 7-25

Let h and k be constants. Replacing x by $x - h$ and y by $y - k$ in an equation shifts the graph horizontally and vertically, with the point (r, s) moving to $(h + r, k + s)$. Therefore

STANDARD EQUATION OF THE PARABOLA

> If p is a nonzero constant, then the graph of $(x - h)^2 = 4p(y - k)$ is a parabola with
>
> vertex: (h, k)
>
> focus: $(h, k + p)$
>
> directrix: the horizontal line $y = k - p$
>
> axis: the vertical line $x = h$
>
> The parabola opens upward if $p > 0$ and downward if $p < 0$.

Example 3 The equation $y = 2(x - 3)^2 - 1$ can be rewritten as

$$y + 1 = 2(x - 3)^2, \quad \text{or equivalently,} \quad (x - 3)^2 = \frac{1}{2}(y - (-1))$$

This is the same form as in the box, with $h = 3$, $k = -1$, and $4p = 1/2$. Hence $p = 1/8$, and the graph is an upward-opening parabola with vertex $(3, -1)$, focus $(3, -1 + 1/8) = (3, -7/8)$, and directrix the horizontal line $y = -1 - 1/8 = -9/8$.

The graph of $y - (-1) = 2(x - 3)^2$ is just the graph of $y = 2x^2$ (Figure 7-23 above) shifted horizontally 3 units to the right and vertically 1 unit downward, as explained on page 313. ∎

Example 4 To graph the equation $y = -4x^2 + 12x - 8$ we first rewrite it as
$$-4(x^2 - 3x) = y + 8$$

Next we complete the square in the expression in parentheses by adding 9/4 (the square of half the coefficient of x). In order not to change the equation we must add $-4(9/4)$ to the right side:

$$-4\left(x^2 - 3x + \frac{9}{4}\right) = y + 8 - 4\left(\frac{9}{4}\right)$$

$$-4\left(x^2 - 3x + \frac{9}{4}\right) = y + 8 - 9$$

$$-4\left(x - \frac{3}{2}\right)^2 = y - 1$$

$$\left(x - \frac{3}{2}\right)^2 = -\frac{1}{4}(y - 1)$$

This is the same form as in the box above, with $h = 3/2$, $k = 1$, and $4p = -1/4$, so that $p = -1/16$. Therefore the graph is a downward-opening parabola with vertex (3/2, 1), as shown in Figure 7-26. ∎

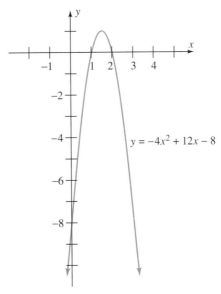

FIGURE 7-26

STANDARD EQUATION OF THE PARABOLA

If p is a nonzero constant, then the graph of $(y - k)^2 = 4p(x - h)$ is a parabola with

vertex: (h, k)

focus: $(h + p, k)$

directrix: the vertical line $x = h - p$

axis: the horizontal line $y = k$

The parabola opens to the right if $p > 0$ and to the left if $p < 0$.

Example 5 To graph $x = 2y^2 + 12y + 14$, we rewrite the equation and complete the square in y:

$$2y^2 + 12y = x - 14$$
$$2(y^2 + 6y) = x - 14$$
$$2(y^2 + 6y + 9) = x - 14 + 2 \cdot 9$$
$$2(y + 3)^2 = x + 4$$
$$(y - (-3))^2 = \frac{1}{2}(x - (-4))$$

The fact in the box above with $h = -4$, $k = -3$, and $4p = 1/2$ shows that the graph is a parabola with vertex $(-4, -3)$ and axis $y = -3$, as shown in Figure 7-27. ∎

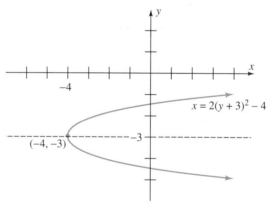

FIGURE 7-27

Applications

Certain laws of physics show that sound waves or light rays from a source at the focus of a parabola will reflect off the parabola in rays parallel to the axis of the parabola, as shown in Figure 7-28. This is the reason that parabolic reflectors are used in automobile headlights and search lights.

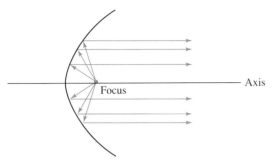

FIGURE 7-28

FIGURE 7-29

Conversely, a light ray coming towards a parabola will be reflected into the focus, as shown in Figure 7-29. This fact is used in the design of radar antennas, satellite dishes, and field microphones used at outdoor sporting events to pick up conversation on the field.

Projectiles follow a parabolic curve, a fact that is used in the design of water slides in which the rider slides down a sharp incline, then up and over a hill, before plunging downward into a pool. At the peak of the parabola shaped hill, the rider shoots up along a parabolic arc several inches above the slide, experiencing a sensation of weightlessness.

EXERCISES

In Exercises 1-4, find the focus and directrix of the parabola.

1. $y = 3x^2$
2. $x = \frac{1}{2} y^2$
3. $y = \frac{1}{4} x^2$
4. $x = -6y^2$

In Exercises 5-16, determine the vertex, focus, and directrix of the parabola *without graphing* **and state whether it opens upward, downward, left, or right.**

5. $x - y^2 = 2$
6. $y - 3 = x^2$
7. $x + (y + 1)^2 = 2$
8. $y + (x + 2)^2 = 3$
9. $3x - 2 = (y + 3)^2$
10. $2x - 1 = -6(y + 1)^2$
11. $x = y^2 - 9y$
12. $x = y^2 + y + 1$
13. $y = 3x^2 + x - 4$
14. $y = -3x^2 + 4x - 1$
15. $y = -3x^2 + 4x + 5$
16. $y = 2x^2 - x - 1$

In Exercises 17-28, sketch the graph of the equation and label the vertex.

17. $y = 4(x - 1)^2 + 2$
18. $y = 3(x - 2)^2 - 3$
19. $x = 2(y - 2)^2$
20. $x = -3(y - 1)^2 - 2$
21. $y = x^2 - 4x - 1$
22. $y = x^2 + 8x + 6$
23. $y = x^2 + 2x$
24. $x = y^2 - 3y$
25. $x^2 = 6x - y - 5$
26. $y^2 = x - 2y - 2$
27. $3y^2 = x - 1 + 2y$
28. $2y^2 = x - 4y - 5$

In Exercises 29-40, find the equation of the parabola satisfying the given conditions.

29. Vertex (0, 0); axis $x = 0$; (2, 12) on graph.
30. Vertex (0, 1); axis $x = 0$; (2, −7) on graph.
31. Vertex (1, 0); axis $x = 1$; (2, 13) on graph.
32. Vertex (−3, 0); axis $y = 0$; (−1, 1) on graph.
33. Vertex (2, 1); axis $y = 1$; (5, 0) on graph.

34. Vertex $(1, -3)$; axis $y = -3$; $(-1, -4)$ on graph.
35. Vertex $(-3, -2)$; focus $(-47/16, -2)$.
36. Vertex $(-5, -5)$; focus $(-5, -99/20)$.
37. Vertex $(1, 1)$; focus $(1, 9/8)$.
38. Vertex $(-4, -3)$; $(-6, -2)$ and $(-6, -4)$ on graph.
39. Vertex $(-1, 3)$; $(8, 0)$ and $(0, 4)$ on graph.
40. Vertex $(1, -3)$; $(0, -1)$ and $(-1, 5)$ on graph.
41. Find the number b such that the vertex of the parabola $y = x^2 + bx + c$ lies on the y-axis.
42. Find the number d such that the parabola $(y + 1)^2 = dx + 4$ passes through $(-6, 3)$.
43. Find the points of intersection of the parabola $4y^2 + 4y = 5x - 12$ and the line $x = 9$.
44. Find the points of intersection of the parabola $4x^2 - 8x = 2y + 5$ and the line $y = 15$.

CHAPTER REVIEW

Important Concepts

Section 7.1 Ellipse 313
 Foci 313
 Center 313
 Major and minor axes 313
 Vertices 313
 Equations of ellipses 314, 315, 316
 Eccentricity of an ellipse 318

Section 7.2 Hyperbola 323
 Foci 323
 Center 323
 Vertices 323
 Asymptotes 323
 Equations of hyperbolas 324, 327
 Eccentricity of a hyperbola 328

Section 7.3 Parabola 333
 Focus 333
 Directrix 333
 Vertex 333
 Equations of parabolas 334, 335, 336

Important Facts and Formulas

- Equation of ellipse with center (h, k) and axes on the lines $x = h, y = k$:

$$\frac{(x - h)^2}{a^2} + \frac{(y - k)^2}{b^2} = 1$$

- Equation of hyperbola with center (h, k) and vertices on the line $y = k$:

$$\frac{(x - h)^2}{a^2} - \frac{(y - k)^2}{b^2} = 1$$

- Equation of hyperbola with center (h, k) and vertices on the line $x = h$:

$$\frac{(y-k)^2}{b^2} - \frac{(x-h)^2}{a^2} = 1$$

- Equation of a parabola with vertex (h, k) and axis $x = h$:
$$(x - h)^2 = 4p(y - k)$$
- Equation of a parabola with vertex (h, k) and axis $y = k$:
$$(y - k)^2 = 4p(x - h)$$

Review Questions

In Questions 1–4, find the foci and vertices of the conic and state whether it is an ellipse or hyperbola.

1. $\dfrac{x^2}{16} + \dfrac{y^2}{20} = 1$

2. $\dfrac{x^2}{9} - \dfrac{y^2}{16} = 1$

3. $\dfrac{(x-1)^2}{7} + \dfrac{(y-3)^2}{16} = 1$

4. $3x^2 = 1 + 2y^2$

5. Find the focus and directrix of the parabola $10y = 7x^2$.

6. Find the focus and directrix of the parabola
$$3y^2 - x - 4y + 4 = 0.$$

7. What is the eccentricity of the ellipse $3x^2 + y^2 = 84$?

8. What is the eccentricity of the ellipse $24x^2 + 30y^2 = 120$?

In Questions 9–20, sketch the graph of the equation. If there are asymptotes, give their equations.

9. $\dfrac{x^2}{4} + \dfrac{y^2}{25} = 1$

10. $25x^2 + 4y^2 = 100$

11. $\dfrac{(x-3)^2}{9} + \dfrac{(y+5)^2}{4} = 1$

12. $\dfrac{x^2}{9} - \dfrac{y^2}{16} = 1$

13. $\dfrac{(y+4)^2}{25} - \dfrac{(x-1)^2}{4} = 1$

14. $4x^2 - 9y^2 = 144$

15. $x^2 + 4y^2 - 10x + 9 = 0$

16. $9x^2 - 4y^2 - 36x + 24y - 36 = 0$

17. $2y = 4(x - 3)^2 + 6$

18. $3y = 6(x + 1)^2 - 9$

19. $x = y^2 + 2y + 2$

20. $y = x^2 - 2x + 3$

21. What is the center of the ellipse $4x^2 + 3y^2 - 32x + 36y + 124 = 0$?

22. Find the equation of the ellipse with center at the origin, one vertex at $(0, 4)$, passing through $(\sqrt{3}, 2\sqrt{3})$.

23. Find the equation of the ellipse with center at $(3, 1)$, one vertex at $(1, 1)$, passing through $(2, 1 + \sqrt{3/2})$.

24. Find the equation of the hyperbola with center at the origin, one vertex at (0, 5), passing through (1, $3\sqrt{5}$).
25. Find the equation of the hyperbola with center at (3, 0), one vertex at (3, 2), passing through (1, $\sqrt{5}$).
26. Find the equation of the parabola with vertex (2, 5), axis $x = 2$, and passing through (3, 12).
27. Find the equation of the parabola with vertex (3/2, $-1/2$), axis $y = -1/2$, and passing through (-3, 1).
28. Find the equation of the parabola with vertex (5, 2) that passes through the points (7, 3) and (9, 6).

CHAPTER 8

Systems of Equations and Inequalities

> ROADMAP: Chapters 8 and 9 are independent of each other and may be read in either order.
>
> Sections 8.5 and 9.6 are independent of Sections 8.2–8.4 and may be read after Section 8.1. Section 8.6 depends on Section 8.5.

This chapter deals with **systems of equations** and **systems of inequalities,** such as:

$$x^2 + y^2 = 25$$
$$x^2 - y = 7$$

Two equations in two variables

$$2x + 5y + z + w = 0$$
$$2y - 4z + 41w = 0$$
$$2x + 17y - 23z + 3w = 0$$

Three equations in four variables

$$6x + y \geq 6$$
$$x + 4y \geq 8$$
$$4x + 5y \leq 20$$

Three inequalities in two variables

A **solution of a system** is a solution that satisfies *all* the equations or inequalities in the system. For instance, in the system of inequalities at the right above, $x = 1$, $y = 3$ is a solution of all three inequalities (check it) and hence is a solution of the system. On the other hand, $x = 2$, $y = 4$ is a solution of the first two inequalities, but not of the third (check it). So $x = 2$, $y = 4$ is *not* a solution of the *system*.

8.1 SYSTEMS OF LINEAR EQUATIONS IN TWO VARIABLES

This section deals with *linear* equations—ones containing only terms of degree 1 (such as x or $3y$) and no terms of higher degree (such as x^2 or y^3 or xy). One method of solving a system of two linear equations in two variables is **substitution.**

Example 1 Any solution of the system

$$x + 2y = 3$$
$$5x - 4y = -6$$

must necessarily be a solution of the first equation. Hence x must satisfy

$$x + 2y = 3, \quad \text{or equivalently,} \quad x = 3 - 2y$$

Substituting this expression for x in the second equation, we have:

$$5x - 4y = -6$$
$$5(3 - 2y) - 4y = -6$$
$$15 - 10y - 4y = -6$$
$$-14y = -21$$
$$y = \frac{-21}{-14} = \frac{3}{2}$$

Therefore every solution of the original system must have $y = 3/2$. But when $y = 3/2$, we see from the first equation that:

$$x + 2y = 3$$
$$x + 2\left(\frac{3}{2}\right) = 3$$
$$x + 3 = 3$$
$$x = 0$$

(We would also have found that $x = 0$ if we had substituted $y = 3/2$ in the second equation.) Consequently, the original system has exactly one solution: $x = 0$, $y = 3/2$. This solution could also have been found by solving the first equation for y instead of x and substituting this value in the second equation. ∎

WARNING

In order to guard against arithmetic mistakes, you should always *check your answers* by substituting them into *all* the equations of the original system. We have in fact checked the answers in all the examples. But these checks are omitted to save space.

The Elimination Method

The **elimination method** of solving systems of linear equations is often more convenient than substitution. It depends on this fact:

Multiplying both sides of an equation by a nonzero constant does not change the solutions of the equation.

8.1 SYSTEMS OF LINEAR EQUATIONS IN TWO VARIABLES

For example, the equation $x + 3 = 5$ has the same solution as $2x + 6 = 10$ (the first equation multiplied by 2). The elimination method also uses this fact from basic algebra:

If $A = B$ and $C = D$, then $A + C = B + D$ and $A - C = B - D$

Example 2 In the system

$$4x - 3y = 8$$
$$2x + 3y = -2$$

the coefficients of the y terms are already negatives of each other. Any solution of this system must also satisfy the equation obtained by adding these two equations:

$$4x - 3y = 8$$
$$\underline{2x + 3y = -2}$$
$$6x = 6 \qquad \text{[The second variable has been eliminated]}$$

Solving this last equation shows that $x = 1$. Substituting $x = 1$ in one of the original equations — say, the first one — shows that

$$4 \cdot 1 - 3y = 8$$
$$-3y = 4$$
$$y = -\frac{4}{3}$$

Therefore the solution of the original system is $x = 1$, $y = -4/3$. ∎

Example 3 To solve the system

$$x - 3y = 4$$
$$2x + y = 1$$

we shall eliminate the x terms. We begin by replacing the first equation by an equivalent one (that is, one with the same solutions):

$$-2x + 6y = -8 \qquad \text{[First equation multiplied by } -2\text{]}$$
$$2x + y = 1$$

The multiplier -2 was chosen so that the coefficients of x in the two equations would be negatives of each other. Any solution of this last system must also be a solution of the sum of the two equations:

$$-2x + 6y = -8$$
$$\underline{2x + y = 1}$$
$$7y = -7 \qquad \text{[The first variable has been eliminated]}$$

Solving this last equation we see that $y = -1$. Substituting this value in the first of the original equations shows that

$$x - 3(-1) = 4$$
$$x = 1$$

Therefore $x = 1$, $y = -1$ is the solution of the original system. ∎

Example 4 Any solution of the system

$$2x - 3y = 1$$
$$3x - 5y = 3$$

must also be a solution of this system:

$$6x - 9y = 3 \quad \text{[First equation multiplied by 3]}$$
$$6x - 10y = 6 \quad \text{[Second equation multiplied by 2]}$$

The multipliers 3 and 2 were chosen so that x would have the same coefficient in both of the new equations. Any solution of this last system must also be a solution of the equation obtained by subtracting the second equation from the first:

$$6x - 9y = 3$$
$$\underline{6x - 10y = 6}$$
$$y = -3 \quad \text{[If this subtraction is confusing, write it out horizontally: } -9y - (-10y) = -9y + 10y = y\text{]}$$

Substituting $y = -3$ in the first of the original equations shows that

$$2x - 3(-3) = 1$$
$$2x = -8$$
$$x = -4$$

Therefore, the solution of the original system is $x = -4$, $y = -3$. ∎

Example 5 To solve the system

$$2x - 3y = 5$$
$$4x - 6y = 1$$

we multiply the first equation by 2 and subtract:

$$4x - 6y = 10$$
$$\underline{4x - 6y = 1}$$
$$0 = 9$$

8.1 SYSTEMS OF LINEAR EQUATIONS IN TWO VARIABLES

Since $0 = 9$ is always false, the original system cannot possibly have any solutions. A system with no solutions is said to be **inconsistent**. ■

Example 6 To solve the system
$$3x - y = 2$$
$$6x - 2y = 4$$
we multiply the first equation by 2 to obtain the system:
$$6x - 2y = 4$$
$$6x - 2y = 4$$
The two equations are identical. So the solutions of this system are the same as the solutions of the single equation $6x - 2y = 4$, which can be rewritten as:
$$2y = 6x - 4$$
$$y = 3x - 2$$
This equation, and hence the original system, has infinitely many solutions. They can be described as follows: Choose any real number for x, say $x = b$. Then $y = 3x - 2 = 3b - 2$. So the solutions of the system are all pairs of numbers of the form
$$x = b, \quad y = 3b - 2 \quad \text{where } b \text{ is any real number}$$
A system such as this is said to be **dependent**. ■

Example 7 575 people attend a ball game and total ticket sales are $2575. If adult tickets cost $5 and children's tickets $3, how many adults attended the game? How many children?

Solution Let x be the number of adults and y the number of children. Then
$$\text{number of adults} + \text{number of children} = \text{total attendance}$$
$$x + y = 575$$
We can obtain a second equation by using the information about ticket sales:

Adult ticket sales + Children ticket sales = Total ticket sales

$$\begin{pmatrix}\text{Price} \\ \text{per} \\ \text{ticket}\end{pmatrix} \times \begin{pmatrix}\text{Number} \\ \text{of} \\ \text{adults}\end{pmatrix} + \begin{pmatrix}\text{Price} \\ \text{per} \\ \text{ticket}\end{pmatrix} \times \begin{pmatrix}\text{Number} \\ \text{of} \\ \text{children}\end{pmatrix} = 2575$$

$$5x + 3y = 2575$$

In order to find x and y we need only solve this system of equations:

$$x + y = 575$$
$$5x + 3y = 2575$$

Multiplying the first equation by -3 and adding we have:

$$-3x - 3y = -1725$$
$$\underline{5x + 3y = 2575}$$
$$2x = 850$$
$$x = 425$$

So 425 adults attended the game. The number of children was $y = 575 - x = 575 - 425 = 150$. ∎

Example 8 How many pounds of tin and how many pounds of copper should be added to 1000 pounds of an alloy that is 10% tin and 30% copper in order to produce a new alloy that is 27.5% tin and 35% copper?

Solution Let x be the number of pounds of tin and y the number of pounds of copper to be added to the 1000 pounds of the old alloy. Then there will be $1000 + x + y$ pounds of the new alloy. We first find the *amounts* of tin and copper in the new alloy:

	Pounds in old alloy	+ Pounds added =	Pounds in new alloy
tin	10% of 1000 +	x	$= 100 + x$
copper	30% of 1000 +	y	$= 300 + y$

Now consider the *percentages* of tin and copper in the new alloy.

	Percentage in new alloy	× Total weight of new alloy	= Pounds in new alloy
tin	27.5%	of $1000 + x + y =$	$.275(1000 + x + y)$
copper	35%	of $1000 + x + y =$	$.35(1000 + x + y)$

The two ways of computing the weight of each metal in the alloy must produce the same result, that is,

$$100 + x = .275(1000 + x + y) \qquad [pounds\ of\ tin]$$
$$300 + y = .35(1000 + x + y) \qquad [pounds\ of\ copper]$$

Multiplying out the right sides and rearranging terms produces this system of equations:

$$.725x - .275y = 175$$
$$-.35x + .65y = 50$$

Multiplying the first equation by .65 and the second by .275 and adding the results, we have:

$$.47125x - .17875y = 113.75$$
$$\underline{-.09625x + .17875y = 13.75}$$
$$.37500x = 127.50$$
$$x = 340$$

Substituting this in the first equation above and solving for y shows that $y = 260$. Therefore 340 pounds of tin and 260 pounds of copper should be added. ∎

Geometric Interpretation

The solution of any system of linear equations in two variables can be seen geometrically by graphing all the equations in the system on the same coordinate plane. As we saw in Section 3.3, the graph of a linear equation is a straight line, and every point on the graph represents a solution of the equation. Therefore a solution of the system will be given by the coordinates of a point that lies on *all* of the lines representing the system.

There are exactly three geometric possibilities for two lines in the plane: They are parallel, or they intersect at a single point, or they coincide, as illustrated in Figure 8–1. Therefore

NUMBER OF SOLUTIONS OF A SYSTEM

A system of two linear equations in two variables must have

no solutions (an inconsistent system) *or*

exactly one solution *or*

an infinite number of solutions (a dependent system).

Lines are parallel

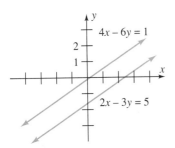

No solutions

Lines intersect at a single point

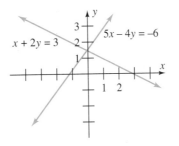

One solution
$x = 0, y = 3/2$

Lines coincide

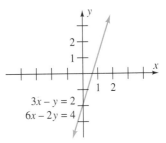

Infinitely many solutions
$x = b, y = 3b - 2$
for any real number b

FIGURE 8–1

EXERCISES

In Exercises 1–6, determine whether the given values of x, y, and z are a solution of the system of equations.

1. $x = -1$, $y = 3$
$2x + y = 1$
$-3x + 2y = 9$

2. $x = 3$, $y = 4$
$2x + 6y = 30$
$x + 2y = 11$

3. $x = 2$, $y = -1$
$\frac{1}{3}x + \frac{1}{2}y = \frac{1}{6}$
$\frac{1}{2}x + \frac{1}{3}y = \frac{2}{3}$

4. $x = .4$, $y = .7$
$3.1x - 2y = -.16$
$5x - 3.5y = -.48$

5. $x = \frac{1}{2}$, $y = 3$, $z = -1$
$2x - y + 4z = -6$
$3y + 3z = 6$
$2z = 2$

6. $x = 2$, $y = \frac{3}{2}$, $z = -\frac{1}{2}$
$3x + 4y - 2z = 13$
$\frac{1}{2}x + 8z = -3$
$x - 3y + 5z = -5$

In Exercises 7–18, use substitution to solve the system.

7. $x - 2y = 5$
$2x + y = 3$

8. $3x - y = 1$
$-x + 2y = 4$

9. $3x - 2y = 4$
$2x + y = -1$

10. $5x - 3y = -2$
$-x + 2y = 3$

11. $-x - 3y = 4$
$3x + 2y = 1$

12. $-2x + 4y = 6$
$5x - 5y = 10$

13. $r + s = 0$
$r - s = 5$

14. $t = 3u + 5$
$t = u + 5$

15. $.4q - p = 3$
$.6q + .5p = 1$

16. $.7w - 1.2z = 2$
$-w + 2.3z = 3$

17. $x + y = c + d$
$x - y = 2c - d$
(where c, d are constants)

18. $x + 3y = c - d$
$2x - y = c + d$
(where c, d are constants)

In Exercises 19–40, use the elimination method to solve the system.

19. $2x - 2y = 12$
$-2x + 3y = 10$

20. $3x + 2y = -4$
$4x - 2y = -10$

21. $x + 3y = -1$
$2x - y = 5$

22. $4x - 3y = -1$
$x + 2y = 19$

23. $2x + 3y = 15$
$8x + 12y = 40$

24. $2x + 5y = 8$
$6x + 15y = 18$

25. $3x - 2y = 4$
$6x - 4y = 8$

26. $2x - 8y = 2$
$3x - 12y = 3$

27. $12x - 16y = 8$
$42x - 56y = 28$

28. $\frac{1}{3}x + \frac{2}{5}y = \frac{1}{6}$
$20x + 24y = 10$

29. $9x - 3y = 1$
$6x - 2y = -5$

30. $8x + 4y = 3$
$10x + 5y = 1$

31. $\frac{x}{3} - \frac{y}{2} = -3$
$\frac{2x}{5} + \frac{y}{5} = -2$

32. $\frac{x}{3} + \frac{3y}{5} = 4$
$\frac{x}{6} - \frac{y}{2} = -3$

33. $\frac{x+y}{4} - \frac{x-y}{3} = 1$
$\frac{x+y}{4} + \frac{x-y}{2} = 9$

34. $\frac{x-y}{4} + \frac{x+y}{3} = 1$
$\frac{x+2y}{3} + \frac{3x-y}{2} = -2$

35. $21.5x + 13.2y = 238.988$
$14.1x - 6.8y = 143.748$

36. $15x - 27y = 20.046$
$21.4x - 568.4y = 27.5392$

37. $3.5x - 2.18y = 2.00782$
$1.92x + 6.77y = -3.86928$

38. $463x - 80y = -13781.6$
$.0375x + .912y = 50.79624$

39. $ax + by = r$ (where a, b, c, d, r, s are
$cx + dy = s$ constants and $ad - bc \neq 0$)

40. $ax + by = ab$ (where a, b are nonzero
$bx - ay = ab$ constants)

41. Let c be any real number. Show that this system has exactly one solution:

$$x + 2y = c$$
$$6x - 3y = 4$$

42. (a) Find the values of c for which this system has an infinite number of solutions.

$$2x - 4y = 6$$
$$-3x + 6y = c$$

(b) Find the values of c for which the system in part (a) has no solutions.

In Exercises 43 and 44, find the values of c and d for which the given points lie on the given straight line.

43. $cx + dy = 2$; (0, 4) and (2, 16)

44. $cx + dy = -6$; (1, 3) and (−2, 12)

45. A 200-seat theater charges $3 for adults and $1.50 for children. If all seats were filled and the total ticket income was $510, how many adults and how many children were in the audience?

46. A theater charges $4 for main floor seats and $2.50 for balcony seats. If all seats are sold, the ticket income is $2100. At one show, 25% of the main floor seats and 40% of the balcony seats were sold and ticket income was $600. How many seats are on the main floor and how many in the balcony?

47. The sum of two numbers is 40. The difference between twice the first number and the second is 11. What are the numbers?

48. The sum of two numbers is 50. The sum of five times one and twice the other is 136. What are the numbers?

49. An investor has part of her money in an account that pays 9% annual interest, and the rest in an account that pays 11% annual interest. If she has $8000 less in the higher paying account than in the lower paying one and her total annual interest income is $2010, how much does she have invested in each account?

50. Joyce has money in two investment funds. Last year the first fund paid a dividend of 8% and the second a dividend of 2% and Joyce received a total of $780. This year the first fund paid a 10% dividend and the second only 1% and Joyce received $810. How much money does she have invested in each fund?

51. At a certain store, cashews cost $4.40/lb and peanuts $1.20/lb. If you want to buy exactly 3 lb of nuts for $6.00, how many pounds of each kind of nuts should you buy? [*Hint:* If you buy x pounds of cashews and y pounds of peanuts, then $x + y = 3$. Find a second equation by considering cost and solve the resulting system.]

52. A store sells deluxe tape recorders for $150. The regular model costs $120. The total tape recorder inventory would sell for $43,800. But during a recent month the store actually sold half of its deluxe models and two thirds of the regular models and took in a total of $26,700. How many of each kind of recorder did they have at the beginning of the month?

53. A plane flies 3000 miles from San Francisco to Boston at a constant speed in 5 hours, flying *with* the wind all the way. The return trip, against the wind, takes 6 hours. Find the speed of the plane and the speed of the wind. [*Hint:* If x is the plane's speed and y the wind speed, then on the trip to Boston (*with* the wind), the plane travels at speed $x + y$ for 5 hours. Since it goes a distance of 3000 miles, we have $5(x + y) = 3000$. Find another equation in x and y and solve the resulting system.]

54. A plane flying into a headwind travels 2000 miles in 4 hours and 24 minutes. The return flight along the same route with a tailwind takes 4 hours. Find the wind speed and the plane's speed (assuming both are constant).

55. A boat made a 4-mile trip upstream against a constant current in 15 minutes. The return trip at the same constant speed with the same current took 12 minutes. What is the speed of the boat and of the current?

56. A boat travels at a constant speed a distance of 57 km downstream in 3 hours, then turns around and travels 55 km upstream in 5 hours. What is the speed of the boat and of the current?

57. A winemaker has two large casks of wine. One wine is 8% alcohol and the other 18% alcohol. How many liters of each wine should be mixed to produce 30 liters of wine that is 12% alcohol?

58. How many cubic centimeters (cm³) of a solution

that is 20% acid and of another solution that is 45% acid should be mixed to produce 100 cm³ of a solution that is 30% acid?

59. How many grams of a 50%-silver alloy should be mixed with a 75%-silver alloy to obtain 40 grams of a 60%-silver alloy?

60. A machine in a pottery factory takes 3 minutes to form a bowl and 2 minutes to form a plate. The material for a bowl costs .25 and the material for a plate costs .20. If the machine runs for 8 hours straight and exactly $44 is spent for material, how many bowls and plates can be produced?

8.1.A EXCURSION: Nonlinear Systems in Two Variables

Substitution or elimination may sometimes be used to solve systems of equations that include nonlinear equations.*

Example 1 Solve the system:

$$x - 8y = -6$$
$$xy = 2$$

Solution You can often determine the number of solutions by graphing both equations in the same coordinate plane. A solution of the system is represented by any point that is on both graphs. Figure 8–2 (in which the techniques of Section 5.3 were used to graph the second equation) shows that this system has two solutions.

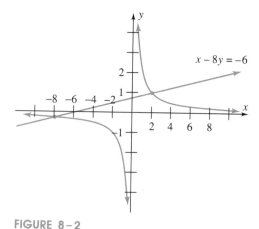

FIGURE 8–2

Unless you use a computer or graphing calculator, however, it may be difficult to read the exact solutions from the graph. To find them algebraically, solve the first equation for x: $x = 8y - 6$. Then substitute this value in the second equation:

---

* Some nonlinear systems may have complex number solutions. Unless stated otherwise, we shall deal only with real number solutions in this section.

$$xy = 2$$
$$(8y - 6)y = 2$$
$$8y^2 - 6y - 2 = 0$$
$$4y^2 - 3y - 1 = 0$$
$$(4y + 1)(y - 1) = 0$$

$4y + 1 = 0$ or $y - 1 = 0$

$y = -\dfrac{1}{4}$ or $y = 1$

Substituting these values in the equation $xy = 2$ we have:

If $y = -\dfrac{1}{4}$, then $x\left(-\dfrac{1}{4}\right) = 2$ If $y = 1$, then $x(1) = 2$

$\qquad\qquad\qquad x = -8 \qquad\qquad\qquad\qquad\qquad\qquad x = 2$

Therefore the solutions are $x = -8$, $y = -\dfrac{1}{4}$ and $x = 2$, $y = 1$. ∎

Example 2 Solve the system:
$$x^2 + y^2 = 8$$
$$x^2 - y = 6$$

Solution The graph of the first equation is a circle with center $(0, 0)$ and radius $\sqrt{8}$ (Section 3.1). The graph of the second is an upward-opening parabola with vertex $(0, -6)$ (Section 5.1 or 7.3). So the rough sketch of the possible graphs in Figure 8–3 shows that there are either two, four, or no solutions.*

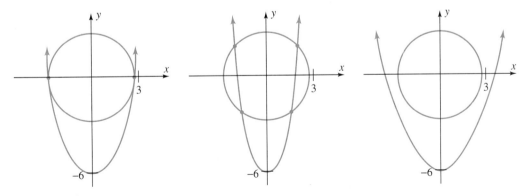

FIGURE 8–3

* If you sketch more carefully or use a graphing calculator, you find that the middle graph is essentially correct.

The system may be solved by substitution (solve the second equation for y and substitute in the first) or by elimination, as follows. Subtracting the second equation from the first yields:

$$y^2 + y = 2$$
$$y^2 + y - 2 = 0$$
$$(y + 2)(y - 1) = 0$$
$$y = -2 \quad \text{or} \quad y = 1$$

Substituting these values of y in the equation $x^2 - y = 6$ produces four possible solutions of the system:

If $y = -2$, then
$$x^2 - (-2) = 6$$
$$x^2 = 4$$
$$x = 2 \quad \text{or} \quad x = -2$$

If $y = 1$, then
$$x^2 - 1 = 6$$
$$x^2 = 7$$
$$x = \sqrt{7} \quad \text{or} \quad x = -\sqrt{7}$$

By checking each of the four possible solutions

$$x = 2 \quad x = -2 \quad x = \sqrt{7} \quad x = -\sqrt{7}$$
$$y = -2 \quad y = -2 \quad y = 1 \quad y = 1$$

in both equations of the original system, you can verify that all four *are* solutions of the system. ■

The solution of some nonlinear systems reduces to solving a related linear system, by either substitution or elimination.

Example 3 To solve the system

$$\frac{1}{x} + \frac{3}{y} = -1$$

$$\frac{2}{x} - \frac{1}{y} = 5$$

we let $u = 1/x$ and $v = 1/y$ so that the system becomes:

$$u + 3v = -1$$
$$2u - v = 5$$

We can solve this system by multiplying the first equation by -2 and adding it to the second equation:

$$-2u - 6v = 2$$
$$\underline{2u - v = 5}$$
$$-7v = 7$$
$$v = -1$$

8.1.A NONLINEAR SYSTEMS IN TWO VARIABLES

Substituting $v = -1$ in the equation $u + 3v = -1$, we see that $u = -3(-1) - 1 = 2$. Consequently, the possible solution of the original system is:

$$x = \frac{1}{u} = \frac{1}{2} \quad \text{and} \quad y = \frac{1}{v} = \frac{1}{(-1)} = -1$$

You should substitute this possible solution in both equations of the original system to check that it is actually a solution of the system. ■

Example 4 A circular tube is to be made by rolling a rectangular sheet of metal and clamping the edges together, as shown in Figure 8–4. If the tube is to have a surface area (excluding ends) of 210 square inches and a volume of 252 cubic inches, what size sheet of metal should be used?

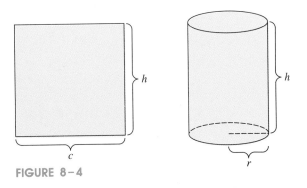

FIGURE 8–4

Solution The area of the metal sheet is given by:

$$\text{height} \cdot \text{width} = \text{area}$$
$$hc = 210$$

The volume of the cylinder is given by

$$\pi \cdot (\text{radius})^2 \cdot \text{height} = \text{volume}$$
$$\pi r^2 h = 252$$

When the sheet is rolled into a tube, the width c of the sheet is the circumference of the circular end of the tube. The relationship between the circumference and radius of this circle is $c = 2\pi r$, or equivalently, $r = c/2\pi$. Substituting this in the last equation above, we have:

$$\pi \left(\frac{c}{2\pi}\right)^2 h = 252$$

$$\frac{c^2 h}{4\pi} = 252$$

$$c^2 h = 1008\pi$$

Therefore we must solve the system

$$hc = 210$$
$$c^2h = 1008\pi$$

Solving the first equation for h and substituting in the second, we have:

$$c^2\left(\frac{210}{c}\right) = 1008\pi$$
$$210c = 1008\pi$$
$$c = 4.8\pi \approx 15.08 \text{ inches}$$

Therefore $h = 210/c = 210/4.8\pi \approx 13.93$ inches. ∎

EXERCISES

In Exercises 1–20, solve the system of equations and check your answers. A rough sketch of the graphs may be helpful.

1. $\begin{aligned} x^2 - y &= 0 \\ -2x + y &= 3 \end{aligned}$
2. $\begin{aligned} x^2 - y &= 0 \\ -3x + y &= -2 \end{aligned}$

3. $\begin{aligned} x^2 - y &= 0 \\ x + 3y &= 6 \end{aligned}$
4. $\begin{aligned} x^2 - y &= 0 \\ x + 4y &= 4 \end{aligned}$

5. $\begin{aligned} x + y &= 10 \\ xy &= 21 \end{aligned}$
6. $\begin{aligned} 2x + y &= 4 \\ xy &= 2 \end{aligned}$

7. $\begin{aligned} xy + 2y^2 &= 8 \\ x - 2y &= 4 \end{aligned}$
8. $\begin{aligned} xy + 4x^2 &= 3 \\ 3x + y &= 2 \end{aligned}$

9. $\begin{aligned} x^2 + y^2 - 4x - 4y &= -4 \\ x - y &= 2 \end{aligned}$

10. $\begin{aligned} x^2 + y^2 - 4x - 2y &= -1 \\ x + 2y &= 2 \end{aligned}$

11. $\begin{aligned} x^2 + y^2 &= 25 \\ x^2 + y &= 19 \end{aligned}$
12. $\begin{aligned} x^2 + y^2 &= 1 \\ x^2 - y &= 5 \end{aligned}$

13. $\begin{aligned} 25x^2 - 16y^2 &= 400 \\ -9x^2 + 4y^2 &= -36 \end{aligned}$
14. $\begin{aligned} 9x^2 + 16y^2 &= 140 \\ -x^2 + 4y^2 &= -4 \end{aligned}$

15. $\dfrac{1}{x} - \dfrac{3}{y} = 2$
16. $\dfrac{5}{x} + \dfrac{2}{y} = 0$

$\dfrac{2}{x} + \dfrac{1}{y} = 3$
$\dfrac{6}{x} + \dfrac{4}{y} = 3$

17. $\dfrac{2}{x} + \dfrac{3}{y} = 8$

$\dfrac{3}{x} - \dfrac{1}{y} = 1$

18. $\dfrac{3}{x^2} + \dfrac{2}{y^2} = 11$ [Hint: Let $u = \dfrac{1}{x^2}$ and $v = \dfrac{1}{y^2}$.]

$\dfrac{1}{x^2} - \dfrac{3}{y^2} = -11$

19. $\dfrac{3}{x+1} - \dfrac{4}{y-2} = 2$

$\dfrac{1}{x+1} + \dfrac{4}{y-2} = 5$

[Hint: Let $u = \dfrac{1}{x+1}$ and $v = \dfrac{1}{y-2}$.]

20. $\dfrac{-5}{x^2+3} - \dfrac{2}{y^2-2} = -12$

$\dfrac{3}{x^2+3} + \dfrac{1}{y^2-2} = 5$

21. Find two real numbers whose sum is -16 and whose product is 48.

22. Find two real numbers whose sum is 34.5 and whose product is 297.

23. Find two positive real numbers whose difference is 1 and whose product is 4.16.

24. Find two real numbers whose difference is 25.75 and whose product is 127.5.

25. Find two real numbers whose sum is 3 such that the sum of their squares is 369.

26. Find two real numbers whose sum is 2 such that the difference of their squares is 60.

27. Find the dimensions of a rectangular room whose perimeter is 58 ft and whose area is 204 sq ft.
28. Find the dimensions of a rectangular room whose perimeter is 53 ft and whose area is 165 sq ft.
29. A rectangle has area 120 square inches and a diagonal of length 17 inches. What are its dimensions?
30. A right triangle has area 225 square centimeters and a hypotenuse of length 35 centimeters. To the nearest tenth of a centimeter, how long are the legs of the triangle?

Unusual Problems

31. A rectangular box (including top) with square ends and a volume of 16 cubic meters is to be constructed from 40 square meters of cardboard. What should the dimensions be? [*Hint:* If a cubic equation occurs, try the Rational Solutions Test.]
32. Find the equation of the straight line that intersects the parabola $y = x^2$ only at the point (3, 9). [*Hint:* What condition on the discriminant guarantees that a quadratic equation has exactly one real solution?]

8.2 LARGER SYSTEMS OF LINEAR EQUATIONS

In order to develop a systematic way of solving systems with three or more equations and variables, we first consider an example of a large system in which the substitution method can be used successfully.

Example 1 Any solution of this system of four equations in four variables

$$3x - 3y + 2z + w = 4$$
$$y - 3z + 2w = -3$$
$$z - 4w = -8$$
$$2w = 6$$

must necessarily be a solution of the last equation. Therefore w must satisfy

$$2w = 6, \quad \text{or equivalently,} \quad w = 3$$

Substituting $w = 3$ in the third equation shows that

$$z - 4w = -8$$
$$z - 4 \cdot 3 = -8$$
$$z = -8 + 12 = 4$$

Thus any solution of the original system must have $w = 3$ and $z = 4$. Substituting these values in the second equation yields:

$$y - 3z + 2w = -3$$
$$y - 3 \cdot 4 + 2 \cdot 3 = -3$$
$$y = -3 + 12 - 6 = 3$$

Finally, substituting $y = 3$, $z = 4$, and $w = 3$ in the first equation shows that

$$3x - 3y + 2z + w = 4$$
$$3x - 3 \cdot 3 + 2 \cdot 4 + 3 = 4$$
$$3x = 4 + 9 - 8 - 3 = 2$$
$$x = \frac{2}{3}$$

Therefore the original system has just one solution: $x = 2/3$, $y = 3$, $z = 4$, $w = 3$. ∎

The substitution process in the preceding example is sometimes called **back substitution** because you begin with the last equation and work back to the first one. Back substitution works in the example because the system there has a very special form, called **echelon form:** The first variable in the first equation, x, does not appear in any subsequent equations; the first variable in the second equation, y, does not appear in any subsequent equations; and in general, the first variable in each equation of the system does not appear in any subsequent equations.

Gaussian Elimination†

In order to see how and why the elimination method works for large systems of equations, it is instructive to examine an example with just two equations.

Example 2 On page 345 we solved the system
$$x - 3y = 4$$
$$2x + y = 1$$
by multiplying the first equation by -2 and adding it to the second in order to eliminate the variable x:

$-2x + 6y = -8$	-2 *times first equation*
$2x + y = 1$	*Second equation*
$7y = -7$	*Sum of second equation and -2 times first equation*

We then solved this last equation for y and substituted the answer $y = -1$ in the original first equation to find that $x = 1$. So what we really did was this: Replace the original system by the following echelon form system and solve by back substitution:

(∗)
$x - 3y = 4$	*First equation*
$7y = -7$	*Sum of second equation and -2 times first equation*

† Named after the great German mathematician K. F. Gauss (1777–1855).

8.2 LARGER SYSTEMS OF LINEAR EQUATIONS

As we saw on page 345, any solution of the original system must be a solution of the first equation and of the sum equation $7y = -7$, and hence of system (*). Conversely, it is easy to check that any solution of system (*) is also a solution of the original system. *Note:* We are not claiming that the second equations in the two systems have the same solutions—they don't—but only that the two *systems* have the same solution, namely, $x = 1$, $y = -1$. ∎

In the preceding example, replacing an equation by the sum of itself and a constant multiple of another equation of the system did not change the solutions of the *system* (although the new equation had different solutions than the one it replaced). A similar argument shows that the same thing is true in *any* system of linear equations and leads to:

THE ELIMINATION METHOD

> **Any system of linear equations can be transformed into a system in echelon form with the same solutions, by using a finite number of operations of these three types:**
>
> 1. **Interchange any two equations in the system.**
> 2. **Replace an equation in the system by a nonzero constant multiple of itself.**
> 3. **Replace an equation in the system by the sum of itself and a constant multiple of another equation in the system.**
>
> **The resulting echelon form system can then be solved by back substitution.**

No formal proof will be given that every system can be put into echelon form by these operations, but the examples will amply illustrate this fact.

Example 3 In order to transform the system

①
$$x + 4y - 3z = 1 \quad \text{Equation A}$$
$$-3x - 6y + z = 3 \quad \text{Equation B}$$
$$2x + 11y - 5z = 0 \quad \text{Equation C}$$

into a system in echelon form, the first step is to replace the second and third equations by equations that do not involve x.

To eliminate x from Equation B we add 3 times Equation A to Equation B:

$$\begin{array}{ll} 3x + 12y - 9z = 3 & \textit{3 times Equation A} \\ \underline{-3x - 6y + z = 3} & \textit{Equation B} \\ 6y - 8z = 6 & \textit{Sum of Equation B and 3 times Equation A} \end{array}$$

Now we replace Equation B by this sum to obtain this system:

② $\quad x + 4y - 3z = 1 \quad$ *Equation A*
$\qquad 6y - 8z = 6 \quad$ *Sum of Equation B and 3 times Equation A*
$\quad 2x + 11y - 5z = 0 \quad$ *Equation C*

In order to eliminate the x term in Equation C, we form a new equation, the sum of Equation C and -2 times Equation A:

$\quad -2x - 8y + 6z = -2 \quad$ *-2 times Equation A*
$\quad \underline{2x + 11y - 5z = 0} \quad$ *Equation C*
$\qquad\quad 3y + z = -2 \quad$ *Sum of Equation C and -2 times Equation A*

Now in the system ② above, replace Equation C by this new equation:

③ $\quad x + 4y - 3z = 1$
$\qquad\quad 6y - 8z = 6$
$\qquad\quad 3y + z = -2 \quad$ *[Sum of Equation C and -2 times Equation A]*

The next step is to eliminate the y term in one of the last two equations. This can be done by replacing the second equation by the sum of itself and -2 times the third equation:

④ $\quad x + 4y - 3z = 1$
$\qquad\qquad\quad -10z = 10 \quad$ *[Sum of second equation and -2 times third equation]*
$\qquad\quad 3y + z = -2$

Finally, interchange the last two equations:

⑤ $\qquad\qquad x + 4y - 3z = 1$
$\qquad\qquad\qquad 3y + z = -2$
$\qquad\qquad\qquad\quad -10z = 10$

We now have a system in echelon form with the same solutions as the system we started with. So we can find the solutions of that original system by using back substitution in this system. Beginning with the last equation, we see that

$$-10z = 10, \quad \text{or equivalently,} \quad z = -1$$

Substituting $z = -1$ in the second equation shows that

$$3y + z = -2$$
$$3y + (-1) = -2$$
$$3y = -1$$
$$y = -\frac{1}{3}$$

8.2 LARGER SYSTEMS OF LINEAR EQUATIONS

Substituting $y = -1/3$ and $z = -1$ in the first equation yields:

$$x + 4y - 3z = 1$$
$$x + 4\left(-\frac{1}{3}\right) - 3(-1) = 1$$
$$x = 1 + \frac{4}{3} - 3 = -\frac{2}{3}$$

Therefore the original system has just one solution: $x = -2/3$, $y = -1/3$, $z = -1$. ∎

Example 4 The first step in solving the system

$$x + 2y + 3z = -2$$
$$2x + 6y + z = 2$$
$$3x + 3y + 10z = -2$$

is to eliminate the x terms from the last two equations by performing suitable operations:

$$x + 2y + 3z = -2$$
$$2y - 5z = 6 \qquad \text{[Sum of second equation and } -2 \text{ times first equation]}$$
$$3x + 3y + 10z = -2$$

$$x + 2y + 3z = -2$$
$$2y - 5z = 6$$
$$-3y + z = 4 \qquad \text{[Sum of third equation and } -3 \text{ times first equation]}$$

In order to eliminate the y term in the last equation, we first arrange for y to have coefficient 1 in the second equation:

$$x + 2y + 3z = -2$$
$$y - \frac{5}{2}z = 3 \qquad \left[\text{Second equation multiplied by } \frac{1}{2}\right]$$
$$-3y + z = 4$$

Now it is easy to eliminate the y term in the last equation:

$$x + 2y + 3z = -2$$
$$y - \frac{5}{2}z = 3$$
$$-\frac{13}{2}z = 13 \qquad \text{[Sum of third equation and 3 times second equation]}$$

This last system is in echelon form and we can solve the third equation: $z = 13(-\frac{2}{13}) = -2$. Substituting $z = -2$ in the second equation shows that

$$y - \frac{5}{2}(-2) = 3$$

$$y = 3 - 5 = -2$$

Substituting $y = -2$ and $z = -2$ in the first equation yields

$$x + 2(-2) + 3(-2) = -2$$

$$x = -2 + 4 + 6 = 8$$

Therefore the only solution of the original system is $x = 8, y = -2, z = -2$. ∎

Example 5 A system such as this one

$$2x + 5y + z + 3w = 0$$
$$2y - 4z + 3w = 0$$
$$2x + 17y - 23z + 41w = 0$$

in which all the constants on the right side are zero is called a **homogeneous system.** It obviously has at least one solution, namely, $x = 0, y = 0, z = 0, w = 0$. This solution is called the **trivial solution.** In order to see if there are any nontrivial solutions, we shall put the system into echelon form. The x term in the third equation can be eliminated by replacing the third equation by the sum of itself and -1 times the first equation:

$$2x + 5y + z + 3w = 0$$
$$2y - 4z + 3w = 0$$
$$12y - 24z + 38w = 0$$

Now we replace the third equation by the sum of itself and -6 times the second equation:

$$2x + 5y + z + 3w = 0$$
$$2y - 4z + 3w = 0$$
$$20w = 0$$

This echelon form system can now be solved. The last equation shows that $w = 0$. Substituting $w = 0$ in the second equation yields:

$$2y - 4z = 0$$
$$2y = 4z$$
$$y = 2z$$

This equation obviously has an infinite number of solutions. For example, $z = 1, y = 2$ is a solution and $z = -\frac{5}{2}, y = -5$ is a solution. More generally, for each real number t, $z = t$ and $y = 2t$ is a solution of the second equation. Substituting $w = 0$, $z = t$, $y = 2t$ in the first equation shows that

$$2x + 5(2t) + t + 3(0) = 0$$
$$2x = -11t$$
$$x = -\frac{11t}{2}$$

Therefore this echelon form system, and hence the original system, has an infinite number of different solutions, one for each real number t:

$$x = -\frac{11t}{2}, \quad y = 2t, \quad z = t, \quad w = 0$$

For example, if $t = 2$, we have the solution

$$x = \frac{-11(2)}{2} = -11, \quad y = 2(2) = 4, \quad z = 2, \quad w = 0$$

If $t = -3$, we have the solution $x = 33/2$, $y = -6$, $z = -3$, $w = 0$; and so on. ∎

A system with infinitely many solutions, as in the last example, is said to be **dependent**. A dependent system may be either homogeneous (as in the example) or nonhomogeneous (Exercises 7–10).

WARNING

> On page 346, we saw an example of a two-equation system with no solutions (an inconsistent system). The same thing can happen with larger systems. If the system is inconsistent, then at some stage in the elimination process the operations used to put the system in echelon form will lead to an equation such as $0x + 0y + 0z = 5$. Since the left side of this equation is always zero and the right side nonzero, the equation has no solution. Any system which contains this equation must be inconsistent.

The preceding examples illustrate the following fact, whose proof is omitted.

NUMBER OF SOLUTIONS OF A SYSTEM

> Any system of linear equations must have
>
> no solutions (an inconsistent system) *or*
>
> exactly one solution *or*
>
> an infinite number of solutions (a dependent system).

EXERCISES

In Exercises 1–22, solve the system.

1. $\begin{aligned} x + y &= 5 \\ -x \phantom{{}+y} + 2z &= 0 \\ 2x + y - z &= 7 \end{aligned}$

2. $\begin{aligned} x + 2y + 4z &= 3 \\ x \phantom{{}+2y} + 2z &= 0 \\ 2x + 4y + z &= 3 \end{aligned}$

3. $\begin{aligned} 2x + y - z &= 4 \\ x \phantom{{}+y} + 2z &= 9 \\ -3x - y + 2z &= 9 \end{aligned}$
   [*Hint:* Begin by interchanging the first and second equations.]

4. $\begin{aligned} 2x - 2y + z &= -6 \\ 3x + y + 2z &= 2 \\ x + y - 2z &= 0 \end{aligned}$

5. $\begin{aligned} -x + 3y + 2z &= 0 \\ 2x - y - z &= 3 \\ x + 2y + 3z &= 0 \end{aligned}$

6. $\begin{aligned} 3x + 7y + 9z &= 0 \\ x + 2y + 3z &= 2 \\ x + 4y + z &= 2 \end{aligned}$

7. $\begin{aligned} x + y + z &= 1 \\ x - 2y + 2z &= 4 \\ 2x - y + 3z &= 5 \end{aligned}$

8. $\begin{aligned} 2x - y + z &= 1 \\ 3x + y + z &= 0 \\ 7x - y + 3z &= 2 \end{aligned}$

9. $\begin{aligned} 11x + 10y + 9z &= 5 \\ x + 2y + 3z &= 1 \\ 3x + 2y + z &= 1 \end{aligned}$

10. $\begin{aligned} -x + 2y - 3z + 4w &= 8 \\ 2x - 4y + z + 2w &= -3 \\ 5x - 4y + z + 2w &= -3 \end{aligned}$

11. $\begin{aligned} x + y &= 3 \\ 5x - y &= 3 \\ 9x - 4y &= 1 \end{aligned}$

12. $\begin{aligned} 2x - y + 2z &= 3 \\ -x + 2y - z &= 0 \\ x + y - z &= 1 \end{aligned}$

13. $\begin{aligned} 2x - y + z &= 1 \\ x + y - z &= 2 \\ -x - y + z &= 0 \end{aligned}$

14. $\begin{aligned} x + 2y + 3z &= 4 \\ 2x - y + z &= 3 \\ 3x + y + 4z &= 7 \end{aligned}$

15. $\begin{aligned} x + y &= 3 \\ -x + 2y &= 3 \\ 2x - y &= 3 \end{aligned}$

16. $\begin{aligned} 3x + y - 2z &= 4 \\ -5x \phantom{{}+y} + 2z &= 5 \\ -7x - y + 3z &= -2 \end{aligned}$

17. $\begin{aligned} x + y + z + w &= 10 \\ x + y + 2z \phantom{{}+w} &= 11 \\ x - 3y \phantom{{}+z} + w &= -14 \\ y + 3z - w &= 7 \end{aligned}$

18. $\begin{aligned} 2x + y + z \phantom{{}+w} &= 3 \\ y + z + w &= 5 \\ 4x \phantom{{}+y} + z + w &= 0 \\ 3y - 2z - w &= 6 \end{aligned}$

19. $\begin{aligned} 2x - 4y + 5z &= 1 \\ x \phantom{{}-4y} - 3z &= 2 \\ 5x - 8y + 7z &= 6 \\ 3x - 4y + 2z &= 3 \\ x - 4y + 8z &= -1 \end{aligned}$

20. $\begin{aligned} 4x \phantom{{}+y} + z + 2w + 24v &= 0 \\ 2x + y \phantom{{}+z+2w} + 12v &= 0 \\ 3x \phantom{{}+y} + z + 2w + 18v &= 0 \\ 4x - y \phantom{{}+z} + w + 24v &= 0 \\ 7x - y + z + 3w + 42v &= 0 \end{aligned}$

21. $\begin{aligned} \frac{3}{x} - \frac{1}{y} + \frac{4}{z} &= -13 \\ \frac{1}{x} + \frac{2}{y} - \frac{1}{z} &= 12 \\ \frac{4}{x} - \frac{1}{y} + \frac{3}{z} &= -7 \end{aligned}$ [*Hint:* Let $u = 1/x$, $v = 1/y$, $w = 1/z$.]

22. $\begin{aligned} \frac{1}{x+1} - \frac{2}{y-3} + \frac{3}{z-2} &= 4 \\ \frac{5}{y-3} - \frac{10}{z-2} &= -5 \\ \frac{-3}{x+1} + \frac{4}{y-3} - \frac{1}{z-2} &= -2 \end{aligned}$
[*Hint:* Let $u = 1/(x+1)$, $v = 1/(y-3)$, $w = 1/(z-2)$.]

In Exercises 23–28, carry out the elimination method far enough to determine whether the system is inconsistent or dependent. (It isn't necessary to solve the systems.)

23. $\begin{aligned} x + 2y &= 0 \\ y - z &= 2 \\ x + y + z &= -2 \end{aligned}$

24. $\begin{aligned} x + 2y + z &= 0 \\ y + 2z &= 0 \\ x + y - z &= 0 \end{aligned}$

25. $\begin{aligned} x + 2y + 4z &= 6 \\ y + z &= 1 \\ x + 3y + 5z &= 10 \end{aligned}$

26. $\begin{aligned} x + y + 2z + 3w &= 1 \\ 2x + y + 3z + 4w &= 1 \\ 3x + y + 4z + 5w &= 2 \end{aligned}$

27. $\begin{aligned} x + 2y &= 1 \\ -x + y &= 0 \\ 2x + 4y &= 3 \end{aligned}$

28. $\begin{aligned} x + 2y + 3z \phantom{{}+w} &= 1 \\ 3x + 2y + 4z \phantom{{}+w} &= -1 \\ 2x + 6y + 8z + w &= 3 \\ 2x \phantom{{}+6y} + 2z - 2w &= 3 \end{aligned}$

Exercises 29–32 deal with systems such as this one:

$$x + y + 4z - w = 1$$
$$y - 2z + 3w = 0$$

Verify that for each *pair* of real numbers, s and t, the system has a solution:

$$w = s, \quad z = t, \quad y = 2t - 3s,$$
$$x = 1 - (2t - 3s) - 4t + s = 1 - 6t + 4s$$

With this model in mind, solve these dependent systems by the elimination method.

29. $x - y + 2z + 3w = 0$
    $x + z + w = 0$
    $3x - 2y + 5z + 7w = 0$

30. $x + 2y + z + 4w = 1$
    $ y + 3z - w = 2$
    $x + 4y + 7z - 2w = 5$
    $3x + 7y + 6z + 11w = 5$

31. $x + y + z - w = 0$
    $2x - 4y - 4z + w = 0$
    $4x - 2y + 2z - 3w = 0$
    $7x - y - z - 3w = 0$

32. $x + 2y + 3z + 4v = 0$
    $2x + 4y + 6z + w + 9v = 0$
    $x + 2y + 3z + w + 5v = 0$

In Exercises 33–38, find constants a, b, c such that the three given points lie on the parabola $y = ax^2 + bx + c$. See the hint for Exercise 33.

33. $(-3, 2)$, $(1, 1)$, $(2, -1)$ [*Hint:* Since $(-3, 2)$ is to be on the graph, we must have $a(-3)^2 + b(-3) + c = 2$, that is, $9a - 3b + c = 2$. In a similar manner, the other two points lead to *linear* equations in a, b, c. Solve this system of three equations for a, b, c.]

34. $(1, -2)$, $(3, 1)$, $(4, -1)$
35. $(1, 0)$, $(-1, 6)$, $(2, 3)$
36. $(1, 1)$, $(0, 0)$, $(-1, 2)$
37. $(-1, 6)$, $(-2, 16)$, $(1, 4)$
38. $(-1, -6)$, $(2, -3)$, $(4, -25)$

39. Find constants a, b, c such that the points $(0, -2)$, $(\ln 2, 1)$, and $(\ln 4, 4)$ lie on the graph of $f(x) = ae^x + be^{-x} + c$. [*Hint:* See Exercise 33.]

40. Find constants a, b, c such that the points $(0, -1)$, $(\ln 2, 4)$, and $(\ln 3, 7)$ lie on the graph of $f(x) = ae^x + be^{-x} + c$.

41. A collection of nickels, dimes, and quarters totals $6.00. If there are 52 coins altogether and twice as many dimes as nickels, how many of each kind of coin are there?

42. A collection of nickels, dimes, and quarters totals $8.20. The number of nickels and dimes together is twice the number of quarters. The value of the nickels is one third of the value of the dimes. How many of each kind of coin are there?

43. An investor has $70,000 invested in a mutual fund, bonds, and a fast food franchise. She has twice as much invested in bonds as in the mutual fund. Last year the mutual fund paid a 2% dividend, the bonds 10%, and the fast food franchise 6%; her dividend income was $4800. How much is invested in each of the three investments?

44. Tickets to a band concert cost $2 for children, $3 for teenagers, and $5 for adults. 570 people attended the concert and total ticket receipts were $1950. Three-fourths as many teenagers as children attended. How many children, adults, and teenagers attended?

45. If Tom, Dick, and Harry work together, they can paint a large room in 4 hours. When only Dick and Harry work together, it takes 8 hours to paint the room. Tom and Dick, working together, take 6 hours to paint the room. How long would it take each of them to paint the room alone? [*Hint:* If x is the amount of the room painted in 1 hour by Tom, y the amount painted by Dick, and z the amount painted by Harry, then $x + y + z = 1/4$.]

46. Pipes R, S, T are connected to the same tank. When all three pipes are running, they can fill the tank in 2 hours. When only pipes S and T are running, they can fill the tank in 4 hours. When only R and T are running, they can fill the tank in 2.4 hours. How long would it take each pipe running alone to fill the tank?

8.2.A EXCURSION: Partial Fractions

In calculus it is sometimes necessary to write a complicated rational expression as the sum of simpler ones. For instance, by using the common denominator $(x - 2)(x^2 + 1)$ on the right-hand side, you can readily verify that

$$\frac{3x^2 - 4x + 1}{x^3 - 2x^2 + x - 2} = \frac{1}{x - 2} + \frac{2x}{x^2 + 1}$$

Each of the fractions on the right-hand side of this equation is called a **partial fraction**, and the entire right-hand side is called the **partial fraction decomposition** of the fraction on the left-hand side. One technique for finding the partial fraction decomposition of a rational expression uses systems of linear equations.

Example 1 Find constants A and B such that

$$\frac{7x - 4}{(x + 3)(x - 2)} = \frac{A}{x + 3} + \frac{B}{x - 2}.$$

Solution Multiplying both sides of this equation by the common denominator $(x + 3)(x - 2)$ shows that

$$7x - 4 = A(x - 2) + B(x + 3)$$
$$7x - 4 = Ax - 2A + Bx + 3B$$
$$7x - 4 = (A + B)x + (-2A + 3B)$$

Since the polynomials on the left and right sides of the last equation are equal, their coefficients must be equal term by term, that is,

$$A + B = 7 \quad \text{[Coefficients of } x \text{ are equal.]}$$
$$-2A + 3B = -4 \quad \text{[Constant terms are equal.]}$$

We now have a system of two equations in the two unknowns A and B. It can be solved by the methods of Section 8.1:

$$2A + 2B = 14 \quad \text{[First equation multiplied by 2]}$$
$$\underline{-2A + 3B = -4}$$
$$5B = 10$$
$$B = 2$$

From the first equation of the original system we have $A = 7 - B = 7 - 2 = 5$. Therefore:

$$\frac{7x - 4}{(x + 3)(x - 2)} = \frac{5}{x + 3} + \frac{2}{x - 2}. \blacksquare$$

Example 2 Find constants A, B, and C such that

$$\frac{2x^2 + 15x + 10}{(x-1)(x+2)^2} = \frac{A}{x-1} + \frac{B}{x+2} + \frac{C}{(x+2)^2}.$$

Solution Multiply both sides of the equation by the common denominator $(x-1)(x+2)^2$ and collect like terms on the right side:

$$\begin{aligned}
2x^2 + 15x + 10 &= A(x+2)^2 + B(x-1)(x+2) + C(x-1) \\
&= A(x^2 + 4x + 4) + B(x^2 + x - 2) + C(x-1) \\
&= Ax^2 + 4Ax + 4A + Bx^2 + Bx - 2B + Cx - C \\
&= (A+B)x^2 + (4A+B+C)x + (4A-2B-C)
\end{aligned}$$

Equating the coefficients of the corresponding powers of x leads to this system:

$$\begin{aligned}
A + B &= 2 \quad &&[\text{Coefficients of } x^2] \\
4A + B + C &= 15 \quad &&[\text{Coefficients of } x] \\
4A - 2B - C &= 10 \quad &&[\text{Constant terms}]
\end{aligned}$$

It can be solved by Gaussian elimination:

$$\begin{aligned}
A + B &= 2 \\
-3B + C &= 7 \quad &&[\text{Sum of second equation and } -4 \text{ times first equation}] \\
4A - 2B - C &= 10
\end{aligned}$$

$$\begin{aligned}
A + B &= 2 \\
-3B + C &= 7 \\
-6B - C &= 2 \quad &&[\text{Sum of third equation and } -4 \text{ times first equation}]
\end{aligned}$$

$$\begin{aligned}
A + B &= 2 \\
-3B + C &= 7 \\
-3C &= -12 \quad &&[\text{Sum of third equation and } -2 \text{ times second equation}]
\end{aligned}$$

The last equation shows that $C = 4$, and hence by the second equation, $-3B + 4 = 7$ so that $B = -1$. Finally, the first equation shows that $A = 2 - B = 2 - (-1) = 3$. Therefore

$$\frac{2x^2 + 15x + 10}{(x-1)(x+2)^2} = \frac{3}{x-1} + \frac{-1}{x+2} + \frac{4}{(x+2)^2}. \blacksquare$$

In the preceding examples the general form of the partial fraction decomposition was given in advance, so we only had to find the constants A, B, C. If the general form is not given, you can determine it by using the following procedure, whose proof will be omitted:

1. Given $f(x)/g(x)$, with $f(x)$, $g(x)$ polynomials and the degree of $f(x)$ less than the degree of $g(x)$, express the denominator $g(x)$ as a product of factors of the form $px + q$ and $ax^2 + bx + c$, where p, q, a, b, c are constants and $ax^2 + bx + c$ has no real roots.*

2. Collect repeated factors and write $g(x)$ as a product of *distinct* factors of the form
$$(px + q)^m \text{ and } (ax^2 + bx + c)^n$$
with m and n nonnegative integers.

3. For each factor of $g(x)$ of the form $(px + q)^m$, with $m \geq 1$, the partial fraction decomposition of $f(x)/g(x)$ includes a sum of fractions of the form
$$\frac{A_1}{px + q} + \frac{A_2}{(px + q)^2} + \frac{A_3}{(px + q)^3} + \cdots + \frac{A_m}{(px + q)^m}$$

4. For each factor of $g(x)$ of the form $(ax^2 + bx + c)^n$, with $n \geq 1$ and $ax^2 + bx + c$ having no real roots, the partial sum decomposition of $f(x)/g(x)$ includes a sum of fractions of the form
$$\frac{B_1 x + C_1}{ax^2 + bx + c} + \frac{B_2 x + C_2}{(ax^2 + bx + c)^2} + \cdots + \frac{B_n x + C_n}{(ax^2 + bx + c)^n}$$

Example 3 Find the form of the partial fraction decomposition of
$$\frac{4x^5 - 12x^4 + 18x^3 - 7x + 8}{(3x^3 - x^2 - 2x)(x^3 - 1)(x^2 + x + 1)^2}$$

Solution First factor the denominator into linear and quadratic factors:
$$(3x^3 - x^2 - 2x)(x^3 - 1)(x^2 + x + 1)^2$$
$$= x(3x^2 - x - 2)(x - 1)(x^2 + x + 1)(x^2 + x + 1)^2$$
$$= x(3x + 2)(x - 1)(x - 1)(x^2 + x + 1)(x^2 + x + 1)^2$$
$$= x(3x + 2)(x - 1)^2(x^2 + x + 1)^3$$

Each of the first three factors is of the form $(px + q)^n$; for example, x is of this form, with $p = 1$, $q = 0$, and $n = 1$. Similarly, $x^2 + x + 1$ is of the form $ax^2 + bx + c$ with $a = 1$, $b = 1$, and $c = 1$. The quadratic formula shows that $x^2 + x + 1$ has no real roots. So this factorization meets the conditions of step 2.

Now form the partial faction sum for each of the factors as specified in steps 3 and 4:

---

* In theory this can always be done; see page 252. For the case when the degree of $f(x)$ is larger than or equal to the degree of $g(x)$, see Exercises 39–42.

8.2.A PARTIAL FRACTIONS

Factor	Sum of partial fractions
x	$\dfrac{A}{x}$
$3x + 2$	$\dfrac{B}{3x + 2}$
$(x - 1)^2$	$\dfrac{C}{x - 1} + \dfrac{D}{(x - 1)^2}$
$(x^2 + x + 1)^3$	$\dfrac{Ex + F}{x^2 + x + 1} + \dfrac{Gx + H}{(x^2 + x + 1)^2} + \dfrac{Jx + K}{(x^2 + x + 1)^3}$

Therefore the partial sum decomposition of the original fraction is of this form:

$$\frac{4x^5 - 12x^4 + 18x^3 - 7x + 8}{(3x^3 - x^2 - 2x)(x^3 - 1)(x^2 + x + 1)^2}$$

$$= \frac{A}{x} + \frac{B}{3x + 2} + \frac{C}{x - 1} + \frac{D}{(x - 1)^2} + \frac{Ex + F}{x^2 + x + 1}$$

$$+ \frac{Gx + H}{(x^2 + x + 1)^2} + \frac{Jx + K}{(x^2 + x + 1)^3}. \quad \blacksquare$$

Example 4 Find the partial fraction decomposition of

$$\frac{5x^4 + x^3 + 5x^2 + 4x + 2}{x(x^2 + 1)^2}$$

Solution Since $x^2 + 1$ has no real roots, the denominator is in the form required by step 2. Steps 3 and 4 show that the partial fraction decomposition is of the form

$$\frac{5x^4 + x^3 + 5x^2 + 4x + 2}{x(x^2 + 1)^2} = \frac{A}{x} + \frac{Bx + C}{x^2 + 1} + \frac{Dx + E}{(x^2 + 1)^2}$$

Multiplying both sides by the common denominator $x(x^2 + 1)^2$ shows that

$$5x^4 + x^3 + 5x^2 + 4x + 2 = A(x^2 + 1)^2 + (Bx + C)x(x^2 + 1) + (Dx + E)x$$

$$= A(x^4 + 2x^2 + 1) + (Bx + C)(x^3 + x) + (Dx + E)x$$

$$= Ax^4 + 2Ax^2 + A + Bx^4 + Cx^3 + Bx^2 + Cx + Dx^2 + Ex$$

$$= (A + B)x^4 + Cx^3 + (2A + B + D)x^2 + (C + E)x + A$$

Consequently, we have this system of equations:
$$A + B = 5$$
$$C = 1$$
$$2A + B + D = 5$$
$$C + E = 4$$
$$A = 2$$

This system can be solved by Gaussian elimination, but it's much faster to note that $A = 2$ and $C = 1$ by the third and fifth equations. Hence by the first equation, $B = 5 - A = 5 - 2 = 3$. Similarly, $D = 5 - 2A - B = 5 - 2 \cdot 2 - 3 = -2$ by the third equation and $E = 4 - C = 4 - 1 = 3$ by the fourth equation. Therefore

$$\frac{5x^4 + x^3 + 5x^2 + 4x + 2}{x(x^2 + 1)^2} = \frac{2}{x} + \frac{3x + 1}{x^2 + 1} + \frac{-2x + 3}{(x^2 + 1)^2}. \blacksquare$$

EXERCISES

In Exercises 1–6, find the constants A, B, C.

1. $\dfrac{x}{(x + 1)(x + 2)} = \dfrac{A}{x + 1} + \dfrac{B}{x + 2}$

2. $\dfrac{1}{(x + 1)(x - 1)} = \dfrac{A}{x + 1} + \dfrac{B}{x - 1}$

3. $\dfrac{2x + 1}{(x + 2)(x - 3)^2} = \dfrac{A}{x + 2} + \dfrac{B}{x - 3} + \dfrac{C}{(x - 3)^2}$

4. $\dfrac{x^2 - x - 21}{(2x - 1)(x^2 + 4)} = \dfrac{A}{2x - 1} + \dfrac{Bx + C}{x^2 + 4}$

5. $\dfrac{5x^2 + 1}{(x + 1)(x^2 - x + 1)} = \dfrac{A}{x + 1} + \dfrac{Bx + C}{x^2 - x + 1}$

6. $\dfrac{x - 2}{(x + 4)(x^2 + 2x + 2)} = \dfrac{A}{x + 4} + \dfrac{Bx + C}{x^2 + 2x + 2}$

In Exercises 7–10, find the form of the partial fraction decomposition of the rational expression, as in Example 3. Do not find the constants.

7. $\dfrac{2x^5 + 5x^4 - 3x^2 + 2x + 1}{x(x - 1)(x - 2)^2(x + 3)^3}$

8. $\dfrac{x^7 - 12x^5 + 2x^4 - x^3 + 9}{(x - 2)^3(2x^2 + 2x + 3)^3}$

9. $\dfrac{3x^8 - 17x^6 + 11x^5 - 7x^3 + 4x - 13}{(x^2 + 1)^2(2x^2 + 2x + 1)^3}$

10. $\dfrac{3x^7 - 5x^5 + 3x + 2}{x^3(x - 3)(x + 2)^2(2x^2 + x + 3)^2(x^2 + 1)^3}$

In Exercises 11–38, find the partial fraction decomposition of the rational expression.

11. $\dfrac{5x - 1}{2x^2 - 5x - 3}$

12. $\dfrac{5x - 5}{3x^2 - 7x + 2}$

13. $\dfrac{x + 7}{x^2 - x - 6}$

14. $\dfrac{1}{x^2 + 4x - 5}$

15. $\dfrac{2x - 3}{x^2 + x}$

16. $\dfrac{4x^2 + 13x - 9}{x^3 + 2x^2 - 3x}$

17. $\dfrac{5x - 2}{x^3 - 4x}$

18. $\dfrac{x^2 + 4x + 1}{(x + 2)(x^2 - 4x + 3)}$

19. $\dfrac{-x^2 + 4x + 3}{(x - 1)^3}$

20. $\dfrac{3x - 10}{(x - 4)^2}$

21. $\dfrac{2x}{(x - 1)^3}$

22. $\dfrac{x^2 + 10x - 36}{x(x - 3)^2}$

23. $\dfrac{5x^2 + 20x + 6}{x^3 + 2x^2 + x}$

24. $\dfrac{8x^2 + 9x + 2}{(x^2 + x)(x + 1)}$

25. $\dfrac{4x^2 - 3x + 5}{(x - 1)(x^2 + x - 2)}$

26. $\dfrac{5x^2 + 15x + 7}{(x + 2)(x^2 + x - 2)}$

27. $\dfrac{4x^2 - x - 9}{(x^2 - 2x + 1)(x + 2)}$

28. $\dfrac{2x + 4}{(x - 5)(x^2 + 6x + 9)}$

29. $\dfrac{5x^3 - 3x^2 + 7x - 3}{(x^2 + 1)^2}$

30. $\dfrac{8x^3 + 13x}{(x^2 + 2)^2}$

31. $\dfrac{x^3 - 2x^2 + 2x - 4}{(x^2 + x + 1)^2}$

32. $\dfrac{2x^2 - x + 5}{(x^2 + x + 2)^2}$

33. $\dfrac{3x^2 + 4x + 4}{x^3 + 4x}$

34. $\dfrac{2x^2 + x + 1}{x^3 + x}$

35. $\dfrac{5x^2 + 2x + 2}{x^3 - 1}$

36. $\dfrac{x^2 - 5x - 2}{(x - 3)(x^2 + 1)}$

37. $\dfrac{2x}{(x^2 + 1)^2(x - 1)}$

38. $\dfrac{x^3 + x^2 + 3}{(x - 1)(x^2 + 2)^2}$

In Exercises 39–42, the degree of the numerator is greater than or equal to the degree of the denominator. A decomposition of a fraction $f(x)/g(x)$ of this type can be obtained as follows. First divide $f(x)$ by $g(x)$; denote the quotient by $q(x)$ and the remainder by $r(x)$. Then $r(x)$ has smaller degree than $g(x)$ and $f(x)/g(x) = q(x) + r(x)/g(x)$.* Use the procedure outlined in the text to find the partial fraction decomposition of $r(x)/g(x)$ and write $f(x)/g(x)$ as $q(x) +$ partial fractions.

39. $\dfrac{6x^3 - 11x^2 + 2x - 3}{3x^2 - 7x + 2}$

40. $\dfrac{6x^3 - 19x + 6x + 5}{2x^2 - 5x - 3}$

41. $\dfrac{x^2 + 4x + 2}{x^2 + x}$

42. $\dfrac{x^3 + 2}{x^2 + 3x + 2}$

In Exercises 43–46, a, b, and c are nonzero constants. Find the partial fraction decomposition of the rational expression.

43. $\dfrac{2x}{x^2 - a^2}$

44. $\dfrac{2x + a}{x^2 + ax}$

45. $\dfrac{(b + c)x + ab}{x^2 + ax}$

46. $\dfrac{(b + c)x + (b - c)a}{x^2 - a^2}$

8.3 MATRIX METHODS

Once you have solved several systems of linear equations by the elimination method, one fact becomes clear. The symbols used for the variables play no real role in the solution process, and a lot of time is wasted copying the x's, y's, z's, and so on, at each stage of the process. This fact suggests a shorthand system for representing a system of equations.

---

* *Reason:* By the Division Algorithm (page 247), $f(x) = g(x)q(x) + r(x)$. Dividing both sides by $g(x)$ shows that $f(x)/g(x) = q(x) + r(x)/g(x)$.

Example 1 This system of equations

$$\begin{aligned} x + 2y - z - 3w &= 2 \\ 6y + 3z - 4w &= 0 \\ 6x + 12y - 3z - 16w &= 3 \\ -5x + 2y + 15z + 10w &= -21 \end{aligned}$$

can be represented by the following rectangular array of numbers, consisting of the coefficients of the variables and the constants on the right side, arranged in the same order they appear in the system:

$$\begin{pmatrix} 1 & 2 & -1 & -3 & \vdots & 2 \\ 0 & 6 & 3 & -4 & \vdots & 0 \\ 6 & 12 & -3 & -16 & \vdots & 3 \\ -5 & 2 & 15 & 10 & \vdots & -21 \end{pmatrix}$$

This array is called the **augmented matrix*** of the system. Note that the second equation above has no x term, meaning that x has coefficient 0 in the equation. This is indicated in the matrix by the 0 at the beginning of the second row.

To solve this system by the elimination method, we begin by eliminating the x term in the third equation. We replace the third equation by the sum of itself and -6 times the first equation. In the matrix shorthand, where a horizontal row represents an equation, this operation is carried out on the rows of the matrix.

$$\begin{array}{rrrrr} -6 & -12 & 6 & 18 & -12 \\ 6 & 12 & -3 & -16 & 3 \\ \hline 0 & 0 & 3 & 2 & -9 \end{array} \quad \begin{array}{l} -6 \text{ times the first row} \\ \text{Third row} \\ \text{Sum of the third row and } -6 \text{ times the first row} \end{array}$$

Replacing the third row of the original matrix by this row yields the matrix

$$\begin{pmatrix} 1 & 2 & -1 & -3 & \vdots & 2 \\ 0 & 6 & 3 & -4 & \vdots & 0 \\ 0 & 0 & 3 & 2 & \vdots & -9 \\ -5 & 2 & 15 & 10 & \vdots & -21 \end{pmatrix}$$

We continue the solution process in the same manner, using the matrix shorthand to represent the system of equations. Replace the fourth row of the last matrix above by the sum of itself and 5 times the first row (this amounts to eliminating the x term from the last equation):

$$\begin{pmatrix} 1 & 2 & -1 & -3 & \vdots & 2 \\ 0 & 6 & 3 & -4 & \vdots & 0 \\ 0 & 0 & 3 & 2 & \vdots & -9 \\ 0 & 12 & 10 & -5 & \vdots & -11 \end{pmatrix}$$

---

* The plural of "matrix" is "matrices."

Next we must make various entries in the last row 0. (This amounts to eliminating the y and z terms from the last equation.)

Replace fourth row by the sum of itself and -2 times the second row:
$$\begin{pmatrix} 1 & 2 & -1 & -3 & \vdots & 2 \\ 0 & 6 & 3 & -4 & \vdots & 0 \\ 0 & 0 & 3 & 2 & \vdots & -9 \\ 0 & 0 & 4 & 3 & \vdots & -11 \end{pmatrix}$$

Multiply third row by $\frac{1}{3}$:
$$\begin{pmatrix} 1 & 2 & -1 & -3 & \vdots & 2 \\ 0 & 6 & 3 & -4 & \vdots & 0 \\ 0 & 0 & 1 & \frac{2}{3} & \vdots & -3 \\ 0 & 0 & 4 & 3 & \vdots & -11 \end{pmatrix}$$

Replace fourth row by the sum of itself and -4 times the third row:
$$\begin{pmatrix} 1 & 2 & -1 & -3 & \vdots & 2 \\ 0 & 6 & 3 & -4 & \vdots & 0 \\ 0 & 0 & 1 & \frac{2}{3} & \vdots & -3 \\ 0 & 0 & 0 & \frac{1}{3} & \vdots & 1 \end{pmatrix}$$

This last matrix is just shorthand notation for this echelon form system:

$$x + 2y - z - 3w = 2$$
$$6y + 3z - 4w = 0$$
$$z + \frac{2}{3}w = -3$$
$$\frac{1}{3}w = 1$$

The solution is now easily found to be $x = -3$, $y = 9/2$, $z = -5$, $w = 3$. ∎

The matrix notation in the example above is certainly more convenient than the equation notation. When matrix notation is used, we usually change our language to suit the situation. Instead of speaking of operations on equations in the system (such as multiplying by a nonzero constant), we speak of **row operations** on the matrix. Similarly, the solution process ends when we obtain an **echelon form matrix** (such as the last matrix shown above).

Example 2 The homogeneous system

$$4x + 12y - 16z = 0$$
$$3x + 4y + 3z = 0$$
$$x + 8y - 19z = 0$$

has augmented matrix

$$\begin{pmatrix} 4 & 12 & -16 & \vdots & 0 \\ 3 & 4 & 3 & \vdots & 0 \\ 1 & 8 & -19 & \vdots & 0 \end{pmatrix}$$

The last vertical column of this matrix consists entirely of zeros. Furthermore, any matrix obtained from this one by the usual row operations will have this same property. Interchanging two rows, or multiplying a row by a constant, or replacing a row by the sum of itself and a constant multiple of another row, will always result in rows with last entry 0. Consequently, when dealing with homogeneous systems such as this, there is no need to write out this last column of zeros at every stage. Instead we need only deal with the **coefficient matrix**

$$\begin{pmatrix} 4 & 12 & -16 \\ 3 & 4 & 3 \\ 1 & 8 & -19 \end{pmatrix}$$

instead of the augmented matrix of the system given above. Using the matrix of coefficients, we proceed as before to reduce it to echelon form:

Multiply first row by $\frac{1}{4}$:
$$\begin{pmatrix} 1 & 3 & -4 \\ 3 & 4 & 3 \\ 1 & 8 & -19 \end{pmatrix}$$

Replace second row by the sum of itself and -3 times the first row:
$$\begin{pmatrix} 1 & 3 & -4 \\ 0 & -5 & 15 \\ 1 & 8 & -19 \end{pmatrix}$$

Replace third row by the sum of itself and -1 times the first row:
$$\begin{pmatrix} 1 & 3 & -4 \\ 0 & -5 & 15 \\ 0 & 5 & -15 \end{pmatrix}$$

Replace third row by the sum of itself and 1 times the second row:
$$\begin{pmatrix} 1 & 3 & -4 \\ 0 & -5 & 15 \\ 0 & 0 & 0 \end{pmatrix}$$

This echelon form matrix represents the following homogeneous system:

$$x + 3y - 4z = 0$$
$$-5y + 15z = 0$$

We don't bother to write out the trivial third equation $0x + 0y + 0z = 0$ since it is satisfied by *all* real numbers. The last equation above is equivalent to $y = 3z$ and hence has infinitely many solutions. For each real number t, there is the solution $z = t$, $y = 3t$. Substituting these values in the first equation shows that $x = -3y + 4z = -9t + 4t = -5t$. Therefore the solutions of this dependent system are given by:

$$x = -5t, \quad y = 3t, \quad z = t \quad (t \text{ any real number}) \blacksquare$$

8.3 MATRIX METHODS

Example 3 To solve the system

$$x + y + 2z = 1$$
$$2x + 4y + 5z = 2$$
$$3x + 5y + 7z = 2$$

we form the augmented matrix and reduce it to echelon form via row operations:

$$\begin{pmatrix} 1 & 1 & 2 & \vdots & 1 \\ 2 & 4 & 5 & \vdots & 2 \\ 3 & 5 & 7 & \vdots & 2 \end{pmatrix}$$

Replace second row by the sum of itself and -2 times first row:

$$\begin{pmatrix} 1 & 1 & 2 & \vdots & 1 \\ 0 & 2 & 1 & \vdots & 0 \\ 3 & 5 & 7 & \vdots & 2 \end{pmatrix}$$

Replace third row by the sum of itself and -3 times first row:

$$\begin{pmatrix} 1 & 1 & 2 & \vdots & 1 \\ 0 & 2 & 1 & \vdots & 0 \\ 0 & 2 & 1 & \vdots & -1 \end{pmatrix}$$

Replace third row by the sum of itself and -1 times second row:

$$\begin{pmatrix} 1 & 1 & 2 & \vdots & 1 \\ 0 & 2 & 1 & \vdots & 0 \\ 0 & 0 & 0 & \vdots & -1 \end{pmatrix}$$

The last row of this matrix is just shorthand for the equation

$$0x + 0y + 0z = -1$$

Since this equation obviously has no solutions, the original system is inconsistent. ∎

EXERCISES

In Exercises 1–6, write the coefficient matrix and the augmented matrix of the system.

1. $2x - 3y + 4z = 1$
   $x + 2y - 6z = 0$
   $3x - 7y + 4z = -3$

2. $x + 2y - 3w + 7z = -5$
   $2x - y + 2z = 4$
   $3x + 7w - 6z = 0$

3. $x - \frac{1}{2}y + \frac{7}{4}z = 0$
   $2x - \frac{3}{2}y + 5z = 0$
   $-2y + \frac{1}{3}z = 0$

4. $2x - \frac{1}{2}y + \frac{7}{2}w - 6z = 1$
   $\frac{1}{4}x - 6y + 2w - z = 2$
   $4y - \frac{1}{2}w + z = 3$
   $2x + 3y + \frac{1}{2}z = 4$

5. $17x - 107y + 5z = \sqrt{2}$
   $7x - 306y + 194z = 0$

6. $7u + 4v = 0$
   $8u - 6v = -1$
   $12u + \frac{1}{2}v = 4$

In Exercises 7–12, the augmented matrix of a system of equations is given. Express the system in ordinary notation.

7. $\begin{pmatrix} 2 & -3 & | & 1 \\ 4 & 7 & | & 2 \end{pmatrix}$ 8. $\begin{pmatrix} 2 & 3 & 5 & | & 2 \\ 1 & 6 & 9 & | & 0 \end{pmatrix}$

9. $\begin{pmatrix} 1 & 0 & 1 & 0 & | & 1 \\ 1 & -1 & 4 & -2 & | & 3 \\ 4 & 2 & 5 & 0 & | & 2 \end{pmatrix}$

10. $\begin{pmatrix} 1 & 7 & 0 & | & 4 \\ 2 & 3 & 1 & | & 6 \\ -1 & 0 & 2 & | & 3 \end{pmatrix}$

11. $\begin{pmatrix} 1 & 2 & 3 & -4 & | & 1 \\ 0 & 3 & 5 & 7 & | & 9 \\ 8 & -7 & 0 & 0 & | & 2 \\ 4 & 3 & 2 & -1 & | & 0 \end{pmatrix}$

12. $\begin{pmatrix} 0 & 1 & -4 & 3 & | & 0 \\ 6 & 2 & 0 & -1 & | & 0 \\ 0 & 0 & 4 & 3 & | & 0 \\ 1 & 0 & 2 & 0 & | & 0 \end{pmatrix}$

In Exercises 13–32, use matrix methods to solve the system.

13. $\begin{aligned} 3x - 2y + 6z &= -6 \\ -3x + 10y + 11z &= 13 \\ x - 2y - z &= -3 \end{aligned}$

14. $\begin{aligned} x + 2y - 3z &= 9 \\ 3x - y - 4z &= 3 \\ 2x - y + 2z &= -8 \end{aligned}$

15. $\begin{aligned} x - y - z &= -5 \\ 2x - 3y - 8z &= -33 \\ x - 2y - 8z &= -32 \end{aligned}$

16. $\begin{aligned} x + 3y + 10z &= -8 \\ -x - 2y - 5z &= 3 \\ 2x + 4y + 5z &= -6 \end{aligned}$

17. $\begin{aligned} 2x - y + z &= 4 \\ 2x + 2y + 3z &= 3 \\ 6x - 9y - 2z &= 17 \end{aligned}$

18. $\begin{aligned} -3x - 2y + z &= 0 \\ x - \tfrac{1}{3}y + 3z &= -2 \\ 3x - y &= -3 \end{aligned}$

19. $\begin{aligned} -8x - 9y &= 11 \\ 24x + 34y &= 2 \\ 16x + 11y &= -57 \end{aligned}$

20. $\begin{aligned} 2x + y &= 7 \\ x - y &= 3 \\ x + 3y &= 4 \end{aligned}$

21. $\begin{aligned} x - 4y - 13z &= 4 \\ x - 2y - 3z &= 2 \\ -3x + 5y + 4z &= 2 \end{aligned}$

22. $\begin{aligned} 2x - 4y + z &= 3 \\ x + 3y - 7z &= 1 \\ -2x + 4y - z &= 10 \end{aligned}$

23. $\begin{aligned} 4x + y + 3z &= 7 \\ x - y + 2z &= 3 \\ 3x + 2y + z &= 4 \end{aligned}$

24. $\begin{aligned} x + 4y + z &= 3 \\ -x + 2y + 2z &= 0 \\ 2x + 2y - z &= 3 \end{aligned}$

25. $\begin{aligned} x + y + z &= 0 \\ 3x - y + z &= 0 \\ -5x - y + z &= 0 \end{aligned}$

26. $\begin{aligned} x + y + z &= 0 \\ x - y - z &= 0 \\ x - y + z &= 0 \end{aligned}$

27. $\begin{aligned} 2x + y + 3z - 2w &= -6 \\ 4x + 3y + z - w &= -2 \\ x + y + z + w &= -5 \\ -2x - 2y + 2z + 2w &= -10 \end{aligned}$

28. $\begin{aligned} x + y + z + w &= -1 \\ -x + 4y + z - w &= 0 \\ x - 2y + z - 2w &= 11 \\ -x - 2y + z + 2w &= -3 \end{aligned}$

29. $\begin{aligned} x - 2y - z - 3w &= -3 \\ -x + y + z &= 0 \\ 4y + 3z - 2w &= -1 \\ 2x - 2y + w &= 1 \end{aligned}$

30. $\begin{aligned} 3x - y + 2z &= 0 \\ -x + 3y + 2z + 5w &= 0 \\ x + 2y + 5z - 4w &= 0 \\ 2x - y + 3w &= 0 \end{aligned}$

31. $\begin{aligned} x - y + 2z + 3w &= 0 \\ 3x - y + 7z + 7w &= 0 \\ 2x - 2y + 5z + 10w &= 0 \\ x - y + 3z + 7w &= 0 \end{aligned}$

32. $\begin{aligned} x - 3y + 4z + w + 2v &= -2 \\ 2x - 2y - z + 2w + 3v &= 7 \\ -x + y - 2z - 2w + v &= 0 \\ -2x - y + 2z + w - v &= 5 \\ 2x - 2y + z + 2w + 2v &= 3 \end{aligned}$

33. Peanuts cost $3 per pound, almonds $4 per pound, and cashews $8 per pound. How many pounds of each should be used to produce 140 pounds of a mixture costing $6 pound, in which there are twice as many peanuts as almonds?

34. A stereo equipment manufacturer produces three models of speakers, R, S, and T, and has three kinds of delivery vehicles: trucks, vans, and station wagons. A truck holds 2 boxes of model R, 1 of model S, and 3 of model T. A van holds 1 box of model R, 3 of model S, and 2 of model T. A station wagon holds 1 box of model R, 3 of model S, and 1 of model T. If 15 boxes of model R, 20 boxes of model S, and 22 boxes of model T are to be delivered, how many vehicles of each type should be used so that all operate at full capacity?

35. A furniture manufacturer has 1950 machine hours available each week in the cutting department, 1490 hours in the assembly department, and 2160 in the finishing department. Manufacturing a chair requires 0.2 hours of cutting, 0.3 hours of assembly, and 0.1 hours of finishing. A chest requires 0.5 hours of cutting, 0.4 hours of assembly, and 0.6 hours of finishing. A table requires 0.3 hours of cutting, 0.1 hours of assembly, and 0.4 hours of finishing. How many chairs, chests, and tables should be produced in order to use all the available production capacity?

8.3.A EXCURSION: Matrix Algebra

Until now, we have used matrices only as a convenient shorthand for dealing with systems of equations. Matrices appear in many other areas of mathematics as well. In this section we present the essential facts about the algebra of matrices.

Let m and n be positive integers. An $m \times n$ **matrix** (read "m by n matrix") is a rectangular array of numbers, with m horizontal rows and n vertical columns. For example,

$$\begin{pmatrix} 3 & 2 & -5 \\ 6 & 1 & 7 \\ -2 & 5 & 0 \end{pmatrix} \qquad \begin{pmatrix} -3 & 4 \\ 2 & 0 \\ 0 & 1 \\ 7 & 3 \\ 1 & -6 \end{pmatrix} \qquad \begin{pmatrix} 3 & 0 & 1 & 0 \\ \sqrt{2} & -\tfrac{1}{2} & 4 & \tfrac{8}{3} \\ 10 & 2 & -\tfrac{3}{4} & 12 \end{pmatrix} \qquad \begin{pmatrix} \sqrt{3} \\ 2 \\ 0 \\ 11 \end{pmatrix}$$

3×3 matrix
3 rows
3 columns

5×2 matrix
5 rows
2 columns

3×4 matrix
3 rows
4 columns

4×1 matrix
4 rows
1 column

In a matrix, the *rows* are horizontal and are numbered from top to bottom. The *columns* are vertical and are numbered from left to right. For example,

$$\begin{array}{r} \text{Row 1} \longrightarrow \\ \text{Row 2} \longrightarrow \\ \text{Row 3} \longrightarrow \end{array} \begin{pmatrix} 11 & 3 & 14 \\ -2 & 0 & -5 \\ \tfrac{1}{3} & 6 & 7 \end{pmatrix}$$

$$\uparrow \qquad \uparrow \qquad \uparrow$$
Column 1 Column 2 Column 3

Each entry in a matrix can be located by stating the row and column in which it appears. For instance, in the 3×3 matrix above, 14 is the entry in row 1, column 3, and 0 is the entry in row 2, column 2.

An arbitrary matrix may be denoted by a capital letter, such as A, or by an array:

$$\begin{pmatrix} a_{11} & a_{12} & a_{13} & \cdots & a_{1n} \\ a_{21} & a_{22} & a_{23} & \cdots & a_{2n} \\ a_{31} & a_{32} & a_{33} & \cdots & a_{3n} \\ \vdots & \vdots & \vdots & & \vdots \\ a_{m1} & a_{m2} & a_{m3} & \cdots & a_{mn} \end{pmatrix}$$

in which a_{ij} denotes the entry in row i and column j. This array notation is usually abbreviated as (a_{ij}).

Two matrices are said to be **equal** if they have the same size (same number of rows and same number of columns) and the corresponding entries are equal. For example,

$$\begin{pmatrix} -3 & (-1)^2 & \frac{6}{8} \\ 6 & 12 & 0 \end{pmatrix} = \begin{pmatrix} -3 & 1 & \frac{3}{4} \\ \sqrt{36} & 3 \cdot 4 & 0 \end{pmatrix}$$

But

$$\begin{pmatrix} 6 & 4 \\ 5 & 1 \end{pmatrix} \neq \begin{pmatrix} 6 & 5 \\ 4 & 1 \end{pmatrix}$$

because the corresponding entries in row 1, column 2, aren't equal ($4 \neq 5$), and similarly in row 2, column 1. More formally,

EQUALITY OF MATRICES

> If $A = (a_{ij})$ and $B = (b_{ij})$ are $m \times n$ matrices,
> then $A = B$ means $a_{ij} = b_{ij}$ for every i and j.

Addition (or **subtraction**) of matrices is defined for any two matrices of the same size by the following rule: Add (or subtract) the corresponding entries. For example,

$$\begin{pmatrix} 1 & 2 \\ 3 & 0 \\ -7 & 4 \end{pmatrix} + \begin{pmatrix} 5 & 8 \\ -3 & 7 \\ 4 & 5 \end{pmatrix} = \begin{pmatrix} 1+5 & 2+8 \\ 3+(-3) & 0+7 \\ -7+4 & 4+5 \end{pmatrix} = \begin{pmatrix} 6 & 10 \\ 0 & 7 \\ -3 & 9 \end{pmatrix}$$

and

$$\begin{pmatrix} 6 & -3 \\ -2 & 8 \end{pmatrix} - \begin{pmatrix} 1 & 4 \\ 5 & -4 \end{pmatrix} = \begin{pmatrix} 6-1 & -3-4 \\ -2-5 & 8-(-4) \end{pmatrix} = \begin{pmatrix} 5 & -7 \\ -7 & 12 \end{pmatrix}$$

In formal terms,

MATRIX ADDITION

> The sum of the $m \times n$ matrices $A = (a_{ij})$ and $B = (b_{ij})$ is the $m \times n$ matrix $A + B$ whose entry in row i, column j, is
>
> $$a_{ij} + b_{ij}$$

The sum of two matrices of different sizes is not defined.

The product of a single number c and a matrix is defined by this rule: Multiply every entry in the matrix by c. For example,

$$2\begin{pmatrix} 3 & -4 & 0 \\ 1 & 5 & 2 \end{pmatrix} = \begin{pmatrix} 2\cdot 3 & 2(-4) & 2\cdot 0 \\ 2\cdot 1 & 2\cdot 5 & 2\cdot 2 \end{pmatrix} = \begin{pmatrix} 6 & -8 & 0 \\ 2 & 10 & 4 \end{pmatrix}$$

The process of multiplying a number by a matrix is called **scalar multiplication**, and the number is sometimes called a **scalar**. Here is the formal definition:

SCALAR MULTIPLICATION

> The product of an $m \times n$ matrix $A = (a_{ij})$ and a scalar c is the $m \times n$ matrix cA whose entry in row i, column j, is ca_{ij}

Example 1 Show that $3(A + B) = 3A + 3B$, where

$$A = \begin{pmatrix} 2 & -1 \\ 0 & 4 \end{pmatrix} \quad \text{and} \quad B = \begin{pmatrix} -3 & 2 \\ 5 & 7 \end{pmatrix}.$$

Solution

$$A + B = \begin{pmatrix} 2 & -1 \\ 0 & 4 \end{pmatrix} + \begin{pmatrix} -3 & 2 \\ 5 & 7 \end{pmatrix} = \begin{pmatrix} 2 + (-3) & -1 + 2 \\ 0 + 5 & 4 + 7 \end{pmatrix} = \begin{pmatrix} -1 & 1 \\ 5 & 11 \end{pmatrix}$$

so that

$$3(A + B) = 3\begin{pmatrix} -1 & 1 \\ 5 & 11 \end{pmatrix} = \begin{pmatrix} 3(-1) & 3\cdot 1 \\ 3\cdot 5 & 3\cdot 11 \end{pmatrix} = \begin{pmatrix} -3 & 3 \\ 15 & 33 \end{pmatrix}.$$

On the other hand,

$$3A = 3\begin{pmatrix} 2 & -1 \\ 0 & 4 \end{pmatrix} = \begin{pmatrix} 3\cdot 2 & 3(-1) \\ 3\cdot 0 & 3\cdot 4 \end{pmatrix} = \begin{pmatrix} 6 & -3 \\ 0 & 12 \end{pmatrix}$$

and

$$3B = 3\begin{pmatrix} -3 & 2 \\ 5 & 7 \end{pmatrix} = \begin{pmatrix} 3(-3) & 3\cdot 2 \\ 3\cdot 5 & 3\cdot 7 \end{pmatrix} = \begin{pmatrix} -9 & 6 \\ 15 & 21 \end{pmatrix}$$

so that

$$3A + 3B = \begin{pmatrix} 6 & -3 \\ 0 & 12 \end{pmatrix} + \begin{pmatrix} -9 & 6 \\ 15 & 21 \end{pmatrix} = \begin{pmatrix} -3 & 3 \\ 15 & 33 \end{pmatrix}.$$

Therefore $3(A + B) = 3A + 3B$. ∎

The preceding example is an illustration of the third property on the following list; the others are illustrated in Exercises 45–52.

PROPERTIES OF MATRIX ADDITION AND SCALAR MULTIPLICATION

For any $m \times n$ matrices A, B, C and numbers c, d:

1. $A + B = B + A$
2. $A + (B + C) = (A + B) + C$
3. $c(A + B) = cA + cB$
4. $(cd)A = c(dA)$
5. $(c + d)A = cA + dA$

The simplest case of matrix multiplication is the product of a single row and a single column that have the same number of entries. This is done by multiplying the corresponding entries (first by first, second by second, and so on) and then adding the results. For example,

$$(1 \quad 2 \quad 3)\begin{pmatrix} 4 \\ 5 \\ 6 \end{pmatrix} = \underbrace{1 \cdot 4}_{\text{First terms}} + \underbrace{2 \cdot 5}_{\text{Second terms}} + \underbrace{3 \cdot 6}_{\text{Third terms}} = 32$$

Note that the product of a row and a column is a *single number*.

Now let A be an $m \times n$ matrix and B an $n \times p$ matrix, so that the number of columns of A is the same as the number of rows of B (namely, n). The product matrix AB is defined to be an $m \times p$ matrix (same number of rows as A and same number of columns as B). The entry in row i, column j, of AB is this number:

the product of row i of A and column j of B.

For example, if

$$A = \begin{pmatrix} 3 & 1 & 2 \\ -1 & 0 & 4 \end{pmatrix} \quad \text{and} \quad B = \begin{pmatrix} 2 & -3 & 0 & 1 \\ 0 & 5 & 2 & 7 \\ 1 & 8 & -4 & 1 \end{pmatrix}$$

then A has 3 columns and B has 3 rows. So the product matrix AB is defined. AB has 2 rows (same as A) and 4 columns (same as B). Its entries are calculated as follows.

Position	Computation of Entry in AB	Result
row 1, column 1	product of row 1 of A and column 1 of B $$(3 \quad 1 \quad 2)\begin{pmatrix} 2 \\ 0 \\ 1 \end{pmatrix} = 3 \cdot 2 + 1 \cdot 0 + 2 \cdot 1 = 8:$$	$\begin{pmatrix} 8 & & & \\ & & & \end{pmatrix}$

Position	Computation of Entry in AB	Result
row 1, column 2	product of row 1 of A and column 2 of B $$(3 \ 1 \ 2)\begin{pmatrix}-3\\5\\8\end{pmatrix} = 3(-3)+1\cdot 5+2\cdot 8 = 12:$$	$\begin{pmatrix}8 & 12 & & \\ & & & \end{pmatrix}$
row 1, column 3	product of row 1 of A and column 3 of B $$(3 \ 1 \ 2)\begin{pmatrix}0\\2\\-4\end{pmatrix} = 3\cdot 0+1\cdot 2+2(-4) = -6:$$	$\begin{pmatrix}8 & 12 & -6 & \\ & & & \end{pmatrix}$
row 1, column 4	product of row 1 of A and column 4 of B $$(3 \ 1 \ 2)\begin{pmatrix}1\\7\\1\end{pmatrix} = 3\cdot 1+1\cdot 7+2\cdot 1 = 12:$$	$\begin{pmatrix}8 & 12 & -6 & 12\\ & & & \end{pmatrix}$
row 2, column 1	product of row 2 of A and column 1 of B $$(-1 \ 0 \ 4)\begin{pmatrix}2\\0\\1\end{pmatrix} = (-1)2+0\cdot 0+4\cdot 1 = 2:$$	$\begin{pmatrix}8 & 12 & -6 & 12\\ 2 & & & \end{pmatrix}$
row 2, column 2	product of row 2 of A and column 2 of B $$(-1 \ 0 \ 4)\begin{pmatrix}-3\\5\\8\end{pmatrix} = (-1)(-3)+0\cdot 5+4\cdot 8 = 35:$$	$\begin{pmatrix}8 & 12 & -6 & 12\\ 2 & 35 & & \end{pmatrix}$
row 2, column 3	product of row 2 of A and column 3 of B $$(-1 \ 0 \ 4)\begin{pmatrix}0\\2\\-4\end{pmatrix} = (-1)\cdot 0+0\cdot 2+4(-4) = -16:$$	$\begin{pmatrix}8 & 12 & -6 & 12\\ 2 & 35 & -16 & \end{pmatrix}$

row 2,
column 4

product of row 2 of A
and column 4 of B

$$(-1 \ 0 \ 4)\begin{pmatrix}1\\7\\1\end{pmatrix} = (-1)\cdot 1 + 0\cdot 7 + 4\cdot 1 = 3:\qquad \begin{pmatrix}8 & 12 & -6 & 12\\ 2 & 35 & -16 & 3\end{pmatrix}$$

Therefore

$$AB = \begin{pmatrix}3 & 1 & 2\\ -1 & 0 & 4\end{pmatrix}\begin{pmatrix}2 & -3 & 0 & 1\\ 0 & 5 & 2 & 7\\ 1 & 8 & -4 & 1\end{pmatrix} = \begin{pmatrix}8 & 12 & -6 & 12\\ 2 & 35 & -16 & 3\end{pmatrix}$$

Row i of a matrix (a_{ij}) consists of the elements a_{i1}, a_{i2}, a_{i3}, etc., and column j of a matrix (b_{ij}) consists of b_{1j}, b_{2j}, b_{3j}, etc. So we have this formal definition:

MATRIX MULTIPLICATION

> The product of an $m \times n$ matrix $A = (a_{ij})$ and an $n \times p$ matrix $B = (b_{ij})$ is the $m \times p$ matrix AB whose entry in row i, column j, is
>
> $$a_{i1}b_{1j} + a_{i2}b_{2j} + a_{i3}b_{3j} + \cdots + a_{in}b_{nj}$$

The product of matrices A and B is not defined if the number of columns of A is different from the number of rows of B.

Matrix multiplication shares some of the properties of ordinary multiplication of numbers. In particular, the associative and distributive laws are valid:*

$$A(BC) = (AB)C \qquad A(B + C) = AB + AC \qquad (B + C)A = BA + CA$$

But there is one very important difference between matrix multiplication and multiplication of numbers. If A and B are matrices such that the product AB is defined, the product BA may not be defined, and even if it is, AB may not be equal to BA.

Example 2

$$\begin{pmatrix}-3 & -9\\ 2 & 6\end{pmatrix}\begin{pmatrix}4 & 6\\ 2 & 3\end{pmatrix} = \begin{pmatrix}(-3)4 + (-9)2 & (-3)6 + (-9)3\\ 2\cdot 4 + 6\cdot 2 & 2\cdot 6 + 6\cdot 3\end{pmatrix} = \begin{pmatrix}-30 & -45\\ 20 & 30\end{pmatrix}.$$

But when the order of the factors is reversed, the answer is different:

$$\begin{pmatrix}4 & 6\\ 2 & 3\end{pmatrix}\begin{pmatrix}-3 & -9\\ 2 & 6\end{pmatrix} = \begin{pmatrix}4(-3) + 6\cdot 2 & 4(-9) + 6\cdot 6\\ 2(-3) + 3\cdot 2 & 2(-9) + 3\cdot 6\end{pmatrix} = \begin{pmatrix}0 & 0\\ 0 & 0\end{pmatrix}.\blacksquare$$

A **zero matrix** is a matrix in which every entry is 0. It is easy to see that for any matrix A and the zero matrix $\mathbf{0}$ of the same size, $A + \mathbf{0} = A = \mathbf{0} + A$.

---

* Provided that the matrices are of appropriate sizes so that the sums and products listed here are defined.

Similarly, the product of a zero matrix with any matrix is another zero matrix. The last example shows that *the product of nonzero matrices may be a zero matrix*. This is quite different from the situation with numbers, in which a product is zero only when one of the factors is zero.

The $n \times n$ **identity matrix** I_n is the matrix with 1's on its main diagonal* and 0's everywhere else; for example:

$$I_2 = \begin{pmatrix} 1 & 0 \\ 0 & 1 \end{pmatrix} \qquad I_3 = \begin{pmatrix} 1 & 0 & 0 \\ 0 & 1 & 0 \\ 0 & 0 & 1 \end{pmatrix} \qquad I_4 = \begin{pmatrix} 1 & 0 & 0 & 0 \\ 0 & 1 & 0 & 0 \\ 0 & 0 & 1 & 0 \\ 0 & 0 & 0 & 1 \end{pmatrix}$$

The number 1 is the multiplicative identity of the number system because $a \cdot 1 = a = 1 \cdot a$ for every number a. The identity matrix I_n is the multiplicative identity for $n \times n$ matrices:

IDENTITY MATRIX

> For any $n \times n$ matrix A, $AI_n = A = I_n A$.

For example, in the 2×2 case

$$\begin{pmatrix} a & b \\ c & d \end{pmatrix} \begin{pmatrix} 1 & 0 \\ 0 & 1 \end{pmatrix} = \begin{pmatrix} a \cdot 1 + b \cdot 0 & a \cdot 0 + b \cdot 1 \\ c \cdot 1 + d \cdot 0 & c \cdot 0 + d \cdot 1 \end{pmatrix} = \begin{pmatrix} a & b \\ c & d \end{pmatrix}$$

Verify that the same answer results if you reverse the order of multiplication.

Every nonzero number c has a multiplicative inverse $c^{-1} = 1/c$ with the property that $cc^{-1} = 1$. The analogous statement for matrix multiplication does not always hold, and special terminology is used when it does. An $n \times n$ matrix A is said to be **invertible** (or **nonsingular**) if there is an $n \times n$ matrix B such that $AB = I_n$. In this case it can be proved that $BA = I_n$ also. The matrix B is called the **inverse** of A and is sometimes denoted A^{-1}.

Example 3 You can readily verify that

$$\begin{pmatrix} 2 & 1 \\ 3 & 1 \end{pmatrix} \begin{pmatrix} -1 & 1 \\ 3 & -2 \end{pmatrix} = \begin{pmatrix} 1 & 0 \\ 0 & 1 \end{pmatrix} = \begin{pmatrix} -1 & 1 \\ 3 & -2 \end{pmatrix} \begin{pmatrix} 2 & 1 \\ 3 & 1 \end{pmatrix}$$

Therefore $A = \begin{pmatrix} 2 & 1 \\ 3 & 1 \end{pmatrix}$ is an invertible matrix with inverse $A^{-1} = \begin{pmatrix} -1 & 1 \\ 3 & -2 \end{pmatrix}$. ∎

Example 4 Find the inverse of the matrix $\begin{pmatrix} 2 & 6 \\ 1 & 4 \end{pmatrix}$.

---

* The **main diagonal** of an $n \times n$ matrix (a_{ij}) runs from the upper left-hand corner to the lower right-hand corner and consists of the numbers $a_{11}, a_{22}, a_{33}, \ldots, a_{nn}$.

Solution We must find numbers x, y, u, v such that

$$\begin{pmatrix} 2 & 6 \\ 1 & 4 \end{pmatrix} \begin{pmatrix} x & u \\ y & v \end{pmatrix} = \begin{pmatrix} 1 & 0 \\ 0 & 1 \end{pmatrix}$$

which is the same as

$$\begin{pmatrix} 2x + 6y & 2u + 6v \\ x + 4y & u + 4v \end{pmatrix} = \begin{pmatrix} 1 & 0 \\ 0 & 1 \end{pmatrix}.$$

Since corresponding entries in these last two matrices are equal, finding x, y, u, v amounts to solving these systems of equations:

$$\begin{matrix} 2x + 6y = 1 \\ x + 4y = 0 \end{matrix} \quad \text{and} \quad \begin{matrix} 2u + 6v = 0 \\ u + 4v = 1 \end{matrix}$$

We shall solve the systems by the matrix methods of Section 8.3. Since the coefficient matrices of the two systems are the same, we shall use the following matrix, whose first three columns form the augmented matrix of the first system and whose first two and last columns form the augmented matrix of the second system:

$$\left(\begin{array}{cc|cc} 2 & 6 & 1 & 0 \\ 1 & 4 & 0 & 1 \end{array} \right)$$

Performing row operations on this matrix amounts to simultaneously doing the operations on the two augmented matrices:

Multiply row 1 by $\frac{1}{2}$:
$$\left(\begin{array}{cc|cc} 1 & 3 & \frac{1}{2} & 0 \\ 1 & 4 & 0 & 1 \end{array} \right)$$

Replace row 2 by the sum of itself and -1 times row 1:
$$\left(\begin{array}{cc|cc} 1 & 3 & \frac{1}{2} & 0 \\ 0 & 1 & -\frac{1}{2} & 1 \end{array} \right)$$

Replace row 1 by the sum of itself and -3 times row 2:
$$\left(\begin{array}{cc|cc} 1 & 0 & 2 & -3 \\ 0 & 1 & -\frac{1}{2} & 1 \end{array} \right)$$

The first three columns of the last matrix show that $x = 2$ and $y = -\frac{1}{2}$. Similarly, the first two and last columns show that $u = -3$ and $v = 1$. Therefore

$$A^{-1} = \begin{pmatrix} 2 & -3 \\ -\frac{1}{2} & 1 \end{pmatrix}$$

Observe that A^{-1} is just the right half of the final form of the augmented matrix above and that the left half is the identity matrix I_2. ∎

The technique used in the preceding example can be used to find the inverse of any matrix that has one. To find the inverse of the $n \times n$ matrix A,

1. Form the matrix $(A|I_n)$ whose left half is A and whose right half is the identity matrix I_n.

2. Use row operations on the entire matrix until the left side has been transformed into the identity matrix I_n.*
3. Then the right side will be the matrix A^{-1}.

If a row of zeros occurs in the left half during step 2, then the left half cannot be transformed into the identity matrix and A does not have an inverse.

Example 5 Find the inverse of $A = \begin{pmatrix} 1 & 1 & 1 \\ 2 & 3 & 0 \\ 1 & 2 & 1 \end{pmatrix}$.

Solution

$$\left(\begin{array}{ccc|ccc} 1 & 1 & 1 & 1 & 0 & 0 \\ 2 & 3 & 0 & 0 & 1 & 0 \\ 1 & 2 & 1 & 0 & 0 & 1 \end{array}\right)$$

Replace row 2 by the sum of itself and -2 times row 1:
Replace row 3 by the sum of itself and -1 times row 1:

$$\left(\begin{array}{ccc|ccc} 1 & 1 & 1 & 1 & 0 & 0 \\ 0 & 1 & -2 & -2 & 1 & 0 \\ 0 & 1 & 0 & -1 & 0 & 1 \end{array}\right)$$

Replace row 1 by the sum of itself and -1 times row 3:
Replace row 2 by the sum of itself and -1 times row 3:

$$\left(\begin{array}{ccc|ccc} 1 & 0 & 1 & 2 & 0 & -1 \\ 0 & 0 & -2 & -1 & 1 & -1 \\ 0 & 1 & 0 & -1 & 0 & 1 \end{array}\right)$$

Multiply row 2 by $-\tfrac{1}{2}$ and interchange rows 2 and 3:

$$\left(\begin{array}{ccc|ccc} 1 & 0 & 1 & 2 & 0 & -1 \\ 0 & 1 & 0 & -1 & 0 & 1 \\ 0 & 0 & 1 & \tfrac{1}{2} & -\tfrac{1}{2} & \tfrac{1}{2} \end{array}\right)$$

Replace row 1 by the sum of itself and -1 times row 3:

$$\left(\begin{array}{ccc|ccc} 1 & 0 & 0 & \tfrac{3}{2} & \tfrac{1}{2} & -\tfrac{3}{2} \\ 0 & 1 & 0 & -1 & 0 & 1 \\ 0 & 0 & 1 & \tfrac{1}{2} & -\tfrac{1}{2} & \tfrac{1}{2} \end{array}\right)$$

Therefore $A^{-1} = \begin{pmatrix} \tfrac{3}{2} & \tfrac{1}{2} & -\tfrac{3}{2} \\ -1 & 0 & 1 \\ \tfrac{1}{2} & -\tfrac{1}{2} & \tfrac{1}{2} \end{pmatrix} = \tfrac{1}{2}\begin{pmatrix} 3 & 1 & -3 \\ -2 & 0 & 2 \\ 1 & -1 & 1 \end{pmatrix}$. ■

---

* Just as in the example, this amounts to simultaneously solving several systems of linear equations (with the same coefficient matrix) to find the entries in the inverse matrix.

EXERCISES

In Exercises 1–8, compute $A + B$, AB, BA, and $2A - 3B$.

1. $A = \begin{pmatrix} 3 & 2 \\ 5 & 1 \end{pmatrix}$, $B = \begin{pmatrix} 7 & -5 \\ -2 & 6 \end{pmatrix}$

2. $A = \begin{pmatrix} -6 & 2 \\ 7 & -1 \end{pmatrix}$, $B = \begin{pmatrix} -8 & 4 \\ 2 & 7 \end{pmatrix}$

3. $A = \begin{pmatrix} \frac{3}{2} & 2 \\ 4 & \frac{7}{2} \end{pmatrix}$, $B = \begin{pmatrix} \frac{1}{2} & -\frac{3}{2} \\ \frac{5}{2} & 1 \end{pmatrix}$

4. $A = \begin{pmatrix} \frac{3}{4} & 7 \\ 6 & -\frac{5}{4} \end{pmatrix}$, $B = \begin{pmatrix} \frac{1}{2} & 3 \\ -5 & \frac{3}{2} \end{pmatrix}$

5. $A = \begin{pmatrix} 0 & 1 \\ 1 & 0 \end{pmatrix}$, $B = \begin{pmatrix} 3 & 5 \\ 7 & 9 \end{pmatrix}$

6. $A = \begin{pmatrix} 1 & -2 \\ -3 & 5 \end{pmatrix}$, $B = \begin{pmatrix} -1 & 2 \\ 3 & -5 \end{pmatrix}$

7. $A = \begin{pmatrix} 5 & 2 \\ 3 & -1 \end{pmatrix}$, $B = \begin{pmatrix} -1 & 2 \\ 3 & -4 \end{pmatrix}$

8. $A = \begin{pmatrix} 1 & -1 \\ 2 & -2 \end{pmatrix}$, $B = \begin{pmatrix} 3 & -1 \\ 3 & -1 \end{pmatrix}$

In Exercises 9–14, determine if the product AB or BA is defined. If a product is defined, state its size (number of rows and columns). Do not calculate any products.

9. $A = \begin{pmatrix} 3 & 6 & 7 \\ 8 & 0 & 1 \end{pmatrix}$, $B = \begin{pmatrix} 2 & 5 & 9 & 1 \\ 7 & 0 & 0 & 6 \\ -1 & 3 & 8 & 7 \end{pmatrix}$

10. $A = \begin{pmatrix} -1 & -2 & -5 \\ 9 & 2 & -1 \\ 10 & 34 & 5 \end{pmatrix}$, $B = \begin{pmatrix} 17 & -9 \\ -6 & 12 \\ 3 & 5 \end{pmatrix}$

11. $A = \begin{pmatrix} 1 & 0 \\ 1 & 1 \\ 0 & 1 \end{pmatrix}$, $B = \begin{pmatrix} 5 & 6 & 11 \\ 7 & 8 & 15 \end{pmatrix}$

12. $A = \begin{pmatrix} 1 & -5 & 7 \\ 2 & 4 & 8 \\ 1 & -1 & 2 \end{pmatrix}$, $B = \begin{pmatrix} -2 & 4 & 9 \\ 13 & -2 & 1 \\ 5 & 25 & 0 \end{pmatrix}$

13. $A = \begin{pmatrix} -4 & 15 \\ 3 & -7 \\ 2 & 10 \end{pmatrix}$, $B = \begin{pmatrix} 1 & 2 \\ 3 & 4 \end{pmatrix}$

14. $A = \begin{pmatrix} 10 & 12 \\ -6 & 0 \\ 1 & 23 \\ -4 & 3 \end{pmatrix}$, $B = \begin{pmatrix} 1 & 2 & 3 \\ 3 & 2 & 1 \end{pmatrix}$

In Exercises 15–20, find AB.

15. $A = \begin{pmatrix} 3 & 2 \\ 2 & 4 \end{pmatrix}$, $B = \begin{pmatrix} 1 & -2 & 3 \\ 0 & 3 & 1 \end{pmatrix}$

16. $A = \begin{pmatrix} -1 & 2 & 3 \\ 0 & -1 & 2 \\ 1 & 2 & 0 \end{pmatrix}$, $B = \begin{pmatrix} 3 & -2 & -1 \\ 1 & 0 & 5 \\ 1 & -1 & -1 \end{pmatrix}$

17. $A = \begin{pmatrix} 1 & 0 & -4 \\ 0 & 2 & -1 \\ 2 & 3 & 4 \end{pmatrix}$, $B = \begin{pmatrix} 1 & 1 \\ 1 & 0 \\ 0 & 1 \end{pmatrix}$

18. $A = \begin{pmatrix} 1 & -2 \\ 3 & 0 \\ 0 & -1 \\ 2 & 1 \end{pmatrix}$, $B = \begin{pmatrix} -1 & 3 & -2 & 0 \\ 6 & 1 & 0 & -2 \end{pmatrix}$

19. $A = \begin{pmatrix} 2 & 0 & -1 \\ 1 & 1 & 2 \\ 0 & 2 & -3 \\ 2 & 3 & 0 \end{pmatrix}$, $B = \begin{pmatrix} 1 & 0 & 1 & 1 \\ 1 & 1 & 0 & 1 \\ 1 & 1 & 1 & 0 \end{pmatrix}$

20. $A = \begin{pmatrix} 10 & 0 & 1 & 0 \\ -1 & 1 & 0 & 1 \end{pmatrix}$,

$B = \begin{pmatrix} 2 & -1 & 0 & 1 \\ -2 & 3 & 1 & -4 \\ 3 & 5 & 2 & -5 \end{pmatrix}$

In Exercises 21–30, perform the indicated operations, where

$A = \begin{pmatrix} 1 & -2 \\ 3 & 4 \end{pmatrix}$, $B = \begin{pmatrix} -1 & 0 \\ 1 & 2 \end{pmatrix}$, $C = \begin{pmatrix} 1 & 0 \\ 0 & 1 \end{pmatrix}$,

$D = \begin{pmatrix} 1 & -2 & 3 \\ 0 & 4 & 1 \end{pmatrix}$, $E = \begin{pmatrix} 1 & 2 \\ -2 & 1 \\ 0 & 5 \end{pmatrix}$,

$F = \begin{pmatrix} 0 & 0 \\ 0 & 0 \\ 0 & 0 \end{pmatrix}$, $G = \begin{pmatrix} 2 & -4 \\ 3 & 0 \\ -1 & 5 \end{pmatrix}$.

21. $B - C$ 22. $2A + C$

23. $-3B + 4C$ 24. $A - 2B$

25. $AD + BD$
26. $EA + EB$
27. $2E + 3G$
28. $B^2 - AB$
29. $DG - B$
30. $DE + 2C$

In Exercises 31–34, find a matrix X satisfying the given equation, where

$$A = \begin{pmatrix} 1 & -2 \\ 4 & 3 \end{pmatrix} \quad \text{and} \quad B = \begin{pmatrix} 2 & -1 \\ 0 & 5 \end{pmatrix}$$

31. $2X = 2A + 3B$
32. $3X = A - 3B$
33. $2X + 3A = 4B$
34. $3X + 2B = 3A$

In Exercises 35–44, find the inverse of the matrix, if it exists.

35. $\begin{pmatrix} 1 & 4 \\ -1 & -3 \end{pmatrix}$

36. $\begin{pmatrix} 3 & -1 \\ -2 & 2 \end{pmatrix}$

37. $\begin{pmatrix} 1 & 2 \\ 3 & 4 \end{pmatrix}$

38. $\begin{pmatrix} 3 & 5 \\ 1 & 4 \end{pmatrix}$

39. $\begin{pmatrix} 3 & -1 \\ -6 & 2 \end{pmatrix}$

40. $\begin{pmatrix} 1 & -1 & 0 \\ 1 & 0 & -1 \\ 6 & -2 & -3 \end{pmatrix}$

41. $\begin{pmatrix} 1 & 2 & 0 \\ 3 & -1 & 2 \\ -2 & 3 & -2 \end{pmatrix}$

42. $\begin{pmatrix} 1 & -3 & 4 \\ 2 & -5 & 7 \\ 0 & -1 & 1 \end{pmatrix}$

43. $\begin{pmatrix} 5 & 0 & 2 \\ 2 & 2 & 1 \\ -3 & 1 & -1 \end{pmatrix}$

44. $\begin{pmatrix} -1 & 3 & 1 \\ 2 & 5 & 0 \\ 3 & 1 & -2 \end{pmatrix}$

In Exercises 45–52, verify that the given statement is true when $c = 2$, $d = 3$, and

$$A = \begin{pmatrix} 1 & 2 \\ 3 & 0 \end{pmatrix}, \quad B = \begin{pmatrix} -1 & 2 \\ 3 & 4 \end{pmatrix}, \quad C = \begin{pmatrix} 2 & 3 \\ 1 & 2 \end{pmatrix}$$

45. $A + B = B + A$
46. $A + (B + C) = (A + B) + C$
47. $c(A + B) = cA + cB$
48. $(c + d)A = cA + dA$
49. $(cd)A = c(dA)$
50. $A(BC) = (AB)C$
51. $A(B + C) = AB + AC$
52. $(B + C)A = BA + CA$

53. If A is an $n \times k$ matrix and B is an $r \times t$ matrix, what conditions must n, k, r, t satisfy in order that both AB and BA be defined?

In Exercises 54–58, verify that the statement is *false* for the given matrices.

54. $AB = BA$; $A = \begin{pmatrix} 1 & 2 \\ 3 & 4 \end{pmatrix}$ and $B = \begin{pmatrix} -1 & 4 \\ 5 & 2 \end{pmatrix}$

55. If $AB = AC$, then $B = C$; $A = \begin{pmatrix} 1 & 2 \\ 2 & 4 \end{pmatrix}$, $B = \begin{pmatrix} 3 & 6 \\ -\frac{3}{2} & -3 \end{pmatrix}$, $C = \begin{pmatrix} 0 & 0 \\ 0 & 0 \end{pmatrix}$

56. $A^2 = \begin{pmatrix} 0 & 0 \\ 0 & 0 \end{pmatrix}$ only if $A = \begin{pmatrix} 0 & 0 \\ 0 & 0 \end{pmatrix}$; $A = \begin{pmatrix} 0 & 2 \\ 0 & 0 \end{pmatrix}$

57. $(A + B)(A - B) = A^2 - B^2$; $A = \begin{pmatrix} 3 & 1 \\ 2 & -4 \end{pmatrix}$ and $B = \begin{pmatrix} 2 & -1 \\ 5 & 3 \end{pmatrix}$

58. $(A + B)(A + B) = A^2 + 2AB + B^2$; $A = \begin{pmatrix} 2 & -1 \\ 3 & 5 \end{pmatrix}$ and $B = \begin{pmatrix} 1 & 2 \\ -3 & 4 \end{pmatrix}$

In Exercises 59–63, the *transpose* of the matrix A is denoted by A^t and is defined by this rule: Row 1 of A is column 1 of A^t; row 2 of A is column 2 of A^t, etc.

59. Find A^t and B^t, when

$$A = \begin{pmatrix} a & b \\ c & d \end{pmatrix} \quad \text{and} \quad B = \begin{pmatrix} r & s \\ u & v \end{pmatrix}$$

In Exercises 60–62, assume A and B are as in Exercise 59 and show that:

60. $(A + B)^t = A^t + B^t$
61. $(A^t)^t = A$
62. $(AB)^t = B^t A^t$ (note order)
63. Show that solving the system of equations

$$a_1 x + b_1 y + c_1 z = d_1$$
$$a_2 x + b_2 y + c_2 z = d_2$$
$$a_3 x + b_3 y + c_3 z = d_3$$

is equivalent to solving the matrix equation $AX = D$, where A is the coefficient matrix of the system and

$$X = \begin{pmatrix} x \\ y \\ z \end{pmatrix} \quad \text{and} \quad D = \begin{pmatrix} d_1 \\ d_2 \\ d_3 \end{pmatrix}.$$

64. Show that every system of n linear equations in n unknowns is equivalent to a matrix equation $AX = D$, where A is the coefficient matrix of the system and X and D are $n \times 1$ matrices.

In Exercises 65–68, solve the system of equations by finding the inverse of the coefficient matrix and using this fact: If A, X, D are as in Exercises 63–64 and A is invertible, then $X = A^{-1}D$ is a solution of the matrix equation because $AX = A(A^{-1}D) = (AA^{-1})D = I_3 D = D$.

65. $\begin{aligned} -x + y &= 1 \\ -x + z &= -2 \\ 6x - 2y - 3z &= 3 \end{aligned}$

66. $\begin{aligned} x + 2y + 3z &= 1 \\ 2x + 5y + 3z &= 0 \\ x + 8z &= -1 \end{aligned}$

67. $\begin{aligned} 2x + y &= 0 \\ -4x - y - 3z &= 1 \\ 3x + y + 2z &= 2 \end{aligned}$

68. $\begin{aligned} -3x - 3y - 4z &= 2 \\ y + z &= 1 \\ 4x + 3y + 4z &= 3 \end{aligned}$

8.4 DETERMINANTS AND CRAMER'S RULE

The **determinant** of the 2×2 matrix $A = \begin{pmatrix} a & b \\ c & d \end{pmatrix}$ is defined to be the number $ad - bc$. An easy way to remember this definition is to note that $ad - bc$ is just the difference of the products of diagonally opposite entries: $\begin{matrix} a & b \\ c & d \end{matrix}$. Here are some examples:

Matrix	Determinant
$\begin{pmatrix} 1 & 2 \\ 6 & 4 \end{pmatrix}$	$1 \cdot 4 - 2 \cdot 6 = 4 - 12 = -8$
$\begin{pmatrix} \frac{1}{2} & 2 \\ 0 & -3 \end{pmatrix}$	$\frac{1}{2}(-3) - 2 \cdot 0 = -\frac{3}{2} - 0 = -\frac{3}{2}$

The determinant of the matrix $A = \begin{pmatrix} a & b \\ c & d \end{pmatrix}$ is denoted by any one of these symbols:

$$\det A \quad \text{or} \quad |A| \quad \text{or} \quad \begin{vmatrix} a & b \\ c & d \end{vmatrix}$$

Note the straight vertical lines in this last symbol. They are *not* the same as the curved parentheses used to denote the matrix. For example,

$$\begin{pmatrix} 3 & 2 \\ 4 & 5 \end{pmatrix} \text{ is a } 2 \times 2 \text{ matrix,}$$

but its determinant $\begin{vmatrix} 3 & 2 \\ 4 & 5 \end{vmatrix}$ is the *number* $3 \cdot 5 - 2 \cdot 4 = 7$.

Before defining determinants of 3×3 matrices, we must introduce a new concept. Observe that if you erase one row and one column of a 3×3 matrix, the remaining entries form a 2×2 matrix. For example, erasing the first row and second column of the 3×3 matrix,

$$\begin{pmatrix} -3 & 4 & -7 \\ 1 & 2 & 0 \\ -4 & 8 & 11 \end{pmatrix} \text{ produces the } 2 \times 2 \text{ matrix } \begin{pmatrix} 1 & 0 \\ -4 & 11 \end{pmatrix}$$

If you now take the determinant of this 2 × 2 matrix, the result is a *number*. The **minor** of any entry in a 3 × 3 matrix is just such a number, namely, the determinant of the 2 × 2 matrix obtained by erasing the row and column in which the given entry appears. For instance, in the 3 × 3 matrix

$$\begin{pmatrix} a_1 & b_1 & c_1 \\ a_2 & b_2 & c_2 \\ a_3 & b_3 & c_3 \end{pmatrix}$$

the minor of c_1 is the number $a_2 b_3 - b_2 a_3$, obtained by erasing the row and column in which c_1 appears and taking the determinant of the result:

$$\begin{pmatrix} a_1 & b_1 & c_1 \\ a_2 & b_2 & c_2 \\ a_3 & b_3 & c_3 \end{pmatrix} \longrightarrow \begin{vmatrix} a_2 & b_2 \\ a_3 & b_3 \end{vmatrix} = a_2 b_3 - b_2 a_3$$

The **determinant of the 3 × 3 matrix**

$$A = \begin{pmatrix} a_1 & b_1 & c_1 \\ a_2 & b_2 & c_2 \\ a_3 & b_3 & c_3 \end{pmatrix}$$

is denoted by any one of these symbols

$$\det A \quad \text{or} \quad |A| \quad \text{or} \quad \begin{vmatrix} a_1 & b_1 & c_1 \\ a_2 & b_2 & c_2 \\ a_3 & b_3 & c_3 \end{vmatrix}$$

and is defined to be the *number*

$$\begin{vmatrix} a_1 & b_1 & c_1 \\ a_2 & b_2 & c_2 \\ a_3 & b_3 & c_3 \end{vmatrix} = a_1 \begin{pmatrix} \text{minor} \\ \text{of } a_1 \end{pmatrix} - b_1 \begin{pmatrix} \text{minor} \\ \text{of } b_1 \end{pmatrix} + c_1 \begin{pmatrix} \text{minor} \\ \text{of } c_1 \end{pmatrix}$$

$$= a_1 \begin{vmatrix} b_2 & c_2 \\ b_3 & c_3 \end{vmatrix} - b_1 \begin{vmatrix} a_2 & c_2 \\ a_3 & c_3 \end{vmatrix} + c_1 \begin{vmatrix} a_2 & b_2 \\ a_3 & b_3 \end{vmatrix}$$

$$= a_1(b_2 c_3 - c_2 b_3) - b_1(a_2 c_3 - c_2 a_3) + c_1(a_2 b_3 - b_2 a_3)$$

$$= a_1 b_2 c_3 - a_1 b_3 c_2 - a_2 b_1 c_3 + a_3 b_1 c_2 + a_2 b_3 c_1 - a_3 b_2 c_1$$

It isn't necessary to memorize the last line of this formula. Just remember the directions given in the top line: to find $|A|$, multiply each entry in the first row of A by its minor, insert the proper signs $(+, -, +)$, and add up the result.

Example 1 The determinant of the matrix

$$A = \begin{pmatrix} 2 & 4 & 3 \\ 0 & 5 & -1 \\ 1 & -1 & 2 \end{pmatrix}$$

is just the number -1 since

$$|A| = 2\begin{vmatrix} 5 & -1 \\ -1 & 2 \end{vmatrix} - 4\begin{vmatrix} 0 & -1 \\ 1 & 2 \end{vmatrix} + 3\begin{vmatrix} 0 & 5 \\ 1 & -1 \end{vmatrix}$$
$$= 2(5\cdot 2 - (-1)(-1)) - 4(0\cdot 2 - (-1)1) + 3(0(-1) - 5\cdot 1)$$
$$= 2(9) - 4(1) + 3(-5) = 18 - 4 - 15 = -1. \blacksquare$$

When the determinant of a 3×3 matrix is defined as above, one says that the determinant is obtained by **expanding along the first row**. Providing that the proper signs are inserted, the determinant can actually be calculated by following an analogous procedure with any row or column. To see how this works, consider the determinant computed before Example 1:

$$\begin{vmatrix} a_1 & b_1 & c_1 \\ a_2 & b_2 & c_2 \\ a_3 & b_3 & c_3 \end{vmatrix} = a_1 b_2 c_3 - a_1 b_3 c_2 - a_2 b_1 c_3 + a_3 b_1 c_2 + a_2 b_3 c_1 - a_3 b_2 c_1$$

If we regroup the terms on the right side and factor, we see that

$$\begin{vmatrix} a_1 & b_1 & c_1 \\ a_2 & b_2 & c_2 \\ a_3 & b_3 & c_3 \end{vmatrix} = (-a_2 b_1 c_3 + a_3 b_1 c_2) + (a_1 b_2 c_3 - a_3 b_2 c_1)$$
$$+ (-a_1 b_3 c_2 + a_2 b_3 c_1)$$
$$= -b_1(a_2 c_3 - a_3 c_2) + b_2(a_1 c_3 - a_3 c_1) - b_3(a_1 c_2 - a_2 c_1)$$
$$= -b_1 \begin{vmatrix} a_2 & c_2 \\ a_3 & c_3 \end{vmatrix} + b_2 \begin{vmatrix} a_1 & c_1 \\ a_3 & c_3 \end{vmatrix} - b_3 \begin{vmatrix} a_1 & c_1 \\ a_2 & c_2 \end{vmatrix}$$
$$= -b_1 \begin{pmatrix} \text{minor} \\ \text{of } b_1 \end{pmatrix} + b_2 \begin{pmatrix} \text{minor} \\ \text{of } b_2 \end{pmatrix} - b_3 \begin{pmatrix} \text{minor} \\ \text{of } b_3 \end{pmatrix}$$

Thus the determinant can also be obtained by multiplying each entry in the second column by its minor, inserting the signs $(-, +, -)$, and adding up the result. This is called **expanding along the second column.**

Example 2 We shall find $|A|$ by expanding along the second column, where A is the matrix

$$A = \begin{pmatrix} 2 & 4 & 3 \\ 0 & 5 & -1 \\ 1 & -1 & 2 \end{pmatrix}$$

$$|A| = -4\begin{vmatrix} 0 & -1 \\ 1 & 2 \end{vmatrix} + 5\begin{vmatrix} 2 & 3 \\ 1 & 2 \end{vmatrix} - (-1)\begin{vmatrix} 2 & 3 \\ 0 & -1 \end{vmatrix}$$
$$= -4(0\cdot 2 - (-1)\cdot 1) + 5(2\cdot 2 - 3\cdot 1) + 1(2\cdot(-1) - 3\cdot 0)$$
$$= -4(1) + 5(1) + 1(-2) = -1$$

As expected, this is the same answer as in Example 1, where the determinant was expanded along the first row. $\blacksquare$

We have seen that the determinant of a 3 × 3 matrix can be found by expanding along the first row or along the second column. It should now seem plausible that the same procedure will work in other cases as well: Choose a particular row or column; multiply each entry by its minor; insert appropriate signs; and add up the result. The proper signs can be found as they were for the second column expansion: Start with the definition of the determinant (first row expansion), regroup the terms, and factor. If you do this, you find that the signs follow this pattern:

$$\begin{pmatrix} + & - & + \\ - & + & - \\ + & - & + \end{pmatrix}$$

meaning that if you expand along the first row, use the signs $+, -, +$; if you expand along the second row, use the signs $-, +, -$; and so on.

Example 3 To compute the determinant of the matrix

$$\begin{pmatrix} 2 & 3 & -5 \\ 0 & 2 & 1 \\ -2 & 4 & 3 \end{pmatrix}$$

we shall expand along the second row. So the signs to be used are $-, +, -$.

$$\begin{vmatrix} 2 & 3 & -5 \\ 0 & 2 & 1 \\ -2 & 4 & 3 \end{vmatrix} = -0 \begin{vmatrix} 3 & -5 \\ 4 & 3 \end{vmatrix} + 2 \begin{vmatrix} 2 & -5 \\ -2 & 3 \end{vmatrix} - 1 \begin{vmatrix} 2 & 3 \\ -2 & 4 \end{vmatrix}$$

$$= 0 + 2(2 \cdot 3 - (-2)(-5)) - 1(2 \cdot 4 - (-2) \cdot 3)$$

$$= 0 + 2(-4) - 1(14) = -8 - 14 = -22$$

You are free to compute a determinant by expanding along any row or column. But the computation is often easier if you choose a row or column with as many 0 entries as possible—that's why we expanded along the second row here. ∎

Determinants of 4 × 4 and larger square matrices are defined in much the same way as was just done for 3 × 3 matrices. See Excursion 8.4.A on page 397 for details.

Determinants and Systems of Linear Equations

The connection between determinants and systems of linear equations can be most easily seen by looking at an arbitrary system of two equations in two unknowns. Suppose a, b, c, d, r, s are fixed real numbers. In order to solve the system

$$ax + by = r$$
$$cx + dy = s$$

we eliminate the y terms by multiplying the first equation by d and the second by $-b$:

$$\begin{array}{ll} adx + bdy = rd & \text{\textit{First equation multiplied by }} d \\ \underline{-bcx - bdy = -bs} & \text{\textit{Second equation multiplied by }} -b \\ adx - bcx = rd - bs & \text{\textit{Sum}} \\ (ad - bc)x = rd - bs & \end{array}$$

Now *if* the number $ad - bc$ is nonzero, we can divide both sides of the last equation by it and conclude that there is only one possible value for x:

$$x = \frac{rd - bs}{ad - bc}$$

Observe that the denominator of x is just the determinant of the **coefficient matrix** $\begin{pmatrix} a & b \\ c & d \end{pmatrix}$, while the numerator is the determinant of the matrix $\begin{pmatrix} r & b \\ s & d \end{pmatrix}$ obtained by replacing the *first* column of the coefficient matrix by the column of constants from the right side of the original equations. Thus when the determinant of the coefficient matrix is nonzero, the only possible value for x is

$$x = \frac{\begin{vmatrix} r & b \\ s & d \end{vmatrix}}{\begin{vmatrix} a & b \\ c & d \end{vmatrix}}$$

A similar argument shows that when the determinant of the coefficient matrix is nonzero, then there is only one possible value for y as well. This leads to:

CRAMER'S RULE FOR TWO EQUATIONS

> The system of equations
>
> $$ax + by = r$$
> $$cx + dy = s$$
>
> has exactly one solution, provided that the determinant of the coefficient matrix is nonzero. In this case the solution is:
>
> $$x = \frac{\begin{vmatrix} r & b \\ s & d \end{vmatrix}}{\begin{vmatrix} a & b \\ c & d \end{vmatrix}} \quad \text{and} \quad y = \frac{\begin{vmatrix} a & r \\ c & s \end{vmatrix}}{\begin{vmatrix} a & b \\ c & d \end{vmatrix}}.$$

Note that the numerator of y in Cramer's Rule is the determinant of the matrix obtained by replacing the *second* column of the coefficient matrix by the column of constants from the right side of the original system of equations.

Example 4 The determinant of the coefficient matrix of the system

$$3x - 4y = 2$$
$$7x + 7y = 3$$

is the number

$$\begin{vmatrix} 3 & -4 \\ 7 & 7 \end{vmatrix} = 3 \cdot 7 - (-4)7 = 21 + 28 = 49$$

Therefore the system has exactly one solution. It can be found by using the formulas in the box on page 392 with $a = 3, b = -4, r = 2$, and $c = 7, d = 7, s = 3$:

$$x = \frac{\begin{vmatrix} 2 & -4 \\ 3 & 7 \end{vmatrix}}{\begin{vmatrix} 3 & -4 \\ 7 & 7 \end{vmatrix}} = \frac{2 \cdot 7 - (-4)3}{49} = \frac{26}{49}$$

and $\quad y = \dfrac{\begin{vmatrix} 3 & 2 \\ 7 & 3 \end{vmatrix}}{\begin{vmatrix} 3 & -4 \\ 7 & 7 \end{vmatrix}} = \dfrac{3 \cdot 3 - 2 \cdot 7}{49} = \dfrac{-5}{49}.$ ■

Analogous but more complicated arguments (which will be omitted here) can be used to prove:

CRAMER'S RULE FOR THREE EQUATIONS

The system of equations

$$a_1 x + b_1 y + c_1 z = r$$
$$a_2 x + b_2 y + c_2 z = s$$
$$a_3 x + b_3 y + c_3 z = t$$

has exactly one solution, provided that the determinant of the coefficient matrix is nonzero. In this case the solution is:

$$x = \frac{\begin{vmatrix} r & b_1 & c_1 \\ s & b_2 & c_2 \\ t & b_3 & c_3 \end{vmatrix}}{\begin{vmatrix} a_1 & b_1 & c_1 \\ a_2 & b_2 & c_2 \\ a_3 & b_3 & c_3 \end{vmatrix}}, \quad y = \frac{\begin{vmatrix} a_1 & r & c_1 \\ a_2 & s & c_2 \\ a_3 & t & c_3 \end{vmatrix}}{\begin{vmatrix} a_1 & b_1 & c_1 \\ a_2 & b_2 & c_2 \\ a_3 & b_3 & c_3 \end{vmatrix}}, \quad z = \frac{\begin{vmatrix} a_1 & b_1 & r \\ a_2 & b_2 & s \\ a_3 & b_3 & t \end{vmatrix}}{\begin{vmatrix} a_1 & b_1 & c_1 \\ a_2 & b_2 & c_2 \\ a_3 & b_3 & c_3 \end{vmatrix}}$$

Don't be put off by all the letters and subscripts in the statement of Cramer's Rule. The pattern here is just the same as before: Each denominator is the determinant of the coefficient matrix. Each numerator has almost the same form; the only difference is that one column has been replaced by the column of constants from the right side of the original system of equations.

Example 5 To solve the system

$$2x - 2y + 3z = 0$$
$$7y - 9z = 1$$
$$6x - 2y + 5z = 2$$

we first compute the determinant of the coefficient matrix by expanding along the first column:

$$\begin{vmatrix} 2 & -2 & 3 \\ 0 & 7 & -9 \\ 6 & -2 & 5 \end{vmatrix} = 2\begin{vmatrix} 7 & -9 \\ -2 & 5 \end{vmatrix} - 0\begin{vmatrix} -2 & 3 \\ -2 & 5 \end{vmatrix} + 6\begin{vmatrix} -2 & 3 \\ 7 & -9 \end{vmatrix}$$
$$= 2(35 - 18) - 0 + 6(18 - 21) = 34 - 18 = 16$$

Since the determinant is nonzero, we can use Cramer's Rule to find the solution:

$$x = \frac{\begin{vmatrix} 0 & -2 & 3 \\ 1 & 7 & -9 \\ 2 & -2 & 5 \end{vmatrix}}{16} = \frac{0 - (-2)\begin{vmatrix} 1 & -9 \\ 2 & 5 \end{vmatrix} + 3\begin{vmatrix} 1 & 7 \\ 2 & -2 \end{vmatrix}}{16} = \frac{2(23) + 3(-16)}{16}$$
$$= \frac{-2}{16} = -\frac{1}{8}$$

$$y = \frac{\begin{vmatrix} 2 & 0 & 3 \\ 0 & 1 & -9 \\ 6 & 2 & 5 \end{vmatrix}}{16} = \frac{2\begin{vmatrix} 1 & -9 \\ 2 & 5 \end{vmatrix} - 0 + 3\begin{vmatrix} 0 & 1 \\ 6 & 2 \end{vmatrix}}{16}$$
$$= \frac{2(23) + 3(-6)}{16} = \frac{28}{16} = \frac{7}{4}$$

$$z = \frac{\begin{vmatrix} 2 & -2 & 0 \\ 0 & 7 & 1 \\ 6 & -2 & 2 \end{vmatrix}}{16} = \frac{2\begin{vmatrix} 7 & 1 \\ -2 & 2 \end{vmatrix} - (-2)\begin{vmatrix} 0 & 1 \\ 6 & 2 \end{vmatrix} + 0}{16} = \frac{2(16) + 2(-6)}{16}$$
$$= \frac{20}{16} = \frac{5}{4} \blacksquare$$

8.4 DETERMINANTS AND CRAMER'S RULE

Example 6 The homogeneous system

$$3x + 2y + z = 0$$
$$7x + 3y + 5z = 0$$
$$-5x + y - z = 0$$

like all homogeneous systems, always has at least one solution, the trivial solution $x = 0$, $y = 0$, $z = 0$. The determinant of the coefficient matrix is

$$\begin{vmatrix} 3 & 2 & 1 \\ 7 & 3 & 5 \\ -5 & 1 & -1 \end{vmatrix} = 3 \begin{vmatrix} 3 & 5 \\ 1 & -1 \end{vmatrix} - 2 \begin{vmatrix} 7 & 5 \\ -5 & -1 \end{vmatrix} + 1 \begin{vmatrix} 7 & 3 \\ -5 & 1 \end{vmatrix}$$
$$= 3(-8) - 2(18) + 22 = -38$$

Since this determinant is nonzero, the system has *exactly one* solution by Cramer's Rule. But we already have one solution, namely, $x = 0$, $y = 0$, $z = 0$. Therefore this trivial solution is the *only* solution of the system. ∎

The argument used in the last example works for any homogeneous system:

> **HOMOGENEOUS SYSTEMS**
>
> If the coefficient matrix of a system of 3 homogeneous equations in 3 variables has a nonzero determinant, then the system has only the trivial solution $x = 0$, $y = 0$, $z = 0$.

This fact is also true for homogeneous systems of 2 equations in 2 variables.

EXERCISES

In Exercises 1–16, compute the determinant.

1. $\begin{vmatrix} 3 & 5 \\ 7 & 2 \end{vmatrix}$

2. $\begin{vmatrix} -2 & 1 \\ 6 & 4 \end{vmatrix}$

3. $\begin{vmatrix} 3 & 5 \\ \frac{5}{2} & 7 \end{vmatrix}$

4. $\begin{vmatrix} 2 & 5 \\ 1 & \frac{5}{2} \end{vmatrix}$

5. $\begin{vmatrix} r & d \\ t & u \end{vmatrix}$

6. $\begin{vmatrix} 1-r & v \\ v & r \end{vmatrix}$

7. $\begin{vmatrix} 1 & 0 & 2 \\ 3 & -1 & 2 \\ 1 & 2 & -3 \end{vmatrix}$

8. $\begin{vmatrix} -1 & 2 & 3 \\ 0 & 1 & 2 \\ 4 & 0 & 5 \end{vmatrix}$

9. $\begin{vmatrix} 0 & 2 & 3 \\ 1 & 7 & 9 \\ 0 & -1 & 5 \end{vmatrix}$

10. $\begin{vmatrix} 3 & 2 & -3 \\ 1 & 0 & 1 \\ 0 & 4 & 0 \end{vmatrix}$

11. $\begin{vmatrix} 3 & 2 & -3 \\ 1 & 0 & 1 \\ 0 & 4 & 1 \end{vmatrix}$

12. $\begin{vmatrix} 1 & 2 & -1 \\ 4 & 3 & -4 \\ -2 & 0 & -3 \end{vmatrix}$

13. $\begin{vmatrix} 4 & 3 & -3 \\ 2 & -2 & 3 \\ -6 & 2 & -5 \end{vmatrix}$

14. $\begin{vmatrix} -1 & 3 & -1 \\ 2 & 5 & 4 \\ 3 & 1 & 4 \end{vmatrix}$

15. $\begin{vmatrix} 2x & 0 & 3 \\ 1 & x^2 & 2 \\ 0 & -1 & x \end{vmatrix}$

16. $\begin{vmatrix} x-1 & 2 & 3 \\ 0 & x+2 & 1 \\ 1 & 0 & x \end{vmatrix}$

In Exercises 17–20, calculate the minor of the given entry.

17. Entry 2: $\begin{pmatrix} 1 & 2 & 3 \\ 4 & 0 & -1 \\ 3 & 1 & 0 \end{pmatrix}$

18. Entry 3: $\begin{pmatrix} 1 & -1 & 0 \\ 2 & 4 & 3 \\ 1 & 0 & 1 \end{pmatrix}$

19. Entry in row 3, column 2: $\begin{pmatrix} 2 & 1 & 3 \\ 3 & 1 & 3 \\ 4 & 1 & 4 \end{pmatrix}$

20. Entry in row 2, column 1: $\begin{pmatrix} 3 & 4 & 2 \\ 3 & 0 & 1 \\ 0 & -1 & -3 \end{pmatrix}$

In Exercises 21–22, verify the given statement by calculating the determinant on the left side, finding the product on the right side, and comparing the two.

21. $\begin{vmatrix} 1 & x & x^2 \\ 1 & y & y^2 \\ 1 & z & z^2 \end{vmatrix} = (x-y)(y-z)(z-x)$

22. $\begin{vmatrix} 1 & 1 & 1 \\ u & v & w \\ u^2 & v^2 & w^2 \end{vmatrix} = (u-v)(v-w)(w-u)$

23. (a) Compute the determinant
$$\begin{vmatrix} 1 & x & y \\ 1 & a & b \\ 1 & c & d \end{vmatrix}$$

(b) Verify that the equation of the straight line through the distinct points (a, b) and (c, d) is
$$\begin{vmatrix} 1 & x & y \\ 1 & a & b \\ 1 & c & d \end{vmatrix} = 0$$

24. Find the equation of the straight line through $(-1, 2)$ and $(3, -3)$.

In Exercises 25–30, solve the equation. (First expand the determinant on the left side.)

25. $\begin{vmatrix} 1-x & 2 \\ 2 & 1-x \end{vmatrix} = 0$

26. $\begin{vmatrix} x-2 & -1 \\ -3 & x-2 \end{vmatrix} = 0$

27. $\begin{vmatrix} x & 2 & 3 \\ 2x & x+4 & 5 \\ 0 & -x & x-2 \end{vmatrix} = 0$

28. $\begin{vmatrix} x & 2 & -1 \\ 0 & x+1 & 3 \\ 3x & 2x+8 & 3 \end{vmatrix} = 0$

29. $\begin{vmatrix} 2-x & 1 & 2 \\ 1 & 2-x & 2 \\ 1 & 1 & 3-x \end{vmatrix} = 0$

[Hint: $x = 5$ is a solution.]

30. $\begin{vmatrix} 1-x & 1 & -1 \\ 2 & 3-x & -4 \\ 4 & 1 & -4-x \end{vmatrix} = 0$

[Hint: $x = 1$ is a solution.]

In Exercises 31–42, use Cramer's Rule to solve the system.

31. $-3x + 5y = 2$
    $2x + 7y = 1$

32. $12x - 7y = 4$
    $5x - 3y = 2$

33. $\frac{3}{2}x + 2y = \frac{5}{2}$
    $5x - 7y = 1$

34. $x - \frac{5}{3}y = 2$
    $6x + \frac{4}{3}y = 1$

35. $7x - 12y = 4$
    $3x - 5y = 2$

36. $\sqrt{5}x - 2\sqrt{3}y = 2$
    $\sqrt{3}x + 2\sqrt{5}y = 3$

37. $-5x + y - 4z = 1$
    $3x + z = -2$
    $2x - 3y - z = 0$

38. $4x - 3y + z = 5$
    $2x + y - 2z = 0$
    $3x - z = 15$

39. $x - 2z = 3$
    $2x + 5z = 0$
    $-y + 3z = 1$

40. $x + y + z = 1$
    $x + y - 2z = 3$
    $2x + y + z = 2$

41. $x + y - z = 4$
    $2x + y + z = 1$
    $3x - 2y - z = 3$

42. $3x + 2y + 4z = 4$
    $3x + 2y - 2z = -\frac{1}{2}$
    $-3x - y + 2z = 0$

In Exercises 43–46, use Cramer's Rule to determine whether the homogeneous system has exactly one solution (namely, the trivial one $x = 0$, $y = 0$, $z = 0$).

43. $2x + y - 2z = 0$
    $3x + 2y - z = 0$
    $4x + y - 3z = 0$

44. $x + 2y - 3z = 0$
    $3x - 5y - 9z = 0$
    $2x + 4y - 6z = 0$

45. $2x + 4y + z = 0$
    $6x - y + 3z = 0$
    $4x + 6y + 2z = 0$

46. $3x + 2y - z = 0$
    $2x + y + z = 0$
    $5x - 2y - z = 0$

52. $(0, 3), (20, 8), (-20, -1)$

53. $(2, -8), (10, -3), (-18, -18)$

54. $(6, 5), (18, 9), (-6, 1)$

In Exercises 47–50, find the area of the triangle whose vertices are given by using this fact: The area of the triangle with vertices (x_1, y_1), (x_2, y_2), (x_3, y_3) is the *absolute value* of

$$\frac{1}{2} \begin{vmatrix} x_1 & y_1 & 1 \\ x_2 & y_2 & 1 \\ x_3 & y_3 & 1 \end{vmatrix}$$

47. $(1, 1), (3, 4), (5, -2)$

48. $(1, 2), (-2, 3), (-4, -1)$

49. $(0, 0), (3, 5), (-4, -7)$

50. $(0, -5), (5, 0), (5, 5)$

In Exercises 51–54, determine whether the three points are collinear (that is, that all three lie on the same straight line).

51. $(1, 1), (4, 10), (-2, -8)$ [*Hint:* Use Exercise 23(b) to check if $(1, 1)$ is on the line through the other two points.]

In Exercises 55–58, all letters represent nonzero real numbers. Verify that the statement is true.

55. $\begin{vmatrix} a+r & b \\ c+s & d \end{vmatrix} = \begin{vmatrix} a & b \\ c & d \end{vmatrix} + \begin{vmatrix} r & b \\ s & d \end{vmatrix}$

56. $\begin{vmatrix} a & b \\ c & d \end{vmatrix} = -\begin{vmatrix} b & a \\ d & c \end{vmatrix}$

57. $\begin{vmatrix} a+kb & b \\ c+kd & d \end{vmatrix} = \begin{vmatrix} a & b \\ c & d \end{vmatrix}$

58. $\begin{vmatrix} ka & b \\ kc & d \end{vmatrix} = k \begin{vmatrix} a & b \\ c & d \end{vmatrix}$

59. Compute AB and find its determinant, where

$$A = \begin{pmatrix} a & b \\ c & d \end{pmatrix} \quad \text{and} \quad B = \begin{pmatrix} r & s \\ t & u \end{pmatrix}$$

60. If A, B are as in Exercise 59, show that $\det AB$ is the product $(\det A) \cdot (\det B)$.

8.4.A EXCURSION: Higher Order Determinants

Determinants of 4×4 or larger square matrices are defined in the same way as determinants of 3×3 matrices: Choose a row or column; multiply each entry in that row or column by its minor; insert the proper signs; and add up the results. The only point needing further explanation here is the choice of "proper signs" for each term.

In order to expand 3×3 determinants we used the signs given by this grid:

$$\begin{matrix} + & - & + \\ - & + & - \\ + & - & + \end{matrix}$$

The sign appearing in row i and column j is the sign of the number $(-1)^{i+j}$. For instance, the sign $-$ appears in row 2 and column 3 and the number $(-1)^{2+3} = (-1)^5 = -1$ is negative. Similarly, the sign $+$ appears in row 3 and column 1, and $(-1)^{3+1}$ is positive.

The **cofactor** of the entry in row i and column j of a square matrix is defined to be the number

$$(-1)^{i+j} \times (\text{minor of the entry})$$

But as we just saw in the 3 × 3 case, the number $(-1)^{i+j}$ gives the correct sign for expanding the determinant. This suggests another definition of the determinant which agrees with the previous one in the 3 × 3 case and makes sense for larger matrices as well:

DETERMINANTS

> **The determinant of a square matrix is the number obtained as follows: Choose a row or column; multiply each entry in it by its cofactor; and add up these products.**

We shall omit the proof that the same number is obtained no matter which row or column of the matrix is used for the expansion of the determinant.

Example 1 We shall calculate the following determinant by expanding along the first column (since it contains a 0 entry):

$$\begin{vmatrix} 3 & 8 & 6 & -3 \\ 2 & 4 & 6 & -2 \\ 2 & 6 & 3 & -3 \\ 0 & 4 & -3 & 1 \end{vmatrix} = 3(\text{cofactor}) + 2(\text{cofactor}) + 2(\text{cofactor}) + 0(\text{cofactor})$$

$$= 3(-1)^{1+1}(\text{minor}) + 2(-1)^{2+1}(\text{minor}) + 2(-1)^{3+1}(\text{minor}) + 0$$

$$= 3\begin{vmatrix} 4 & 6 & -2 \\ 6 & 3 & -3 \\ 4 & -3 & 1 \end{vmatrix} - 2\begin{vmatrix} 8 & 6 & -3 \\ 6 & 3 & -3 \\ 4 & -3 & 1 \end{vmatrix} + 2\begin{vmatrix} 8 & 6 & -3 \\ 4 & 6 & -2 \\ 4 & -3 & 1 \end{vmatrix}$$

Expanding each of these 3 × 3 determinants along the first row produces:

$$\begin{vmatrix} 3 & 8 & 6 & -3 \\ 2 & 4 & 6 & -2 \\ 2 & 6 & 3 & -3 \\ 0 & 4 & -3 & 1 \end{vmatrix} = 3\left[4\begin{vmatrix} 3 & -3 \\ -3 & 1 \end{vmatrix} - 6\begin{vmatrix} 6 & -3 \\ 4 & 1 \end{vmatrix} - 2\begin{vmatrix} 6 & 3 \\ 4 & -3 \end{vmatrix}\right]$$

$$-2\left[8\begin{vmatrix} 3 & -3 \\ -3 & 1 \end{vmatrix} - 6\begin{vmatrix} 6 & -3 \\ 4 & 1 \end{vmatrix} - 3\begin{vmatrix} 6 & 3 \\ 4 & -3 \end{vmatrix}\right]$$

$$+2\left[8\begin{vmatrix} 6 & -2 \\ -3 & 1 \end{vmatrix} - 6\begin{vmatrix} 4 & -2 \\ 4 & 1 \end{vmatrix} - 3\begin{vmatrix} 4 & 6 \\ 4 & -3 \end{vmatrix}\right]$$

$$= 3[4(-6) - 6(18) - 2(-30)]$$
$$- 2[8(-6) - 6(18) - 3(-30)]$$
$$+ 2[8(0) - 6(12) - 3(-36)]$$

$$= -12. \blacksquare$$

It should be clear from this example that calculating the determinant of a large matrix directly from the definition can be quite long and cumbersome. For instance, the determinant of a 6 × 6 matrix involves six 5 × 5 minors. Each of these in turn involves five 4 × 4 minors, each of which involves four

3×3 minors—for a total of 120 3×3 minors to be calculated. Consequently other methods are often used to compute determinants. These methods depend on the following:

PROPERTIES OF DETERMINANTS

1. **If a matrix contains a row of zeros or a column of zeros, then its determinant is 0.**
2. **If a matrix has two identical rows or two identical columns, then its determinant is 0.**
3. **Interchanging two rows or two columns of a matrix changes the sign of its determinant.**
4. **Replacing one row of a matrix by the sum of itself and a constant multiple of another row does not change the determinant, and similarly for columns.**
5. **Multiplying one row or one column of a matrix by a constant c multiplies the determinant by c.**

It is easy to see why property 1 holds (just expand the determinant along the row or column of zeros). Proofs of the other properties will be omitted. We shall illustrate these properties in examples.

Example 2 Here is a matrix with two identical rows. Expanding the determinant along the first column yields:

$$\begin{vmatrix} 1 & 2 & 3 \\ 1 & 2 & 3 \\ 7 & 4 & 5 \end{vmatrix} = 1 \begin{vmatrix} 2 & 3 \\ 4 & 5 \end{vmatrix} - 1 \begin{vmatrix} 2 & 3 \\ 4 & 5 \end{vmatrix} + 7 \begin{vmatrix} 2 & 3 \\ 2 & 3 \end{vmatrix} = (-2) - (-2) + 7(0) = 0. \blacksquare$$

Example 3 Expanding the following determinant on the first row, we have:

$$\begin{vmatrix} 1 & 0 & 2 \\ 2 & 1 & 3 \\ 5 & 2 & 6 \end{vmatrix} = 1 \begin{vmatrix} 1 & 3 \\ 2 & 6 \end{vmatrix} - 0 + 2 \begin{vmatrix} 2 & 1 \\ 5 & 2 \end{vmatrix} = 0 - 2 = -2$$

Interchanging the first two columns changes the sign of this determinant:

$$\begin{vmatrix} 0 & 1 & 2 \\ 1 & 2 & 3 \\ 2 & 5 & 6 \end{vmatrix} = 0 - 1 \begin{vmatrix} 1 & 3 \\ 2 & 6 \end{vmatrix} + 2 \begin{vmatrix} 1 & 2 \\ 2 & 5 \end{vmatrix} = 0 + 2 = 2$$

If we replace the third row of this last matrix by the sum of itself and -2 times the second row, the new matrix has the same determinant:

$-2 \times$ (row 2) + (row 3)
$$\begin{vmatrix} 0 & 1 & 2 \\ 1 & 2 & 3 \\ 0 & 1 & 0 \end{vmatrix} = 0 - 1 \begin{vmatrix} 0 & 2 \\ 1 & 3 \end{vmatrix} + 0 = 2. \blacksquare$$

Example 4 In order to evaluate the following 4 × 4 determinant with a minimum of calculation, we try to arrange things so that there are as many zeros as possible in a single row or column. We note that we can get an additional zero in column 2 by replacing row 3 by the sum of itself and 4 times row 1. According to property 4 this does not change the determinant:

$$\begin{vmatrix} 1 & 2 & -1 & 3 \\ -4 & 0 & 2 & -4 \\ -4 & -8 & -5 & -2 \\ 2 & 0 & 2 & 1 \end{vmatrix} = \begin{vmatrix} 1 & 2 & -1 & 3 \\ -4 & 0 & 2 & -4 \\ 0 & 0 & -9 & 10 \\ 2 & 0 & 2 & 1 \end{vmatrix} \quad 4 \times (row\ 1) + (row\ 3)$$

Expanding the determinant along the second column produces:

$$\begin{vmatrix} 1 & 2 & -1 & 3 \\ -4 & 0 & 2 & -4 \\ 0 & 0 & -9 & 10 \\ 2 & 0 & 2 & 1 \end{vmatrix} = -2 \begin{vmatrix} -4 & 2 & -4 \\ 0 & -9 & 10 \\ 2 & 2 & 1 \end{vmatrix}$$

In this 3 × 3 determinant we can replace row 1 by the sum of itself and 2 times row 3. (Once again the value of the determinant doesn't change by property 4.)

$$-2 \begin{vmatrix} -4 & 2 & -4 \\ 0 & -9 & 10 \\ 2 & 2 & 1 \end{vmatrix} = -2 \begin{vmatrix} 0 & 6 & -2 \\ 0 & -9 & 10 \\ 2 & 2 & 1 \end{vmatrix} \quad 2 \times (row\ 3) + (row\ 1)$$

Finally, we expand this last determinant along the first column:

$$-2 \begin{vmatrix} 0 & 6 & -2 \\ 0 & -9 & 10 \\ 2 & 2 & 1 \end{vmatrix} = (-2)2 \begin{vmatrix} 6 & -2 \\ -9 & 10 \end{vmatrix} = -4(42) = -168. \quad \blacksquare$$

By following the obvious pattern it is possible to use Cramer's Rule to solve systems of 4 linear equations in 4 variables, 5 equations in 5 variables, and so on. But the number of calculations needed to compute large determinants makes Cramer's Rule an extremely inefficient method of solving large systems —even when a computer is used. Consequently almost all computer programs for solving large systems of linear equations are based on the elimination method presented in Sections 8.1 and 8.2.

EXERCISES

In Exercises 1–4, calculate the determinant by using the definition.

1. $\begin{vmatrix} 1 & 2 & 0 & 2 \\ -1 & 0 & 1 & 3 \\ 0 & 2 & -1 & 4 \\ 1 & 3 & -2 & 0 \end{vmatrix}$

2. $\begin{vmatrix} 2 & -1 & 2 & 0 \\ 1 & 2 & 0 & 1 \\ 2 & 1 & 2 & -1 \\ 0 & 1 & 2 & 1 \end{vmatrix}$

3. $\begin{vmatrix} 1 & 0 & 2 & 0 & 3 \\ 2 & -1 & 3 & 1 & 2 \\ 1 & 2 & -1 & 0 & 2 \\ 0 & 1 & -2 & 1 & 3 \\ 3 & 0 & 1 & 2 & -2 \end{vmatrix}$

4. $\begin{vmatrix} -1 & 2 & 1 & 3 & -1 \\ 2 & 0 & -3 & 1 & 2 \\ 0 & 1 & 0 & 2 & -1 \\ 3 & 0 & 1 & 2 & 0 \\ 4 & 1 & 2 & -1 & 1 \end{vmatrix}$

In Exercises 5–10, explain why each statement is true. [*Hint:* Use the properties of determinants listed in the text.]

5. $\begin{vmatrix} 1 & 2 & 4 & 2 \\ 3 & -2 & 7 & -2 \\ 5 & 4 & 9 & 4 \\ 7 & 3 & 11 & 3 \end{vmatrix} = 0$

6. $\begin{vmatrix} 3 & -2 & 1 & -3 \\ 4 & 2 & -3 & 2 \\ 1 & 0 & 5 & 0 \\ -6 & 2 & 0 & 1 \end{vmatrix} = - \begin{vmatrix} -6 & 2 & 0 & 1 \\ 4 & 2 & -3 & 2 \\ 1 & 0 & 5 & 0 \\ 3 & -2 & 1 & -3 \end{vmatrix}$

7. $\begin{vmatrix} 1 & 3 & 0 & 4 \\ 2 & 6 & 1 & -1 \\ 3 & -12 & 4 & 10 \\ 7 & 0 & 5 & 11 \end{vmatrix} = 3\begin{vmatrix} 1 & 1 & 0 & 4 \\ 2 & 2 & 1 & -1 \\ 3 & -4 & 4 & 10 \\ 7 & 0 & 5 & 11 \end{vmatrix}$

8. $\begin{vmatrix} 2 & 1 & -3 & 2 \\ 4 & 3 & 5 & 7 \\ 6 & 0 & 7 & 1 \\ 0 & 2 & -1 & 5 \end{vmatrix} = \begin{vmatrix} 2 & 1 & -3 & 2 \\ 0 & 2 & -1 & 5 \\ 4 & 3 & 5 & 7 \\ 6 & 0 & 7 & 1 \end{vmatrix}$

9. $\begin{vmatrix} 1 & 2 & -1 & 3 \\ 4 & 0 & 5 & 2 \\ -3 & 2 & 0 & 4 \\ 6 & 5 & 3 & -2 \end{vmatrix} = \begin{vmatrix} 1 & 2 & -1 & 3 \\ 4 & 0 & 5 & 2 \\ 0 & 8 & -3 & 13 \\ 6 & 5 & 3 & -2 \end{vmatrix}$

10. $\begin{vmatrix} 1 & 3 & 5 & 4 \\ -2 & 2 & 7 & 3 \\ 4 & 1 & 2 & 2 \\ 0 & 6 & 7 & -1 \end{vmatrix} = - \begin{vmatrix} 1 & 0 & 4 & 5 \\ -2 & 8 & 3 & 7 \\ 4 & -11 & 2 & 2 \\ 0 & 6 & -1 & 7 \end{vmatrix}$

In Exercises 11–14, calculate the determinant. [*Hint:* First use the properties of determinants to replace the given one by an equal one that's easier to compute, as in Example 4 in the text.]

11. $\begin{vmatrix} 1 & 3 & 5 & 3 \\ 2 & 1 & 10 & 0 \\ 1 & 2 & 3 & 4 \\ -1 & 0 & 2 & 1 \end{vmatrix}$

12. $\begin{vmatrix} 2 & 3 & 5 & 3 \\ 4 & 3 & 1 & 3 \\ -6 & -9 & -12 & -9 \\ 1 & 2 & 2 & 0 \end{vmatrix}$

13. $\begin{vmatrix} 1 & 0 & -1 & 2 & 3 \\ 2 & 1 & 5 & 2 & 1 \\ 3 & 0 & -3 & 6 & 1 \\ 0 & 1 & 2 & 1 & 0 \\ 1 & 0 & 2 & 1 & -1 \end{vmatrix}$

14. $\begin{vmatrix} 1 & 2 & 1 & 2 & 1 \\ 3 & 1 & 3 & 6 & 2 \\ -5 & 0 & 1 & -8 & 1 \\ 2 & 1 & 0 & 4 & -3 \\ 0 & 2 & -3 & 0 & 2 \end{vmatrix}$

8.5 SYSTEMS OF INEQUALITIES

A graph is often the best way to describe the solutions of an inequality in two variables. The graph of an inequality in two variables consists of all points (x, y) whose coordinates are a solution of the inequality. For instance, $(3, 1)$ is on the graph of $y \leq 2x + 1$, because $x = 3$, $y = 1$ is a solution. For convenience, we sometimes say that the *point* $(3, 1)$ is a solution of the inequality.

Example 1 To graph $y \leq 2x + 1$, we first graph the line $y = 2x + 1$ (Figure 8–5). It consists of all points $(x, 2x + 1)$, and every such point is a solution of $y \leq 2x + 1$. A point not on the line must lie above or below it. If (x, y) is below the line, then it must be below the point $(x, 2x + 1)$ (Figure 8–5):

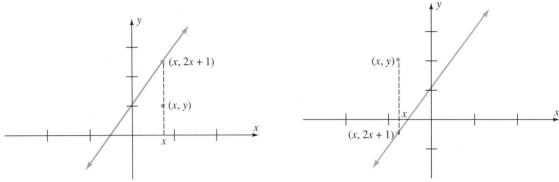

FIGURE 8-5

FIGURE 8-6

Since these points have the same first coordinate, the lower one must have a smaller second coordinate, that is, $y < 2x + 1$. Therefore every point *below* the line is a solution of $y \leq 2x + 1$. If (x, y) is a point above the line, then it lies above the point $(x, 2x + 1)$ and hence it has a larger second coordinate (Figure 8-6, above). Hence $y > 2x + 1$, and the point (x, y) is *not* a solution of $y \leq 2x + 1$. Therefore the graph of $y \leq 2x + 1$ is the half-plane consisting of *all points on or below the line $y = 2x + 1$* (Figure 8-7). ∎

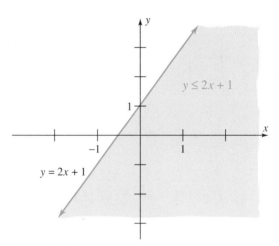

FIGURE 8-7

Example 2 The preceding argument also shows that the graph of $y > 2x + 1$ consists of *all points above the line $y = 2x + 1$*, as shown in Figure 8-8. The line $y = 2x + 1$ is dotted to indicate that it is *not* part of the graph of $y > 2x + 1$. ∎

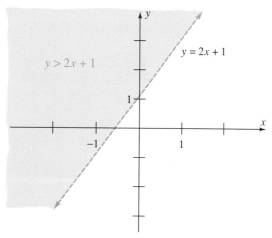

FIGURE 8-8

The argument in these examples can be used with any function—just replace $2x + 1$ by $f(x)$:

GRAPHS OF INEQUALITIES

> **The graph of $y < f(x)$ consists of all points below the graph of $y = f(x)$.**
>
> **The graph of $y > f(x)$ consists of all points above the graph of $y = f(x)$.***

Example 3 To graph $-x^2 + 2x + y - 1 > 0$, we first rewrite the inequality as $y > x^2 - 2x + 1$. The graph consists of all points above the graph of $f(x) = x^2 - 2x + 1 = (x - 1)^2$. From Section 5.1 we know that the graph of f is a parabola with vertex $(1, 0)$. The graph of the inequality is shown in Figure 8–9. ∎

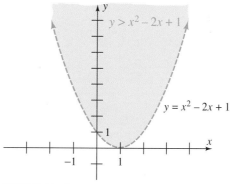

FIGURE 8-9

---

* If $<$ is replaced by $\leq$ (or $>$ by $\geq$), then the graph of $y = f(x)$ is also included in the graph of the inequality.

When an inequality cannot easily be described in terms of a function, other *ad hoc* techniques are necessary.

Example 4 The graph of $x > 3$* consists of all points (x, y) with first coordinate greater than 3. So the graph is the half-plane of all points to the right of the vertical line $x = 3$, as shown in Figure 8–10. ∎

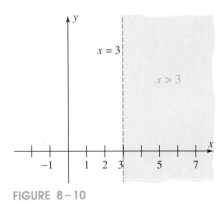

FIGURE 8–10

Example 5 If (c, d) is a solution of $(x - 1)^2 + y^2 \geq 4$, then $(c - 1)^2 + (d - 0)^2 \geq 4$. Taking square roots shows that

$$\sqrt{(c - 1)^2 + (d - 0)^2} \geq 2$$

According to the distance formula, this says that

The distance from (c, d) to $(1, 0)$ is greater than or equal to 2.

Therefore the graph of $(x - 1)^2 + y^2 \geq 4$ consists of all points on or outside the circle with center $(1, 0)$ and radius 2, as shown in Figure 8–11.

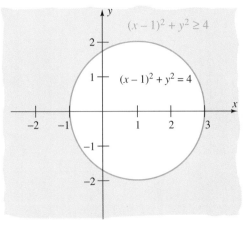

FIGURE 8–11

* All inequalities here are assumed to involve two variables, one of which may have zero coefficient.

Similarly, the graph of $(x - 1)^2 + y^2 < 4$ consists of all points inside the circle (the white disc in Figure 8–11). ∎

Systems of Inequalities

The graph of a system of inequalities in two variables consists of all points that are solutions of the system. A point (x, y) is a solution of the *system* if (x, y) is a solution of *every* inequality of the system.

Example 6 Sketch the graph of the system

$$x^2 + y^2 \leq 9$$
$$x - y < 2$$

Solution Since the graph of $x^2 + y^2 = 9$ is a circle with center $(0, 0)$ and radius 3, the graph of the first inequality consists of all points on or inside this circle (Figure 8–12). The second inequality can be rewritten as $y > x - 2$. So its graph is the set of all points above the line $y = x - 2$ (Figure 8–13). Therefore the graph of the system (the set of solutions of the system) consists of all points that are on or inside the circle *and* above this line (Figure 8–14 on the next page). ∎

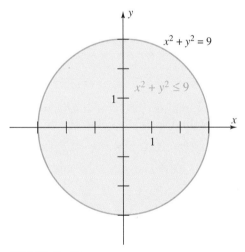

FIGURE 8–12

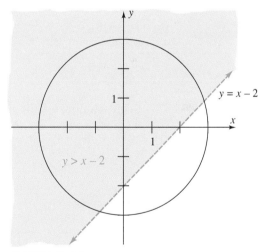

FIGURE 8–13

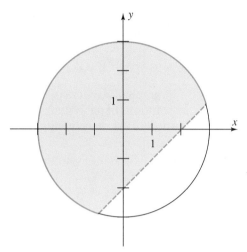

FIGURE 8-14

Example 7 Sketch the graph of the system

$$2x + y > 0$$
$$x - y < -3$$

Solution By rearranging each inequality the system becomes:

$$y > -2x$$
$$y < x + 3$$

The solutions of the first inequality are the points *above* the line $y = -2x$ (Figure 8-15). The solutions of the second are the points *below* the line $y =$

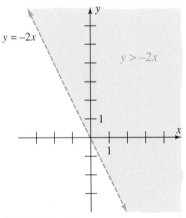

FIGURE 8-15

$x + 3$ (Figure 8–16). So the solutions of the *system* are the points that satisfy both of these conditions (Figure 8–17). ∎

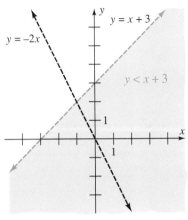

FIGURE 8–16

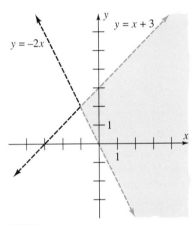

FIGURE 8–17

Example 8   To graph the system

$$x + y > 11$$
$$2x - y < 10$$
$$x - 2y > -16$$

we rewrite it as

$$y > -x + 11$$
$$y > 2x - 10$$
$$y < \frac{1}{2}x + 8$$

The solutions of the system consist of all points that lie *above both* of the lines $y = -x + 11$ and $y = 2x - 10$, and also *below* the line $y = \frac{1}{2}x + 8$, as shown in Figure 8–18. ∎

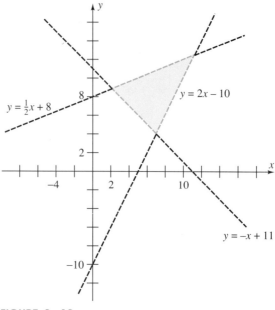

FIGURE 8–18

Example 9 A candy store sells two kinds of 4-pound gift boxes. The Regular Mix contains 2 pounds of assorted chocolates and 2 pounds of assorted caramels. The Chewey Mix has 1 pound of chocolates and 3 pounds of caramels. If 100 pounds of chocolates and 180 pounds of caramels are available, what are the possible numbers of both mixes that can be made up?

Solution Let x be the number of Regular Mixes to be made and y the number of Chewey Mixes. We know that

$$\begin{pmatrix} \text{Pounds of chocolates} \\ \text{in } x \text{ boxes of Regular} \\ \text{(2 pounds per box)} \end{pmatrix} + \begin{pmatrix} \text{Pounds of chocolates} \\ \text{in } y \text{ boxes of Chewey} \\ \text{(1 pound per box)} \end{pmatrix} \leq 100$$

$$2x \quad + \quad 1y \quad \leq 100$$

Similarly,

$$\begin{pmatrix} \text{Pounds of caramels} \\ \text{in } x \text{ boxes of Regular} \\ \text{(2 pounds per box)} \end{pmatrix} + \begin{pmatrix} \text{Pounds of caramels} \\ \text{in } y \text{ boxes of Chewey} \\ \text{(3 pounds per box)} \end{pmatrix} \leq 180$$

$$2x \quad + \quad 3y \quad \leq 180$$

Furthermore, both x and y must be nonnegative (you can't have a negative number of boxes). So the possible numbers of the two mixes must be solutions of this system of inequalities:

$$2x + y \le 100$$
$$2x + 3y \le 180$$
$$x \ge 0$$
$$y \ge 0$$

We can picture the possibilities by graphing this system. To do this we rewrite it as:

$$y \le -2x + 100$$
$$y \le -\frac{2}{3}x + 60$$
$$x \ge 0, y \ge 0$$

So the graph consists of the points that satisfy *all* of these conditions:

on or below the lines $y = -2x + 100$ and $y = -\frac{2}{3}x + 60$;

on or above the line $y = 0$ (the *x*-axis);

on or to the right of the vertical line $x = 0$ (the *y*-axis):

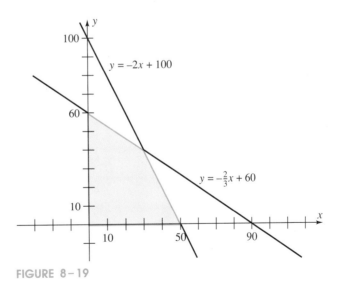

FIGURE 8-19

Since we can't have fractional boxes, the points on the graph with integer coordinates represent the possible numbers of Regular and Chewey Mixes that can be made from the available supplies. ■

EXERCISES

In Exercises 1–16, sketch the graph of the inequality.

1. $y < x + 3$
2. $y > 5 - 2x$
3. $2x - y \leq 4$
4. $4x - 3y \leq 24$
5. $3x - 2y \geq 18$
6. $2x + 5y \geq 10$
7. $y \geq -3$
8. $x \leq -1$
9. $y > x^2 - 2$
10. $y < x^2 + 3$
11. $x^2 + y^2 \geq 4$
12. $(x + 3)^2 + (y - 5)^2 \leq 9$
13. $(x - 2)^2 + y^2 \leq 4$
14. $(x - 1)^2 + (y + 2)^2 < 5$
15. $y - x^3 \leq 2$
16. $(x - 3)^2 + y \geq 25$

In Exercises 17–38, sketch the graph of the system.

17. $2x + y \leq 5$
    $x + 2y \leq 5$
18. $4x + y \geq 9$
    $2x + 3y \leq 7$
19. $2x + y > 8$
    $4x - y < 3$
20. $x + y > 5$
    $x - 2y < 2$
21. $3x + y \geq 6$
    $x + 2y \geq 7$
    $x \geq 0, y \geq 0$
22. $2x + 3y \geq 12$
    $x + y \geq 4$
    $x \geq 0, y \geq 0$
23. $3x + 2y < 18$
    $x + 2y < 10$
    $x \geq 0, y \geq 0$
24. $x + 2y > 5$
    $3x + 4y > 11$
    $x \geq 0, y \geq 0$
25. $2x + y \geq 8$
    $2x - y \geq 0$
    $x \leq 10, y \geq 2$
26. $3x + y \leq -4$
    $2x - y \leq 6$
    $x \geq -5, y \geq -8$
27. $3x + 2y \geq 12$
    $3x + 2y \leq 18$
    $3x - y \geq 0$
28. $4x + 5y \leq 20$
    $x + 4y \geq 8$
    $6x + y \geq 6$
29. $x^2 + y^2 > 25$
    $7x - y < 25$
30. $x^2 + y^2 \leq 25$
    $2x - y \geq 0$
31. $x^2 + y^2 \leq 16$
    $x - 2y \geq -4$
    $3x + 4y \geq -12$
32. $x^2 + y^2 \geq 9$
    $x + y \leq 7$
    $x \geq -1$
33. $x^2 + y^2 \leq 16$
    $x^2 + y^2 \geq 9$
34. $(x + 3)^2 + (y - 1)^2 \geq 1$
    $(x + 3)^2 + (y - 1)^2 \leq 9$
35. $x^2 - y < 2$
    $x - y > -2$
36. $x^2 - 4y < 8$
    $x^2 - y > 0$
37. $y \leq \sqrt{x}$
    $y \geq 2x - 3$
    $x \geq 0$
38. $y \geq 1/x$
    $y \geq x - 4$
    $x \geq 1$

In Exercises 39–42, find a system of inequalities whose graph is the interior of the given figure.

39. Rectangle with vertices $(2, 3)$, $(2, -1)$, $(7, 3)$, $(7, -1)$.
40. Parallelogram with vertices $(2, 4)$, $(2, -2)$, $(5, 6)$, $(5, 0)$.
41. Triangle with vertices $(2, 4)$, $(-4, 0)$, $(2, -1)$.
42. Quadrilateral with vertices $(-2, 3)$, $(2, 4)$, $(5, 0)$, $(0, -3)$.

43. Graph the inequality $xy \leq 1$. [*Hint:* When $x > 0$, it is equivalent to $y \leq 1/x$, but when $x < 0$, it is equivalent to $y \geq 1/x$ (why?).]

44. Graph the inequality $xy - y \leq 1$.

In Exercises 45–50, find a system of inequalities that describes all the possibilities and sketch its graph.

45. In order for an oil company to make 1 storage tank of regular gasoline, it takes 2 hours of processing and 3 hours of refining. One tank of premium gas takes 3 hours of processing and 3.5 hours of refining. The processing plant can be operated for at most 12 hours a day and the refinery for at most 16 hours a day. What are the possible outputs of regular and premium gas?

46. A typewriter manufacturer makes both manual and electric models. The demand for manual typewriters is never more than half of that for electric ones. The factory's production cannot exceed 1200 typewriters per month. What are the possibilities for making x manual and y electric typewriters each month?

47. Jack takes 4 hours to assemble a toy chest, and Jill takes 3 hours to decorate it. To make a silverware case takes Jack 2 hours for assembly and Jill 4 hours for decorating. Neither Jack nor Jill wants to

work more than 20 hours a week. What are the possible production levels for toy and silverware chests?

48. A builder plans to construct a building containing both one-bedroom apartments, each with 1000 square feet of space, and two-bedroom apartments, each with 1500 square feet. The maximum building size is 45,000 square feet. If there can be at most 40 apartments, what are the possibilities for the number of one- and two-bedroom units?

49. A pet food is to be made of grain and meat by-products. One serving must provide at most 12 units of fat, at least 2 units of carbohydrates, and at least 1 unit of protein. Each gram of grain contains 2 units of fat, 2 units of carbohydrates, and no units of protein. A gram of meat by-products has 3 units of fat, 1 unit of carbohydrate, and 1 unit of protein. What are the possible ways of combining x grams of grain and y grams of meat by-products in one serving?

50. An airline dietician is planning a snack package of fruit and nuts. Each ounce of fruit will supply 1 unit of protein, 2 units of carbohydrates, and 1 unit of fat. Each ounce of nuts will supply 1 unit of protein, 1 unit of carbohydrates, and 1 unit of fat. Every package must provide at least 7 units of protein, at least 10 units of carbohydrates, and no more than 9 units of fat. What are the possible ways of mixing x ounces of fruit and y ounces of nuts in each package?

8.6 INTRODUCTION TO LINEAR PROGRAMMING

Many business problems are concerned with finding the optimal value of a function (for instance, the maximum value of the profit function or the minimum value of the cost function), subject to various constraints (such as transportation costs, environmental protection laws, interest rates, availability of parts, etc.). An example from the last section will give you an idea of what is involved.

Example 1 A candy store sells two kinds of 4-pound gift boxes, Regular and Chewey. A box of Regular has 2 pounds of chocolates and 2 pounds of caramels. A box of Chewey has 1 pound of chocolates and 3 pounds of caramels. 100 pounds of chocolates and 180 pounds of caramels are available. If the profit is $3 on each box of Regular and $4 on each box of Chewey, how many boxes of each kind should be made in order to maximize the profit?

Solution Let x be the number of boxes of Regular and y the number of boxes of Chewey. Then the profit P is given by

$$P = \begin{pmatrix} \text{profit on } x \text{ boxes} \\ \text{of Regular at \$3} \\ \text{per box} \end{pmatrix} + \begin{pmatrix} \text{profit on } y \text{ boxes} \\ \text{of Chewey at \$4} \\ \text{per box} \end{pmatrix} = 3x + 4y$$

The value of the profit function $P = 3x + 4y$ can be made as large as you want by taking x and y large enough. But as we saw on pages 408–409, the *constraints* on available supplies imply that the only *feasible solutions* are those pairs of integers (x, y) that are also solutions of the system of inequalities in Figure 8–20:*

---

* If you haven't read Example 9 on pages 408–410, do so now.

$y \leq -2x + 100$
$y \leq -\frac{2}{3}x + 60$
$x \geq 0, \quad y \geq 0$

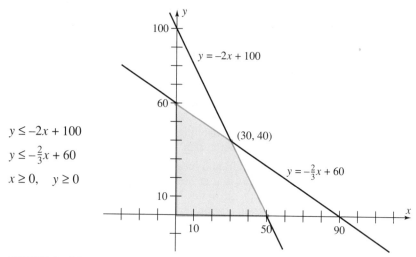

FIGURE 8-20

If you want to take the time, you can substitute each solution (x, y) of this system in the profit function $P = 3x + 4y$ to see which ones produce the highest value of P.* Fortunately, however, mathematicians have found a shortcut by proving this fact: *The maximum value of P must occur at a vertex (corner point) of the graph.* So we need only evaluate $P = 3x + 2y$ at the four vertices to find the maximum profit.

The vertices $(0, 0)$, $(0, 60)$ and $(50, 0)$ can be read off the graph. The other vertex is the intersection of the lines $y = -2x + 100$ and $y = -\frac{2}{3}x + 60$, that is, the solution of the system

$$y = -2x + 100$$
$$y = -\frac{2}{3}x + 60$$

Solving this system as in Section 8.1 (do it), we find that $(30, 40)$ is the other vertex. So we have:

Vertex	Profit = 3x + 4y
$(0, 0)$	$P = 3 \cdot 0 + 4 \cdot 0 \quad = 0$
$(0, 60)$	$P = 3 \cdot 0 + 4 \cdot 60 \; = 240$
$(50, 0)$	$P = 3 \cdot 50 + 4 \cdot 0 \; = 150$
$(30, 40)$	$P = 3 \cdot 30 + 4 \cdot 40 = 250$

* Since only the solutions with integer coordinates apply in our situation, there are only a finite number of them to check.

Therefore, making 30 boxes of Regular and 40 of Chewey will produce the maximum profit. ∎

The preceding example is a **linear programming problem.** Such a problem consists of a linear function of the form $F = ax + by$ (called the **objective function**) subject to certain **constraints** that are expressed by a system of linear inequalities. Any pair (x, y) that satisfies the constraints is called a **feasible solution.** The goal is to find an **optimal solution,** that is, a feasible solution that produces the maximum or minimum value of the objective function.

As illustrated in the last example, the set of feasible solutions (the graph of the system of inequalities given by the constraints) is a region of the plane whose edges are straight line segments. This region may be either bounded (as in the example) or unbounded (as in Figure 8–21 below). The key to solving linear programming problems is this fact, whose proof is omitted:

LINEAR PROGRAMMING THEOREM

> If the feasible solutions form a bounded region, then the objective function has both a maximum and a minimum value. These occur at vertices of the set of feasible solutions.
>
> If the feasible solutions form an unbounded region in the first quadrant and both coefficients of the objective function are positive,* then the objective function has a minimum, but no maximum, value. The minimum value occurs at a vertex.

Example 2 Find the maximum and minimum values of $F = 3x + 2y$ subject to the constraints:

$$2x + y \geq 9$$
$$x + y \geq 5$$
$$y \geq 0$$
$$x \geq 1.5$$

Solution By rewriting the system of constraints, we can easily find its graph (the set of feasible solutions):

---

* This is the only case of an unbounded region of feasible solutions that occurs in most applications.

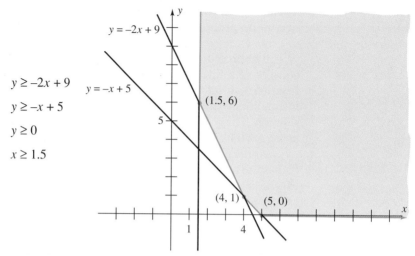

FIGURE 8-21

The vertices were found by solving each of these systems of equations:

$$y = -x + 5 \qquad y = -2x + 9 \qquad y = -2x + 9$$
$$y = 0 \qquad x = 1.5 \qquad y = -x + 5$$
$$\text{vertex } (5, 0) \qquad \text{vertex } (1.5, 6) \qquad \text{vertex } (4, 1)$$

The set of feasible solutions is unbounded, and F has no maximum value. Its minimum value will occur at one of the vertices, so we test each of them:

Vertex	$F = 3x + 2y$	
(5, 0)	$F = 3 \cdot 5 + 0$	$= 15$
(1.5, 6)	$F = 3(1.5) + 2 \cdot 6$	$= 16.5$
(4, 1)	$F = 3 \cdot 4 + 2 \cdot 1$	$= 14$

Therefore the minimum value is 14, and it occurs when $x = 4$, $y = 1$. ∎

Example 3 An animal food is to be made from grain and silage. Each pound of grain provides 2400 calories and 20 grams of protein and costs 18¢. Each pound of silage provides 600 calories and 2 grams of protein and costs 3¢. An animal must have at least 18,000 calories and 120 grams of protein each day. It is unhealthy for an animal to have more than 12 pounds of grain or 20 pounds of silage per day. What combination of grain and silage should be used to keep costs as low as possible?

8.6 INTRODUCTION TO LINEAR PROGRAMMING

Solution Let x be the number of pounds of grain an animal eats per day, and y the number of pounds of silage. Then the cost function to be minimized is given by $C = .18x + .03y$. The health limits on the amount of food imply these constraints:

$$0 \le x \le 12 \quad \text{and} \quad 0 \le y \le 20.$$

The minimal calorie requirement also leads to a constraint:

$$\begin{pmatrix} \text{Calories in } x \\ \text{pounds of grain} \end{pmatrix} + \begin{pmatrix} \text{Calories in } y \\ \text{pounds of silage} \end{pmatrix} \ge 18{,}000$$

$$2400x + 600y \ge 18{,}000$$

A final constraint comes from the minimal protein requirement:

$$\begin{pmatrix} \text{Protein in } x \\ \text{pounds of grain} \end{pmatrix} + \begin{pmatrix} \text{Protein in } y \\ \text{pounds of silage} \end{pmatrix} \ge 120$$

$$20x + 2y \ge 120$$

So the entire system of constraints is given by:

$$2400x + 600y \ge 18{,}000$$
$$20x + 2y \ge 120$$
$$x \le 12, \ y \le 20$$
$$x \ge 0, \ y \ge 0$$

which is equivalent to

$$4x + y \ge 30$$
$$10x + y \ge 60$$
$$x \le 12, \ y \le 20$$
$$x \ge 0, \ y \ge 0$$

The graph of this system (set of feasible solutions) and its vertices are found as above:

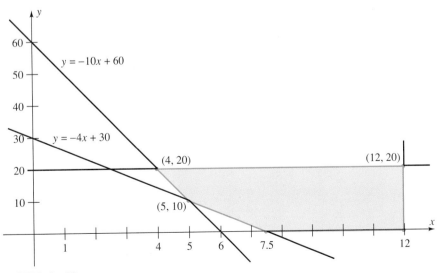

FIGURE 8–22

Since the set of feasible solutions is bounded, the cost function has both a maximum and a minimum value, each of which occurs at a vertex. The table below shows that the minimum is $1.20 and occurs when 5 pounds of grain and 10 pounds of silage are used.

Vertex	$C = .18x + .03y$
(4, 20)	$C = .18(4) + .03(20) = \$1.32$
(5, 10)	$C = .18(5) + .03(10) = \$1.20$
(7.5, 0)	$C = .18(7.5) + .03(0) = \$1.35$
(12, 0)	$C = .18(12) + .03(0) = \$2.16$
(12, 20)	$C = .18(12) + .03(20) = \$2.76$

EXERCISES

In Exercises 1–8, find the minimum and maximum values (if any) of the function F, subject to the given constraints.

1. $F = 4x + 3y$; constraints:
$3x + y \leq 6$
$2x + y \leq 5$
$x \geq 0, y \geq 0$

2. $F = 3x + 7y$; constraints:
$6x + 5y \leq 100$
$8x + 5y \leq 120$
$x \geq 0, y \geq 0$

3. $F = 3x + 2y$; constraints:
$11x + 5y \geq 75$
$4x + 15y \geq 80$
$x \geq 0, y \geq 0$

4. $F = 6x + 3y$; constraints:
$-2x + 3y \leq 9$
$-x + 3y \leq 12$
$x \geq 0, y \geq 0$

5. $F = 3x + y$; constraints:
$2x + y \leq 20$
$10x + y \geq 36$
$2x + 5y \geq 36$

6. $F = 2x + 6y$; constraints:
$x + y \geq 6$
$-x + y \leq 2$
$2x - y \leq 8$

7. $F = x + 3y$; constraints:
$2x + 3y \leq 30$
$-x + y \leq 5$
$x + y \geq 5$
$x \leq 10$
$x \geq 0, y \geq 0$

8. $F = 1100 - 2x - y$; constraints:
$x + y \leq 80$
$x + y \geq 30$
$0 \leq x \leq 40$
$0 \leq y \leq 70$

9. Wonderboat Company makes 4-person and 2-person inflatable rafts. A 4-person raft requires 6 hours of fabrication and 1 hour of finishing. A 2-person raft requires 4 hours of fabrication and 1 hour of finishing. Each week 108 hours of fabrication time and 24 hours of finishing time are available. If $40 is the profit on a 4-person raft and $30 on a 2-person one, how many of each kind should be made to maximize weekly profit?

10. The Coop sells summer-weight and winter-weight sleeping bags. A summer bag requires .9 hour of labor for cutting and .8 hour for assembly. A winter bag requires 1.8 hours for cutting and 1.2 hours for assembly. There are at most 864 hours of labor available in the cutting department each month and at most 672 hours in the assembly department. If the profit is $25 on a summer bag and

$40 on a winter bag, how many of each kind should be made to maximize profit?

11. 60 pounds of chocolates and 100 pounds of mints are available to make up 5-pound boxes of candy. A box of Choco-Mix has 4 pounds of chocolates and 1 pound of mints and sells for $10. A box of Minto-Mix has 2 pounds of chocolates and 3 pounds of mints and sells for $16. How many boxes of each kind should be made to maximize revenue?

12. CalenCo makes large wall calendars and small desk calendars. Printing a wall calendar takes 2 minutes and binding it 1 minute. A desk calendar requires 1 minute for printing and 3 minutes for binding. The printing machine is available for 3 hours a day and the binding machine for 5 hours. The profit is $1.00 on a wall calendar and $1.20 on a desk calendar. How many of each type should be made in order to maximize daily profit?

13. A person who is suffering from nutritional deficiencies is told to take at least 2400 mg of iron, 2100 mg of vitamin B-1, and 1500 mg of vitamin B-2. One Supervites pill contains 40 mg of iron, 10 mg of B-1, and 5 mg of B-2, and costs 6¢. One Vitahealth pill provides 10 mg of iron, 15 mg of B-1, and 15 mg of B-2, and costs 8¢. What combination of Supervites and Vitahealth pills will meet the requirements at lowest cost?

14. A fertilizer is to contain two major ingredients. Each pound of Ingredient I contains 4 oz of nitrogen and 2 oz of phosphates, and costs $1.00. Each pound of Ingredient II contains 3 oz of nitrogen and 5 oz of phosphates, and costs $2.50. A bag of fertilizer must contain at least 70 oz of nitrogen and 110 oz of phosphates. How much of each ingredient should be used to produce a bag of fertilizer at minimum cost? How many pounds of each ingredient does it contain?

15. A pound of chicken feed A contains 3000 units of nutrigood and 1000 units of fasgrow, and costs 20¢. A pound of chicken feed B contains 4000 units of nutrigood and 4000 units of fasgrow, and costs 40¢. The minimum daily requirement for one chicken farm is 36,000 units of nutrigood and 20,000 units of fasgrow. How many pounds of each chicken feed should be used each day to minimize food costs while meeting or exceeding the daily nutritional requirements?

16. A company has warehouses in Titusville and Rockland. It has 80 stereo systems stored in Titusville and 70 in Rockland. Superstore orders 35 systems and Giantval orders 60. It costs $8 to ship a system from Titusville to Superstore and $12 to ship one to Giantval. It costs $10 to ship a system from Rockland to Superstore and $13 to ship one to Giantval. How should the orders be filled to keep shipping costs as low as possible?

17. A cereal company plans to combine bran and oats to manufacture at least 28 tons of a new cereal. The cereal must contain at least twice as much oats as bran. If bran costs $350 per ton and oats cost $225 per ton, how much of each should be used to minimize costs?

18. Newsmag publishes a U.S. and a Canadian edition each month. There are 30,000 subscribers in the U.S. and 20,000 in Canada. Other copies are sold at newsstands. Shipping costs average $80 per thousand copies for U.S. newsstands and $60 per thousand copies for Canadian newsstands. Surveys show that no more than 120,000 copies of each issue can be sold (including subscriptions) and that the number of copies of the Canadian edition should not exceed twice the number of copies of the U.S. edition. The publisher can spend at most $8400 a month on shipping costs to newsstands. If the profit is $200 for each thousand copies of the U.S. edition and $150 for each thousand copies of the Canadian edition, how many copies of each version should be printed to earn as large a profit as possible?

19. Students at Zonker College are required to take at least 3 humanities and 4 science courses. The maximum allowable number of science courses is 12. Each humanities course carries 4 credits and each science course 5 credits. The total number of credits in humanities and science cannot exceed 80. Quality points for each course are assigned in the usual way: the number of credit hours times 4 for an A grade; times 3 for a B grade; times 2 for a C grade. A student expects to get B's in all her science courses. She expects to get C's in half her humanities courses, B's in one fourth of them, and A's in the rest. Under these assumptions how many courses of each kind should she take in order to earn the maximum possible number of quality points?

20. An investor has $20,000 to invest in bonds. He expects an annual yield of 10% for grade B bonds and 6% for grade AAA bonds. He decides that he shouldn't invest more than half his money in B bonds, and should invest at least one fourth in AAA bonds. He wants the amount invested in B bonds to be at least two thirds of the amount invested in AAA bonds. How much should he invest in each type of bond to maximize his annual yield?

Unusual Problems

21. Explain why it is impossible to maximize the function $P = 3x + 4y$, subject to the constraints: $x + y \geq 8$; $2x + y \leq 10$; $x + 2y \leq 8$; $x \geq 0$; $y \geq 0$.

22. The Linear Programming Theorem on page 413 asserts that the optimal solution (when it exists) must occur at a vertex of the set of feasible solutions. This allows the possibility that the optimal solution might occur at *two* vertices.

   (a) Show that this is indeed the case here:

   Maximize $P = 3x + 2y$, subject to the constraints:

   $3x + 2y \leq 22$; $x \leq 6$; $y \leq 5$; $x \geq 0$; $y \geq 0$

   (b) Show that the optimal value of P in part (a) also occurs at every point on the line segment joining the two vertices where the optimal value occurs. [*Hint:* The maximum value of P is 22; so any optimal solution satisfies $3x + 2y = 22$. Where does the line $3x + 2y = 22$ intersect the set of feasible solutions?]

23. In Exercise 19, find the student's grade point average (total number of quality points divided by total number of credit hours) at each vertex of the set of feasible solutions. Does the distribution of courses that produces the highest number of quality points also yield the highest grade point average? Is this a contradiction?

CHAPTER REVIEW

Important Concepts

Section 8.1	System of equations	343
	Solution of a system	343
	Substitution method	344
	Elimination method	344
	Inconsistent system	347
	Dependent system	347
	Number of solutions of a system	349
Excursion 8.1.A	Nonlinear systems	352
Section 8.2	Back substitution	358
	Echelon form	358
	Elimination method	358
	Homogeneous system	362
	Number of solutions of a system	363
Excursion 8.2.A	Partial fraction decomposition	366
Section 8.3	Matrix	372
	Augmented matrix	372
	Row operations	373
	Echelon form matrix	373
	Coefficient matrix	374

Excursion 8.3.A	$m \times n$ matrix 377	
	Rows and columns 377	
	Equality of matrices 378	
	Addition and subtraction of matrices 378	
	Scalar multiplication 379	
	Multiplication of matrices 380	
	Zero matrix 382	
	Identity matrix 383	
	Invertible matrix 383	
	Inverse 383	
Section 8.4	Determinant 388	
	Minor 389	
	Expansion along a row or column 390	
	Cramer's Rule 392, 393	
Excursion 8.4.A	Cofactor 397	
	Properties of determinants 399	
Section 8.5	Graphical solution of inequalities 392	
	Systems of inequalities 605	
Section 8.6	Linear programming problems 613	
	Objective function 613	
	Feasible solution 613	
	Optimal solution 613	
	Constraints 613	
	Linear Programming Theorem 613	

Review Questions

In Questions 1–10, solve the system of linear equations by any means you want.

1. $\begin{aligned} -5x + 3y &= 4 \\ 2x - y &= -3 \end{aligned}$

2. $\begin{aligned} 3x - y &= 6 \\ 2x + 3y &= 7 \end{aligned}$

3. $\begin{aligned} 3x - 5y &= 10 \\ 4x - 3y &= 6 \end{aligned}$

4. $\begin{aligned} \frac{1}{4}x - \frac{1}{3}y &= -\frac{3}{12} \\ \frac{1}{10}x + \frac{2}{5}y &= \frac{2}{5} \end{aligned}$

5. $\begin{aligned} 3x + y - z &= 13 \\ x + 2z &= 9 \\ -3x - y + 2z &= 9 \end{aligned}$

6. $\begin{aligned} x + 2y + 3z &= 1 \\ 4x + 4y + 4z &= 2 \\ 10x + 8y + 6z &= 4 \end{aligned}$

7. $\begin{aligned} 4x + 3y - 3z &= 2 \\ 5x - 3y + 2z &= 10 \\ 2x - 2y + 3z &= 14 \end{aligned}$

8. $\begin{aligned} x + y - 4z &= 0 \\ 2x + y - 3z &= 2 \\ -3x - y + 2z &= -4 \end{aligned}$

9. $\begin{aligned} x - 2y - 3z &= 1 \\ 5y + 10z &= 0 \\ 8x - 6y - 4z &= 8 \end{aligned}$

10. $\begin{aligned} 4x - y - 2z &= 4 \\ x - y - \tfrac{1}{2}z &= 1 \\ 2x - y - z &= 8 \end{aligned}$

11. The sum of one number and three times a second number is -20. The sum of the second number and two times the first number is 55. Find the two numbers.

12. You are given $144 in $1, $5, and $10 bills. There are 35 bills. There are two more $10 bills than $5 bills. How many bills of each type do you have?

13. Let L be the line with equation $4x - 2y = 6$ and M the line with equation $-10x + 5y = -15$. Which of the following statements is true?
    (a) L and M do not intersect.
    (b) L and M intersect at a single point.
    (c) L and M are the same line.
    (d) All of the above are true.
    (e) None of the above are true.

14. Which of the following statements about this system of equations is *false*?

$$x + z = 2$$
$$6x + 4y + 14z = 24$$
$$2x + y + 4z = 7$$

   (a) $x = 2, y = 3, z = 0$ is a solution.
   (b) $x = 1, y = 1, z = 1$ is a solution.
   (c) $x = 1, y = -3, z = 3$ is a solution.
   (d) The system has an infinite number of solutions.
   (e) $x = 2, y = 5, z = -1$ is not a solution.

15. Which of these systems are inconsistent (that is, have no solutions)?

   (i) $x + y = 5$
   $2x + 2y = 6$

   (ii) $x - y = 0$
   $7x - 2y = 0$

   (iii) $2x + y = 1$
   $3x - 4y = 2$

   (iv) $3x + 2y = 1$
   $6x + 4y = 0$

   (a) (i) only
   (b) (i) and (ii)
   (c) (ii) and (iv)
   (d) (ii) and (iii)
   (e) (i) and (iv)

In Questions 16–18, solve the nonlinear system.

16. $x^2 - y = 0$
    $y - 2x = 3$

17. $x^2 + y^2 = 25$
    $x^2 + y = 19$

18. $x^2 + y^2 = 16$
    $x + y = 2$

19. Tickets to a lecture cost $1 for students, $1.50 for faculty, and $2 for others. Total attendance at the lecture was 460, and the total income from tickets was $570. Three times as many students as faculty attended. How many faculty members attended the lecture?

20. An alloy containing 40% gold and an alloy containing 70% gold are to be mixed to produce 50 pounds of an alloy containing 60% gold. How much of each alloy is needed?

In Questions 21–22, find the *form* of the partial fraction decomposition of the rational expression. Do not find the constants.

21. $\dfrac{3x^6 - 7x^5 + 2x^4 - 3x^2 + 8x - 5}{(x^2 - 1)(x^2 + 1)(x^3 + x)}$

22. $\dfrac{x^7 - x^6 + x^5 - x^4 + x^3 - x^2 + x - 1}{(x^2 + 4x + 4)(x^2 - 4)(x^2 + 1)^2}$

In Questions 23–26, find the partial fraction decomposition of the rational expression.

23. $\dfrac{4x - 7}{x^2 - x - 6}$ **24.** $\dfrac{6x^2 + 6x - 6}{(x^2 - 1)(x + 2)}$

25. $\dfrac{x + 1}{x^2 - 4x + 4}$ **26.** $\dfrac{4x^2 - 3x + 1}{x^3 - x^2 + x}$

27. Write the augmented matrix of the system

$$x - 2y + 3z = 4$$
$$2x + y - 4z = 3$$
$$-3x + 4y - z = -2$$

28. Use matrix methods to solve the system in Question 27.

29. Write the coefficient matrix of the system

$$2x - y - 2z + 2u = 0$$
$$x + 3y - 2z + u = 0$$
$$-x + 4y + 2z - 3u = 0$$

30. Use matrix methods to solve the system in Question 29.

In Questions 31–38, perform the indicated matrix operations, if they are defined, or state that the operation is not defined. Use these matrices:

$$A = \begin{pmatrix} -1 & 0 \\ 0 & -1 \end{pmatrix}, \quad B = \begin{pmatrix} 2 & -3 \\ 4 & 1 \end{pmatrix}, \quad C = \begin{pmatrix} 3 & 2 \\ 2 & 4 \end{pmatrix},$$

$$D = \begin{pmatrix} -3 & 1 & 2 \\ 1 & 0 & 4 \end{pmatrix}, \quad E = \begin{pmatrix} 1 & 2 \\ -3 & 4 \\ 0 & 5 \end{pmatrix}, \quad F = \begin{pmatrix} 2 & 3 \\ 6 & 3 \\ 6 & 1 \end{pmatrix}$$

31. $3F - 2E$ **32.** $3A + B - C$

33. $AB + AC$ **34.** AD

35. $DA - A$ 36. AE
37. $DF + FE$ 38. $DF - DE$

In Questions 39–42, find the inverse of the matrix, if it exists.

39. $\begin{pmatrix} 3 & -7 \\ 4 & -9 \end{pmatrix}$ 40. $\begin{pmatrix} 2 & 6 \\ 1 & 3 \end{pmatrix}$

41. $\begin{pmatrix} 3 & 2 & 6 \\ 1 & 1 & 2 \\ 2 & 2 & 5 \end{pmatrix}$ 42. $\begin{pmatrix} 1 & -1 & 1 \\ 2 & -3 & 2 \\ -4 & 6 & 1 \end{pmatrix}$

In Questions 43–46, evaluate the determinant.

43. $\begin{vmatrix} 3 & -1 \\ 6 & 7 \end{vmatrix}$ 44. $\begin{vmatrix} \frac{5}{3} & -\frac{7}{3} \\ -\frac{2}{3} & \frac{4}{3} \end{vmatrix}$

45. $\begin{vmatrix} 1 & -2 & 3 \\ 0 & 2 & 4 \\ -1 & 0 & 5 \end{vmatrix}$ 46. $\begin{vmatrix} 3 & 1 & -4 \\ 2 & 0 & 1 \\ -3 & 1 & 0 \end{vmatrix}$

47. $\begin{vmatrix} 1 & 1 & 1 \\ x & y & z \\ x^3 & y^3 & z^3 \end{vmatrix} = ?$

   (a) $(x - y)(y - z)(z - x)$
   (b) $(z - y)(y - z)(z - x)(x + y + z)$
   (c) $(x - y)(x - z)(y - z)(x + y)$
   (d) all of the above
   (e) none of the above

48. $\begin{vmatrix} 1 & x & y \\ 1 & a & b \\ 1 & c & d \end{vmatrix} = 0$ is the equation of

   (a) the circle with center (a, b) and radius cd
   (b) the line through (a, b) with slope c/d
   (c) the line through (a, c) and (b, d)
   (d) the line through (a, b) and (c, d)
   (e) none of the above

49. $\begin{vmatrix} 0 & 0 & 0 \\ 5 & 7 & 9 \\ 6 & 8 & 11 \end{vmatrix} \cdot \begin{vmatrix} -1 & 2 & 3 \\ -3 & 4 & 5 \\ 6 & -7 & 8 \end{vmatrix} = ?$

50. $2\begin{vmatrix} 3 & 5 \\ 1 & -1 \end{vmatrix} - 3\begin{vmatrix} 7 & 5 \\ -5 & -1 \end{vmatrix} + 4\begin{vmatrix} 7 & 3 \\ -5 & 1 \end{vmatrix} = ?$

51. $\begin{vmatrix} 1 & 0 & 2 & -1 \\ 3 & 1 & -2 & 3 \\ 2 & 0 & 1 & 4 \\ 0 & 2 & -1 & 0 \end{vmatrix} = ?$ 52. $\begin{vmatrix} 3 & -1 & 3 & 0 \\ 1 & 3 & 0 & 1 \\ 3 & 1 & 3 & -1 \\ 0 & 1 & 3 & 1 \end{vmatrix} = ?$

In Questions 53–56, use Cramer's Rule to solve the system.

53. $2x - 3y = -4$
    $5x + 7y = 1$

54. $4x + 5y = 13$
    $3x + y = -4$

55. $x - 2z = 3$
    $2x + 5z = 0$
    $-y + 3z = 1$

56. $5x + 2y - z = -7$
    $3y + z = 10$
    $x - 2y + 2z = 0$

57. Solve graphically: $2x - y \geq 5$
58. Solve graphically: $(x - 2)^2 + (y - 3)^2 \leq 9$

In Questions 59–61, sketch the graph of the system:

59. $x + y \geq 0$
    $-x + y \leq 6$
    $x \leq 5$

60. $2x + y \geq 10$
    $-x + y \leq 1$
    $x \geq 0, y \geq 0$

61. $x^2 + y^2 \leq 4$
    $x - y \geq 1$
    $y \leq 0$

In Questions 62–64, find the minimum and maximum values (if any) of F subject to the given constraints. State the values of x and y that produce each optimal value of F.

62. $F = 5x + 7y$; constraints:
    $10x + 20y \leq 480$
    $10x + 30y \leq 570$
    $x \geq 0, y \geq 0$

63. $F = .08x + .05y$; constraints:
    $3x + y \geq 20$
    $3x + 6y \geq 60$
    $x \geq 0, y \geq 0$

64. $F = 2x - 3y$; constraints:
    $2x + 3y \geq 12$
    $3x + 2y \leq 18$
    $3x - y \leq 0$
    $x \geq 0, y \geq 0$

65. A private trash collection firm charges $260 for each container it removes. A container weighs 20 pounds and is 5 cubic feet in volume. The firm charges $200 for each barrel it removes. A barrel weighs 20 pounds and is 3 cubic feet in volume. If the truck can carry at most 1000 pounds and 180 cubic feet of material, what combination of containers and barrels will produce the largest possible revenue? Will there be any unused space in the truck?

66. A grass seed mixture contains bluegrass seeds costing 30¢ an ounce and rye seeds costing 15¢ an ounce. What is the cheapest way to obtain a mixture of 500 pounds that is at least 60% bluegrass?

CHAPTER 9

Discrete Algebra and Probability

This chapter deals with a variety of subjects involving the natural numbers 0, 1, 2, 3, ... and counting processes.

> **ROADMAP:** The sections in this chapter are independent of each other and may be read in any order, with one exception: Section 9.1 is a prerequisite for Sections 9.2 and 9.3. Also, a few examples in Section 9.6 depend on Section 9.5.

9.1 SEQUENCES AND SUMS

A **sequence** is an ordered list of numbers. We usually write them horizontally, with the ordering understood to be from left to right. The same number may appear several times on the list. Each number on the list is called a **term** of the sequence. We are primarily interested in infinite sequences, such as

$$2, 4, 6, 8, 10, 12, \ldots$$

$$1, -3, 5, -7, 9, -11, 13, \ldots$$

$$2, 1, \frac{2}{3}, \frac{2}{4}, \frac{2}{5}, \frac{2}{6}, \frac{2}{7}, \ldots$$

where the dots indicate that the same pattern continues forever.*

When the pattern in an ordered list of numbers isn't obvious, the sequence is usually described as follows: The first term is denoted a_1, the second term a_2, and so on. Then a formula is given for the nth term a_n.

---

*Such a list defines a function f whose domain is the set of positive integers. The rule is $f(1)$ = first number on the list, $f(2)$ = second number on the list, and so on. Conversely, any function g whose domain is the set of positive integers leads to an ordered list of numbers, namely, $g(1)$, $g(2)$, $g(3)$, ... So a sequence is formally defined to be a function whose domain is the set of positive integers.

9.1 SEQUENCES AND SUMS

Example 1 Consider the sequence $a_1, a_2, a_3, \ldots, a_n, \ldots$ where a_n is given by the formula
$$a_n = \frac{n^2 - 3n + 1}{2n + 5}$$

To find a_1 we substitute $n = 1$ in the formula for a_n; to find a_2 we substitute $n = 2$ in the formula; and so on:

$$a_1 = \frac{1^2 - 3 \cdot 1 + 1}{2 \cdot 1 + 5} = -\frac{1}{7}$$

$$a_2 = \frac{2^2 - 3 \cdot 2 + 1}{2 \cdot 2 + 5} = -\frac{1}{9}$$

$$a_3 = \frac{3^2 - 3 \cdot 3 + 1}{2 \cdot 3 + 5} = \frac{1}{11}$$

Thus the sequence begins $-1/7, -1/9, 1/11, \ldots$. The 39th term is

$$a_{39} = \frac{39^2 - 3 \cdot 39 + 1}{2 \cdot 39 + 5} = \frac{1405}{83}. \blacksquare$$

Example 2 It is easy to list the first few terms of the sequence
$$a_1, a_2, a_3, \ldots \quad \text{where} \quad a_n = \frac{(-1)^n}{n + 2}$$

Substituting $n = 1$, $n = 2$, etc., in the formula for a_n shows that:

$$a_1 = \frac{(-1)^1}{1 + 2} = -\frac{1}{3}, \quad a_2 = \frac{(-1)^2}{2 + 2} = \frac{1}{4}, \quad a_3 = \frac{(-1)^3}{3 + 2} = -\frac{1}{5}.$$

Similarly,

$$a_{41} = \frac{(-1)^{41}}{41 + 2} = -\frac{1}{43} \quad \text{and} \quad a_{206} = \frac{(-1)^{206}}{206 + 2} = \frac{1}{208}. \blacksquare$$

Example 3 Here are some other sequences whose nth term can be described by a formula:

Sequence	nth Term	First 5 Terms
$a_1, a_2, a_3, \ldots$	$a_n = n^2 + 1$	$2, 5, 10, 17, 26$
$b_1, b_2, b_3, \ldots$	$b_n = \dfrac{1}{n}$	$1, \dfrac{1}{2}, \dfrac{1}{3}, \dfrac{1}{4}, \dfrac{1}{5}$
$c_1, c_2, c_3, \ldots$	$c_n = \dfrac{(-1)^{n+1}2n}{(n+1)(n+2)}$	$\dfrac{1}{3}, -\dfrac{1}{3}, \dfrac{3}{10}, -\dfrac{4}{15}, \dfrac{5}{21}$
$x_1, x_2, x_3, \ldots$	$x_n = 3 + \dfrac{1}{10^n}$	$3.1, 3.01, 3.001, 3.0001, 3.00001.$ $\blacksquare$

A **constant sequence** is a sequence in which every term is the same number, such as the sequence 7, 7, 7, 7, . . . or the sequence $a_1, a_2, a_3, \ldots$ where $a_n = -18$ for every $n \geq 1$.

The subscript notation for sequences is sometimes abbreviated by writing $\{a_n\}$ in place of $a_1, a_2, a_3, \ldots$.

Example 4 $\{1/2^n\}$ denotes the sequence whose first four terms are

$$a_1 = \frac{1}{2^1}, \quad a_2 = \frac{1}{2^2} = \frac{1}{4}, \quad a_3 = \frac{1}{2^3} = \frac{1}{8}, \quad a_4 = \frac{1}{2^4} = \frac{1}{16}.$$

Similarly, $\{(-1)^n n^2\}$ denotes the sequence with first three terms

$$a_1 = (-1)^1 \cdot 1^2 = -1, \quad a_2 = (-1)^2 \cdot 2^2 = 4, \quad a_3 = (-1)^3 \cdot 3^2 = -9$$

and 23rd term $a_{23} = (-1)^{23} \cdot 23^2 = -529$. ∎

A sequence is said to be defined **recursively** (or **inductively**) if the first term is given (or the first several terms) and there is a method of determining the nth term by using the terms that precede it.

Example 5 Consider the sequence whose first two terms are

$$a_1 = 1 \quad \text{and} \quad a_2 = 1$$

and whose nth term (for $n \geq 3$) is the sum of the two preceding terms:

$$a_3 = a_2 + a_1 = 1 + 1 = 2$$
$$a_4 = a_3 + a_2 = 2 + 1 = 3$$
$$a_5 = a_4 + a_3 = 3 + 2 = 5$$

For each integer n, the two preceding integers are $n - 1$ and $n - 2$. So

$$a_n = a_{n-1} + a_{n-2} \quad (n \geq 3)$$

This sequence 1, 1, 2, 3, 5, 8, 13, . . . is called the **Fibonacci sequence**, and the numbers that appear in it are called **Fibonacci numbers**. Fibonacci numbers have many surprising and interesting properties. See Exercises 54–60 for details. ∎

Example 6 The sequence given by

$$a_1 = -7 \quad \text{and} \quad a_n = a_{n-1} + 3 \quad \text{for } n \geq 2$$

is defined recursively. Its first three terms are:

$$a_1 = -7, \quad a_2 = a_1 + 3 = -7 + 3 = -4,$$
$$a_3 = a_2 + 3 = -4 + 3 = -1. \ \blacksquare$$

Sometimes it is convenient or more natural to begin numbering the terms of a sequence with a number other than 1. So we may consider sequences such as

$$b_4, b_5, b_6, \ldots \quad \text{or} \quad c_0, c_1, c_2, \ldots$$

Example 7 The sequence 4, 5, 6, 7, ... can be conveniently described by saying $b_n = n$, with $n \geq 4$. In the brackets notation, we write $\{n\}_{n \geq 4}$. Similarly, the sequence

$$2^0, 2^1, 2^2, 2^3, \ldots$$

may be described as $\{2^n\}_{n \geq 0}$ or by saying $c_n = 2^n$, with $n \geq 0$. ■

Summation Notation and Partial Sums

It is sometimes necessary to find the sum of various terms in a squence. For instance, we might want to find the sum of the first nine terms of the sequence $\{a_n\}$. Mathematicians often use the Greek letter sigma (Σ) to abbreviate such a sum:*

$$\sum_{k=1}^{9} a_k = a_1 + a_2 + a_3 + a_4 + a_5 + a_6 + a_7 + a_8 + a_9$$

Similarly, for any positive integer m and numbers $c_1, c_2, \ldots, c_m$

SUMMATION NOTATION

$$\sum_{k=1}^{m} c_k \text{ means } c_1 + c_2 + c_3 + \cdots + c_m.$$

Example 8 $\sum_{k=1}^{5} k^2$ denotes the sum of all terms of the form k^2, as k takes values from 1 to 5, that is,

$$\sum_{k=1}^{5} k^2 = 1^2 + 2^2 + 3^2 + 4^2 + 5^2 = 55. \quad ■$$

Example 9 To find $\sum_{k=1}^{4} k^2(k-2)$, we successively substitute 1, 2, 3, 4 for k in the expression $k^2(k-2)$ and add up the result:

$$\sum_{k=1}^{4} k^2(k-2) = 1^2(1-2) + 2^2(2-2) + 3^2(3-2) + 4^2(4-2)$$

$$= 1(-1) + 4(0) + 9(1) + 16(2) = 40. \quad ■$$

---

* Σ is the letter S in the Greek alphabet, the first letter in *Sum*.

Example 10 The sum $\sum_{k=1}^{6} (-1)^k k$ is

$$(-1)^1 \cdot 1 + (-1)^2 \cdot 2 + (-1)^3 \cdot 3 + (-1)^4 \cdot 4 + (-1)^5 \cdot 5 + (-1)^6 \cdot 6$$
$$= -1 + 2 - 3 + 4 - 5 + 6 = 3. \blacksquare$$

In sums such as $\sum_{k=1}^{5} k^2$ and $\sum_{k=1}^{6} (-1)^k k$, the letter k is called the **summation index**. Any letter may be used for the summation index, just as the variable of a function f may be denoted by $f(x)$ or $f(t)$ or $f(k)$, etc. For example, $\sum_{n=1}^{5} n^2$ means: Take the sum of the terms n^2 as n takes values from 1 to 5. In other words, $\sum_{n=1}^{5} n^2 = \sum_{k=1}^{5} k^2$. Similarly,

$$\sum_{k=1}^{4} k^2(k-2) = \sum_{j=1}^{4} j^2(j-2) = \sum_{n=1}^{4} n^2(n-2)$$

The Σ notation for sums can also be used for sums that don't begin with $k = 1$. For instance,

$$\sum_{k=4}^{10} k^2 = 4^2 + 5^2 + 6^2 + 7^2 + 8^2 + 9^2 + 10^2 = 371$$

$$\sum_{j=0}^{3} j^2(2j+5) = 0^2(2 \cdot 0 + 5) + 1^2(2 \cdot 1 + 5) + 2^2(2 \cdot 2 + 5) + 3^2(2 \cdot 3 + 5)$$
$$= 142$$

Suppose $\{a_n\}$ is a sequence and k is a positive integer. The sum of the first k terms of the sequence is called the **kth partial sum** of the sequence. Thus

PARTIAL SUMS

$$k\text{th partial sum of } \{a_n\} = \sum_{n=1}^{k} a_n = a_1 + a_2 + a_3 + \cdots + a_k.$$

Example 11 Here are some partial sums of the sequence $\{n^3\}$:

First partial sum: $\sum_{n=1}^{1} n^3 = 1^3 = 1$

Second partial sum: $\sum_{n=1}^{2} n^3 = 1^3 + 2^3 = 9$

Sixth partial sum: $\sum_{n=1}^{6} n^3 = 1^3 + 2^3 + 3^3 + 4^3 + 5^3 + 6^3 = 441. \blacksquare$

9.1 SEQUENCES AND SUMS

Example 12 The sequence $\{2^n\}_{n \geq 0}$ begins with the 0th term, so the fourth partial sum (the sum of the first four terms) is

$$2^0 + 2^1 + 2^2 + 2^3 = \sum_{n=0}^{3} 2^n$$

Similarly, the fifth partial sum of the sequence $\left\{\dfrac{1}{n(n-2)}\right\}_{n \geq 3}$ is the sum of the first five terms:

$$\frac{1}{3(3-2)} + \frac{1}{4(4-2)} + \frac{1}{5(5-2)} + \frac{1}{6(6-2)} + \frac{1}{7(7-2)}$$

$$= \sum_{n=3}^{7} \frac{1}{n(n-2)}. \blacksquare$$

Certain calculations can be written very compactly in summation notation. For example, the distributive law shows that

$$ca_1 + ca_2 + ca_3 + \cdots + ca_r = c(a_1 + a_2 + a_3 + \cdots + a_r)$$

In summation notation this becomes

$$\sum_{n=1}^{r} ca_n = c\left(\sum_{n=1}^{r} a_n\right).$$

This proves the first of the following statements.

PROPERTIES OF SUMS

1. $\sum_{n=1}^{r} ca_n = c\left(\sum_{n=1}^{r} a_n\right)$ for any number c.

2. $\sum_{n=1}^{r} (a_n + b_n) = \sum_{n=1}^{r} a_n + \sum_{n=1}^{r} b_n$

3. $\sum_{n=1}^{r} (a_n - b_n) = \sum_{n=1}^{r} a_n - \sum_{n=1}^{r} b_n$

To prove statement 2, use the commutative and associative laws repeatedly to show that:

$$(a_1 + b_1) + (a_2 + b_2) + (a_3 + b_3) + \cdots + (a_r + b_r)$$
$$= (a_1 + a_2 + a_3 + \cdots + a_r) + (b_1 + b_2 + b_3 + \cdots + b_r)$$

which can be written in summation notation as

$$\sum_{n=1}^{r} (a_n + b_n) = \sum_{n=1}^{r} a_n + \sum_{n=1}^{r} b_n$$

The last statement is proved similarly.

EXERCISES

In Exercises 1–10, find the first five terms of the sequence $\{a_n\}$.

1. $a_n = 2n + 6$
2. $a_n = 2^n - 7$
3. $a_n = \dfrac{1}{n^3}$
4. $a_n = \dfrac{1}{(n+3)(n+1)}$
5. $a_n = (-1)^n \sqrt{n+2}$
6. $a_n = (-1)^{n+1} n(n-1)$
7. $a_n = 4 + (-.1)^n$
8. $a_n = 5 - (.1)^n$
9. $a_n = (-1)^n + 3n$
10. $a_n = (-1)^{n+2} - (n+1)$

In Exercises 11–14, express the sum in Σ notation.

11. $1 + 2 + 3 + 4 + 5 + 6 + 7 + 8 + 9 + 10 + 11$
12. $1^1 + 2^2 + 3^3 + 4^4 + 5^5$
13. $\dfrac{1}{2^7} + \dfrac{1}{2^8} + \dfrac{1}{2^9} + \dfrac{1}{2^{10}} + \dfrac{1}{2^{11}} + \dfrac{1}{2^{12}} + \dfrac{1}{2^{13}}$
14. $(-6)^{11} + (-6)^{12} + (-6)^{13} + (-6)^{14} + (-6)^{15}$

In Exercises 15–20, find the sum.

15. $\displaystyle\sum_{i=1}^{5} 3i$
16. $\displaystyle\sum_{i=1}^{4} \dfrac{1}{2^i}$
17. $\displaystyle\sum_{n=1}^{6} (2n - 3)$
18. $\displaystyle\sum_{n=1}^{7} (-1)^n (3n + 1)$
19. $\displaystyle\sum_{n=3}^{6} (n^2 - 8)$
20. $\displaystyle\sum_{k=0}^{5} (2n^2 - 5n + 1)$

In Exercises 21–26, find a formula for the nth term of the sequence whose first few terms are given.

21. $-1, 1, -1, 1, -1, 1, \ldots$
22. $2, -2, 2, -2, 2, -2, \ldots$
23. $\dfrac{1}{2}, \dfrac{2}{3}, \dfrac{3}{4}, \dfrac{4}{5}, \dfrac{5}{6}, \ldots$
24. $\dfrac{1}{2 \cdot 3}, \dfrac{1}{3 \cdot 4}, \dfrac{1}{4 \cdot 5}, \dfrac{1}{5 \cdot 6}, \dfrac{1}{6 \cdot 7}, \ldots$
25. $2, 7, 12, 17, 22, 27, \ldots$
26. $8, -5, 2, -11, -4, -17, -10, \ldots$

In Exercises 27–34, find the first five terms of the given sequence.

27. $a_1 = 4$ and $a_n = 2a_{n-1} + 3$ for $n \geq 2$
28. $a_1 = -3$ and $a_n = (-1)^n 4 a_{n-1} - 5$ for $n \geq 2$
29. $a_1 = 1, a_2 = -2, a_3 = 3$, and $a_n = a_{n-1} + a_{n-2} + a_{n-3}$ for $n \geq 4$
30. $a_1 = 1, a_2 = 3$, and $a_n = 2a_{n-1} + 3a_{n-2}$ for $n \geq 3$
31. $a_0 = 2, a_1 = 3$, and $a_n = (a_{n-1})\left(\dfrac{1}{2} a_{n-2}\right)$ for $n \geq 2$
32. $a_0 = 1, a_1 = 1$, and $a_n = na_{n-1}$ for $n \geq 2$
33. a_n is the nth digit in the decimal expansion of π.
34. a_n is the nth digit in the decimal expansion of $1/13$.

In Exercises 35–38, find the third and the sixth partial sums of the sequence.

35. $\{n^2 - 5n + 2\}$
36. $\{(2n - 3n^2)^2\}$
37. $\{(-1)^{n+1} 5\}$
38. $\{2^n (2 - n^2)\}_{n \geq 0}$

In Exercises 39–42, express the given sum in Σ notation.

39. $\dfrac{1}{3} + \dfrac{1}{5} + \dfrac{1}{7} + \dfrac{1}{9} + \dfrac{1}{11} + \dfrac{1}{13}$
40. $2 + 1 + \dfrac{4}{5} + \dfrac{5}{7} + \dfrac{2}{3} + \dfrac{7}{11} + \dfrac{8}{13}$
41. $\dfrac{1}{8} - \dfrac{2}{9} + \dfrac{3}{10} - \dfrac{4}{11} + \dfrac{5}{12}$
42. $\dfrac{2}{3 \cdot 5} + \dfrac{4}{5 \cdot 7} + \dfrac{8}{7 \cdot 9} + \dfrac{16}{9 \cdot 11} + \dfrac{32}{11 \cdot 13} + \dfrac{64}{13 \cdot 15} + \dfrac{128}{15 \cdot 17}$

In Exercises 43–48, use a calculator to approximate the required term or sum.

43. a_{12} where $a_n = \left(1 + \dfrac{1}{n}\right)^n$

44. a_{50} where $a_n = \dfrac{\ln n}{n^2}$

45. a_{102} where $a_n = \dfrac{n^3 - n^2 + 5n}{3n^2 + 2n - 1}$

46. a_{125} where $a_n = \sqrt[n]{n}$

47. $\displaystyle\sum_{k=1}^{14} \dfrac{1}{k^2}$ 48. $\displaystyle\sum_{n=8}^{22} \dfrac{1}{n}$

Unusual Problems

Exercises 49–53 deal with prime numbers. A positive integer greater than 1 is *prime* if its only positive integer factors are itself and 1. For example, 7 is prime because its only factors are 7 and 1, but 15 is not prime because it has factors other than 15 and 1 (namely, 3 and 5).

49. (a) Let $\{a_n\}$ be the sequence of prime integers in their usual ordering. Verify that the first ten terms are 2, 3, 5, 7, 11, 13, 17, 19, 23, 29.
    (b) Find $a_{17}, a_{18}, a_{19}, a_{20}$.

In Exercises 50–53, find the first five terms of the sequence.

50. a_n is the nth prime integer larger than 10. [*Hint:* $a_1 = 11$.]
51. a_n is the square of the nth prime integer.
52. a_n is the number of prime integers less than n.
53. a_n is the largest prime integer less than $5n$.

Exercises 54–60 deal with the Fibonacci sequence $\{a_n\}$ that was discussed in Example 5.

54. Leonardo Fibonacci discovered the sequence in the 13th century in connection with this problem: A rabbit colony begins with one pair of adult rabbits (one male, one female). Each adult pair produces one pair of babies (one male, one female) every month. Each pair of baby rabbits becomes adult and produces the first offspring at age two months. Assuming that no rabbits die, how many adult pairs of rabbits are in the colony at the end of n months ($n = 1, 2, 3, \ldots$)? [*Hint:* It may be helpful to make up a chart listing for each month the number of adult pairs, the number of one-month-old pairs, and the number of baby pairs.]

55. (a) List the first ten terms of the Fibonacci sequence.
    (b) List the first ten partial sums of the sequence.
    (c) Do the partial sums follow an identifiable pattern?

56. Verify that every positive integer less than or equal to 15 can be written as a sum of Fibonacci numbers, with none used more than once.

57. Verify that $5(a_n)^2 + 4(-1)^n$ is always a perfect square for $n = 1, 2, \ldots, 10$.

58. Verify that $(a_n)^2 = a_{n+1} a_{n-1} + (-1)^{n-1}$ for $n = 2, \ldots, 10$.

59. Show that $\displaystyle\sum_{n=1}^{k} a_n = a_{k+2} - 1$. [*Hint:* $a_1 = a_3 - a_2$; $a_2 = a_4 - a_3$; etc.]

60. Show that $\displaystyle\sum_{n=1}^{k} a_{2n-1} = a_{2k}$, that is, the sum of the first k odd-numbered terms is the kth even-numbered term. [*Hint:* $a_3 = a_4 - a_2$; $a_5 = a_6 - a_4$; etc.]

9.2 ARITHMETIC SEQUENCES

In this section and the next we consider two types of sequences that arise frequently. Both are easy to deal with because there are simple formulas for their nth terms and various partial sums.

An **arithmetic sequence** (sometimes called an **arithmetic progression**) is a sequence in which the difference between each term and the preceding one is always the same constant.

Example 1 In the sequence 3, 8, 13, 18, 23, 28, . . . the difference between each term and the preceding one is always 5. So this is an arithmetic sequence. ∎

Example 2 In the sequence 14, 10, 6, 2, -2, -6, -10, -14, . . . the difference between each term and the preceding one is -4 (for example, $10 - 14 = -4$ and $-6 - (-2) = -4$). Hence this sequence is arithmetic. ■

If $\{a_n\}$ is an arithmetic sequence, then for each $n \geq 2$, the term preceding a_n is a_{n-1} and the difference $a_n - a_{n-1}$ is some constant—call it d. Therefore $a_n - a_{n-1} = d$, or equivalently,

ARITHMETIC SEQUENCES

> In an arithmetic sequence $\{a_n\}$
> $$a_n = a_{n-1} + d$$
> for some constant d and all $n \geq 2$.

The number d is called the **common difference** of the arithmetic sequence.

Example 3 If $\{a_n\}$ is an arithmetic sequence with $a_1 = 3$ and $a_2 = 4.5$, then the common difference is $d = a_2 - a_1 = 4.5 - 3 = 1.5$. So the sequence begins 3, 4.5, 6, 7.5, 9, 10.5, 12, 13.5, ... ■

Example 4 The sequence $\{-7 + 4n\}$ is an arithmetic sequence because for each $n \geq 2$,
$$a_n - a_{n-1} = (-7 + 4n) - (-7 + 4(n - 1))$$
$$= (-7 + 4n) - (-7 + 4n - 4) = 4.$$
Therefore the common difference is $d = 4$. ■

If $\{a_n\}$ is an arithmetic sequence with common difference d, then for each $n \geq 2$ we know that $a_n = a_{n-1} + d$. Applying this fact repeatedly shows that
$$a_2 = a_1 + d$$
$$a_3 = a_2 + d = (a_1 + d) + d = a_1 + 2d$$
$$a_4 = a_3 + d = (a_1 + 2d) + d = a_1 + 3d$$
$$a_5 = a_4 + d = (a_1 + 3d) + d = a_1 + 4d$$
and in general

n-TH TERM OF AN ARITHMETIC SEQUENCE

> In an arithmetic sequence $\{a_n\}$ with common difference d
> $$a_n = a_1 + (n - 1)d \quad \text{for every } n \geq 1$$

Example 5 Find the nth term of the arithmetic sequence with first term -5 and common difference 3.

Solution Since $a_1 = -5$ and $d = 3$, the formula in the box shows that
$$a_n = a_1 + (n-1)d = -5 + (n-1)3 = 3n - 8. \blacksquare$$

Example 6 What is the 45th term of the arithmetic sequence whose first three terms are 5, 9, 13?

Solution The first three terms show that $a_1 = 5$ and that the common difference d is 4. Applying the formula in the box with $n = 45$, we have
$$a_{45} = a_1 + (45-1)d = 5 + (44)4 = 181. \blacksquare$$

Example 7 If $\{a_n\}$ is an arithmetic sequence with $a_6 = 57$ and $a_{10} = 93$, find a_1 and a formula for a_n.

Solution Apply the formula $a_n = a_1 + (n-1)d$ with $n = 6$ and $n = 10$:
$$a_6 = a_1 + (6-1)d \quad \text{and} \quad a_{10} = a_1 + (10-1)d$$
$$57 = a_1 + 5d \qquad\qquad\qquad 93 = a_1 + 9d$$

We can find a_1 and d by solving this system:
$$a_1 + 9d = 93$$
$$a_1 + 5d = 57$$

Subtracting the second equation from the first shows that $4d = 36$, and hence $d = 9$. Substituting $d = 9$ in the second equation shows that $a_1 = 12$. So the formula for a_n is
$$a_n = a_1 + (n-1)d = 12 + (n-1)9 = 9n + 3. \blacksquare$$

Partial Sums

It's easy to compute partial sums of arithmetic sequences by using the following formulas.

PARTIAL SUMS OF AN ARITHMETIC SEQUENCE

If $\{a_n\}$ is an arithmetic sequence with common difference d, then for each positive integer k the kth partial sum can be found by using *either* of these formulas:

1. $\sum_{n=1}^{k} a_n = \dfrac{k}{2}(a_1 + a_k)$ or

2. $\sum_{n=1}^{k} a_n = ka_1 + \dfrac{k(k-1)}{2}d$

Proof Let S denote the kth partial sum $a_1 + a_2 + \cdots + a_k$. For reasons that will become apparent later we shall calculate the number $2S$:

$$2S = S + S = (a_1 + a_2 + \cdots + a_k) + (a_1 + a_2 + \cdots + a_k)$$

Now we rearrange the terms on the right by grouping the first and last terms together, then the first and last of the remaining terms, and so on:

$$2S = (a_1 + a_k) + (a_2 + a_{k-1}) + (a_3 + a_{k-2}) + \cdots + (a_k + a_1)$$

Since adjacent terms of the sequence differ by d we have:

$$a_2 + a_{k-1} = (a_1 + d) + (a_k - d) = a_1 + a_k.$$

Using this fact,

$$a_3 + a_{k-2} = (a_2 + d) + (a_{k-1} - d) = a_2 + a_{k-1} = a_1 + a_k.$$

Continuing in this manner we see that every pair in the sum for $2S$ is equal to $a_1 + a_k$. Therefore

$$2S = (a_1 + a_k) + (a_2 + a_{k-1}) + (a_3 + a_{k-2}) + \cdots + (a_k + a_1)$$
$$= (a_1 + a_k) + (a_1 + a_k) + (a_1 + a_k) + \cdots + (a_1 + a_k) \quad (k \text{ terms})$$
$$= k(a_1 + a_k)$$

Dividing both sides of this last equation by 2 shows that $S = \dfrac{k}{2}(a_1 + a_k)$.

This proves the first formula. To obtain the second one, note that

$$a_1 + a_k = a_1 + (a_1 + (k-1)d) = 2a_1 + (k-1)d$$

Substituting the right side of this equation in the first formula for S shows that

$$S = \frac{k}{2}(a_1 + a_k) = \frac{k}{2}(2a_1 + (k-1)d) = ka_1 + \frac{k(k-1)}{2}d.$$

This proves the second formula. ∎

Example 8 To find the 12th partial sum of the arithmetic sequence that begins $-8, -3, 2, 7, \ldots$ we first note that the common difference d is 5. Since $a_1 = -8$ and $d = 5$, the second formula in the box with $k = 12$ shows that

$$\sum_{n=1}^{12} a_n = 12(-8) + \frac{12(11)}{2} 5 = -96 + 330 = 234. \quad\blacksquare$$

Example 9 The sum of all multiples of 3 from 3 to 333 can be found by noting that it is just a partial sum of the arithmetic sequence $3, 6, 9, 12, \ldots$. Since this sequence can be written in the form

$$3 \cdot 1, 3 \cdot 2, 3 \cdot 3, 3 \cdot 4, 3 \cdot 5, 3 \cdot 6, \ldots$$

we see that $333 = 3 \cdot 111$ is the 111th term. The 111th partial sum of this sequence can be found by using the first formula in the box with $k = 111$, $a_1 = 3$, and $a_{111} = 333$:

$$\sum_{n=1}^{111} a_n = \frac{111}{2}(3 + 333) = \frac{111}{2}(336) = 18,648. \blacksquare$$

Example 10 If the starting salary for a job is $20,000 and you get a $2000 raise at the beginning of each subsequent year, what will your salary be during the tenth year? How much will you earn during the first ten years?

Solution Your yearly salary rates form a sequence: 20000, 22000, 24000, 26000, and so on. It is an arithmetic sequence with $a_1 = 20000$ and $d = 2000$. Your tenth-year salary is

$$a_{10} = a_1 + (10 - 1)d = 20000 + 9 \cdot 2000 = \$38,000.$$

Your ten-year total earnings are the 10th partial sum of the sequence:

$$\frac{10}{2}(a_1 + a_{10}) = \frac{10}{2}(20000 + 38000) = 5(58000) = \$290,000. \blacksquare$$

Arithmetic Means

To find the **average** or **mean** of numbers a and b, you add them and divide by 2, obtaining $(a + b)/2$. In the three term sequence

$$a, \frac{a + b}{2}, b$$

the difference of the first two terms is

$$\frac{a + b}{2} - a = \frac{a + b}{2} - \frac{2a}{2} = \frac{b - a}{2}$$

and the difference of the last two terms is the same number:

$$b - \frac{a + b}{2} = \frac{2b}{2} - \frac{a + b}{2} = \frac{b - a}{2}.$$

Thus $a, (a + b)/2, b$ is a finite arithmetic sequence, whose middle term is the mean of a and b.

If $a, b,$ and $m_1, m_2, \ldots, m_k$ are real numbers such that

$$a, m_1, m_2, \ldots, m_k, b$$

is an arithmetic sequence, we say that $m_1, m_2, \ldots, m_k$ are k **arithmetic means** between a and b.

Example 11 Insert four arithmetic means between 6 and 14.

Solution We must find numbers m_1, m_2, m_3, m_4 such that

$$6, m_1, m_2, m_3, m_4, 14$$

is an arithmetic sequence. Thus the sequence $\{a_n\}$ we are looking for has $a_1 = 6, a_2 = m_1, a_3 = m_2, \ldots, a_6 = 14$. If the common difference of this sequence is d, then the formula for the n-th term (with $n = 6$) shows that

$$a_6 = a_1 + (6-1)d$$
$$14 = 6 + 5d$$
$$d = \frac{8}{5}$$

Since each term in an arithmetic sequence is the sum of the preceding term and the common difference d, the four arithmetic means are:

$$m_1 = a_1 + d = 6 + \frac{8}{5} = 7\frac{3}{5}$$

$$m_2 = m_1 + d = 7\frac{3}{5} + \frac{8}{5} = 9\frac{1}{5}$$

$$m_3 = m_2 + d = 9\frac{1}{5} + \frac{8}{5} = 10\frac{4}{5}$$

$$m_4 = m_3 + d = 10\frac{4}{5} + \frac{8}{5} = 12\frac{2}{5}. \blacksquare$$

EXERCISES

In Exercises 1–6, the first term a_1 and the common difference d of an arithmetic sequence are given. Find the fifth term and the formula for the nth term.

1. $a_1 = 5, d = 2$
2. $a_1 = -4, d = 5$
3. $a_1 = 4, d = \frac{1}{4}$
4. $a_1 = -6, d = \frac{2}{3}$
5. $a_1 = 10, d = -\frac{1}{2}$
6. $a_1 = \pi, d = \frac{1}{5}$

In Exercises 7–12, find the kth partial sum of the arithmetic sequence $\{a_n\}$ with common difference d.

7. $k = 6, a_1 = 2, d = 5$
8. $k = 8, a_1 = \frac{2}{3}, d = -\frac{4}{3}$
9. $k = 7, a_1 = \frac{3}{4}, d = -\frac{1}{2}$
10. $k = 9, a_1 = 6, a_9 = -24$
11. $k = 6, a_1 = -4, a_6 = 14$
12. $k = 10, a_1 = 0, a_{10} = 30$

In Exercises 13–18, show that the sequence is arithmetic and find its common difference.

13. $\{3 - 2n\}$
14. $\left\{4 + \frac{n}{3}\right\}$
15. $\left\{\frac{5 + 3n}{2}\right\}$
16. $\left\{\frac{\pi - n}{2}\right\}$
17. $\{c + 2n\}$ (c constant)
18. $\{2b + 3nc\}$ (b, c constants)

In Exercises 19–24, use the given information about the arithmetic sequence with common difference d to find a_5 and a formula for a_n.

19. $a_4 = 12, d = 2$
20. $a_7 = -8, d = 3$
21. $a_2 = 4, a_6 = 32$
22. $a_7 = 6, a_{12} = -4$
23. $a_5 = 0, a_9 = 6$
24. $a_5 = -3, a_9 = -18$

In Exercises 25–28, find the sum.

25. $\sum_{n=1}^{20} (3n + 4)$
26. $\sum_{n=1}^{25} \left(\frac{k}{4} + 5\right)$
27. $\sum_{n=1}^{40} \frac{n+3}{6}$
28. $\sum_{n=1}^{30} \frac{4-6n}{3}$

29. Find the sum of all the even integers from 2 to 100.
30. Find the sum of all the integer multiples of 7 from 7 to 700.
31. Find the sum of the first 200 positive integers.
32. Find the sum of the positive integers from 101 to 200 (inclusive). [*Hint:* What's the sum from 1 to 100? Use it and Exercise 31.]
33. Insert 5 arithmetic means between 4 and 14.
34. Insert 4 arithmetic means between -3 and 7.
35. A business makes a $10,000 profit during its first year. If the yearly profit increases by $7500 in each subsequent year, what will the profit be in the tenth year and what will the total profit for the first ten years be?
36. If a man's starting salary is $15,000 and he receives a $1000 increase every six months, what will his salary be during the last six months of the sixth year? How much will he earn during the first six years?
37. A lecture hall has 6 seats in the first row, 8 in the second, 10 in the third, and so on. But rows 12 through 20 all have the same number of seats. Find the number of seats in the lecture hall.
38. A monument is constructed by laying a row of 60 bricks at ground level. A second row, with 2 fewer bricks, is centered on that; a third row, with 2 fewer bricks, is centered on the second; and so on. The top row contains 10 bricks. How many bricks are there in the monument?
39. A ladder with nine rungs is to be built, with the bottom rung 24 inches wide and the top rung 18 inches wide. If the lengths of the rungs decreases uniformly from bottom to top, how long should each of the seven intermediate rungs be?
40. Find the first eight numbers in an arithmetic sequence in which the sum of the first and seventh term is 40 and the product of the first and fourth terms is 160.

9.3 GEOMETRIC SEQUENCES

A **geometric sequence** (sometimes called a **geometric progression**) is a sequence in which the quotient of each term and the preceding one is the same constant r. This constant r is called the **common ratio** of the geometric sequence.

Example 1 The sequence $3, 9, 27, \ldots, 3^n, \ldots$ is geometric with common ratio 3. For instance $a_2/a_1 = 9/3 = 3$ and $a_3/a_2 = 27/9 = 3$. If 3^n is any term ($n \geq 2$), then the preceding term is 3^{n-1} and

$$\frac{3^n}{3^{n-1}} = \frac{3 \cdot 3^{n-1}}{3^{n-1}} = 3. \blacksquare$$

Example 2 The sequence $\{5/2^n\}$ which begins $5/2, 5/4, 5/8, 5/16, \ldots$ is geometric with common ratio $r = 1/2$ because for each $n \geq 1$

$$\frac{5/2^n}{5/2^{n-1}} = \frac{5}{2^n} \cdot \frac{2^{n-1}}{5} = \frac{2^{n-1}}{2^n} = \frac{2^{n-1}}{2^{n-1} \cdot 2} = \frac{1}{2}. \blacksquare$$

If $\{a_n\}$ is a geometric sequence with common ratio r, then for each $n \geq 2$ the term preceding a_n is a_{n-1} and

$$a_n/a_{n-1} = r, \quad \text{or equivalently,} \quad a_n = ra_{n-1}$$

Applying this last formula for $n = 2, 3, 4, \ldots$ we have:

$$a_2 = ra_1$$
$$a_3 = ra_2 = r(ra_1) = r^2 a_1$$
$$a_4 = ra_3 = r(r^2 a_1) = r^3 a_1$$
$$a_5 = ra_4 = r(r^3 a_1) = r^4 a_1$$

and in general

nTH TERM OF A GEOMETRIC SEQUENCE

If $\{a_n\}$ is a geometric sequence with common ratio r, then for all $n \geq 1$,

$$a_n = r^{n-1} a_1$$

Example 3 To find a formula for the nth term of the geometric sequence $\{a_n\}$ where $a_1 = 7$ and $r = 2$, we use the equation in the box:

$$a_n = r^{n-1} a_1 = 2^{n-1} \cdot 7.$$

So the sequence is $\{7 \cdot 2^{n-1}\}$. ■

Example 4 If the first two terms of a geometric sequence are 2 and $-2/5$, then the common ratio must be

$$r = \frac{a_2}{a_1} = \frac{-2/5}{2} = \frac{-2}{5} \cdot \frac{1}{2} = -\frac{1}{5}.$$

Using the equation in the box, we now see that the formula for the nth term is

$$a_n = r^{n-1} a_1 = \left(-\frac{1}{5}\right)^{n-1}(2) = \frac{(1)^{n-1}}{(-5)^{n-1}}(2) = \frac{2}{(-5)^{n-1}}$$

So the sequence begins $2, -2/5, 2/5^2, -2/5^3, 2/5^4, \ldots$. ■

Example 5 If $\{a_n\}$ is a geometric sequence with $a_2 = 20/9$ and $a_5 = 160/243$, then by the equation in the box above,

$$\frac{160/243}{20/9} = \frac{a_5}{a_2} = \frac{r^4 a_1}{r a_1} = r^3.$$

Consequently,

$$r = \sqrt[3]{\frac{160/243}{20/9}} = \sqrt[3]{\frac{160}{243} \cdot \frac{9}{20}} = \sqrt[3]{\frac{8 \cdot 9}{243}} = \sqrt[3]{\frac{8}{27}} = \frac{2}{3}.$$

Since $a_2 = ra_1$ we see that
$$a_1 = \frac{a_2}{r} = \frac{20/9}{2/3} = \frac{20}{9} \cdot \frac{3}{2} = \frac{10}{3}.$$

Therefore,
$$a_n = r^{n-1}a_1 = \left(\frac{2}{3}\right)^{n-1} \cdot \frac{10}{3} = \frac{2^{n-1} \cdot 2 \cdot 5}{3^{n-1} \cdot 3} = \frac{2^n \cdot 5}{3^n} = 5\left(\frac{2}{3}\right)^n. \blacksquare$$

Partial Sums

If the common ratio r of a geometric sequence is the number 1, then we have
$$a_n = 1^{n-1}a_1 \quad \text{for every } n \geq 1.$$

Therefore the sequence is just the constant sequence $a_1, a_1, a_1, \ldots$. For any positive integer k, the kth partial sum of this constant sequence is
$$\underbrace{a_1 + a_1 + \cdots + a_1}_{k \text{ terms}} = ka_1.$$

In other words, the kth partial sum of a constant sequence is just k times the constant. If a geometric sequence is not constant (that is, $r \neq 1$), then its partial sums are given by the following formula.

PARTIAL SUMS OF A GEOMETRIC SEQUENCE

> The kth partial sum of the geometric sequence $\{a_n\}$ with common ratio $r \neq 1$ is
> $$\sum_{n=1}^{k} a_n = a_1\left(\frac{1 - r^k}{1 - r}\right)$$

Proof If S denotes the k-th partial sum, then the formula for the n-th term of a geometric sequence shows that
$$S = a_1 + a_2 + \cdots + a_k = a_1 + a_1 r + a_1 r^2 + a_1 r^3 + \cdots + a_1 r^{k-1}.$$

Use this equation to compute $S - rS$:
$$\begin{aligned} S &= a_1 + a_1 r + a_1 r^2 + a_1 r^3 + \cdots + a_1 r^{k-1} \\ rS &= \phantom{a_1 +{}} a_1 r + a_1 r^2 + a_1 r^3 + \cdots + a_1 r^{k-1} + a_1 r^k \\ \hline S - rS &= a_1 \phantom{{}+ a_1 r + a_1 r^2 + a_1 r^3 + \cdots + a_1 r^{k-1}} - a_1 r^k \\ (1 - r)S &= a_1(1 - r^k) \end{aligned}$$

Since $r \neq 1$, we can divide both sides of this last equation by $1 - r$ to complete the proof:
$$S = \frac{a_1(1 - r^k)}{1 - r} = a_1\left(\frac{1 - r^k}{1 - r}\right). \blacksquare$$

Example 6 To find the sum

$$-\frac{3}{2}+\frac{3}{4}-\frac{3}{8}+\frac{3}{16}-\frac{3}{32}+\frac{3}{64}-\frac{3}{128}+\frac{3}{256}-\frac{3}{512}$$

we note that this is the 9th partial sum of the geometric sequence $\left\{3\left(\frac{-1}{2}\right)^n\right\}$. The common ratio is $r = -1/2$. The formula in the box shows that

$$\sum_{n=1}^{9} 3\left(\frac{-1}{2}\right)^n = a_1\left(\frac{1-r^9}{1-r}\right) = \left(\frac{-3}{2}\right)\left(\frac{1-(-1/2)^9}{1-(-1/2)}\right)$$

$$= \left(\frac{-3}{2}\right)\left(\frac{1+1/2^9}{3/2}\right) = \left(\frac{-3}{2}\right)\left(\frac{2}{3}\right)\left(1+\frac{1}{2^9}\right)$$

$$= -1 - \frac{1}{2^9} = -1 - \frac{1}{512} = -\frac{513}{512}. \blacksquare$$

Example 7 A superball is dropped from a height of 9 ft. It hits the ground and bounces to a height of 6 ft. It continues to bounce up and down. On each bounce it rises to 2/3 of the height of the previous bounce. How far has the ball traveled (both up and down) when it hits the ground for the seventh time?

Solution We first consider how far the ball travels on each bounce. On the first bounce it rises 6 ft and falls 6 ft for a total of 12 ft. On the second bounce it rises and falls 2/3 of the previous height, and hence travels 2/3 of 12 ft. If a_n denotes the distance traveled on the nth bounce, then

$$a_1 = 12 \qquad a_2 = \left(\frac{2}{3}\right)a_1 \qquad a_3 = \left(\frac{2}{3}\right)a_2 = \left(\frac{2}{3}\right)^2 a_1$$

and in general

$$a_n = \left(\frac{2}{3}\right)a_{n-1} = \left(\frac{2}{3}\right)^{n-1} a_1$$

So $\{a_n\}$ is a geometric sequence with common ratio $r = 2/3$. When the ball hits the ground for the seventh time it has completed six bounces. Therefore the total distance it has traveled is the distance it was originally dropped (9 ft) plus the distance traveled in six bounces, namely,

$$9 + a_1 + a_2 + a_3 + a_4 + a_5 + a_6 = 9 + \sum_{n=1}^{6} a_n = 9 + a_1\left(\frac{1-r^6}{1-r}\right)$$

$$= 9 + 12\left(\frac{1-(2/3)^6}{1-(2/3)}\right)$$

$$= 9 + 36(1-(2/3)^6) \approx 41.84 \text{ ft.} \blacksquare$$

Geometric Means

If a, b, and $m_1, m_2, \ldots, m_k$ are real numbers such that

$$a, m_1, m_2, \ldots, m_k, b$$

is a geometric sequence, we say that $m_1, m_2, \ldots, m_k$ are k **geometric means** between a and b.

Example 8 Insert two geometric means between 8 and 27.

Solution We must find numbers m_1 and m_2 such that

$$8, m_1, m_2, 27$$

is a geometric sequence. Thus the sequence $\{a_n\}$ we are looking for has $a_1 = 8$, $a_2 = m_1$, $a_3 = m_2$, $a_4 = 27$. If the common ratio of this sequence is r, then the formula for the n-th term (with $n = 4$) shows that

$$a_4 = a_1 r^3$$
$$27 = 8r^3$$
$$r^3 = \frac{27}{8} = \left(\frac{3}{2}\right)^3$$
$$r = \frac{3}{2}$$

Since each term in a geometric sequence is r times the preceding one, the two geometric means are:

$$m_1 = ra_1 = \frac{3}{2} \cdot 8 = 12$$

$$m_2 = rm_1 = \frac{3}{2} \cdot 12 = 18. \blacksquare$$

EXERCISES

In Exercises 1–8, determine whether the sequence is arithmetic, geometric, or neither.

1. 2, 7, 12, 17, 22, . . .
2. 2, 6, 18, 54, 162, . . .
3. 13, 13/2, 13/4, 13/8, . . .
4. $-1, -\frac{1}{2}, 0, \frac{1}{2}, \ldots$
5. 50, 48, 46, 44, . . .
6. 2, −3, 9/2, −27/4, −81/8, . . .
7. 3, −3/2, 3/4, −3/8, 3/16, . . .
8. −6, −3.7, −1.4, .9, 3.2, . . .

In Exercises 9–14, the first term a_1 and the common ratio r of a geometric sequence are given. Find the sixth term and a formula for the nth term.

9. $a_1 = 5, r = 2$
10. $a_1 = 1, r = -2$

11. $a_1 = 4, r = \dfrac{1}{4}$ 12. $a_1 = -6, r = \dfrac{2}{3}$

13. $a_1 = 10, r = -\dfrac{1}{2}$ 14. $a_1 = \pi, r = \dfrac{1}{5}$

In Exercises 15–18, find the kth partial sum of the geometric sequence $\{a_n\}$ with common ratio r.

15. $k = 6, a_1 = 5, r = \dfrac{1}{2}$

16. $k = 8, a_1 = 9, r = \dfrac{1}{3}$

17. $k = 7, a_2 = 6, r = 2$

18. $k = 9, a_2 = 6, r = \dfrac{1}{4}$

In Exercises 19–22, show that the given sequence is geometric and find the common ratio.

19. $\left\{\left(-\dfrac{1}{2}\right)^n\right\}$ 20. $\{2^{3n}\}$

21. $\{5^{n+2}\}$ 22. $\{3^{n/2}\}$

In Exercises 23–28, use the given information about the geometric sequence $\{a_n\}$ to find a_5 and a formula for a_n.

23. $a_1 = 256, a_2 = -64$

24. $a_1 = 1/6, a_2 = -1/18$

25. $a_2 = 4, a_5 = 1/16$

26. $a_3 = 4, a_6 = -32$

27. $a_4 = -4/5, r = 2/5$

28. $a_2 = 6, a_7 = 192$

In Exercises 29–34, find the sum.

29. $\displaystyle\sum_{n=1}^{7} 2^n$ 30. $\displaystyle\sum_{k=1}^{6} 3\left(\dfrac{1}{2}\right)^k$

31. $\displaystyle\sum_{n=1}^{9} \left(-\dfrac{1}{3}\right)^n$ 32. $\displaystyle\sum_{n=1}^{5} 5 \cdot 3^{n-1}$

33. $\displaystyle\sum_{j=1}^{6} 4\left(\dfrac{3}{2}\right)^{j-1}$ 34. $\displaystyle\sum_{t=1}^{8} 6(.9)^{t-1}$

35. Insert two geometric means between 4 and 256.

36. Insert four geometric means between 96 and 3.

37. Insert three geometric means between 25 and 50.

38. Insert one geometric mean between 3 and 108. Insert one arithmetic mean between 3 and 108. Which is larger?

Unusual Problems

39. Suppose $\{a_n\}$ is a geometric sequence with common ratio $r > 0$ and each $a_n > 0$. Show that the sequence $\{\log a_n\}$ is an arithmetic sequence with common difference $\log r$.

40. Suppose $\{a_n\}$ is an arithmetic sequence with common difference d. Let C be any positive number. Show that the sequence $\{C^{a_n}\}$ is a geometric sequence with common ratio C^d.

41. A ball is dropped from a height of 8 ft. On each bounce it rises to half its previous height. When the ball hits the ground for the seventh time, how far has it traveled?

42. A ball is dropped from a height of 10 ft. On each bounce it rises to 45% of its previous height. When it hits the ground for the tenth time, how far has it traveled?

43. In the geometric sequence 1, 2, 4, 8, 16, . . . , show that each term is 1 plus the sum of all preceding terms.

44. In the geometric sequence 2, 6, 18, 54, . . . , show that each term is twice the sum of 1 and all preceding terms.

45. If you are paid a salary of 1¢ on the first day of March, 2¢ on the second day, and your salary continues to double each day, how much will you earn in the month of March?

46. Starting with your parents, how many ancestors do you have for the past ten generations?

47. A car that sold for $8000 new depreciates in value 25% each year. What is it worth after five years?

48. A vacuum pump removes 60% of the air in a container at each stroke. What percentage of the original amount of air remains after 6 strokes?

49. The minimum monthly payment for a certain bank credit card is the larger of 1/25 of the outstanding balance or $5. If the balance is less than $5, the entire balance is due. If you make only the minimum payment each month, how long will it take to pay off a balance of $200 (excluding any interest that might be due)?

50. If a and b are nonnegative, show that $\sqrt{ab}$ is a single geometric mean between a and b.

9.3.A EXCURSION: Infinite Series

We now introduce a topic that is closely related to infinite sequences and has some very useful applications. We can only give a few highlights here; complete coverage requires calculus. Consider the sequence $\{3/10^n\}$ and let S_k denote its k-th partial sum; then

$$S_1 = \frac{3}{10}$$

$$S_2 = \frac{3}{10} + \frac{3}{10^2} = \frac{33}{100}$$

$$S_3 = \frac{3}{10} + \frac{3}{10^2} + \frac{3}{10^3} = \frac{333}{1000}$$

$$S_4 = \frac{3}{10} + \frac{3}{10^2} + \frac{3}{10^3} + \frac{3}{10^4} = \frac{3333}{10,000}.$$

These partial sums $S_1, S_2, S_3, S_4, \ldots$ themselves form a sequence:

$$\frac{3}{10}, \frac{33}{100}, \frac{333}{1000}, \frac{3333}{10,000}, \ldots$$

The terms in the sequence of partial sums appear to be getting closer and closer to 1/3. In other words, as k gets larger and larger, the corresponding partial sum S_k gets closer and closer to 1/3. Consequently, we write

$$\frac{3}{10} + \frac{3}{10^2} + \frac{3}{10^3} + \frac{3}{10^4} + \cdots = \frac{1}{3}$$

and say that 1/3 is the *sum* of the *infinite series*

$$\frac{3}{10} + \frac{3}{10^2} + \frac{3}{10^3} + \frac{3}{10^4} + \cdots.$$

In the general case, an **infinite series** (or simply **series**) is defined to be an expression of the form

$$a_1 + a_2 + a_3 + a_4 + a_5 + \cdots$$

in which each a_n is a real number. This series is also denoted by the symbol

$$\sum_{n=1}^{\infty} a_n.$$

Example 1

(a) $\displaystyle\sum_{n=1}^{\infty} 2(.6)^n$ denotes the series

$$2(.6) + 2(.6)^2 + 2(.6)^3 + 2(.6)^4 + \cdots$$

(b) $\sum_{n=1}^{\infty} \left(\frac{-1}{2}\right)^n$ denotes the series

$$-\frac{1}{2} + \left(\frac{-1}{2}\right)^2 + \left(\frac{-1}{2}\right)^3 + \left(\frac{-1}{2}\right)^4 + \cdots = -\frac{1}{2} + \frac{1}{4} - \frac{1}{8} + \frac{1}{16} + \cdots . \blacksquare$$

The **partial sums** of the series $a_1 + a_2 + a_3 + a_4 + \cdots$ are

$$S_1 = a_1$$
$$S_2 = a_1 + a_2$$
$$S_3 = a_1 + a_2 + a_3$$

and in general, for any $k \geq 1$

$$S_k = a_1 + a_2 + a_3 + a_4 + \cdots + a_k.$$

If it happens that the terms $S_1, S_2, S_3, S_4, \ldots$ of the *sequence* of partial sums get closer and closer to a particular real number S in such a way the partial sum S_k is arbitrarily close to S when k is large enough, then we say that the series **converges** and that S is the **sum of the convergent series.** For example, we just saw that the series

$$\frac{3}{10} + \frac{3}{10^2} + \frac{3}{10^3} + \frac{3}{10^4} + \cdots$$

converges and that its sum is 1/3.

A sequence is a *list* of numbers $a_1, a_2, a_3, \ldots$. Intuitively, you can think of a convergent series $a_1 + a_2 + a_3 + \cdots$ as an "infinite sum" of numbers. But be careful: not every series has a sum. For instance, the partial sums of the series

$$1 + 2 + 3 + 4 + \cdots$$

get larger and larger (compute some) and do not get closer and closer to a single real number. So this series is not convergent.

Example 2 Although no proof will be given here, it is intuitively clear that every infinite decimal may be thought of as the sum of a convergent series. For instance,

$$\pi = 3.1415926 \cdots = 3 + .1 + .04 + .001 + .0005 + .00009 + \cdots .$$

Note that the third partial sum is $3 + .1 + .04 = 3.14$, which is π to 2 decimal places. Similarly, the k-th partial sum of this series is just π to $k - 1$ decimal places. $\blacksquare$

Infinite Geometric Series

If $\{a_n\}$ is a geometric sequence with common ratio r, then the corresponding infinite series

$$a_1 + a_2 + a_3 + a_4 + a_5 + \cdots$$

is called an **infinite geometric series.** By using the formula for the n-th term of a geometric sequence, we can also express the corresponding geometric series in the form

$$a_1 + ra_1 + r^2 a_1 + r^3 a_1 + r^4 a_1 + \cdots$$

Under certain circumstances, an infinite geometric series is convergent and has a sum:

SUM OF AN INFINITE GEOMETRIC SERIES

If $|r| < 1$, then the infinite geometric series

$$a_1 + ra_1 + r^2 a_1 + r^3 a_1 + r^4 a_1 + \cdots$$

converges and its sum is

$$\frac{a_1}{1-r}.$$

Although we cannot prove this fact rigorously here, we can make it highly plausible. The k-th partial sum S_k of the geometric series $a_1 + a_2 + a_3 + \cdots$ is the same as the k-th partial sum of the geometric sequence $\{a_n\}$ and hence

$$S_k = a_1 \left(\frac{1 - r^k}{1 - r} \right).$$

As k gets larger and larger, the number r^k gets very close to 0 (because $|r| < 1$; for instance $.7^{20} \approx .000798$ and $(-.3)^9 \approx .0000197$). Consequently, when k is very large $1 - r^k$ is very close to $1 - 0$ so that

$$S_k = a_1 \left(\frac{1 - r^k}{1 - r} \right) \text{ is very close to } a_1 \left(\frac{1 - 0}{1 - r} \right) = \frac{a_1}{1 - r}.$$

Example 3 $\sum_{n=1}^{\infty} \left(\frac{-1}{2} \right)^n = -\frac{1}{2} + \frac{1}{4} - \frac{1}{8} + \frac{1}{16} + \cdots$ is an infinite geometric series with $a_1 = -1/2$ and $r = -1/2$. Since $|r| < 1$, this series converges and its sum is

$$\frac{a_1}{1-r} = \frac{-\frac{1}{2}}{1 - \left(-\frac{1}{2}\right)} = \frac{-\frac{1}{2}}{\frac{3}{2}} = -\frac{1}{3}. \blacksquare$$

Infinite geometric series provide another way of writing an infinite repeating decimal as a rational number.

Example 4 To express $6.8573573573\ldots$ as a rational number, we first write it as $6.8 + .0573573573\ldots$. Consider $.0573573573\ldots$ as an infinite series:

$$.0573 + .0000573 + .0000000573 + .0000000000573 + \cdots$$

which is the same as

$$.0573 + (.001)(.0573) + (.001)^2(.0573) + (.001)^3(.0573) + \cdots$$

This is a convergent geometric series with $a_1 = .0573$ and $r = .001$. Its sum is

$$\frac{a_1}{1-r} = \frac{.0573}{1-.001} = \frac{.0573}{.999} = \frac{573}{9990}.$$

Therefore

$$\begin{aligned} 6.8573573573\cdots &= 6.8 + [.0573 + .0000573 + \cdots] \\ &= 6.8 + \frac{573}{9990} \\ &= \frac{68}{10} + \frac{573}{9990} \\ &= \frac{68{,}505}{9990} = \frac{4567}{666}. \blacksquare \end{aligned}$$

EXERCISES

In Exercises 1–8, find the sum of the infinite series, if it has one.

1. $\sum_{n=1}^{\infty} \frac{1}{2^n}$
2. $\sum_{n=1}^{\infty} \left(-\frac{3}{4}\right)^n$
3. $\sum_{n=1}^{\infty} (.06)^n$
4. $1 - .5 + .25 - .375 + .0625 - \cdots$
5. $500 + 200 + 80 + 32 + \cdots$
6. $9 - 3\sqrt{3} + 3 - \sqrt{3} + 1 - \frac{1}{\sqrt{3}} + \cdots$
7. $2 + \sqrt{2} + 1 + \frac{1}{\sqrt{2}} + \frac{1}{2} + \cdots$
8. $\sum_{n=1}^{\infty} \left(\frac{1}{2^n} - \frac{1}{3^n}\right)$

In Exercises 9–15, express the repeating decimal as a rational number.

9. $.22222\ldots$
10. $.37373737\ldots$
11. $5.4272727\ldots$
12. $85.131313\ldots$
13. $2.1425425425\ldots$
14. $3.7165165165\ldots$
15. $1.74241241241\ldots$

16. If $\{a_n\}$ is an arithmetic sequence with common difference $d > 0$ and each $a_i > 0$, explain why the infinite series $a_1 + a_2 + a_3 + a_4 + \cdots$ is not convergent.

9.4 THE BINOMIAL THEOREM

The Binomial Theorem provides a formula for calculating the product $(x + y)^n$ for any positive integer n. Before we state the theorem, some preliminaries are needed.

9.4 THE BINOMIAL THEOREM

Let n be a positive integer. The symbol $n!$ (read **n factorial**) denotes the product of all the integers from 1 to n. For example,

$$2! = 1 \cdot 2 = 2, \quad 3! = 1 \cdot 2 \cdot 3 = 6, \quad 4! = 1 \cdot 2 \cdot 3 \cdot 4 = 24,$$

$$5! = 1 \cdot 2 \cdot 3 \cdot 4 \cdot 5 = 120, \quad 10! = 1 \cdot 2 \cdot 3 \cdot 4 \cdot 5 \cdot 6 \cdot 7 \cdot 8 \cdot 9 \cdot 10 = 3{,}628{,}800$$

and in general,

n FACTORIAL

$$n! = 1 \cdot 2 \cdot 3 \cdot 4 \cdots (n-2)(n-1)n$$

We *define* $0!$ to be the number 1.

If r and n are integers with $0 \le r \le n$, then the symbol $\binom{n}{r}$ is called a **binomial coefficient** and denotes this number:

BINOMIAL COEFFICIENTS

$$\binom{n}{r} = \frac{n!}{r!(n-r)!}$$

For example,

$$\binom{5}{3} = \frac{5!}{3!(5-3)!} = \frac{5!}{3!2!} = \frac{1 \cdot 2 \cdot 3 \cdot 4 \cdot 5}{(1 \cdot 2 \cdot 3)(1 \cdot 2)} = \frac{4 \cdot 5}{2} = 10$$

$$\binom{4}{2} = \frac{4!}{2!(4-2)!} = \frac{4!}{2!2!} = \frac{1 \cdot 2 \cdot 3 \cdot 4}{(1 \cdot 2)(1 \cdot 2)} = \frac{3 \cdot 4}{2} = 6$$

The examples illustrate the fact that *every binomial coefficient* $\binom{n}{r}$ *is an integer.* We shall assume this fact for now and prove it in Section 9.5. For any nonnegative integer n,

$$\binom{n}{0} = 1 \quad \text{and} \quad \binom{n}{n} = 1$$

because

$$\binom{n}{0} = \frac{n!}{0!(n-0)!} = \frac{n!}{0!n!} = \frac{n!}{n!} = 1 \text{ and } \binom{n}{n} = \frac{n!}{n!(n-n)!} = \frac{n!}{n!0!} = \frac{n!}{n!} = 1.$$

If we list the binomial coefficients for each value of n in this manner:

$n = 0$ $\binom{0}{0}$

$n = 1$ $\binom{1}{0}$ $\binom{1}{1}$

$n = 2$ $\binom{2}{0}$ $\binom{2}{1}$ $\binom{2}{2}$

$n = 3$ $\binom{3}{0}$ $\binom{3}{1}$ $\binom{3}{2}$ $\binom{3}{3}$

$n = 4$ $\binom{4}{0}$ $\binom{4}{1}$ $\binom{4}{2}$ $\binom{4}{3}$ $\binom{4}{4}$

$\vdots$ $\iddots$ $\ddots$

and then calculate each of them, we obtain the following array of numbers:

row 0			1		
row 1		1		1	
row 2		1	2	1	
row 3	1	3		3	1
row 4	1	4	6	4	1

$\vdots$ $\iddots$ $\ddots$

This array is called **Pascal's triangle**. Its pattern is easy to remember: Each entry (except the 1's at the beginning or end of a row) is the sum of the two closest entries in the row above it. In the fourth row, for instance, 6 is the sum of the two 3's above it, and each 4 is the sum of the 1 and 3 above it. See Exercise 45 for a proof.

In order to develop a formula for calculating $(x + y)^n$, we first calculate these products for small values of n to see if we can find some kind of pattern:

$n = 0$ $\quad (x + y)^0 = \quad 1$

$n = 1$ $\quad (x + y)^1 = \quad x + y$

$n = 2$ $\quad (x + y)^2 = \quad x^2 + 2xy + y^2$

$n = 3$ $\quad (x + y)^3 = \quad x^3 + 3x^2y + 3xy^2 + y^3$

$n = 4$ $\quad (x + y)^4 = \quad x^4 + 4x^3y + 6x^2y^2 + 4xy^3 + y^4$

Some parts of the pattern are already clear. For each positive n, the first term is x^n and the last term is y^n. Beginning with the second term,

The successive exponents of y are 1, 2, 3, . . . , n.

In each term before the last one, the exponent of x is 1 less than the preceding term. Suppose this pattern holds true for larger values of n as well. Then for a fixed n, the expansion of $(x + y)^n$ would have first term x^n. In the second term, the exponent of x would be 1 less than n, namely, $n - 1$, and the exponent of y would be 1. So the second term would be of the form (constant)$x^{n-1}y$. In the next term, the exponent of x would be 1 less again, namely, $n - 2$, and the exponent of y would be 2. Continuing in this fashion, we would have

$$(x + y)^n = x^n + (*)x^{n-1}y + (*)x^{n-2}y^2 + (*)x^{n-3}y^3 + \cdots (*)xy^{n-1} + y^n$$

where the symbols $(*)$ indicate the various constant coefficients.

In order to determine the constant coefficients in the expansion of $(x + y)^n$, we return to the computations made above for $n = 0, 1, 2, 3, 4$. Each of the terms x, y, x^2, y^2, and so on, has coefficient 1. If we omit the x's and y's and just list the coefficients that appear in the computations above, we obtain this array of numbers:

$$
\begin{array}{cccccccccc}
n = 0 & & & & & 1 & & & & \\
n = 1 & & & & 1 & & 1 & & & \\
n = 2 & & & 1 & & 2 & & 1 & & \\
n = 3 & & 1 & & 3 & & 3 & & 1 & \\
n = 4 & 1 & & 4 & & 6 & & 4 & & 1
\end{array}
$$

But this is just the top of Pascal's triangle. In the case $n = 4$, it means that the coefficients of the expansion of $(x + y)^4$ are just the binomial coefficients $\binom{4}{0}$, $\binom{4}{1}$, $\binom{4}{2}$, $\binom{4}{3}$, $\binom{4}{4}$; and similarly for the other small values of n. If this pattern holds true for larger n as well, then the coefficients of the expansion of $(x + y)^n$ are just the binomial coefficients

$$\binom{n}{0}, \binom{n}{1}, \binom{n}{2}, \binom{n}{3}, \ldots, \binom{n}{n-1}, \binom{n}{n}$$

Since $\binom{n}{0} = 1$ and $\binom{n}{n} = 1$ for every n, the first and last coefficients on this list are 1. This is consistent with the fact that the first and last terms are x^n and y^n.

The preceding discussion suggests that the following result is true:

THE BINOMIAL THEOREM

For each positive integer n,

$$(x + y)^n = x^n + \binom{n}{1}x^{n-1}y + \binom{n}{2}x^{n-2}y^2 +$$

$$\binom{n}{3}x^{n-3}y^3 + \cdots + \binom{n}{n-1}xy^{n-1} + y^n$$

Using summation notation and the fact that $\binom{n}{0} = 1 = \binom{n}{n}$, the Binomial Theorem can be written compactly as $(x + y)^n = \sum_{j=0}^{n} \binom{n}{j} x^{n-j} y^j$. The Binomial Theorem will be proved in Section 9.5 by means of mathematical induction. We shall assume its truth for now and illustrate some of its uses.

Example 1 In order to compute $(x + y)^8$ we apply the Binomial Theorem in the case $n = 8$:

$$(x + y)^8 = x^8 + \binom{8}{1} x^7 y + \binom{8}{2} x^6 y^2 + \binom{8}{3} x^5 y^3$$
$$+ \binom{8}{4} x^4 y^4 + \binom{8}{5} x^3 y^5 + \binom{8}{6} x^2 y^6 + \binom{8}{7} xy^7 + y^8$$

Now verify that

$$\binom{8}{1} = \frac{8!}{1!7!} = 8, \quad \binom{8}{2} = \frac{8!}{2!6!} = 28,$$

$$\binom{8}{3} = \frac{8!}{3!5!} = 56, \quad \binom{8}{4} = \frac{8!}{4!4!} = 70$$

Using these facts, we see that

$$\binom{8}{5} = \frac{8!}{5!3!} = \binom{8}{3} = 56, \quad \binom{8}{6} = \frac{8!}{6!2!} = \binom{8}{2} = 28,$$

$$\binom{8}{7} = \frac{8!}{7!1!} = \binom{8}{1} = 8$$

Substituting these values in the expansion above, we have

$(x + y)^8 = x^8 + 8x^7 y + 28x^6 y^2$
$\qquad + 56x^5 y^3 + 70x^4 y^4 + 56x^3 y^5 + 28x^2 y^6 + 8xy^7 + y^8.$ ■

Example 2 To find $(1 - z)^6$, we note that $1 - z = 1 + (-z)$ and apply the Binomial Theorem with $x = 1$, $y = -z$, and $n = 6$:

$$(1 - z)^6 = 1^6 + \binom{6}{1} 1^5(-z) + \binom{6}{2} 1^4(-z)^2 + \binom{6}{3} 1^3(-z)^3$$
$$+ \binom{6}{4} 1^2(-z)^4 + \binom{6}{5} 1(-z)^5 + (-z)^6$$
$$= 1 - \binom{6}{1} z + \binom{6}{2} z^2 - \binom{6}{3} z^3 + \binom{6}{4} z^4 - \binom{6}{5} z^5 + z^6$$
$$= 1 - 6z + 15z^2 - 20z^3 + 15z^4 - 6z^5 + z^6. ■$$

9.4 THE BINOMIAL THEOREM

Example 3 To expand $(x^2 + x^{-1})^4$, use the Binomial Theorem with x^2 in place of x and x^{-1} in place of y:

$$(x^2 + x^{-1})^4 = (x^2)^4 + \binom{4}{1}(x^2)^3(x^{-1}) + \binom{4}{2}(x^2)^2(x^{-1})^2$$

$$+ \binom{4}{3}(x^2)(x^{-1})^3 + (x^{-1})^4$$

$$= x^8 + 4x^6 x^{-1} + 6x^4 x^{-2} + 4x^2 x^{-3} + x^{-4}$$

$$= x^8 + 4x^5 + 6x^2 + 4x^{-1} + x^{-4}. \blacksquare$$

Example 4 To show that $(1.001)^{1000} > 2$ without using a calculator, we write 1.001 as $1 + .001$ and apply the Binomial Theorem with $x = 1$, $y = .001$, and $n = 1000$:

$$(1.001)^{1000} = (1 + .001)^{1000} = 1^{1000} + \binom{1000}{1} 1^{999}(.001) + \text{other positive terms}$$

$$= 1 + \binom{1000}{1}(.001) + \text{other positive terms}$$

But $\binom{1000}{1} = \dfrac{1000!}{1!999!} = \dfrac{1000 \cdot 999!}{999!} = 1000.$ Therefore $\binom{1000}{1}(.001) = 1{,}000(.001) = 1$ and

$$(1.001)^{1000} = 1 + 1 + \text{other positive terms} = 2 + \text{other positive terms}.$$

Hence $(1.001)^{1000} > 2$. $\blacksquare$

Sometimes we need to know only one term in the expansion of $(x + y)^n$. If you examine the expansion given by the Binomial Theorem, you will see that in the second term y has exponent 1, in the third term y has exponent 2, and so on. Thus

PROPERTIES OF THE BINOMIAL EXPANSION

In the binomial expansion of $(x + y)^n$,

The exponent of y is always one less than the number of the term.

Furthermore, in each of the middle terms of the expansion,

The coefficient of the term containing y^r is $\binom{n}{r}$.

The sum of the x exponent and the y exponent is n.

For instance, in the *ninth* term of the expansion of $(x + y)^{13}$, y has exponent 8, the coefficient is $\binom{13}{8}$, and x must have exponent 5 (since $8 + 5 = 13$). Thus the ninth term is $\binom{13}{8} x^5 y^8$.

Example 5 What is the ninth term of the expansion of $\left(2x^2 + \dfrac{\sqrt[4]{y}}{\sqrt{6}}\right)^{13}$? We shall use the Binomial Theorem with $n = 13$ and with $2x^2$ in place of x and $\sqrt[4]{y}/\sqrt{6}$ in place of y. The remarks above show that the ninth term is

$$\binom{13}{8}(2x^2)^5 \left(\dfrac{\sqrt[4]{y}}{\sqrt{6}}\right)^8$$

Since $\sqrt[4]{y} = y^{1/4}$ and $\sqrt{6} = \sqrt{3}\sqrt{2} = 3^{1/2}2^{1/2}$, we can simplify as follows:

$$\binom{13}{8}(2x^2)^5 \left(\dfrac{\sqrt[4]{y}}{\sqrt{6}}\right)^8 = \binom{13}{8} 2^5(x^2)^5 \dfrac{(y^{1/4})^8}{(3^{1/2})^8(2^{1/2})^8} = \binom{13}{8} 2^5 x^{10} \dfrac{y^2}{3^4 \cdot 2^4}$$

$$= \binom{13}{8} \dfrac{2}{3^4} x^{10} y^2 = \dfrac{13 \cdot 12 \cdot 11 \cdot 10 \cdot 9}{5 \cdot 4 \cdot 3 \cdot 2} \cdot \dfrac{2}{3^4} x^{10} y^2 = \dfrac{286}{9} x^{10} y^2. \blacksquare$$

EXERCISES

In Exercises 1–10, evaluate the expression.

1. $6!$
2. $\dfrac{11!}{8!}$
3. $\dfrac{12!}{9!3!}$
4. $\dfrac{9! - 8!}{7!}$
5. $\binom{5}{3} + \binom{5}{2} - \binom{6}{3}$
6. $\binom{12}{11} - \binom{11}{10} + \binom{7}{0}$
7. $\binom{6}{0} + \binom{6}{1} + \binom{6}{2} + \binom{6}{3} + \binom{6}{4} + \binom{6}{5} + \binom{6}{6}$
8. $\binom{6}{0} - \binom{6}{1} + \binom{6}{2} - \binom{6}{3} + \binom{6}{4} - \binom{6}{5} + \binom{6}{6}$
9. $\binom{100}{96}$
10. $\binom{75}{72}$

In Exercises 11–16, expand the expression.

11. $(x + y)^5$
12. $(a + b)^7$
13. $(a - b)^5$
14. $(c - d)^8$
15. $(2x + y^2)^5$
16. $(3u - v^3)^6$

In Exercises 17–26, use the Binomial Theorem to expand and (where possible) simplify the expression.

17. $(\sqrt{x} + 1)^6$
18. $(2 - \sqrt{y})^5$
19. $(1 - c)^{10}$
20. $\left(\sqrt{c} + \dfrac{1}{\sqrt{c}}\right)^7$
21. $(x^{-3} + x)^4$
22. $(3x^{-2} - x^2)^6$
23. $(1 + \sqrt{3})^4 + (1 - \sqrt{3})^4$
24. $(\sqrt{3} + 1)^6 - (\sqrt{3} - 1)^6$
25. $(1 + i)^6$, where $i^2 = -1$
26. $(\sqrt{2} - i)^4$, where $i^2 = -1$

In Exercises 27–32, find the indicated term of the expansion of the given expression.

27. third, $(x + y)^5$
28. fourth, $(a + b)^6$
29. fifth, $(c - d)^7$
30. third, $(a + 2)^8$

31. fourth, $\left(u^{-2} + \dfrac{u}{2}\right)^7$ 32. fifth, $(\sqrt{x} - \sqrt{2})^7$

33. Find the coefficient of $x^5 y^8$ in the expansion of $(2x - y^2)^9$.

34. Find the coefficient of $x^{12} y^6$ in the expansion of $(x^3 - 3y)^{10}$.

35. Find the coefficient of $1/x^3$ in the expansion of $\left(2x + \dfrac{1}{x^2}\right)^6$.

36. Find the constant term in the expansion of $\left(y - \dfrac{1}{2y}\right)^{10}$.

37. (a) Verify that $\dbinom{9}{1} = 9$ and $\dbinom{9}{8} = 9$.

   (b) Prove that for each positive integer n,
   $\dbinom{n}{1} = n$ and $\dbinom{n}{n-1} = n$. [Note: Part (a) is just the case when $n = 9$ and $n - 1 = 8$.]

38. (a) Verify that $\dbinom{7}{2} = \dbinom{7}{5}$.

   (b) Let r and n be integers with $0 \le r \le n$. Prove that $\dbinom{n}{r} = \dbinom{n}{n-r}$. [Note: Part (a) is just the case when $n = 7$ and $r = 2$.]

39. Prove that for any positive integer n,
$$2^n = \binom{n}{0} + \binom{n}{1} + \binom{n}{2} + \cdots + \binom{n}{n}.$$
[Hint: $2^n = (1 + 1)^n$.]

40. Prove that for any positive integer n,
$$\binom{n}{0} - \binom{n}{1} + \binom{n}{2} - \binom{n}{3} + \binom{n}{4} - \cdots$$
$$+ (-1)^k \binom{n}{k} + \cdots + (-1)^n \binom{n}{n} = 0.$$

41. (a) Let f be the function given by $f(x) = x^5$. Let h be a nonzero number and compute $f(x + h) - f(x)$ (but leave all binomial coefficients in the form $\dbinom{5}{r}$ here and below).

   (b) Use part (a) to show that h is a factor of $f(x + h) - f(x)$ and find $\dfrac{f(x+h) - f(x)}{h}$.

   (c) If h is *very* close to 0, find a simple approximation of the quantity $\dfrac{f(x+h) - f(x)}{h}$. (See part (b).)

42. Do Exercise 41 with $f(x) = x^8$ in place of $f(x) = x^5$.

43. Do Exercise 41 with $f(x) = x^{12}$ in place of $f(x) = x^5$.

44. Let n be a fixed positive integer. Do Exercise 41 with $f(x) = x^n$ in place of $f(x) = x^5$.

Unusual Problems

45. Let r and n be integers such that $0 \le r \le n$.

   (a) Verify that $(n - r)! = (n - r)(n - (r + 1))!$
   (b) Verify that $(n - r)! = ((n + 1) - (r + 1))!$
   (c) Prove that $\dbinom{n}{r+1} + \dbinom{n}{r} = \dbinom{n+1}{r+1}$ for any $r \le n - 1$. [Hint: Write out the terms on the left side and use parts (a) and (b) to express each of them as a fraction with denominator $(r + 1)!(n - r)!$. Then add these two fractions, simplify the numerator, and compare the result with $\dbinom{n+1}{r+1}$.]
   (d) Use (c) to explain why each entry in Pascal's triangle (except the 1's at the beginning or end of a row) is the sum of the two closest entries in the row above it.

46. (a) Find these numbers and write them one *below* the next: 11^0, 11^1, 11^2, 11^3, 11^4.
   (b) Compare the list in part (a) with rows 0 to 4 of Pascal's triangle. What's the explanation?
   (c) What can be said about 11^5 and row 5 of Pascal's triangle?
   (d) Calculate all integer powers of 101 from 101^0 to 101^8, list the results one under the other, and compare the list with rows 0 to 8 of Pascal's triangle. What's the explanation? What happens with 101^9?

9.5 MATHEMATICAL INDUCTION

Mathematical induction is a method of proof that can be used to prove a wide variety of mathematical facts, including the Binomial Theorem and statements such as:

The sum of the first n positive integers is the number $\dfrac{n(n+1)}{2}$.

$2^n > n$ for every positive integer n.

For each positive integer n, 4 is a factor of $7^n - 3^n$.

All of the statements above have a common property. For example, a statement such as

The sum of the first n positive integers is the number $\dfrac{n(n+1)}{2}$

or, in symbols,

$$1 + 2 + 3 + \cdots + n = \frac{n(n+1)}{2}$$

is really an infinite sequence of statements, one for each possible value of n:

$$n = 1: \qquad 1 = \frac{1(2)}{2}$$

$$n = 2: \qquad 1 + 2 = \frac{2(3)}{2}$$

$$n = 3: \qquad 1 + 2 + 3 = \frac{3(4)}{2}$$

and so on. Obviously, there isn't time enough to verify every one of the statements on this list, one at a time. But we can find a workable method of proof by examining how each statement on the list is *related* to the *next* statement on the list.

For instance, for $n = 50$, the statement is:

$$1 + 2 + 3 + \cdots + 50 = \frac{50(51)}{2}$$

At the moment, we don't know whether or not this statement is true. But just *suppose* that it were true. What could then be said about the next statement, the one for $n = 51$:

$$1 + 2 + 3 + \cdots + 50 + 51 = \frac{51(52)}{2}?$$

Well, *if* it is true that

$$1 + 2 + 3 + \cdots + 50 = \frac{50(51)}{2}$$

then adding 51 to both sides and simplifying the right side would yield these equalities:

$$1 + 2 + 3 + \cdots + 50 + 51 = \frac{50(51)}{2} + 51$$

$$1 + 2 + 3 + \cdots + 50 + 51 = \frac{50(51)}{2} + \frac{2(51)}{2} = \frac{50(51) + 2(51)}{2}$$

$$1 + 2 + 3 + \cdots + 50 + 51 = \frac{(50 + 2)51}{2}$$

$$1 + 2 + 3 + \cdots + 50 + 51 = \frac{51(52)}{2}$$

Since this last equality is just the original statement for $n = 51$, we conclude that

If the statement is true for $n = 50$, *then* it is also true for $n = 51$.

We have *not* proved that the statement actually *is* true for $n = 50$, but only that *if* it is, then it is also true for $n = 51$.

We claim that this same conditional relationship holds for any two consecutive values of n. In other words, we claim that for any positive integer k,

① ***If* the statement is true for $n = k$, *then* it is also true for $n = k + 1$.**

The proof of this claim is the same argument used above (with k and $k + 1$ in place of 50 and 51): *If* it is true that

$$1 + 2 + 3 + \cdots + k = \frac{k(k + 1)}{2} \quad \text{[\textit{Original statement for } n = k]}$$

then adding $k + 1$ to both sides and simplifying the right side produces these equalities:

$$1 + 2 + 3 + \cdots + k + (k + 1) = \frac{k(k + 1)}{2} + (k + 1)$$

$$1 + 2 + 3 + \cdots + k + (k + 1) = \frac{k(k + 1)}{2} + \frac{2(k + 1)}{2} = \frac{k(k + 1) + 2(k + 1)}{2}$$

$$1 + 2 + 3 + \cdots + k + (k + 1) = \frac{(k + 2)(k + 1)}{2}$$

$$1 + 2 + 3 + \cdots + k + (k + 1) = \frac{(k + 1)((k + 1) + 1)}{2}$$

[Original statement for $n = k + 1$]

We have proved that claim ① is valid for each positive integer k. We have *not* proved that the original statement is true for any value of n, but only that *if* it is true for $n = k$, then it is also true for $n = k + 1$. Applying this fact when $k = 1, 2, 3, \ldots$, we see that

②
$$\begin{cases} \textit{if}\text{ the statement is true for } n = 1, & \textit{then}\text{ it is also true for } n = 1 + 1 = 2; \\ \textit{if}\text{ the statement is true for } n = 2, & \textit{then}\text{ it is also true for } n = 2 + 1 = 3; \\ \textit{if}\text{ the statement is true for } n = 3, & \textit{then}\text{ it is also true for } n = 3 + 1 = 4; \\ \quad\vdots \\ \textit{if}\text{ the statement is true for } n = 50, & \textit{then}\text{ it is also true for } n = 50 + 1 = 51; \\ \textit{if}\text{ the statement is true for } n = 51, & \textit{then}\text{ it is also true for } n = 51 + 1 = 52; \\ \quad\vdots \end{cases}$$

and so on.

We are finally in a position to *prove* the original statement: $1 + 2 + 3 + \cdots + n = n(n + 1)/2$. Obviously, it *is true* for $n = 1$ since $1 = 1(2)/2$. Now apply in turn each of the propositions on list ② above. Since the statement *is* true for $n = 1$, it must also be true for $n = 2$, and hence for $n = 3$, and hence for $n = 4$, and so on, for every value of n. Therefore the original statement is true for *every* positive integer n.

The preceding proof is an illustration of the following principle:

PRINCIPLE OF MATHEMATICAL INDUCTION

> **Suppose there is given a statement involving the positive integer n and that:**
>
> **(i) The statement is true for $n = 1$.**
>
> **(ii) If the statement is true for $n = k$ (where k is a positive integer), then the statement is also true for $n = k + 1$.**
>
> **Then the statement is true for every positive integer n.**

Property (i) is simply a statement of fact. To verify that it holds, you must prove the given statement is true for $n = 1$. This is usually easy, as in the preceding example.

Property (ii) is a *conditional* property. It does not assert that the given statement *is* true for $n = k$, but only that *if* it is true for $n = k$, then it is also true for $n = k + 1$. So to verify that property (ii) holds, you need only prove this conditional proposition:

If the statement is true for $n = k$, *then* it is also true for $n = k + 1$.

In order to prove this, or any conditional proposition, you must proceed as in the example above: Assume the "if" part and use this assumption to prove the "then" part. As we saw above, the same argument will usually work for any possible k. Once this conditional proposition has been proved, you can use it *together* with property (i) to conclude that the given statement is necessarily true for every n, just as in the preceding example.

Thus proof by mathematical induction reduces to two steps:

Step 1: Prove that the given statement is true for $n = 1$.

Step 2: Let k be a positive integer. Assume that the given statement is true for $n = k$. Use this assumption to prove that the statement is true for $n = k + 1$.

Step 2 may be performed before step 1 if you wish. Step 2 is sometimes referred to as the **inductive step.** The assumption that the given statement is true for $n = k$ in this inductive step is called the **induction hypothesis.**

Example 1 Prove that $2^n > n$ for every positive integer n.

Solution Here the statement involving n is: $2^n > n$.

Step 1 When $n = 1$, we have the statement $2^1 > 1$. This is obviously true.

Step 2 Let k be any positive integer. We assume that the statement is true for $n = k$, that is, we assume that $2^k > k$. We shall use this assumption to prove that the statement is true for $n = k + 1$, that is, that $2^{k+1} > k + 1$. We begin with the induction hypothesis:* $2^k > k$. Multiplying both sides of this inequality by 2 yields:

$$2 \cdot 2^k > 2k$$

③ $$2^{k+1} > 2k$$

Since k is a positive integer, we know that $k \geq 1$. Adding k to each side of the inequality $k \geq 1$, we have

$$k + k \geq k + 1$$
$$2k \geq k + 1$$

Combining this result with inequality ③ above, we see that

$$2^{k+1} > 2k \geq k + 1$$

The first and last terms of this inequality show that $2^{k+1} > k + 1$. Therefore the statement is true for $n = k + 1$. This argument works for any positive integer k. Thus we have completed the inductive step. By the Principle of Mathematical Induction, we conclude that $2^n > n$ for every positive integer n. ■

Example 2 Simple arithmetic shows that:

$$7^2 - 3^2 = 49 - 9 = 40 = 4 \cdot 10 \quad \text{and} \quad 7^3 - 3^3 = 343 - 27 = 316 = 4 \cdot 79$$

---

* This is the point at which you usually must do some work. Remember that what follows is the "finished proof." It does not include all the thought, scratch work, false starts, and so on, that were done before this proof was actually found.

In each case, 4 is a factor. These examples suggest that

For each positive integer n, 4 is a factor of $7^n - 3^n$.

This conjecture can be proved by induction as follows.

Step 1 When $n = 1$, the statement is: 4 is a factor of $7^1 - 3^1$. Since $7^1 - 3^1 = 4 = 4 \cdot 1$, the statement is true for $n = 1$.

Step 2 Let k be a positive integer and assume that the statement is true for $n = k$, that is, that 4 is a factor of $7^k - 3^k$. Let us denote the other factor by D, so that the induction hypothesis is: $7^k - 3^k = 4D$. We must use this assumption to prove that the statement is true for $n = k + 1$, that is, that 4 is a factor of $7^{k+1} - 3^{k+1}$. Here is the proof:

$$7^{k+1} - 3^{k+1} = 7^{k+1} - 7 \cdot 3^k + 7 \cdot 3^k - 3^{k+1} \quad [\textit{Since } -7 \cdot 3^k + 7 \cdot 3^k = 0]$$
$$= 7(7^k - 3^k) + (7 - 3)3^k \quad [\textit{Factor}]$$
$$= 7(4D) + (7 - 3)3^k \quad [\textit{Induction hypothesis}]$$
$$= 7(4D) + 4 \cdot 3^k \quad [7 - 3 = 4]$$
$$= 4(7D + 3^k) \quad [\textit{Factor out 4}]$$

From this last line, we see that 4 is a factor of $7^{k+1} - 3^{k+1}$. Thus the statement is true for $n = k + 1$, and the inductive step is complete. Therefore by the Principle of Mathematical Induction the conjecture is actually true for every positive integer n. ∎

Another example of mathematical induction, the proof of the Binomial Theorem, is given at the end of this section.

Sometimes a statement involving the integer n may be false for $n = 1$ and (possibly) other small values of n, but true for all values of n beyond a particular number. For instance, the statement $2^n > n^2$ is false for $n = 1, 2, 3, 4$. But it is true for $n = 5$ and all larger values of n. A variation on the Principle of Mathematical Induction can be used to prove this fact and similar statements. See Exercise 28 for details.

A Common Mistake with Induction

It is sometimes tempting to omit step 2 of an inductive proof when the given statement can easily be verified for small values of n, especially if a clear pattern seems to be developing. As the next example shows, however, *omitting step 2 may lead to error.*

Example 3 An integer (> 1) is said to be *prime* if its only positive integer factors are itself and 1. For instance, 11 is prime since its only positive integer factors are 11 and 1. But 15 is not prime because it has factors other than 15 and 1 (namely, 3 and 5). For each positive integer n, consider the number

$$f(n) = n^2 - n + 11$$

You can readily verify that

$$f(1) = 11, \quad f(2) = 13, \quad f(3) = 17, \quad f(4) = 23, \quad f(5) = 31$$

and that *each of these numbers is prime*. Furthermore, there is a clear pattern: The first two numbers (11 and 13) differ by 2; the next two (13 and 17) differ by 4; the next two (17 and 23) differ by 6; and so on. On the basis of this evidence, we might conjecture:

For each positive integer n, the number $f(n) = n^2 - n + 11$ is prime.

We have seen that this conjecture is true for $n = 1, 2, 3, 4, 5$. Unfortunately, however, it is *false* for some values of n. For instance, when $n = 11$,

$$f(11) = 11^2 - 11 + 11 = 11^2 = 121$$

But 121 is obviously *not* prime since it has a factor other than 121 and 1, namely, 11. You can verify that the statement is also false for $n = 12$ but true for $n = 13$. ■

In the preceding example, the proposition

If the statement is true for $n = k$, then it is true for $n = k + 1$

is false when $k = 10$ and $k + 1 = 11$. If you were not aware of this and tried to complete step 2 of an inductive proof, you would not have been able to find a valid proof for it. Of course, the fact that you can't find a proof of a proposition doesn't always mean that no proof exists. But when you are unable to complete step 2, you are warned that there is a possibility that the given statement may be false for some values of n. This warning should prevent you from drawing any wrong conclusions.

Proof of the Binomial Theorem

We shall use induction to prove that for every positive integer n,

$$(x + y)^n = x^n + \binom{n}{1} x^{n-1} y + \binom{n}{2} x^{n-2} y^2 + \binom{n}{3} x^{n-3} y^3 + \cdots + \binom{n}{n-1} xy^{n-1} + y^n$$

This theorem was discussed and its notation explained in Section 9.4.

Step 1 When $n = 1$, there are only two terms on the right side of the equation above, and the statement reads $(x + y)^1 = x^1 + y^1$. This is certainly true.

Step 2 Let k be any positive integer and assume that the theorem is true for $n = k$, that is, that

$$(x + y)^k = x^k + \binom{k}{1} x^{k-1}y + \binom{k}{2} x^{k-2}y^2 + \cdots$$
$$+ \binom{k}{r} x^{k-r}y^r + \cdots + \binom{k}{k-1} xy^{k-1} + y^k$$

[On the right side above, we have included a typical middle term $\binom{k}{r} x^{k-r}y^r$. The sum of the exponents is k, and the bottom part of the binomial coefficient is the same as the y exponent.] We shall use this assumption to prove that the theorem is true for $n = k + 1$, that is, that

$$(x + y)^{k+1} = x^{k+1} + \binom{k+1}{1} x^k y + \binom{k+1}{2} x^{k-1}y^2 + \cdots$$
$$+ \binom{k+1}{r+1} x^{k-r}y^{r+1} + \cdots + \binom{k+1}{k} xy^k + y^{k+1}$$

We have simplified some of the terms on the right side; for instance, $(k + 1) - 1 = k$ and $(k + 1) - (r + 1) = k - r$. But this is the correct statement for $n = k + 1$: The coefficients of the middle terms are $\binom{k+1}{1}$, $\binom{k+1}{2}$, $\binom{k+1}{3}$, and so on; the sum of the exponents of each middle term is $k + 1$, and the bottom part of each binomial coefficient is the same as the y exponent.

In order to prove the theorem for $n = k + 1$, we shall need this fact about binomial coefficients: For any integers r and k with $0 \leq r < k$,

④ $$\binom{k}{r+1} + \binom{k}{r} = \binom{k+1}{r+1}$$

A proof of this fact is outlined in Exercise 45 on page 453.

To prove the theorem for $n = k + 1$, we first note that

$$(x + y)^{k+1} = (x + y)(x + y)^k$$

Applying the induction hypothesis to $(x + y)^k$, we see that

$$(x + y)^{k+1} = (x + y)\left[x^k + \binom{k}{1} x^{k-1}y + \binom{k}{2} x^{k-2}y^2 + \cdots + \binom{k}{r} x^{k-r}y^r \right.$$
$$\left. + \binom{k}{r+1} x^{k-(r+1)}y^{r+1} + \cdots + \binom{k}{k-1} xy^{k-1} + y^k\right]$$
$$= x\left[x^k + \binom{k}{1} x^{k-1}y + \cdots + y^k\right] + y\left[x^k + \binom{k}{1} x^{k-1}y + \cdots + y^k\right]$$

Next we multiply out the right-hand side. Remember that multiplying by x increases the x exponent by 1 and multiplying by y increases the y exponent by 1.

$$(x+y)^{k+1} = \left[x^{k+1} + \binom{k}{1} x^k y + \binom{k}{2} x^{k-1} y^2 + \cdots + \binom{k}{r} x^{k-r+1} y^r \right.$$
$$\left. + \binom{k}{r+1} x^{k-r} y^{r+1} + \cdots + \binom{k}{k-1} x^2 y^{k-1} + xy^k \right]$$
$$+ \left[x^k y + \binom{k}{1} x^{k-1} y^2 + \binom{k}{2} x^{k-2} y^3 + \cdots + \binom{k}{r} x^{k-r} y^{r+1} \right.$$
$$\left. + \binom{k}{r+1} x^{k-(r+1)} y^{r+2} + \cdots + \binom{k}{k-1} xy^k + y^{k+1} \right]$$
$$= x^{k+1} + \left[\binom{k}{1} + 1 \right] x^k y + \left[\binom{k}{2} + \binom{k}{1} \right] x^{k-1} y^2 + \cdots$$
$$+ \left[\binom{k}{r+1} + \binom{k}{r} \right] x^{k-r} y^{r+1} + \cdots + \left[1 + \binom{k}{k-1} \right] xy^k + y^{k+1}$$

Now apply statement ④ above to each of the coefficients of the middle terms. For instance, with $r = 1$, statement ④ shows that $\binom{k}{2} + \binom{k}{1} = \binom{k+1}{2}$. Similarly, with $r = 0$, $\binom{k}{1} + 1 = \binom{k}{1} + \binom{k}{0} = \binom{k+1}{1}$, and so on. Then the expression above for $(x+y)^{k+1}$ becomes:

$$(x+y)^{k+1} = x^{k+1} + \binom{k+1}{1} x^k y + \binom{k+1}{2} x^{k-1} y^2 + \cdots$$
$$+ \binom{k+1}{r+1} x^{k-r} y^{r+1} + \cdots + \binom{k+1}{k} xy^k + y^{k+1}$$

Since this last statement says the theorem is true for $n = k + 1$, the inductive step is complete. By the Principle of Mathematical Induction the theorem is true for every positive integer n.

EXERCISES

In Exercises 1–18, use mathematical induction to prove that each of the given statements is true for every positive integer n.

1. $1 + 2 + 2^2 + 2^3 + 2^4 + \cdots + 2^{n-1} = 2^n - 1$

2. $1 + 3 + 3^2 + 3^3 + 3^4 + \cdots + 3^{n-1} = \dfrac{3^n - 1}{2}$

3. $1 + 3 + 5 + 7 + \cdots + (2n - 1) = n^2$

4. $2 + 4 + 6 + 8 + \cdots + 2n = n^2 + n$

5. $1^2 + 2^2 + 3^2 + \cdots + n^2 = \dfrac{n(n+1)(2n+1)}{6}$

6. $\dfrac{1}{2} + \dfrac{1}{4} + \dfrac{1}{8} + \cdots + \dfrac{1}{2^n} = 1 - \dfrac{1}{2^n}$

7. $\dfrac{1}{1 \cdot 2} + \dfrac{1}{2 \cdot 3} + \dfrac{1}{3 \cdot 4} + \cdots + \dfrac{1}{n(n+1)} = \dfrac{n}{n+1}$

8. $\left(1 + \dfrac{1}{1}\right)\left(1 + \dfrac{1}{2}\right)\left(1 + \dfrac{1}{3}\right) \cdots \left(1 + \dfrac{1}{n}\right) = n + 1$

9. $n + 2 > n$ 10. $2n + 2 > n$

11. $3^n \geq 3n$ 12. $3^n \geq 1 + 2n$

13. $3n > n + 1$ 14. $\left(\dfrac{3}{2}\right)^n > n$

15. 3 is a factor of $2^{2n+1} + 1$

16. 5 is a factor of $2^{4n-2} + 1$

17. 64 is a factor of $3^{2n+2} - 8n - 9$

18. 64 is a factor of $9^n - 8n - 1$

19. Let c and d be fixed real numbers. Prove that
$$c + (c + d) + (c + 2d) + (c + 3d) + \cdots + (c + (n-1)d) = \dfrac{n(2c + (n-1)d)}{2}.$$

20. Let r be a fixed real number with $r \neq 1$. Prove that $1 + r + r^2 + r^3 + \cdots + r^{n-1} = \dfrac{r^n - 1}{r - 1}$.
(Remember that $1 = r^0$; so when $n = 1$ the left side reduces to $r^0 = 1$.)

21. (a) Write *each* of $x^2 - y^2, x^3 - y^3,$ and $x^4 - y^4$ as a product of $x - y$ and another factor.
    (b) Make a conjecture as to how $x^n - y^n$ can be written as a product of $x - y$ and another factor. Use induction to prove your conjecture.

22. Let $x_1 = \sqrt{2}$; $x_2 = \sqrt{2 + \sqrt{2}}$; $x_3 = \sqrt{2 + \sqrt{2 + \sqrt{2}}}$; and so on. Prove that $x_n < 2$ for every positive integer n.

In Exercises 23–27, if the given statement is true, prove it. If it is false, give a counterexample.

23. Every odd positive integer is prime.

24. The number $n^2 + n + 17$ is prime for every positive integer n.

25. $(n + 1)^2 > n^2 + 1$ for every positive integer n.

26. 3 is a factor of the number $n^3 - n + 3$ for every positive integer n.

27. 4 is a factor of the number $n^4 - n + 4$ for every positive integer n.

28. Let q be a *fixed* integer. Suppose a statement involving the integer n has these two properties:
    (i) The statement is true for $n = q$.
    (ii) *If* the statement is true for $n = k$ (where k is an integer with $k \geq q$), then the statement is also true for $n = k + 1$.

    Then we claim that the statement is true for every integer n greater than or equal to q.
    (a) Give an informal explanation that shows why the claim above should be valid. Note that when $q = 1$, this claim is precisely the Principle of Mathematical Induction.
    (b) The claim made above will be called the *Extended Principle of Mathematical Induction*. State the two steps necessary to use this principle to prove that a given statement is true for all $n \geq q$. (See discussion on pages 456–457.)

In Exercises 29–34, use the Extended Principle of Mathematical Induction (Exercise 28) to prove the given statement.

29. $2n - 4 > n$ for every $n \geq 5$. (Use 5 for q here.)

30. Let r be a fixed real number with $r > 1$. Then $(1 + r)^n > 1 + nr$ for every integer $n \geq 2$. (Use 2 for q here.)

31. $n^2 > n$ for all $n \geq 2$

32. $2^n > n^2$ for all $n \geq 5$

33. $3^n > 2^n + 10n$ for all $n \geq 4$

34. $2n < n!$ for all $n \geq 4$

Unusual Problems

35. Let n be a positive integer. Suppose that there are three pegs and on one of them n rings are stacked, with each ring being smaller in diameter than the one below it (see Figure 9–1). We want to transfer the stack of rings to another peg according to these rules: (i) Only one ring may be moved at a time; (ii) a ring can be moved to any peg, provided it is never placed on top of a smaller ring; (iii) the final order of the rings on the new peg must be the same as the original order on the first peg.

FIGURE 9-1

(a) What is the smallest possible number of moves when $n = 2$? $n = 3$? $n = 4$?
(b) Make a conjecture as to the smallest possible number of moves required for any n. Prove your conjecture by induction.

36. The basic formula for compound interest $T(x) = P(1 + r)^x$ was discussed on page 278. Prove by induction that the formula is valid whenever x is a positive integer. [*Note: P* and *r* are assumed to be constant.]

9.6 PERMUTATIONS AND COMBINATIONS

Many problems involve systematic counting of all the possibilities that might occur.

Example 1 Anne, Bill, Charlie, Dana, Elsie, and Fred enter a short-story competition in which three prizes will be awarded (and there will be no ties). If they are the only contestants, in how many possible ways can the prizes be awarded?

Solution For convenience we use initials. There are 6 possibilities for first place: *A, B, C, D, E, F*. For each possible first-place winner there are 5 possible second-place winners:

1. 2.	1. 2.	1. 2.	1. 2.	1. 2.	1. 2.
A B	B A	C A	D A	E A	F A
A C	B C	C B	D B	E B	F B
A D	B D	C D	D C	E C	F C
A E	B E	C E	D E	E D	F D
A F	B F	C F	D F	E F	F E

So the total number of ways first and second prize can be awarded is $6 \cdot 5 = 30$. If, for instance, *A* and *B* take first and second, then there are four possibilities for third:

$$A, B, C; \quad A, B, D; \quad A, B, E; \quad A, B, F$$

Similarly, for *each* of the 30 ways that first and second prizes can be awarded, there are 4 ways to award third prize. So there are $30 \cdot 4 = 120$ possible ways to award all three prizes. ■

The argument in the example may be summarized like this:

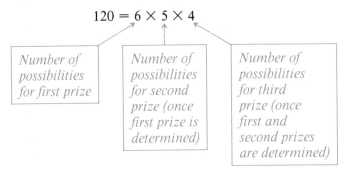

When thought of this way, the example is an illustration of the following principle (in the case where $k = 3$, each event is the awarding of a prize, and $n_1 = 6, n_2 = 5, n_3 = 4$).

FUNDAMENTAL COUNTING PRINCIPLE

Consider a list of k events. Suppose the first event can occur in n_1 ways; the second event can occur in n_2 ways (once the first event has occurred); the third event can occur in n_3 ways (once the first two events have occurred); and so on. Then the total number of ways that all k events can occur is the product $n_1 n_2 n_3 \cdots n_k$.

Example 2 A committee consisting of 8 Democrats and 6 Republicans must choose a chairperson, a vice-chairperson, a secretary, and a treasurer. The chairperson must be a Democrat and the vice-chairperson a Republican. The secretary and treasurer may belong to either party. If no person can hold more than one office, in how many different ways can these four officers be chosen?

Solution Consider the choice of a chairperson as the first event. Since there are 8 Democrats, there are 8 possible ways for this event to occur ($n_1 = 8$). The second event is the choice of a vice-chairperson. It can occur in 6 ways because there are 6 Republicans ($n_2 = 6$). Once the chairperson and vice-chairperson are determined, any of the remaining 12 people may be chosen secretary. So that event can occur in 12 ways ($n_3 = 12$). The fourth and last event ($k = 4$) is the choice of a treasurer, who can be any of the 11 people remaining ($n_4 = 11$). By the Fundamental Counting Principle the number of ways that all four officers can be chosen is the product $n_1 n_2 n_3 n_4 = 8 \cdot 6 \cdot 12 \cdot 11 = 6336$. ∎

Example 3 How many license plates can be made consisting of 3 letters followed by 3 one-digit numbers, subject to these conditions: No plate may begin with I or O, and the first number may not be a zero?

Solution Consider the filling of each of the six positions on a license plate as an event. Since O and I are excluded, there are 24 possible letters for the first position ($n_1 = 24$). There are 26 possible letters for each of the second and third positions ($n_2 = 26, n_3 = 26$). Since 0 cannot appear in the fourth position (the first position for a number), there are 9 possibilities ($n_4 = 9$). Any digit from 0 through 9 may appear in the last two positions, so there are 10 possible digits for each position ($n_5 = 10, n_6 = 10$). By the Fundamental Counting Principle, the total number of license plates is the product

$$24 \cdot 26 \cdot 26 \cdot 9 \cdot 10 \cdot 10 = 14{,}601{,}600. \blacksquare$$

Permutations

A **permutation** of a set of n elements is an ordering of the elements of the set. We can use the Fundamental Counting Principle to determine the number of possible permutations of a set.

Example 4 The number of possible batting orders for a 9-person baseball team is the number of permutations of a set of 9 elements. There are 9 possibilities for first batter, 8 possibilities for second (after the first one is determined), 7 possibilities for third (after the first two are determined), and so on, down to the eighth (2 possibilities) and ninth batter (1 possibility). So the number of possible batting orders is

$$9 \cdot 8 \cdot 7 \cdot 6 \cdot 5 \cdot 4 \cdot 3 \cdot 2 \cdot 1 = 9! = 362{,}880.* \blacksquare$$

The argument in the example carries over to the general case with n in place of 9: To determine a permutation of n elements, there are n possible choices for first position, $n - 1$ choices for second position, and so on. Therefore

PERMUTATIONS OF n ELEMENTS

The total number of permutations of a set of n elements is

$$n! = n(n - 1)(n - 2) \cdots 3 \cdot 2 \cdot 1*$$

Example 5 The number of possible ways that a deck of 52 cards can be shuffled (the number of permutations of a 52-element set) is 52! A calculator with scientific notation and an $\boxed{n!}$ key shows that 52! is approximately 8.066×10^{67}. $\blacksquare$

In certain cases we are interested in the order of *some* (not necessarily all) of the elements in a set. For instance, in Example 1 (the short-story contest), we considered possible ways of ordering three of the six contestants (the prize winners). Such an ordering is called a permutation of 6 elements taken 3 at a

---

* Factorial notation ($n!$) is explained on page 447.

time. More generally, if $r \leq n$, an ordering of r elements of an n-element set is called a **permutation of n elements taken r at a time.***

In determining the number of permutations of n elements taken r at a time, there are n choices for first position, $n - 1$ for second position, and so on. Since we want only an ordered list of r elements, the process stops after r steps. Then the total number of orderings is the product of the number of possibilities for each of the first r positions. For instance, in the short-story contest we saw that the number of permutations of 6 things taken 3 at a time was $6 \cdot 5 \cdot 4$, which is the product of 3 consecutive integers from 6 down. In the general case

> **PERMUTATIONS OF n ELEMENTS TAKEN r AT A TIME**
>
> The number of permutations of n elements taken r at a time is the product of r consecutive integers, beginning with n and working downward: $n(n-1)(n-2) \cdots (n-(r-1))$. This number is denoted $_nP_r$.

Example 6 The number of permutations of 8 elements taken 5 at a time is the product of 5 consecutive integers, going down from 8:

$$_8P_5 = 8 \cdot 7 \cdot 6 \cdot 5 \cdot 4 = 6720.$$

Similarly, $_{20}P_4$ is the product of 4 consecutive integers:

$$_{20}P_4 = 20 \cdot 19 \cdot 18 \cdot 17 = 116{,}280. \blacksquare$$

Since $3!/3! = 1$, the number $_8P_5$ can be rewritten like this:

$$_8P_5 = 8 \cdot 7 \cdot 6 \cdot 5 \cdot 4 = \frac{8 \cdot 7 \cdot 6 \cdot 5 \cdot 4}{1} \cdot \frac{3!}{3!} = \frac{8!}{3!} = \frac{8!}{(8-5)!}.$$

A similar computation works in the general case (with n in place of 8 and r in place of 5) and leads to this description of $_nP_r$:

> **FACTORIAL FORM FOR $_nP_r$**
>
> $$_nP_r = \frac{n!}{(n-r)!}$$

Example 7

$$_9P_3 = \frac{9!}{(9-3)!} = \frac{9 \cdot 8 \cdot 7 \cdot 6 \cdot 5 \cdot 4 \cdot 3 \cdot 2 \cdot 1}{6 \cdot 5 \cdot 4 \cdot 3 \cdot 2 \cdot 1} = 9 \cdot 8 \cdot 7 = 504$$

$$_{16}P_{16} = \frac{16!}{(16-16)!} = \frac{16!}{0!} = \frac{16!}{1} = 16! \blacksquare$$

---

* When $r = n$, a permutation of n things taken n at a time is just a permutation of the entire set, as discussed above.

Combinations

To play the state lottery you select 6 different numbers from 1 to 44. The order in which you list them doesn't matter. If these six numbers are drawn in *any* order, you win. How many choices of six numbers do you have?

We can answer this question by considering a more general situation. A **combination of *n* elements taken *r* at a time** is any collection of *r* distinct objects in a set of *n* objects, without regard to the order in which the *r* objects might be chosen.

Example 8 To play the state lottery you select a combination of 44 elements taken 6 at a time. A 3-person committee selected from a group of 10 people is a combination of 10 elements taken 3 at a time. A 5-card poker hand is a combination of 52 elements (the entire deck of cards) taken 5 at a time. ■

Example 9 The alphabet has 26 letters. So the letters B, Q, T form a combination of 26 elements taken 3 at a time. If we list these three letters as B, T, Q, or Q, B, T, or Q, T, B, or T, B, Q, or T, Q, B, we still have the *same combination* because the order doesn't matter. The situation with permutations is quite different. There are six different permutations of these three letters (BQT, BTQ, QBT, QTB, TBQ, TQB) because order *does* matter for permutations. ■

The following fact is the key to dealing with combinations.

COMBINATIONS OF *n* ELEMENTS TAKEN *r* AT A TIME

The number of combinations of *n* elements taken *r* at a time is $_nC_r$, where

$$_nC_r = \frac{n!}{r!(n-r)!} *$$

Proof We shall prove the statement by analyzing the relationship between permutations and combinations.

A *permutation* of *n* elements taken *r* at a time is determined by two events:

1. A choice of *r* elements in the *n*-element set, that is, a *combination* of *n* elements taken *r* at a time; and
2. A specific ordering of these *r* elements, that is, a *permutation* of *r* elements.

So the number $_nP_r$ of *permutations* of *n* elements taken *r* at a time is the same as the total number of ways that events 1 and 2 can occur. By the Fundamental Counting Principle and the last box on page 466:

---

* If you have read Section 9.4, note that the number $_nC_r$ is precisely the binomial coefficient $\binom{n}{r}$.

$$_nP_r = \begin{pmatrix} \text{Number of ways event 1} \\ \text{can occur} \end{pmatrix} \cdot \begin{pmatrix} \text{Number of ways event 2} \\ \text{can occur} \end{pmatrix}$$

$$_nP_r = \begin{pmatrix} \text{Number of combinations} \\ \text{of } n \text{ things taken } r \\ \text{at a time} \end{pmatrix} \cdot \begin{pmatrix} \text{Number of permutations} \\ \text{of } r \text{ things} \end{pmatrix}$$

$$\frac{n!}{(n-r)!} = {_nC_r} \cdot r!$$

Dividing both sides by $r!$ shows that $_nC_r = \dfrac{n!}{r!(n-r)!}$. ∎

Example 10 The number of 3-person committees that can be selected from a group of 10 people is the number of combinations of 10 elements taken 3 at a time:

$$_{10}C_3 = \frac{10!}{3!(10-3)!} = \frac{10!}{3! \cdot 7!} = \frac{10 \cdot 9 \cdot 8 \cdot 7!}{3 \cdot 2 \cdot 1 \cdot 7!} = \frac{10 \cdot 9 \cdot 8}{3 \cdot 2 \cdot 1} = 120. \;\blacksquare$$

Example 11 The number of ways of playing the state lottery (in which you select 6 different numbers from 1 to 44) is the number of combinations of 44 elements taken 6 at a time:

$$_{44}C_6 = \frac{44!}{6!(44-6)!} = \frac{44!}{6! \cdot 38!} = 7{,}059{,}052.$$

Since there are more than 7 million choices, it's not surprising that you probably haven't won the lottery. ∎

Example 12 The number of possible 5-card poker hands is the number of combinations of 52 elements taken 5 at a time:

$$_{52}C_5 = \frac{52!}{5!(52-5)!} = \frac{52!}{5! \cdot 47!} = 2{,}598{,}960. \;\blacksquare$$

Example 13 The city council consists of 18 Democrats, 20 Republicans, and 7 Independents. In how many ways can you select a committee of 3 Democrats, 4 Republicans, and 2 Independents?

Solution The number of ways of choosing 3 of the 18 Democrats is $_{18}C_3$, the number of combinations of 18 elements taken 3 at a time. Similarly, the number of ways of choosing 4 Republicans is $_{20}C_4$, and the number of ways of choosing 2 Independents is $_7C_2$. By the Fundamental Counting Principle, the number of ways all these choices can be made (the number of ways of selecting the committee) is the product

$$_{18}C_3 \cdot {_{20}C_4} \cdot {_{7}C_2} = \frac{18!}{3!(18-3)!} \cdot \frac{20!}{4!(20-4)!} \cdot \frac{7!}{2!(7-2)!}$$

$$= \frac{18!}{3! \cdot 15!} \cdot \frac{20!}{4! \cdot 16!} \cdot \frac{7!}{2! \cdot 5!} = 83{,}023{,}920. \blacksquare$$

EXERCISES

In Exercises 1–16, compute the number.

1. $_4P_3$
2. $_7P_5$
3. $_8P_8$
4. $_5P_1$
5. $_{12}P_2$
6. $_{11}P_3$
7. $_{14}C_2$
8. $_{20}C_3$
9. $_{99}C_{95}$
10. $_{65}C_{60}$
11. $_nP_2$
12. $_nP_{n-1}$
13. $_nC_{n-1}$
14. $_nC_1$
15. $_nC_2$
16. $_nP_1$

In Exercises 17–22, find the number n that makes the statement true.

17. $_nP_4 = 8(_nP_3)$
18. $_nP_5 = 7(_nP_4)$
19. $_nP_6 = 9(_{n-1}P_5)$
20. $_nP_7 = 11(_{n-1}P_6)$
21. $_nP_5 = 21(_{n-1}P_3)$
22. $_nP_6 = 30(_{n-2}P_4)$

23. In how many different orders can 8 people be seated in a row?

24. In how many orders can 6 pictures be arranged in a vertical line?

25. How many outcomes are possible in a 4-team tournament (in which there are no ties)?

26. How many 5-digit numbers can be formed from the numerals 2, 3, 4, 5, 6 if
    (a) no repeated digits are allowed?
    (b) repeated digits are allowed?

27. In how many ways can the first 4 people in the batting order of a 9-person baseball team be chosen?

28. In how many different ways can first through third prizes be awarded in a 10-person race (assume no ties occur)?

29. How many 5-letter identification codes can be formed from the first 20 letters of the alphabet if no repetitions are allowed?

30. How many different numbers can be formed from the digits 3, 5, 7, 9 if no repetitions are allowed? [*Hint:* How many 1-digit numbers can be formed? How many 2-digit ones?]

31. How many 7-digit phone numbers can be formed from the numerals 0 through 9 if none of the first three digits can be 0?

32. How many batting orders are possible for a 9-player baseball team if the 4 infielders bat first and the pitcher bats last?

33. A man has 4 pairs of pants, 6 sports coats, and 8 ties. In how many different ways can he wear one of each?

34. In how many different ways can an 8-question true-false test be answered?

35. An exam consists of 10 multiple-choice questions, with 4 choices for each question. In how many ways can the exam be answered?

36. In how many ways can 6 men and 6 women be seated in a row of 12 chairs if the women sit in the even-numbered seats?

37. In how many ways can 4 men and 8 women be seated in a row of 12 chairs if men sit at both ends of the row?

38. A student must answer 5 questions on an 8-question exam. In how many ways can this be done?

39. How many different straight lines are determined by 8 points in the plane, no 3 of which lie on the same straight line? (Remember that a line is determined by 2 points.)

40. How many games must be played in an 8-team league if each team is to play every other team exactly twice?

41. How many different 6-card cribbage hands are possible from a 52-card deck?

42. How many 5-card poker hands are flushes (5 cards of the same suit)? [*Hint:* The number of club flushes is the number of ways 5 cards can be chosen from the 13 clubs.]

43. How many 8-digit numbers can be formed using three 6's and five 7's?

44. How many different 5-person committees consisting of 3 women and 2 men can be chosen from a group of 10 women and 8 men?

45. How many ways can two committees, one of 4 people and one of 3 people, be chosen from a group of 11 people if no person serves on both committees?

46. How many 6-digit numbers can be formed using only the numerals 5 and 9?

47. (a) A basketball squad consists of 5 people who can play center or forward and 7 people who can play only guard. In how many ways can a team consisting of a center, 2 forwards, and 2 guards be chosen?
    (b) How many teams are possible if the squad consists of 2 centers, 4 forwards, and 6 people who can play either guard or forward?

48. (a) How many different pairs are possible from a standard 52-card deck? [*Hint:* How many pairs can be formed from the 4 aces? How many from the 4 kings?]
    (b) How many 5-card poker hands consist of a pair of aces and 3 other cards, none of which are aces?

(c) How many 5-card poker hands contain at least 2 aces?
(d) How many 5-card poker hands are full houses (3 of a kind and a pair of another kind)?

Unusual Problems

49. (a) In how many different orders can 3 objects be placed around a circle? (Note that ABC, BCA, and CAB are different orders when the 3 letters are in a line, but the *same* order in a circular arrangement. However, ABC is not the same circular arrangement as ACB, because B is to the left of A in one case and to the right of A in the other.)
    (b) Do part (a) for 4 objects.
    (c) Do part (a) for 5 objects.
    (d) Let n be a positive integer and do part (a) for n objects.

50. In how many ways can 6 different foods be arranged around the edge of a circular table? (See Exercise 49.)

51. In how many different orders could King Arthur and 12 knights be seated at the Round Table? (See Exercise 49.)

52. All students at a certain college are required to take year-long courses in economics, mathematics, and history during their first year. The economics classes meet at 9, 11, 1, and 3 o'clock; mathematics classes meet at 10, 12, 2, and 4 o'clock; history classes meet at 8, 10, 12, and 5 o'clock. How many different schedules are possible for a student? Assume each class lasts 50 minutes and that a student has the same schedule for the entire year.

53. A promoter wants to make up a program consisting of 10 acts. He has 7 singing acts and 9 instrumental acts available. If singing and instrumental acts are to be alternated, how many different program orders are possible? [*Hint:* First consider the possibilities when a singing act begins the program.]

9.6.A EXCURSION: Distinguishable Permutations

Suppose you have 3 identical red marbles, 3 identical white marbles, and 2 identical blue ones. How many different color patterns can be formed by placing the 8 marbles in a row? Each way of placing the marbles in a row is a permutation, and we know that there are 8! permutations of 8 marbles. But the *same* color pattern may result from *different* permutations. To see how this can happen, mentally label the red marbles R_1, R_2, R_3, the white ones W_1, W_2, W_3, and the blue ones B_1, B_2. Here are several different permutations of the marbles that all produce the *same* color pattern:

$$R_3\ W_1\ B_1\ R_1\ W_3\ B_2\ R_2\ W_2; \quad R_2\ W_2\ B_1\ R_3\ W_1\ B_2\ R_1\ W_3;$$
$$R_1\ W_3\ B_2\ R_3\ W_2\ B_1\ R_2\ W_1$$

Permutations that produce the same color pattern, such as the three listed above, are said to be *indistinguishable*. Observe that any one of the three indistinguishable permutations shown above may be obtained from any of the others simply by rearranging the red marbles among themselves, the white marbles among themselves, and the blue marbles among themselves.

Permutations that produce different color patterns are said to be *distinguishable*. If two permutations are distinguishable, then one *cannot* be obtained from the other simply by rearranging the marbles of the same color. Finding the total number of different color patterns is the same as finding the total number of distinguishable permutations of the 8 marbles.

In order to find the total number of different color patterns (distinguishable permutations), we shall proceed indirectly. We shall examine the way in which a specific permutation of the marbles is determined. Any *permutation* of the marbles (for instance, $R_1\ W_3\ B_2\ R_3\ W_2\ B_1\ R_2\ W_1$) is determined by these four things:

(i) A color pattern (in the example, the pattern R W B R W B R W).

(ii) The order in which the red marbles appear in the red positions of the pattern (in the example, $R_1\ R_3\ R_2$), that is, a permutation of the 3 red marbles.

(iii) The order in which the white marbles appear in the white positions of the pattern (in the example, $W_3\ W_2\ W_1$), that is, a permutation of the 3 white marbles.

(iv) The order in which the blue marbles appear in the blue positions of the pattern (in the example, $B_2\ B_1$), that is, a permutation of the 2 blue marbles.

A specific choice of one possibility for *each* of items (i) to (iv) will lead to exactly one permutation of the 8 marbles. Conversely, every permutation of the 8 marbles uniquely determines a choice in each one of items (i) to (iv), as shown by the example above. So the total number of permutations of the 8 marbles, namely, 8!, is the same as the total number of ways that items (i) to (iv) can all occur. According to the Fundamental Counting Principle, this number is the product of the numbers of ways each of the four items can occur. Therefore

$$8! = \begin{pmatrix}\text{Number of}\\\text{color}\\\text{patterns}\end{pmatrix} \times \begin{pmatrix}\text{Number of}\\\text{permutations of}\\\text{3 red marbles,}\\\text{namely, 3!}\end{pmatrix} \times \begin{pmatrix}\text{Number of}\\\text{permutations of}\\\text{3 white marbles,}\\\text{namely, 3!}\end{pmatrix} \times \begin{pmatrix}\text{Number of}\\\text{permutations of}\\\text{2 blue marbles,}\\\text{namely, 2!}\end{pmatrix}$$

If we let N denote the number of color patterns, this statement becomes:

$$8! = N \cdot 3! \cdot 3! \cdot 2!$$

Solving this equation for N we see that

$$N = \frac{8!}{3! \cdot 3! \cdot 2!} = 560$$

More generally, suppose that n and $k_1, k_2, \ldots, k_t$ are positive integers such that $n = k_1 + k_2 + \cdots + k_t$. Suppose that a set consists of n objects and that k_1 of these objects are all of one kind (such as red marbles), that k_2 of the objects are all of another kind (such as white marbles), that k_3 of the objects are all of a third kind, and so on. We say that two permutations of this set are **distinguishable** if one cannot be obtained from the other simply by rearranging objects of the same kind. The total number of distinguishable permutations can be found by using the same method that was used to find the total number of color patterns in the marble example above (where we had $n = 8$ and $k_1 = 3$, $k_2 = 3$, $k_3 = 2$):

DISTINGUISHABLE PERMUTATIONS

Given a set of n objects in which k_1 are of one kind, k_2 are of a second kind, k_3 are of a third kind, and so on, then the number of distinguishable permutations of the set is

$$\frac{n!}{k_1! \cdot k_2! \cdot k_3! \cdots k_t!}$$

Example 1 The number of distinguishable ways that the letters in the word TENNESSEE can be arranged is just the number of distinguishable permutations of the set consisting of the 9 symbols T, E, N, N, E, S, S, E, E. There are 4 E's, 2 N's, 2 S's, and 1 T. So we apply the formula in the box above with $n = 9$; $k_1 = 4$; $k_2 = 2$; $k_3 = 2$; $k_4 = 1$; and find that the number of distinguishable permutations is

$$\frac{9!}{4! \, 2! \, 2! \, 1!} = \frac{9 \cdot 8 \cdot 7 \cdot 6 \cdot 5 \cdot 4 \cdot 3 \cdot 2 \cdot 1}{4 \cdot 3 \cdot 2 \cdot 1 \cdot 2 \cdot 1 \cdot 2 \cdot 1 \cdot 1} = 3780. \blacksquare$$

EXERCISES

In Exercises 1–4, determine the number of distinguishable ways the letters in the word can be arranged.

1. LOOK
2. MISSISSIPPI
3. CINCINNATI
4. BOOKKEEPER

5. How many color patterns can be obtained by placing 5 red, 7 black, 3 white, and 4 orange discs in a row?

6. In how many different ways can you write the algebraic expression $x^4 y^2 z^3$ without using exponents or fractions?

9.7 INTRODUCTION TO PROBABILITY

Probability theory, which was first developed in the seventeenth century to analyze games of chance, now plays a significant role in business and the sciences. Virtually everyone has seen examples of this in statements such as:

The chance of rain tomorrow is 40%.
There is a high probability of another serious earthquake in California.
The vaccine is effective with 87% of the patients.

Probability theory provides mathematical tools for dealing with situations like these that involve uncertainty.

In the study of probability, an **experiment** is any activity or occurrence with an observable result; this result is called an **outcome** of the experiment. For instance, flipping a coin is an experiment with two possible outcomes, heads or tails. Drawing a card from a standard deck is an experiment with 52 possible outcomes. An experiment with the outcomes "satisfactory" or "defective" occurs when an item coming off a factory assembly line is tested by the quality control department.

We shall restrict our study to experiments with **equally likely** outcomes. Flipping a fair coin (one that is not weighted or shaved to favor one side over the other) is an example of such an experiment, as is rolling a fair die (one that isn't loaded so that one number comes up more often than the others).* A well-run factory, however, will produce far more satisfactory than defective items, so that the two outcomes of a quality control test will not be equally likely.

The set of all possible outcomes of an experiment is called the **sample space** of the experiment.

Example 1

(a) In the experiment of rolling a die, the sample space is

$$\{1, 2, 3, 4, 5, 6\}$$

since these numbers are the possible outcomes of rolling the die.

(b) If the experiment consists of flipping two coins, one after the other, then there are four possible outcomes:

> heads on both coins (HH);
> heads on the first coin and tails on the second (HT);
> tails on the first coin and heads on the second (TH);
> tails on both coins (TT).

Using the labels above, the sample space is the set

$$\{HH, HT, TH, TT\}. \blacksquare$$

---

*Hereafter all dice and coins in the discussion are assumed to be fair.

Any subset of the sample space (that is, any collection of some of the possible outcomes) is called an **event**. In the experiment of rolling a die, the subset {2, 4, 6} is the event "rolling an even number" and the subset {3, 4, 5, 6} is the event "rolling a number greater than 2." The event "rolling a number less than 7" is the entire sample space {1, 2, 3, 4, 5, 6}.

The basic idea of probability theory is to assign to each event a number between 0 and 1 that indicates its likelihood, with 0 meaning the event *never* occurs and 1 meaning that it *always* occurs.

PROBABILITY OF AN EVENT

If S is the sample space of an experiment with equally likely outcomes and E is an event, then the *probability* of E is denoted $P(E)$ and defined by

$$P(E) = \frac{\text{number of outcomes in } E}{\text{number of outcomes in } S}$$

We shall denote the *number* of outcomes in an event E by $n(E)$ and the number of outcomes in the sample space S by $n(S)$. In this notation,

$$P(E) = \frac{n(E)}{n(S)}.$$

Example 2 In the experiment of rolling a die, find the probability of rolling:

(a) 5;   (b) an even number;
(c) a number greater than 2;   (d) a number less than 7.

Solution

(a) The sample space $S = \{1, 2, 3, 4, 5, 6\}$ contains 6 possible outcomes, so $n(S) = 6$. The event "rolling a 5" is the one-element subset $E = \{5\}$, so $n(E) = 1$. Therefore the probability of rolling a 5 is

$$P(E) = \frac{n(E)}{n(S)} = \frac{1}{6}.$$

(b) The event "rolling an even number" is the subset $E = \{2, 4, 6\}$, with $n(E) = 3$. Its probability is

$$P(E) = \frac{n(E)}{n(S)} = \frac{3}{6} = \frac{1}{2}.$$

(c) The probability of the event {3, 4, 5, 6} is

$$P(E) = \frac{n(E)}{n(S)} = \frac{4}{6} = \frac{2}{3}.$$

(d) Rolling a number less than 7 is the event $E = \{1, 2, 3, 4, 5, 6\}$, which is the entire sample space S. The probability of this event is

$$P(E) = \frac{n(E)}{n(S)} = \frac{6}{6} = 1. \blacksquare$$

As Example 2 illustrates,

$$0 \leq P(E) \leq 1 \quad \text{for any event } E.$$

This is true because $n(E)$, the number of outcomes in E, is always less than or equal to $n(S)$, the number of outcomes in the entire sample space, so that $P(E) = n(E)/n(S)$ is a number between 0 and 1.

An event E that *always* occurs (such as rolling a number less than 7 with a single die) has probability 1. An event E that *never* occurs (for instance, rolling an 8 with a single die) has probability 0.*

Example 3 Assuming that the probability of a girl being born is the same as that of a boy, find the probability that a family with three children has

(a) At least two girls.   (b) Exactly two girls.

Solution The possible ways for three children to be born lead to the following sample space, in which b stands for boy and g for girl:

$$S = \{bbb, bbg, bgb, bgg, gbb, gbg, ggb, ggg\}.$$

(a) The event of having at least two girls is the subset

$$E = \{bgg, gbg, ggb, ggg\}$$

and

$$P(E) = \frac{n(E)}{n(S)} = \frac{4}{8} = \frac{1}{2}.$$

(b) The event of having exactly two girls is the subset

$$F = \{bgg, gbg, ggb\}.$$

Therefore

$$P(F) = \frac{n(F)}{n(S)} = \frac{3}{8}. \blacksquare$$

Example 4 If a pair of dice are rolled, find the probability that the total showing is (a) 7; (b) 11.

Solution Denote the result of rolling the dice by an ordered pair of numbers, with the first coordinate being the number showing on the first die and the second coordinate the number showing on the second. The sample space consists of all such pairs. There are 6 possibilities for the first coordinate and 6

---

* E is a set that contains *no* outcomes, so that $n(E) = 0$ and hence $P(E) = 0$.

possibilities for the second coordinate. By the Fundamental Counting Principle, there are $6 \cdot 6 = 36$ possible outcomes of rolling two dice.

(a) A total of 7 can occur in six ways:

$$(1, 6), (2, 5), (3, 4), (4, 3), (5, 2), (6, 1).$$

So the probability of a total of 7 is $6/36 = 1/6$.

(b) A total of 11 can occur in only two ways: $(5, 6)$ and $(6, 5)$. So the probability of a total of 11 is $2/36 = 1/18$. ∎

WARNING

The set

$$\{2, 3, 4, 5, 6, 7, 8, 9, 10, 11, 12\}$$

consists of the possible totals when rolling two dice, but it cannot be used as the sample space in Example 4 because the outcomes in this set are *not* equally likely. For instance, the total 2 only occurs in one way (namely, $(1, 1)$) and therefore is less likely than a total of 7, which can occur in six ways.

The solution of many probability problems depends on the counting techniques of Section 9.6.

Example 5 A committee of three people is chosen at random* from a group of 4 Republicans and 6 Democrats. Find the probability that the committee consists of

(a) Three Democrats; (b) One Republican and two Democrats.

Solution The sample space S consists of all the possible three-person committees. The number of outcomes in S is the number of ways of choosing 3 people from a group of 10, namely,

$$_{10}C_3 = \binom{10}{3} = \frac{10!}{3!7!} = 120.$$

(a) The number of ways of choosing 3 Democrats from the 6 available is

$$\binom{6}{3} = \frac{6!}{3!3!} = 20$$

so the probability that the committee consists entirely of Democrats is $20/120 = 1/6$.

---

*This means that each person is equally likely to be chosen.

(b) There are 4 choices for the Republican member. The number of ways of choosing 2 Democrats is $\binom{6}{2} = \frac{6!}{2!4!} = 15$. By the Fundamental Counting Principle, there are $4 \cdot 15 = 60$ ways of choosing the committee. The probability of such a committee being chosen is $60/120 = 1/2$. ∎

Mutually Exclusive Events

Two events in the same sample space are said to be **mutually exclusive** if they cannot both occur simultaneously. For example, when rolling a die, the events "rolling a 3 or 4" and "rolling a 1" cannot both occur on the same roll, so these events are mutually exclusive. In terms of sets, this means that the sets {3, 4} and {1} have no elements in common.* On the other hand, the events "rolling a 3 or 4" and "rolling an even number" are *not* mutually exclusive because both occur when a 4 is rolled. In other words, the sets {3, 4} and {2, 4, 6} have an element in common.

PROBABILITY OF MUTUALLY EXCLUSIVE EVENTS

> If E and F are mutually exclusive events in the same sample space, then the probability that E *or* F will occur, denoted $P(E \text{ or } F)$, is
> $$P(E) + P(F).$$

Although we won't give a formal proof of this fact, the following example indicates why it is true.

Example 6 In the experiment of rolling a die, with sample space $S = \{1, 2, 3, 4, 5, 6\}$, consider the following events:

$$\text{rolling a 3 or 4:} \quad E = \{3, 4\}$$
$$\text{rolling a 1:} \quad F = \{1\}$$

It is easy to see that

$$P(E) = \frac{2}{6} = \frac{1}{3} \quad \text{and} \quad P(F) = \frac{1}{6}.$$

E and F are mutually exclusive and the event "E or F" is the set $\{1, 3, 4\}$ consisting of the possible outcomes when E or F occurs. Note that $P(E \text{ or } F) = 3/6 = 1/2$. Thus

$$P(E) + P(F) = \frac{1}{3} + \frac{1}{6} = \frac{1}{2} = P(E \text{ or } F). \blacksquare$$

---

* Sets with no elements in common are said to be **disjoint**.

> **WARNING**
>
> The formula in the box above is *not* valid when the events are not mutually exclusive. In the experiment of rolling a die, for instance, the events "rolling a 3 or 4" and "rolling an even number," that is, $E = \{3, 4\}$ and $G = \{2, 4, 6\}$, are not mutually exclusive because both contain the outcome 4. The event "E or G" is the set $\{2, 3, 4, 6\}$, consisting of the possible outcomes when E or G occurs; its probability is $4/6 = 2/3$. In this case, $P(E \text{ or } G)$ is *not* the sum of the probabilities of E and of G because
>
> $$P(E \text{ or } G) = \frac{2}{3} \quad \text{but} \quad P(E) + P(G) = \frac{1}{3} + \frac{1}{2} = \frac{5}{6}.$$

Example 7 If a 5-card hand is dealt from a well-shuffled standard 52-card deck, what is the probability that it will consist of 5 hearts or of 5 spades?

Solution The sample space consists of the possible 5-card hands; the number of such hands is

$$\binom{52}{5} = \frac{52!}{5!47!} = 2{,}598{,}960.$$

The event E of getting 5 hearts can occur in as many ways as 5 cards can be chosen from the 13 hearts in the deck, namely,

$$\binom{13}{5} = \frac{13!}{5!8!} = 1287.$$

Similarly, the event F of getting 5 spades can occur in 1287 ways. Since E and F are mutually exclusive,

$$P(E \text{ or } F) = P(E) + P(F) = \frac{1287}{2{,}598{,}960} + \frac{1287}{2{,}598{,}960} = \frac{2574}{2{,}598{,}960}$$

$$\approx .00099$$

Rounded to three decimal places, this probability is .001. So the chances of being dealt 5 hearts or 5 spades are about 1 in 1000. ∎

The **complement** of an event consists of all the outcomes in the sample space that are *not* in the event. In rolling a die, for instance, "rolling a 3 or 4" is the set $\{3, 4\}$. The complement of this event is the set $\{1, 2, 5, 6\}$, which is the event "rolling a 1, 2, 5, or 6." Note that either the event $\{3, 4\}$ or its complement *must* occur whenever the die is rolled.

The complement of an event E is denoted E'. In every case, E or E' must occur (since every possible outcome is in one or the other). Thus $P(E \text{ or } E') = 1$. Since E and E' are mutually exclusive (no outcome is in both E and E'), we have:

$$P(E \text{ or } E') = P(E) + P(E')$$
$$1 = P(E) + P(E')$$

Rearranging this last equation, we have

$$P(E') = 1 - P(E).$$

Example 8 If a committee of three people is chosen at random from a group of 4 Republicans and 6 Democrats, what is the probability that it will have at least one Republican member?

Solution The event "at least one Republican member" is the complement of the event E, "all three committee members are Democrats". In Example 5(a) we saw that $P(E) = 1/6$. Therefore

$$P(E') = 1 - P(E) = 1 - \frac{1}{6} = \frac{5}{6}. \blacksquare$$

Probability Distributions

In many real-life situations it isn't possible to determine the probability of particular outcomes with the kind of theoretical analysis that we have used with dice, cards, etc. In such cases probabilities are determined empirically, by running experiments, constructing models, or analyzing previous data. To predict the chance of rain tomorrow, for example, a weather forecaster might note that there were 3800 days in the past century with atmospheric conditions similar to those today and that 1600 of these days were followed by rain. It would be reasonable to conclude that the probability of rain tomorrow is approximately $1600/3800 \approx .42$.

In a sample space with n equally likely outcomes, such as those studied above, each outcome has a probability of $1/n$ and the sum of these n probabilities is 1. This idea can be generalized in order to deal with experiments in which the outcomes are not all equally likely. Each outcome in the sample space is assigned a number (its probability) by some reasonable means (using theoretical analysis or an empirical approach), subject to the following conditions:

1. The probability of each outcome is a number between 0 and 1.

2. The sum of the probabilities of all the outcomes is 1.

Such an assignment of probabilities is called a **probability distribution**. It can be shown that the formulas developed above for mutually exclusive events and complements are valid for any probability distribution.

Example 9 Based on survey data, a pollster concludes that Smith has a 42% chance of winning an upcoming election and that Jones has a 38% chance of winning. Assuming these figures are accurate, find the probability that

(a) Smith or Jones will win.

(b) The third candidate, Brown, will win.

Solution The sample space consists of three outcomes:

$$S = \text{Smith wins}, \quad J = \text{Jones wins}, \quad B = \text{Brown wins}.$$

We are given that $P(S) = .42$ and $P(J) = .38$.

(a) Since Smith and Jones can't both win, S and J are mutually exclusive events; hence

$$P(S \text{ or } J) = P(S) + P(J) = .42 + .38 = .80.$$

Thus there is an 80% chance that Smith or Jones will win.

(b) The complement of the event "S or J" is B (the only other outcome in the sample space). Consequently,

$$P(B) = 1 - P(S \text{ or } J) = 1 - .80 = .20.$$

Alternatively, $P(B) = .20$ because the sum $P(S) + P(J) + P(B)$ must be 1. In any case, we conclude that Brown has a 20% chance of winning the election. ∎

EXERCISES

In Exercises 1–4, list a sample space for the experiment.

1. Three coins are flipped.
2. A day in March is chosen at random.
3. A marble is drawn from a jar containing 2 red and 2 blue marbles.
4. A coin is flipped and a die is rolled.
5. If a single die is rolled, find the probability that the result is
   (a) an odd number.
   (b) a number greater than 3.
   (c) a number other than 5.
   (d) any number except 5 or 6.
6. If three coins are flipped, find the probability that
   (a) exactly one shows heads.
   (b) at least one shows heads.

In Exercises 7–10, a letter is chosen at random from the word ABRACADABRA. Find the probability that the chosen letter is:

7. A
8. B
9. C or D
10. E

In Exercises 11–14, two dice are rolled. Find the probability that the sum is

11. less than 9.
12. greater than 7.
13. 2 or 11.
14. 8, 9 or 10.

In Exercises 15–20, a marble is drawn from an urn containing 3 red, 4 white, and 8 blue marbles. Find the probability that this marble is

15. red.
16. white.
17. blue or white.
18. black.
19. not blue.
20. not white.

In Exercises 21–28, a card is drawn from a well-shuffled standard 52-card deck. Find the probability that this card is

21. an ace.
22. a heart.
23. a red queen.

24. a face card (king, queen, or jack).
25. a red face card.
26. a red card that is not a face card.
27. not a face card and greater than 5 (ace is 1).
28. a black card less than 6 (ace is 1).

In Exercises 29–34, a 5-card hand is dealt from a well-shuffled standard 52-card deck. Find the probability that it contains

29. four aces.
30. four of a kind.
31. two aces and three kings
32. five face cards.
33. at least one ace.
34. at least one spade.

35. If a student taking a ten question true-false quiz guesses randomly, find the probability that the student will get
   (a) 10 questions correct.
   (b) 5 questions correct and 5 wrong.
   (c) At least 7 questions correct.

36. A committee of six people is to be chosen from a group of 5 men and 10 women. Find the probability that the committee consists of:
   (a) 3 women and 3 men.
   (b) 4 women and 2 men.
   (c) 6 women.

37. A shipment of 20 stereos contains 5 defective ones. If 5 stereos in this shipment are chosen at random, what is the probability that exactly 2 of them will be defective?

38. A package contains 14 fasteners, 2 of which are defective. If 3 fasteners are chosen at random from the package, what is the probability that all 3 are not defective?

39. In a certain state lottery, the bettor chooses six numbers between 1 and 44. If these six numbers are drawn in the lottery, the player wins a prize of several million dollars. What is the probability that a bettor who buys only one ticket will win the lottery?

40. In another state lottery (see Exercise 39), the bettor chooses six numbers from 1 to 50, but gets two such choices for each ticket bought. What is the probability that a bettor who buys only one ticket will win this lottery?

In Exercises 41–44, use the following results of a survey of 800 consumers, which show the number of people in each income class who watch TV for the stated number of hours each week:

Annual Income	Hours of TV Watched per Week		
	<10	10–20	>20
Less than $20,000	40	120	100
$20,000–$39,999	70	160	50
$40,000–$59,999	80	60	20
$60,000 or more	40	50	10

If a person from the survey group is chosen at random, what is the probability that this person

41. has an income of at least $40,000?
42. watches TV at least 10 hours per week?
43. has an income under $40,000 and watches TV at least 10 hours per week?
44. has an income of at least $20,000 and watches TV less than 10 hours per week?

CHAPTER REVIEW

Important Concepts

Section 9.1

Sequence 424
Term 424
Constant sequence 426
Recursively defined sequence 426
Fibonacci sequence 426
Summation notation 427
Summation index 428
Partial sum 428

Section 9.2	Arithmetic sequence 431	
	Common difference 432	
	Partial sum formulas 433	
Section 9.3	Geometric sequence 437	
	Common ratio 437	
	Partial sum formula 439	
Excursion 9.3.A	Infinite series 443	
	Infinite geometric series 444	
Section 9.4	n factorial ($n!$) 447	
	Binomial coefficient 447	
	Pascal's triangle 448	
	Binomial Theorem 449	
Section 9.5	Principle of Mathematical Induction 456	
	Inductive step 457	
	Induction hypothesis 457	
Section 9.6	Fundamental Counting Principle 464	
	Permutation 465	
	Permutation of n things taken r at a time 466	
	Combination of n things taken r at a time 467	
Excursion 9.6.A	Distinguishable permutations 470	
Section 9.7	Experiment 473	
	Outcome 473	
	Sample space 473	
	Event 474	
	Probability of an event 474	
	Mutually exclusive events 477	
	Complement of an event 478	
	Probability Distribution 479	

Important Facts and Formulas

- In an arithmetic sequence $\{a_n\}$ with common difference d:

$$a_n = a_1 + (n-1)d \qquad \sum_{n=1}^{k} a_n = \frac{k}{2}(a_1 + a_k)$$

$$\sum_{n=1}^{k} a_n = ka_1 + \frac{k(k-1)}{2}d$$

- In a geometric sequence $\{a_n\}$ with common ratio r:

$$a_n = r^{n-1}a_1 \qquad \sum_{n=1}^{k} a_n = a_1\left(\frac{1-r^k}{1-r}\right)$$

- $n! = 1 \cdot 2 \cdot 3 \cdots (n-2)(n-1)n$

- $\binom{n}{r} = \dfrac{n!}{r!(n-r)!}$
- *The Binomial Theorem:*

$$(x+y)^n = x^n + \binom{n}{1}x^{n-1}y + \binom{n}{2}x^{n-2}y^2 + \binom{n}{3}x^{n-3}y^3 + \cdots + \binom{n}{n-1}xy^{n-1} + y^n$$

$$= \sum_{j=0}^{n} \binom{n}{j} x^{n-j}y^j$$

- The number of permutations of a set of n elements is $n!$.
- $_nP_r = \dfrac{n!}{(n-r)!} = \left(\begin{array}{l}\text{the product of } r \text{ consecutive integers,}\\ \text{starting with } n \text{ and working down}\end{array}\right)$
- $_nC_r = \dfrac{n!}{r!(n-r)!} = \binom{n}{r}$

Review Questions

In Questions 1–4, find the first four terms of the sequence $\{a_n\}$.

1. $a_n = 2n - 5$
2. $a_n = 3n - 27$
3. $a_n = \left(\dfrac{-1}{n}\right)^2$
4. $a_n = (-1)^{n+1}(n-1)$

5. Find the 5th partial sum of the sequence $\{a_n\}$, where $a_1 = -4$ and $a_n = 3a_{n-1} + 2$.
6. Find the 4th partial sum of the sequence $\{a_n\}$, where $a_1 = 1/9$ and $a_n = 3a_{n-1}$.
7. $\sum_{n=0}^{4} 2^n(n+1) = ?$
8. $\sum_{n=2}^{4} (3n^2 - n + 1) = ?$

In Questions 9–12, find a formula for a_n; assume that the sequence is arithmetic.

9. $a_1 = 3$ and the common difference is -6.
10. $a_2 = 4$ and the common difference is 3.
11. $a_1 = -5$ and $a_3 = 7$.
12. $a_3 = 2$ and $a_7 = -1$.

In Questions 13–16, find a formula for a_n; assume that the sequence is geometric.

13. $a_1 = 2$ and the common ratio is 3.
14. $a_1 = 5$ and the common ratio is $-1/2$.
15. $a_2 = 192$ and $a_7 = 6$.
16. $a_3 = 9/2$ and $a_6 = -243/16$.
17. Find the 11th partial sum of the arithmetic sequence with $a_1 = 5$ and common difference -2.
18. Find the 12th partial sum of the arithmetic sequence with $a_1 = -3$ and $a_{12} = 16$.
19. Find the 5th partial sum of the geometric sequence with $a_1 = 1/4$ and common ratio 3.
20. Find the 6th partial sum of the geometric sequence with $a_1 = 5$ and common ratio $1/2$.
21. Insert three arithmetic means between 4 and 23.
22. Insert two geometric means between 4 and 62.5.
23. Is it better to be paid $5 per day for 100 days or to be paid 5¢ the first day, 10¢ the second day, 20¢ the third day, and have your salary increase in this fashion every day for 100 days?
24. Tuition at Bigstate University is now $1000 per year and will increase $150 per year in succeeding years. If a student starts school now, spends four years as an undergraduate, three years in law school, and five years getting a PhD, how much tuition will she have paid?

Find the following sums, if they exist.

25. $\sum_{n=1}^{\infty} \frac{1}{2^{n-1}}$ 26. $\sum_{n=1}^{\infty} \left(\frac{-1}{4^n}\right)$

27. Use the Binomial Theorem to show that $(1.02)^{51} > 2.5$.
28. What is the coefficient of u^3v^2 in the expansion of $(u + 5v)^5$?

29. $\binom{15}{12} = ?$ 30. $\binom{18}{3} = ?$

31. Let n be a positive integer. Simplify $\binom{n+1}{n}$.

32. Use the Binomial Theorem to expand $(\sqrt{x} + 1)^5$. Simplify your answer.

33. $\frac{20!5!}{6!17!} = ?$ 34. $\frac{7! - 5!}{4!} = ?$

35. Find the coefficient of x^2y^4 in the expansion of $(2y + x^2)^5$.

36. Prove that for every positive integer n,
$$1^3 + 2^3 + 3^3 + \cdots + n^3 = \frac{n^2(n+1)^2}{4}$$

37. Prove that for every positive integer n,
$$1 + 5 + 5^2 + 5^3 + \cdots + 5^{n-1} = \frac{5^n - 1}{4}$$

38. Prove that $2^n \geq 2n$ for every positive integer n.

39. If x is a real number with $|x| < 1$, then prove that $|x^n| < 1$ for all $n \geq 1$.

40. Prove that for any positive integer n, $1 + 5 + 9 + \cdots + (4n - 3) = n(2n - 1)$.

41. Prove that for any positive integer n, $1 + 4 + 4^2 + 4^3 + \cdots + 4^{n-1} = \frac{1}{3}(4^n - 1)$.

42. Prove that $3n < n!$ for every $n \geq 4$.

43. Prove that for every positive integer n, 8 is a factor of $9^n - 8n - 1$.

In Questions 44–48, compute the number.

44. $_7P_3$ 45. $_{24}P_6$

46. $_7C_3$ 47. $_{24}C_{20}$

48. $_kC_t$

49. How many 3-digit numbers are there in which the first digit is even and the second is odd?

50. How many possible finishes (first, second, third place) are there in an 8-horse race?

51. A woman has three skirts, four blouses, and two scarves. How many different outfits can she wear (an outfit being a skirt, blouse, and scarf)?

52. A bridge hand consists of 13 cards from a standard 52-card deck. How many different bridge hands are there?

53. How many different committees consisting of 2 men and 2 women can be formed from a group of 8 women and 5 men?

54. List a sample space for the experiment of flipping a coin four times in succession.

55. If two dice are rolled, what is the probability that the sum is 8?

56. Which event is more likely: rolling a 3 with a single die or rolling a total of 6 with two dice?

57. Assuming that the probability of a girl being born is the same as that of a boy, find the probability that a family with four children has 3 girls and a boy.

58. If a card is drawn from a well-shuffled standard 52-card deck, what is the probability that it is a 6 or a king?

59. A committee of four people is to be randomly chosen from a group of 8 mathematicians and 12 chemists. What is the probability that the committee will consist of 2 mathematicians and 2 chemists?

60. Two stale candy bars are inadvertently dropped in a box with 18 fresh bars of the same brand. If you take 3 bars from the box, what is the probability that all of them will be fresh?

APPENDIX 1

Geometry Review

A **triangle** has three sides (straight line segments) and three angles, formed at the points where the various sides meet. When angles are measured in degrees, the sum of the measures of all three angles of a triangle is *always* 180°. For instance, see Figure A–1.

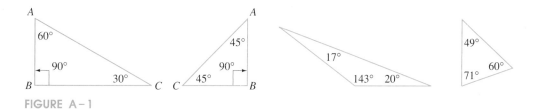

FIGURE A–1

A **right angle** is an angle that measures 90°. A right triangle is a triangle, one of whose angles is a right angle, such as the first two triangles shown in Figure A–1. The side of a right triangle that lies opposite the right angle is called the **hypotenuse**. In each of the right triangles in Figure A–1, side AC is the hypotenuse.

PYTHAGOREAN THEOREM

If the sides of a right triangle have lengths a and b and the hypotenuse has length c, then

$$c^2 = a^2 + b^2$$

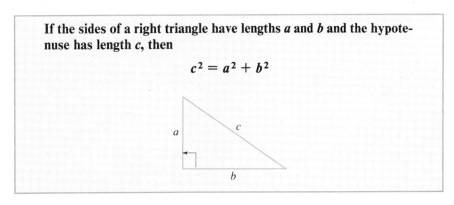

Example 1 Consider the right triangle with sides of lengths 5 and 12, as shown in Figure A–2.

487

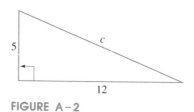

FIGURE A–2

According to the Pythagorean Theorem the length c of the hypotenuse satisfies the equation: $c^2 = 5^2 + 12^2 = 25 + 144 = 169$. Since $169 = 13^2$, we see that c must be 13. ■

THEOREM I | If two angles of a triangle are equal, then the two sides opposite these angles have the same length.

Example 2 Suppose the hypotenuse of the right triangle shown in Figure A–3 has length 1 and that angles B and C measure 45° each.

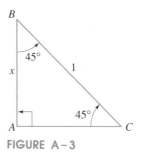

FIGURE A–3

Then by Theorem I, sides AB and AC have the same length. If x is the length of side AB, then by the Pythagorean Theorem:

$$x^2 + x^2 = 1^2$$
$$2x^2 = 1$$
$$x^2 = \frac{1}{2}$$
$$x = \sqrt{\frac{1}{2}} = \frac{1}{\sqrt{2}} = \frac{\sqrt{2}}{2}$$

(We ignore the other solution of this equation, namely, $x = -\sqrt{1/2}$, since x represents a length here and thus must be nonnegative.) Therefore the sides of a 90°-45°-45° triangle with hypotenuse 1 are each of length $\sqrt{2}/2$. ■

THEOREM II | In a right triangle that has an angle of 30°, the length of the side opposite the 30° angle is one-half the length of the hypotenuse.

Example 3 Suppose that in the right triangle shown in Figure A–4 angle B is 30° and the length of hypotenuse BC is 2.

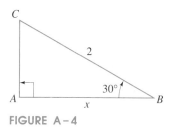

FIGURE A–4

By Theorem II the side opposite the 30° angle, namely, side AC, has length 1. If x denotes the length of side AB, then by the Pythagorean Theorem:

$$1^2 + x^2 = 2^2$$
$$x^2 = 3$$
$$x = \sqrt{3}.$$ ■

Example 4 The right triangle shown in Figure A–5 has a 30° angle at C, and side AC has length $\sqrt{3}/2$.

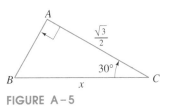

FIGURE A–5

Let x denote the length of the hypotenuse BC. By Theorem II, side AB has length $\frac{1}{2}x$. By the Pythagorean Theorem:

$$\left(\frac{1}{2}x\right)^2 + \left(\frac{\sqrt{3}}{2}\right)^2 = x^2$$
$$\frac{x^2}{4} + \frac{3}{4} = x^2$$
$$\frac{3}{4} = \frac{3}{4}x^2$$
$$x^2 = 1$$
$$x = 1$$

Therefore the triangle has hypotenuse of length 1 and sides of lengths 1/2 and $\sqrt{3}/2$. ■

Two triangles, as in Figure A–6,

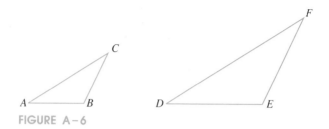

FIGURE A–6

are said to be **similar** if their corresponding angles are equal (that is, $\angle A = \angle D$; $\angle B = \angle E$; and $\angle C = \angle F$). Thus similar triangles have the same *shape* but not necessarily the same *size*.

THEOREM III

Suppose triangle *ABC* with sides *a, b, c* is similar to triangle *DEF* with sides *d, e, f* (that is, $\angle A = \angle D$; $\angle B = \angle E$; $\angle C = \angle F$),

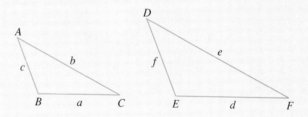

then,

$$\frac{a}{d} = \frac{b}{e} = \frac{c}{f}$$

These equalities are equivalent to:

$$\frac{a}{b} = \frac{d}{e}, \quad \frac{b}{c} = \frac{e}{f}, \quad \frac{a}{c} = \frac{d}{f}$$

The equivalence of the equalities in the conclusion of the theorem is easily verified. For example, since

$$\frac{a}{d} = \frac{b}{e}$$

we have

$$ae = db$$

Dividing both sides of this equation by *be* yields:

$$\frac{ae}{be} = \frac{db}{be}$$

$$\frac{a}{b} = \frac{d}{e}$$

The other equivalences are proved similarly.

Example 5 Suppose the triangles in Figure A-7

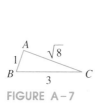
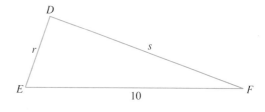

FIGURE A-7

are similar and that the sides have the lengths indicated. Then by Theorem III,

$$\frac{\text{length } AC}{\text{length } DF} = \frac{\text{length } BC}{\text{length } EF}$$

In other words,

$$\frac{\sqrt{8}}{s} = \frac{3}{10}$$

so that

$$3s = 10\sqrt{8}$$

$$s = \left(\frac{10}{3}\right)\sqrt{8}$$

Similarly, by Theorem III,

$$\frac{\text{length } AB}{\text{length } DE} = \frac{\text{length } BC}{\text{length } EF}$$

so that

$$\frac{1}{r} = \frac{3}{10}$$

$$3r = 10$$

$$r = \frac{10}{3}$$

Therefore the sides of triangle DEF are of lengths 10, $\frac{10}{3}$, and $\frac{10}{3}\sqrt{8}$. ■

APPENDIX 2

Logarithm Tables

A typical table of common (base 10) logarithms appears on the next two pages. The following examples illustrate its use.

Example 1 To find the logarithm of 4.63, we first look in the *left*-hand column for the first two digits of our number, namely, 4.6. Now on the same line as 4.6 we look at the entry in the column labeled 3; it is .6656. So the logarithm of 4.63 is approximately .6656. ∎

Example 2 We cannot find log 5.527 immediately since our table does *not* include 5.527. However, the table does include 5.52 = 5.520 and 5.53 = 5.530 and

$$5.520 < 5.527 < 5.530$$

Since 27 lies $\frac{7}{10}$ of the way from 20 to 30, we see that 5.527 lies $\frac{7}{10}$ of the way from 5.520 to 5.530. Therefore it seems reasonable that log 5.527 lies approximately $\frac{7}{10}$ of the way from log 5.520 to log 5.530. Using the table, we find that

$$\log 5.530 = \log 5.53 \approx .7427$$
$$\log 5.520 = \log 5.52 \approx .7419$$

difference .0008

Therefore log 5.527 lies $\frac{7}{10}$ of the way between .7419 and .7427. The distance from .7419 to .7427 is .0008, and $\frac{7}{10}$ of this distance is $(.7)(.0008) = .00056$. Since we are using four-place logarithms, we round off this number to four places: $.00056 \approx .0006$. Then we have:

$$\log 5.527 \approx \log 5.52 + .0006 \approx .7419 + .0006 = .7425. \blacksquare$$

The process used to find log 5.527 in the preceding example is called **linear interpolation.** Owing to the approximating involved there may be a slight error in the results. But for most purposes this error is insignificant.

Example 3 In order to find the logarithm of 573.2, we note that

$$573.2 = 5.732 \times 10^2$$

Consequently,

$$\log 573.2 = \log(10^2 \cdot 5.732) = \log 10^2 + \log 5.732.$$

x	0	1	2	3	4	5	6	7	8	9
1.0	.0000	.0043	.0086	.0128	.0170	.0212	.0253	.0294	.0334	.0374
1.1	.0414	.0453	.0492	.0531	.0569	.0607	.0645	.0682	.0719	.0755
1.2	.0792	.0828	.0864	.0899	.0934	.0969	.1004	.1038	.1072	.1106
1.3	.1139	.1173	.1206	.1239	.1271	.1303	.1335	.1367	.1399	.1430
1.4	.1461	.1492	.1523	.1553	.1584	.1614	.1644	.1673	.1703	.1732
1.5	.1761	.1790	.1818	.1847	.1875	.1903	.1931	.1959	.1987	.2014
1.6	.2041	.2068	.2095	.2122	.2148	.2175	.2201	.2227	.2253	.2279
1.7	.2304	.2330	.2355	.2380	.2405	.2430	.2455	.2480	.2504	.2529
1.8	.2553	.2577	.2601	.2625	.2648	.2672	.2695	.2718	.2742	.2765
1.9	.2788	.2810	.2833	.2856	.2878	.2900	.2923	.2945	.2967	.2989
2.0	.3010	.3032	.3054	.3075	.3096	.3118	.3139	.3160	.3181	.3201
2.1	.3222	.3243	.3263	.3284	.3304	.3324	.3345	.3365	.3385	.3404
2.2	.3424	.3444	.3464	.3483	.3502	.3522	.3541	.3560	.3579	.3598
2.3	.3617	.3636	.3655	.3674	.3692	.3711	.3729	.3747	.3766	.3784
2.4	.3802	.3820	.3838	.3856	.3874	.3892	.3909	.3927	.3945	.3962
2.5	.3979	.3997	.4014	.4031	.4048	.4065	.4082	.4099	.4116	.4133
2.6	.4150	.4166	.4183	.4200	.4216	.4232	.4249	.4265	.4281	.4298
2.7	.4314	.4330	.4346	.4362	.4378	.4393	.4409	.4425	.4440	.4456
2.8	.4472	.4487	.4502	.4518	.4533	.4548	.4564	.4579	.4594	.4609
2.9	.4624	.4639	.4654	.4669	.4683	.4698	.4713	.4728	.4742	.4757
3.0	.4771	.4786	.4800	.4814	.4829	.4843	.4857	.4871	.4886	.4900
3.1	.4914	.4928	.4942	.4955	.4969	.4983	.4997	.5011	.5024	.5038
3.2	.5051	.5065	.5079	.5092	.5105	.5119	.5132	.5145	.5159	.5172
3.3	.5185	.5198	.5211	.5224	.5237	.5250	.5263	.5276	.5289	.5302
3.4	.5315	.5328	.5340	.5353	.5366	.5378	.5391	.5403	.5416	.5428
3.5	.5441	.5453	.5465	.5478	.5490	.5502	.5514	.5527	.5539	.5551
3.6	.5563	.5575	.5587	.5599	.5611	.5623	.5635	.5647	.5658	.5670
3.7	.5682	.5694	.5705	.5717	.5729	.5740	.5752	.5763	.5775	.5786
3.8	.5798	.5809	.5821	.5832	.5843	.5855	.5866	.5877	.5888	.5899
3.9	.5911	.5922	.5933	.5944	.5955	.5966	.5977	.5988	.5999	.6010
4.0	.6021	.6031	.6042	.6053	.6064	.6075	.6085	.6096	.6107	.6117
4.1	.6128	.6138	.6149	.6159	.6170	.6180	.6191	.6201	.6212	.6222
4.2	.6232	.6243	.6253	.6263	.6274	.6284	.6294	.6304	.6314	.6325
4.3	.6335	.6345	.6355	.6365	.6375	.6385	.6395	.6405	.6415	.6425
4.4	.6435	.6444	.6454	.6464	.6474	.6484	.6493	.6503	.6513	.6522
4.5	.6532	.6542	.6551	.6561	.6571	.6580	.6590	.6599	.6609	.6618
4.6	.6628	.6637	.6646	.6656	.6665	.6675	.6684	.6693	.6702	.6712
4.7	.6721	.6730	.6739	.6749	.6758	.6767	.6776	.6785	.6794	.6803
4.8	.6812	.6821	.6830	.6839	.6848	.6857	.6866	.6875	.6884	.6893
4.9	.6902	.6911	.6920	.6928	.6937	.6946	.6955	.6964	.6972	.6981
5.0	.6990	.6998	.7007	.7016	.7024	.7033	.7042	.7050	.7059	.7067
5.1	.7076	.7084	.7093	.7101	.7110	.7118	.7126	.7135	.7143	.7152
5.2	.7160	.7168	.7177	.7185	.7193	.7202	.7210	.7218	.7226	.7235
5.3	.7243	.7251	.7259	.7267	.7275	.7284	.7292	.7300	.7308	.7316
5.4	.7324	.7332	.7340	.7348	.7356	.7364	.7372	.7380	.7388	.7396
x	0	1	2	3	4	5	6	7	8	9

x	0	1	2	3	4	5	6	7	8	9
5.5	.7404	.7412	.7419	.7427	.7435	.7443	.7451	.7459	.7466	.7474
5.6	.7482	.7490	.7497	.7505	.7513	.7520	.7528	.7536	.7543	.7551
5.7	.7559	.7566	.7574	.7582	.7589	.7597	.7604	.7612	.7619	.7627
5.8	.7634	.7642	.7649	.7657	.7664	.7672	.7679	.7686	.7694	.7701
5.9	.7709	.7716	.7723	.7731	.7738	.7745	.7752	.7760	.7767	.7774
6.0	.7782	.7789	.7796	.7803	.7810	.7818	.7825	.7832	.7839	.7846
6.1	.7853	.7860	.7868	.7875	.7882	.7889	.7896	.7903	.7910	.7917
6.2	.7924	.7931	.7938	.7945	.7952	.7959	.7966	.7973	.7980	.7987
6.3	.7993	.8000	.8007	.8014	.8021	.8028	.8035	.8041	.8048	.8055
6.4	.8062	.8069	.8075	.8082	.8089	.8096	.8102	.8109	.8116	.8122
6.5	.8129	.8136	.8142	.8149	.8156	.8162	.8169	.8176	.8182	.8189
6.6	.8195	.8202	.8209	.8215	.8222	.8228	.8235	.8241	.8248	.8254
6.7	.8261	.8267	.8274	.8280	.8287	.8293	.8299	.8306	.8312	.8319
6.8	.8325	.8331	.8338	.8344	.8351	.8357	.8363	.8370	.8376	.8382
6.9	.8388	.8395	.8401	.8407	.8414	.8420	.8426	.8432	.8439	.8445
7.0	.8451	.8457	.8463	.8470	.8476	.8482	.8488	.8494	.8500	.8506
7.1	.8513	.8519	.8525	.8531	.8537	.8543	.8549	.8555	.8561	.8567
7.2	.8573	.8579	.8585	.8591	.8597	.8603	.8609	.8615	.8621	.8627
7.3	.8633	.8639	.8645	.8651	.8657	.8663	.8669	.8675	.8681	.8686
7.4	.8692	.8698	.8704	.8710	.8716	.8722	.8727	.8733	.8739	.8745
7.5	.8751	.8756	.8762	.8768	.8774	.8779	.8785	.8791	.8797	.8802
7.6	.8808	.8814	.8820	.8825	.8831	.8837	.8842	.8848	.8854	.8859
7.7	.8865	.8871	.8876	.8882	.8887	.8893	.8899	.8904	.8910	.8915
7.8	.8921	.8927	.8932	.8938	.8943	.8949	.8954	.8960	.8965	.8971
7.9	.8976	.8982	.8987	.8993	.8998	.9004	.9009	.9015	.9020	.9025
8.0	.9031	.9036	.9042	.9047	.9053	.9058	.9063	.9069	.9074	.9079
8.1	.9085	.9090	.9096	.9101	.9106	.9112	.9117	.9122	.9128	.9133
8.2	.9138	.9143	.9149	.9154	.9159	.9165	.9170	.9175	.9180	.9186
8.3	.9191	.9196	.9201	.9206	.9212	.9217	.9222	.9227	.9232	.9238
8.4	.9243	.9248	.9253	.9258	.9263	.9269	.9274	.9279	.9284	.9289
8.5	.9294	.9299	.9304	.9309	.9315	.9320	.9325	.9330	.9335	.9340
8.6	.9345	.9350	.9355	.9360	.9365	.9370	.9375	.9380	.9385	.9390
8.7	.9395	.9400	.9405	.9410	.9415	.9420	.9425	.9430	.9435	.9440
8.8	.9445	.9450	.9455	.9460	.9465	.9469	.9474	.9479	.9484	.9489
8.9	.9494	.9499	.9504	.9509	.9513	.9518	.9523	.9528	.9533	.9538
9.0	.9542	.9547	.9552	.9557	.9562	.9566	.9571	.9576	.9581	.9586
9.1	.9590	.9595	.9600	.9605	.9609	.9614	.9619	.9624	.9628	.9633
9.2	.9638	.9643	.9647	.9652	.9657	.9661	.9666	.9671	.9675	.9680
9.3	.9685	.9689	.9694	.9699	.9703	.9708	.9713	.9717	.9722	.9727
9.4	.9731	.9736	.9741	.9745	.9750	.9754	.9759	.9763	.9768	.9773
9.5	.9777	.9782	.9786	.9791	.9795	.9800	.9805	.9809	.9814	.9818
9.6	.9823	.9827	.9832	.9836	.9841	.9845	.9850	.9854	.9859	.9863
9.7	.9868	.9872	.9877	.9881	.9886	.9890	.9894	.9899	.9903	.9908
9.8	.9912	.9917	.9921	.9926	.9930	.9934	.9939	.9943	.9948	.9952
9.9	.9956	.9961	.9965	.9969	.9974	.9978	.9983	.9987	.9991	.9996
x	0	1	2	3	4	5	6	7	8	9

But log $10^2 = 2$ and 5.732 is a number between 1 and 10. Using the tables and interpolation we find that log $5.732 \approx .7583$. Therefore

$$\log 573.2 = \log 10^2 + \log 5.732 \approx 2 + .7583 = 2.7583 \;\blacksquare$$

Example 4 In order to find log .00563, we write .00563 as follows:

$$.00563 = 5.63 \times .001 = 5.63 \times 10^{-3}$$

Consequently,

$$\log .00563 = \log(10^{-3} \cdot 5.63) = \log 10^{-3} + \log 5.63$$

We know that log $10^{-3} = -3$. The tables show that log $5.63 \approx .7505$. Therefore,

$$\log .00563 = \log 10^{-3} + \log 5.63 \approx -3 + .7505 = -2.2495. \;\blacksquare$$

The preceding examples show that

The logarithm of any positive number can always be approximated by the sum of an integer and a number between 0 and 1.

For instance, log $5.75 \approx .7597 = 0 + .7597$ and log $573.2 \approx 2.7583 = 2 + .7583$ and log $.00563 \approx -3 + .7505$. A logarithm written in this way is said to be in **standard form**. The integer part of a logarithm in standard form is called the **characteristic**. The decimal fraction part of a logarithm in standard form (that is, the number between 0 and 1) is called the **mantissa**.

Antilogarithms

The number whose logarithm is u is called the **antilogarithm** of u. To find the antilogarithm of u, recall that log $10^u = u$ for every real number u. In other words,

The antilogarithm of u is 10^u.

Antilogarithms can be found by using the logarithm tables "in reverse."

Example 5 The antilogarithm of .7435 is the number y such that log $y = .7435$. In order to find y, look through the logarithm table until you find the entry .7435. As shown on page 486, this entry lies on the same line as 5.5 (left column) and in the column labeled 4 at the top. This means that log $5.54 \approx .7435$, so that $y \approx 5.54$. $\blacksquare$

Example 6 The antilogarithm of 2.7435 is $10^{2.7435}$. Simple arithmetic and the laws of exponents show that

$$10^{2.7435} = 10^{2+.7435} = (10^2)(10^{.7435}).$$

Now $10^{.7435}$ is just the antilogarithm of .7435. So once we know this number, we can multiply by $10^2 = 100$ to get the antilogarithm of 2.7435. But .7435 *is* a number between 0 and 1, and we have just seen how to use the tables to find antilogarithms of such numbers. In fact, in Example 5 we found that

$$10^{.7435} = \text{antilogarithm of } .7435 \approx 5.54$$

Therefore the antilogarithm of 2.7435 is just

$$10^2(\text{antilogarithm of } .7435) \approx 10^2(5.54) = 100(5.54) = 554. \blacksquare$$

Example 7 Suppose $\log x = -3.2505$. In order to use the table to find x, it is first necessary to write $\log x = -3.2505$ in standard form (that is, as the sum of an integer and a number between 0 and 1). Be careful—it is *not* true that $-3.2505 = -3 + .2505$. It is true that $-3.2505 = -3 - .2505$, but $-.2505$ does not lie between 0 and 1. Here's how to write -3.2505 in standard form:

$$\log x = -3.2505 = (-4 + 4) - 3.2505$$
$$= -4 + (4 - 3.2505) = -4 + .7495$$

Now we can proceed as before. Since the antilogarithm of -3.2505 is known to be $10^{-3.2505}$ and since

$$10^{-3.2505} = 10^{-4+.7495} = (10^{-4})(10^{.7495})$$

we need only find $10^{.7495}$, the antilogarithm of .7495. For once we have this, we just multiply by $10^{-4} = .0001$ to obtain the antilogarithm of $-4 + .7495 = -3.2505$. Since .7495 lies between 0 and 1, this can be done via tables and interpolation:

$$10^{.7495} = \text{antilogarithm of } .7495 \approx 5.617$$

Therefore the antilogarithm of $-3.2505 = -4 + .7495$ is

$$10^{-4}(\text{antilogarithm of } .7495) \approx (10^{-4})(5.617)$$
$$= (.0001)(5.617) = .0005617. \blacksquare$$

Here is a summary of the procedure used in the preceding examples to find antilogarithms.

ANTILOGARITHMS

Given u, find the antilogarithm of u as follows:

(i) Write u in standard form, as the sum of an integer k and a number v between 0 and 1: $u = k + v$. [*Note: k may be positive, negative, or zero.*]

(ii) Use the tables (and interpolation, if necessary) to find the antilogarithm of v (that is, the number y with $\log y = v$).

(iii) Then the antilogarithm of u is $(10^k)y$.

Answers to Odd-Numbered Exercises

CHAPTER 1

Section 1.1, page 7

1. [number line showing points at −7, −5, −4.75, −1, 0, 1/2, 2.25, 8/3, 10]

3. $-4 > -8$ **5.** $\pi < 100$ **7.** $y \le 7.5$ **9.** $t > 0$ **11.** $\pi < d < 7.93$
13. $-8 \le y < 0$ **15.** $c \le 3$ **17.** $c < 4 \le d$ **19.** $<$ **21.** $=$ **23.** $>$ **25.** $>$
27. $>$ **29.** $\ne$ **31.** $\ne$ **33.** $\ne$ **35.** -1 **37.** 67.43 **39.** 35 **41.** -10.77
43. 7 **45.** 26 **47.** 70 **49.** -50 **51.** -19 **53.** 0 **55.** 10

57. Yes: If r and s are rational numbers, then, by definition, $r = a/b$ and $s = c/d$ for some integers a, b, c, d, with $b \ne 0, d \ne 0$. Hence $rs = \frac{a}{b} \cdot \frac{c}{d} = \frac{ac}{bd}$. Since the product of two integers is an integer, ac and bd are integers. Also, $bd \ne 0$ since both b and d are nonzero. Therefore rs is rational.

59. (a) Many correct answers, including $19/63, 20/63, \ldots, 34/63$. (b) Half the distance between r/s and a/b is $\frac{1}{2}\left(\frac{a}{b} - \frac{r}{s}\right)$, a rational number. Thus $\frac{r}{s} + \frac{1}{2}\left(\frac{a}{b} - \frac{r}{s}\right) = \frac{2br + as - rb}{2bs} = \frac{br + as}{2bs}$ is a rational number that lies halfway between r/s and a/b.

61. $\le$ **63.** $\ge$ **65.** If $-b \le a \le 0$, then $a^2 \le b^2$; if $a \le -b \le 0$, then $a^2 \ge b^2$.
67. $b + c = a$ **69.** a lies to the right of b. **71.** $a < b$ **73.** $\sqrt{2}$

75. (a) If $\frac{0}{0} = 1$, then $1 = \frac{0}{0} = \frac{2 \cdot 0}{0} = 2 \cdot \frac{0}{0} = 2 \cdot 1 = 2$, which is a contradiction. (b) The argument in part (a), with c in place of 1, shows that $c = 2c$ and hence $1 = 2$, a contradiction. (c) $0/0$ ought to be close to d/d when d is a number very close to 0. But $d/d = 1$ for every nonzero d.

Excursion 1.1.A, page 12

1. $.7777\ldots$ **3.** $1.6428571428571\ldots$ **5.** $.052631578947368421052\ldots$

7. no; $\frac{2}{3} = .6666\ldots$ **9.** yes; $\frac{1}{64} = .015625$ **11.** no **13.** yes; $\frac{1}{.625} = 1.6$

15. $\frac{37}{99}$ **17.** $\frac{758679}{9900} = \frac{252893}{3300}$ **19.** $\frac{5}{37}$ **21.** $\frac{517896}{9900} = \frac{14386}{275}$

A-1

23. If $d = .74999\ldots$, then $10{,}000d - 100d = (7499.999\ldots) - (74.999\ldots) = 7425$. Hence $9900d = 7425$ so that $d = \dfrac{7425}{9900} = \dfrac{3}{4}$. Also $.75000\ldots = .75 = \dfrac{75}{100} = \dfrac{3}{4}$.

25. $\dfrac{6}{17} = .35294117647058823529\ldots$

27. $\dfrac{1}{29} = .0344827586206896551724137931 0344\ldots$

29. $\dfrac{283}{47} = 6.021276595744680851063829787234042553191489361 70212\ldots$

31. (a) One of many possible ways is to use the nonrepeating decimal expansion of π. For instance, with .75, associate .7531415926 . . . ; with 6.593 associate 6.59331415926 . . . , etc. Thus different terminating decimals correspond to different nonrepeating ones.
(b) As suggested in the *Hint*, associate with .134134134 . . . the irrational number .134013400134000134 . . . . With 6.17398419841 associate 6.173984109841000984100098410009841 . . . , etc. Thus different repeating decimals correspond to different nonrepeating ones.

Section 1.2, page 18

1. 36   **3.** 73   **5.** -5 **7.** $-125/64$ **9.** 1/3   **11.** -112 **13.** 81/16
15. $-211/216$ **17.** 129/8   **19.** x^{10} **21** $.03y^9$ **23.** $24x^7$ **25.** $9x^4y^2$
27. $-21a^6$ **29.** $384w^6$ **31.** ab^3 **33.** $8x^{-1}y^3$ **35.** a^8x^{-3} **37.** $3xy$
39. 7.9327×10^4 **41.** 2×10^{-3} **43.** 5.963×10^{12} **45.** 740,000
47. .0000000000038   **49.** 34,570,000,000   **51.** 2^{12} **53.** 2^{-12} **55.** x^7 **57.** ce^9
59. $b^2c^2d^6$ **61.** $a^{12}b^8$ **63.** $1/(c^{10}d^6)$ **65.** $1/(108x)$ **67.** a^7c/b^6 **69.** c^3d^6
71. $a + \dfrac{1}{a}$ **73.** negative   **75.** negative   **77.** negative   **79.** 3^s **81.** $a^{6t}b^{4t}$
83. b^{rs+st}/c^{2rt} **85.** 1.36×10^5 **87.** 2×10^{-12} **89.** 1.34456×10^{50}
91. (a) $983.58   (b) $3379.93   (c) approximately 10.24   **93.** 1.584×10^9 inches
95. 500 seconds $= 8\dfrac{1}{3}$ minutes.
97. Many possible examples, including $3^2 + 4^2 = 9 + 16 = 25$, but $(3 + 4)^2 = 7^2 = 49$.
99. Many possible examples, including $3^2 \cdot 2^3 = 9 \cdot 8 = 72$; but $(3 \cdot 2)^{2+3} = 6^5 = 7776$.
101. Many possible examples, including $2^6/2^3 = 64/8 = 8$, but $2^{6/3} = 2^2 = 4$.
103. False for all nonzero a; for instance, $(-3)^2 = (-3)(-3) = 9$, but $-3^2 = -9$.

Section 1.3, page 24

1. .09   **3.** $.08^6$ **5.** $6\sqrt{2}$ **7.** 1/2   **9.** $-1 + \sqrt{3}$ **11.** $14 + 3\sqrt{3}$ **13.** 100
15. 3   **17.** .00001   **19.** 343/8   **21.** 326.158   **23.** 34.164   **25.** $>$ **27.** $<$

29. ab^2 **31.** $2xy^4$ **33.** $3xy\sqrt{2y}$ **35.** $3r^2s^3$ **37.** $4x+2y$ **39.** c^2 **41.** $3a^2b$
43. $2c^2d^3$ **45.** $1/(2x^2)$ **47.** $3c^2/(2d^4)$ **49.** $(a^2+b^2)^{1/3}$ **51.** $a^{3/16}$ **53.** $4t^{27/10}$
55. $4a^4/b$ **57.** $d^5/(2\sqrt{c})$ **59.** $15\sqrt{5}$ **61.** $(4x+2y)^2$ **63.** 1 **65.** $x^{9/2}$
67. $c^{42/5}d^{10/3}$ **69.** $\dfrac{a^{1/2}}{49b^{5/2}}$ **71.** $\dfrac{2^{9/2}a^{12/5}}{3^4b^4}$ **73.** a^x **75.** $\dfrac{1}{x^{1/5}y^{2/5}}$
77. 1 **79.** $x^{7/6} - x^{11/6}$ **81.** $x - y$ **83.** $x + y - (x+y)^{3/2}$
85. $(x^{1/3}+3)(x^{1/3}-2)$ **87.** $(x^{1/2}+3)(x^{1/2}+1)$ **89.** $(x^{2/5}+9)(x^{1/5}+3)(x^{1/5}-3)$
91. 3π

93. (a) (i) $\sqrt{10} \approx 3\dfrac{1}{7} \approx 3.1429$; (ii) $\sqrt{10} \approx 3\dfrac{1}{6} \approx 3.1667$; (iii) $\sqrt{10} \approx 3\dfrac{1}{6} - \dfrac{1}{216} \approx 3.1620$;

calculator: $\sqrt{10} \approx 3.1623$ **(b)** Using $a = 4$ and $b = 3$: (i) $\sqrt{19} \approx 4\dfrac{1}{3} \approx 4.3333$;

(ii) $\sqrt{19} \approx 4\dfrac{3}{8} = 4.375$; (iii) $\sqrt{19} \approx 4\dfrac{3}{8} - \dfrac{9}{512} \approx 4.3574$; calculator: $\sqrt{19} \approx 4.3589$

95. Many possible examples, including: $\sqrt[3]{8+27} = \sqrt[3]{35} \approx 3.27$, but $\sqrt[3]{8} + \sqrt[3]{27} = 2 + 3 = 5$.
97. Many possible examples, including: $\sqrt{8 \cdot 3} = \sqrt{24} \approx 4.9$, but $4\sqrt{3} \approx 6.9$.

Section 1.4, page 31

1. 11 **3.** 0 **5.** 10 **7.** 169 **9.** $\pi - \sqrt{2}$ **11.** π **13.** 1 **15.** $<$ **17.** $>$
19. $<$ **21.** $=$ **23.** [number line from -2 to 8, bracket at 0, parenthesis at 8]
25. [number line from -3 to 3, bracket at -2, parenthesis at 2] **27.** [number line, parenthesis at -4, bracket at 1]
29. $[5, 8]$ **31.** $(-3, 14)$ **33.** $[-3.7, -2.4)$ **35.** $[-8, \infty)$ **37.** $(-6.7, \infty)$
39. $(-\infty, 15]$
41. Many answers, including: true for $x = 0, y = 1$ and $x = -1, y = 0$; false for $x = 1, y = 1$ and $x = 2, y = -3$.
43. Many answers, including: true for $x = 1, y = 2$ and $x = 3, y = 4$; false for $x = -3, y = 1$ and $x = -2, y = 5$.
45. Many answers, including: true for $x = 5, y = 6$ and $x = 2, y = 7$; false for $x = 2, y = 1$ and $x = 3, y = 0$.
47. 7 **49.** $14\dfrac{1}{2}$ **51.** $\pi - 3$ **53.** $\sqrt{3} - \sqrt{2}$ **55.** t^2
57. $(-3-y)^2 = (-1)^2(3+y)^2 = (3+y)^2$ **59.** $b - 3$ **61.** $-(c-d) = d - c$
63. 0 **65.** $|(c-d)^2| = (c-d)^2 = c^2 - 2cd + d^2$ **67.** $|x - 5| < 4$
69. $|x + 4| \leq 17$ **71.** $|c| < |b|$ **73.** $|x| > |x + 6|$ **75.** $|x| > 2$
77. The distance from x to 3 is less than 2 units. **79.** The distance from x to -7 is at most 3 units.
81. The distance from b to 0 is less than the distance from c to 3.
83. The distance from b to -3 is less than the distance from b to 3.

85. There is no number that is within 2 units of 1 and at the same time within 3 units of 12.
87. All real numbers.   89. Zero and all negative real numbers.   91. $x = 1$ or -1
93. $x = 1$ or 3 95. $x = -\pi + 4$ or $-\pi - 4$ 97. $-7 < x < 7$ 99. $-4 < x < -2$
101. $x \le -5$ or $x \ge 5$
103. Since $|a| \ge 0, |b| \ge 0$, and $|c| \ge 0$, the sum $|a| + |b| + |c|$ is positive only when one or more of $|a|, |b|, |c|$ is positive. But $|a|$ is positive only when $a \ne 0$; similarly for b, c.
105. (a) By the first inequality property in the box on page 7, $c \le |c|$ implies that $c + d \le |c| + d$. Similarly, $d \le |d|$ implies that $|c| + d \le |c| + |d|$. By the last fact in the box on page 2, $c + d \le |c| + d$ and $|c| + d \le |c| + |d|$ imply that $c + d \le |c| + |d|$. A similar argument shows that $(-c) + (-d) \le |c| + |d|$. Since $-(c + d) = (-c) + (-d)$, we have $-(c + d) \le |c| + |d|$.   (b) If $c + d$ is nonnegative, then by the definition of absolute value and part (a), $|c + d| = c + d \le |c| + |d|$. If $c + d$ is negative, then by the definition of absolute value and part (a), $|c + d| = -(c + d) \le |c| + |d|$.

Section 1.5, page 40

1. $8x$ 3. $-2a^2b$ 5. $-x^3 + 4x^2 + 2x - 3$ 7. $5u^3 + u - 4$
9. $4z - 12z^2w + 6z^3w^2 - zw^3 + 8$ 11. $-3x^3 + 15x + 8$ 13. $-5xy - x$
15. $15y^3 - 5y$ 17. $12a^2x^2 - 6a^3xy + 6a^2xy$ 19. $12z^4 + 30z^3$
21. $12a^2b - 18ab^2 + 6a^3b^2$ 23. $x^2 - x - 2$ 25. $2x^2 + 2x - 12$
27. $y^2 + 7y + 12$ 29. $-6x^2 + x + 35$ 31. $3y^3 - 9y^2 + 4y - 12$ 33. $x^2 - 16$
35. $16a^2 - 25b^2$ 37. $y^2 - 22y + 121$ 39. $25x^2 - 10bx + b^2$
41. $16x^6 - 8x^3y^4 + y^8$ 43. $9x^4 - 12x^2y^4 + 4y^8$ 45. $2y^3 + 9y^2 + 7y - 3$
47. $-15w^3 + 2w^2 + 9w - 18$ 49. $24x^3 - 4x^2 - 4x$ 51. $x^3 - 6x^2 + 11x - 6$
53. $-3x^3 - 5x^2y + 26xy^2 - 8y^3$
55. yes; leading coefficient 1; constant term 1; degree 3
57. yes; leading coefficient 1; constant term -1; degree 3
59. yes; leading coefficient 1; constant term -3; degree 2   61. no
63. quotient $3x^3 - 3x^2 + 5x - 11$; remainder 12
65. quotient $x^2 + 2x - 6$; remainder $-7x + 7$
67. quotient $5x^2 + 5x + 5$; remainder zero   69. no   71. yes   73. 3   75. -6
77. 6   79. 1   81. 5   83. $x - 25$ 85. $9 + 6\sqrt{y} + y$ 87. $\sqrt{3}x^2 + 4x + \sqrt{3}$
89. $3ax^2 + (3b + 2a)x + 2b$ 91. $abx^2 + (a^2 + b^2)x + ab$
93. $x^3 - (a + b + c)x^2 + (ab + ac + bc)x - abc$ 95. 3^{4+r+t}
97. $x^{m+n} + 2x^n - 3x^m - 6$ 99. $2x^{4n} - 5x^{3n} + 8x^{2n} - 18x^n - 5$
101. example: if $y = 4$, then $3(4 + 2) \ne (3 \cdot 4) + 2$; correct statement: $3(y + 2) = 3y + 6$
103. example: if $x = 2, y = 3$, then $(2 + 3)^2 \ne 2 + 3^2$; correct statement: $(x + y)^2 = x^2 + 2xy + y^2$
105. example: if $x = 2, y = 3$, then $(7 \cdot 2)(7 \cdot 3) \ne 7 \cdot 2 \cdot 3$; correct statement: $(7x)(7y) = 49xy$

107. example: if $y = 2$, then $2 + 2 + 2 \neq 2^3$; correct statement: $y + y + y = 3y$
109. example: if $x = 4$, then $(4 - 3)(4 - 2) \neq 4^2 - 5 \cdot 4 - 6$; correct statement: $(x - 3)(x - 2) = x^2 - 5x + 6$
111. If x is the chosen number, then adding one and squaring the result gives $(x + 1)^2$. Subtracting one from the original number x and squaring the result gives $(x - 1)^2$. Subtracting the second of these squares from the first yields: $(x + 1)^2 - (x - 1)^2 = (x^2 + 2x + 1) - (x^2 - 2x + 1) = 4x$. Dividing by the original number x now gives $\dfrac{4x}{x} = 4$. So the answer is always 4, no matter what number x is chosen.
113. many correct answers

Excursion 1.5.A, page 45

1. $2 \mid$ 3 −8 0 9 5
    6 −4 −8 2
    3 −2 −4 1 $\mid$ 7
   quotient $3x^3 - 2x^2 - 4x + 1$;
   remainder 7

3. $-3 \mid$ 2 5 0 −2 −8
    −6 3 −9 33
    2 −1 3 −11 $\mid$ 25
   quotient $2x^3 - x^2 + 3x - 11$;
   remainder 25

5. $7 \mid$ 5 0 −3 −4 6
    35 245 1,694 11,830
    5 35 242 1,690 $\mid$ 11,836
   quotient $5x^3 + 35x^2 + 242x + 1690$;
   remainder 11,836

7. $2 \mid$ 1 −6 4 2 −7
    2 −8 −8 −12
    1 −4 −4 −6 $\mid$ −19
   quotient $x^3 - 4x^2 - 4x - 6$;
   remainder −19

9. quotient $3x^3 + \dfrac{3}{4}x^2 - \dfrac{29}{16}x - \dfrac{29}{64}$; remainder $\dfrac{483}{256}$
11. quotient $2x^3 - 6x^2 + 2x + 2$; remainder 1
13. $g(x) = (x + 4)(3x^2 - 3x + 1)$ 15. $g(x) = \left(x - \dfrac{1}{2}\right)(2x^4 - 6x^3 + 12x^2 - 10)$
17. quotient $x^2 - 2.15x + 4$; remainder 2.25   19. $c = -4$

Section 1.6, page 48

1. $(x + 2)(x - 2)$ 3. $(3y + 5)(3y - 5)$ 5. $(9x + 2)^2$ 7. $(\sqrt{5} + x)(\sqrt{5} - x)$
9. $(7 + 2z)^2$ 11. $(x^2 + y^2)(x + y)(x - y)$ 13. $(x + 3)(x - 2)$ 15. $(z + 3)(z + 1)$
17. $(y + 9)(y - 4)$ 19. $(x - 3)^2$ 21. $(x + 5)(x + 2)$ 23. $(x + 9)(x + 2)$
25. $(3x + 1)(x + 1)$ 27. $(2z + 3)(z + 4)$ 29. $9x(x - 8)$ 31. $2(x - 1)(5x + 1)$
33. $(4u - 3)(2u + 3)$ 35. $(2x + 5y)^2$ 37. $(x - 5)(x^2 + 5x + 25)$ 39. $(x + 2)^3$
41. $(2 + x)(4 - 2x + x^2)$ 43. $(-x + 5)^3$ 45. $(x + 1)(x^2 - x + 1)$
47. $(2x - y)(4x^2 + 2xy + y^2)$
49. $(x^3 + 2^3)(x^3 - 2^3) = (x + 2)(x^2 - 2x + 4)(x - 2)(x^2 + 2x + 4)$
51. $(y^2 + 5)(y^2 + 2)$ 53. $(9 + y^2)(3 + y)(3 - y)$

55. $(z+1)(z^2-z+1)(z-1)(z^2+z+1)$ **57.** $(x^2+3y)(x^2-y)$
59. $(x+z)(x-y)$ **61.** $(a+2b)(a^2-b)$
63. $(x^2-8)(x+4) = (x+\sqrt{8})(x-\sqrt{8})(x+4)$
65. $(2x-y)(x+3y) + 3(2x-y) = (2x-y)(x+3y+3)$
67. $(x-3y)(x^2+3xy+9y^2+1)$ **69.** $(x+\frac{1}{8})(x-\frac{1}{8})$ **71.** $(y+\frac{1}{6})(y-\frac{5}{6})$
73. $(z+\frac{7}{4})(z+\frac{5}{4})$
75. If $x^2+1 = (x+c)(x+d) = x^2+(c+d)x+cd$, then $c+d=0$ and $cd=1$. But $c+d=0$ implies that $c=-d$ and hence that $1-cd = (-d)d = -d^2$, or equivalently, that $d^2 = -1$. Since there is no real number with this property, x^2+1 cannot possibly factor in this way.

Section 1.7, page 55

1. $\dfrac{9}{7}$ **3.** $\dfrac{195}{8}$ **5.** $\dfrac{x-2}{x+1}$ **7.** $\dfrac{a+b}{a^2+ab+b^2}$ **9.** $1/x$ **11.** $\dfrac{29}{35}$ **13.** $\dfrac{121}{42}$

15. $\dfrac{ce+3cd}{de}$ **17.** $\dfrac{b^2-c^2}{bc}$ **19.** $\dfrac{-1}{x(x+1)}$ **21.** $\dfrac{x+3}{(x+4)^2}$ **23.** $\dfrac{2x-4}{x(3x-4)}$

25. $\dfrac{x^2-xy+y^2+x+y}{x^3+y^3}$ **27.** $\dfrac{-6x^5-38x^4-84x^3-71x^2-14x+1}{4x(x+1)^3(x+2)^3}$ **29.** 2

31. $2/(3c)$ **33.** $3y/x^2$ **35.** $\dfrac{12x}{x-3}$ **37.** $\dfrac{5y^2}{3(y+5)}$ **39.** $\dfrac{u+1}{u}$

41. $\dfrac{(u+v)(4u-3v)}{(2u-v)(2u-3v)}$ **43.** $\dfrac{35}{24}$ **45.** $u^2/(vw)$ **47.** $\dfrac{x+3}{2x}$ **49.** $\dfrac{x^2y^2}{(x+y)(x+2y)}$

51. $\dfrac{cd(c+d)}{c-d}$ **53.** $\dfrac{y-x}{xy}$ **55.** $\dfrac{-3y+3}{y}$ **57.** $\dfrac{-1}{x(x+h)}$ **59.** $\dfrac{xy}{x+y}$

61. $2\sqrt{5}/5$ **63.** $\sqrt{70}/10$ **65.** $\sqrt{x}/x$ **67.** $\dfrac{(x+1)(\sqrt{x}-1)}{x-1}$ **69.** $\dfrac{\sqrt{a}+2\sqrt{b}}{a-4b}$

71. example: if $a=1$, $b=2$, then $\dfrac{1}{1}+\dfrac{1}{2} \neq \dfrac{1}{1+2}$; correct statement: $\dfrac{1}{a}+\dfrac{1}{b}=\dfrac{b+a}{ab}$

73. example: if $a=4$, $b=9$, then $\left(\dfrac{1}{\sqrt{4}+\sqrt{9}}\right)^2 \neq \dfrac{1}{4+9}$; correct statement: $\left(\dfrac{1}{\sqrt{a}+\sqrt{b}}\right)^2 = \dfrac{1}{a+2\sqrt{ab}+b}$

75. example: if $u=1$, $v=2$, then $\dfrac{1}{2}+\dfrac{2}{1} \neq 1$; correct statement: $\dfrac{u}{v}+\dfrac{v}{u}=\dfrac{u^2+v^2}{vu}$

77. example: if $x=4$, $y=9$, then $(\sqrt{4}+\sqrt{9}) \cdot \dfrac{1}{\sqrt{4}+\sqrt{9}} \neq 4+9$; correct statement: $(\sqrt{x}+\sqrt{y}) \cdot \dfrac{1}{\sqrt{x}+\sqrt{y}} = 1$

Chapter 1 Review, page 57

1. (a) > (b) < (c) < (d) > (e) =   3. 28/99
5. (a) $-10 < y < 0$ (b) $0 \le x \le 10$ 7. (a) $|x + 7| < 3$ (b) $|y| > |x - 3|$
9. 3.3×10^{12} 11. $x = 2$ or 8 13. $x = -11/2$ or $-1/2$ 15. $-4 \le x \le 0$
17. (a) $7 - \pi$ (b) $\sqrt{23} - \sqrt{3}$ 19. (a) $(-8, \infty)$ (b) $(-\infty, 5]$ 21. $c^{15}/(d^{15}e^5)$
23. 1   25. $-1/27$ 27. $7 + 2\sqrt{10}$ 29. $8x^6 + 60x^4y + 150x^2y^2 + 125y^3$
31. $u^4 + 2u^2v^2 - 4u^2w^2 - 4v^2w^2 + v^4 + 4w^4$ 33. $\dfrac{3x^2 - x - 6}{x^2 - 4}$ 35. $\dfrac{c - 1}{c + 1}$
37. $2/u$ 39. $c^2d^4/2$ 41. $2x + 1$ 43. $(x + 1)(x + 4)$ 45. $(x - 4)(x + 2)$
47. $(2x + 3)(2x - 3)$ 49. $(2x - 5)(2x + 3)$ 51. $x(3x - 1)(x + 2)$
53. $(x^2 + 1)(x + 1)(x - 1)$ 55. $7\sqrt{3}/3$ 57. $\dfrac{\sqrt{3} - 2\sqrt{c}}{3 - 4c}$ 59. (d)   61. (d)   63. (d)

CHAPTER 2

Section 2.1, page 66

1. $x = 3/2$ 3. $x = -5$ 5. $x = 8$ 7. $x = -5/6$ 9. $y = -32$
11. $z = -1/13$ 13. $x = -3$ 15. $x = -1$ 17. $x = 5/2$ 19. $x = 26$
21. $x = 5/6$ 23. no solution   25. $x = 1$ 27. no solution   29. $x = -5/2$
31. $x = -3/2$ 33. $z = -2/3$ 35. $y = 1$ 37. $x = -22/3$ 39. $x = 0$
41. $x = -1/7$ 43. identity   45. not an identity   47. identity   49. not an identity
51. $x \approx -1.239697$ 53. $y = 3.765$ 55. $x \approx .24361$ 57. $x \approx -1065.11596$
59. $x \approx 67.85484$ 61. $y = \dfrac{x + 5}{3}$ 63. $y = \dfrac{3x}{4} - 18$ 65. $b = \dfrac{2A - hc}{h}$ with $h \ne 0$
67. $h = \dfrac{4V}{\pi d^2}$ with $d \ne 0$ 69. $v = 1 - \dfrac{b}{S}$ with $v \ne 1, S \ne 0$
71. $x = \dfrac{3b^2 - 12ab - a}{5a - b - 2}$ with $5a - b - 2 \ne 0$
73. $z = \dfrac{14}{(a + c)(a - b)} - b$ with $a + c \ne 0, a - b \ne 0$ 75. $c = 1/2$ 77. $c = -3$

Section 2.2, page 74

1. 91   3. $17.88   5. $1475   7. 3.6 hours   9. average 79; 91 on fourth exam
11. $366.67 at 12% and $733.33 at 6%   13. 14%   15. 5
17. $29\frac{1}{6}$ lb of peanuts; $15\frac{5}{6}$ lb of cashews   19. 112.5 ounces   21. 30 lb   23. $2\frac{2}{3}$ quarts
25. Lionel 177.5 lb; boa 22.5 lb   27. 4:40 P.M.   29. 65 mph   31. 110 miles
33. 60 mph   35. 45 by 90 ft   37. 60 by 20 meters   39. 45 hours   41. 30 minutes

43. 90 minutes ["That old National School," said Mrs. Morgan. "There is silly the sums are with them. Filling up an old bath with holes in it, indeed. Who would be such a fool?"]
45. $1437.50   **47.** 1.2 ounces

Section 2.3, page 81

1. $x = 3$ or 5 **3.** $x = -2$ or 7 **5.** $y = \frac{1}{2}$ or -3 **7.** $t = -2$ or $-\frac{1}{4}$
9. $u = 1$ or $-\frac{4}{3}$ **11.** $x = \frac{1}{4}$ or $-\frac{4}{3}$ **13.** $x = -3$ or 5
15. $x = (1 + \sqrt{5})/2$ or $(1 - \sqrt{5})/2$ **17.** $x = 2 \pm \sqrt{3}$ **19.** $x = -3 \pm \sqrt{2}$
21. no real number solutions   **23.** $x = \frac{1}{2} \pm \sqrt{2}$ **25.** $x = \frac{2 \pm \sqrt{3}}{2}$ **27.** $u = \frac{-4 \pm \sqrt{6}}{5}$
29. 2   **31.** 2   **33.** 1   **35.** $x = -5$ or 8 **37.** $x = \frac{-5 \pm \sqrt{57}}{8}$
39. $x = -3$ or -6 **41.** $x = \frac{-1 \pm \sqrt{2}}{2}$ **43.** $x = 5$ or $-\frac{3}{2}$
45. no real number solutions   **47.** no real number solutions   **49.** $x \approx 1.824$ or $.47$
51. $x = 13.79$ **53.** $c = \pm\sqrt{E/m}$; $m \neq 0$, $E/m \geq 0$
55. $r = \frac{-\pi h \pm \sqrt{\pi^2 h^2 + 4A\pi}}{2\pi}$; $\pi h^2 + 4A \geq 0$ **57.** $x = \frac{-y \pm \sqrt{49y^2 + 108}}{6}$
59. $x = 3$ or $-\frac{2}{k}$; $k \neq 0$ **61.** $k = 10$ or -10 **63.** $k = 16$ **65.** $k = 4$

Section 2.4, page 85

1. $-13, -12$ or $12, 13$ **3.** 3 cm, 4 cm, 5 cm   **5.** 4 cm   **7.** approx. 1.753 ft
9. 2 meters   **11.** 2.5 yd   **13.** 2 in. by 2 in.   **15.** 5   **17.** 12 hr
19. 9 mph; 16 mph   **21.** Red Riding Hood, 54 mph; wolf, 48 mph
23. $4\sqrt{5}/5 \approx 1.79$ hours   **25. (a)** approximately 6.3 sec   **(b)** approximately 4.9 sec
27. (a) approximately 4.4 sec   **(b)** after 50 sec   **29.** 2000 meters

Section 2.5, page 92

1. $8 + 2i$ **3.** $-2 - 10i$ **5.** $-\frac{1}{2} - 2i$ **7.** $\left(\frac{\sqrt{2} - \sqrt{3}}{2}\right) + 2i$ **9.** $1 + 13i$
11. $-10 + 11i$ **13.** $-21 - 20i$ **15.** 4   **17.** $-i$ **19.** i **21.** i
23. $\frac{5}{29} + \frac{2}{29}i$ **25.** $-\frac{1}{3}i$ **27.** $\frac{12}{41} - \frac{15}{41}i$ **29.** $\frac{-5}{41} - \frac{4}{41}i$ **31.** $\frac{10}{17} - \frac{11}{17}i$

33. $\frac{7}{10} + \frac{11}{10}i$ **35.** $-\frac{113}{170} + \frac{41}{170}i$ **37.** $6i$ **39.** $\sqrt{14}i$ **41.** $-4i$ **43.** $11i$
45. $(\sqrt{15} - 3\sqrt{2})i$ **47.** $\frac{2}{3}$ **49.** $-41 - i$ **51.** $(2 + 5\sqrt{2}) + (\sqrt{5} - 2\sqrt{10})i$
53. $\frac{1}{3} - \frac{\sqrt{2}}{3}i$ **55.** $x = 2, y = -2$ **57.** $x = -3/4, y = 3/2$ **59.** $x = \frac{1}{3} \pm \frac{\sqrt{14}}{3}i$
61. $x = -\frac{1}{2} \pm \frac{\sqrt{7}}{2}i$ **63.** $x = \frac{1}{4} \pm \frac{\sqrt{31}}{4}i$ **65.** $x = \frac{3 \pm \sqrt{3}}{2}$
67. $x = 2, -1 + \sqrt{3}i, -1 - \sqrt{3}i$ **69.** $x = 1, -1, i, -i$ **71.** -1
73. $z + w = (a + bi) + (c + di) = (a + c) + (b + d)i$ and hence
$\overline{z + w} = (a + c) - (b + d)i = a + c - bi - di$
$= (a - bi) + (c - di) = \overline{z} + \overline{w}$
75. We first express z/w in standard form: $\frac{z}{w} = \frac{a + bi}{c + di} = \frac{a + bi}{c + di} \cdot \frac{c - di}{c - di} =$
$\frac{(ac + bd) + (bc - ad)i}{c^2 + d^2}$. Hence $\overline{\left(\frac{z}{w}\right)} = \frac{(ac + bd) - (bc - ad)i}{c^2 + d^2} = \frac{ac + bd - bci + adi}{c^2 + d^2}$.
On the other hand, $\frac{\overline{z}}{\overline{w}} = \frac{a - bi}{c - di} = \frac{a - bi}{c - di} \cdot \frac{c + di}{c + di} = \frac{ac + bd - bci + adi}{c^2 + d^2}$.
77. If $z = a + bi$, with a, b real numbers, then $z - \overline{z} = (a + bi) - (a - bi) = 2bi$. If $z = a + bi$ is real, then $b = 0$ and hence $z - \overline{z} = 2bi = 0$. Therefore $z = \overline{z}$. Conversely, if $z = \overline{z}$, then $0 = z - \overline{z} = 2bi$, which implies that $b = 0$. Hence $z = a$ is real.
79. $\frac{1}{z} = \left(\frac{a}{a^2 + b^2}\right) + \left(\frac{-b}{a^2 + b^2}\right)i$

Section 2.6, page 98

1. $x = 3$ **3.** $x = 0$ or 5 **5.** $x = 4$ **7.** $x = 0$ or $\pm .2$ **9.** $y = \pm 1$ or $\pm \sqrt{6}$
11. $x = \pm \sqrt{5}$ **13.** $y = \pm 2$ or $\pm 1/\sqrt{2}$ **15.** $x = \pm 1/\sqrt{5}$ **17.** $x = 7$ **19.** $x = -2$
21. $x = -1$ or 2 **23.** $x = -1$ or -4 **25.** $x = -1$ or 2 **27.** $x = 9$ **29.** $x = \frac{1}{2}$
31. $x = \frac{1}{2}$ or -4 **33.** $x = 6$ **35.** no solutions   **37.** $b = \sqrt{\frac{a^2}{A^2 - 1}}$
39. $u = \sqrt{\frac{x^2}{1 - K^2}}$ **41.** $x = 4$ **43.** $x = 4$ **45.** $x = -1$ or -8
47. $x = -64$ or 8 **49.** $x = 16$ **51.** $x = -\frac{1}{2}$ or $\frac{1}{3}$ **53.** $x = -1$ or 2
55. $x = -6$ or 3 **57.** $x = 3/2$ **59.** $x = -5$ or 1 or -3 or -1
61. $x = 1$ or 4 or $\frac{5 \pm \sqrt{33}}{2}$ **63.** 50 by 120 cm   **65.** $r = \sqrt{-8 + \sqrt{388}} \approx 3.42$

Excursion 2.6.A, page 102

1. $x = \pm 1$ or -3
3. $x = \pm 1$ or -5
5. $x = -4, 1$, or $\frac{1}{2}$
7. $x = -3$ or 2
9. $x = 2$
11. not a factor
13. factor
15. not a factor
17. $x = 1$ or 2 or $-\frac{1}{2}$
19. $x = 1$ or $\frac{1}{2}$ or $\frac{1}{3}$
21. $x = -1$ or 2
23. $y = \frac{2}{3}$
25. $k = 1$
27. $k = 1$
29. If $x - c$ were a factor of $x^4 + x^2 + 1$, then c would be a solution of $x^4 + x^2 + 1 = 0$, that is, c would satisfy $c^4 + c^2 = -1$. But $c^4 \geq 0$ and $c^2 \geq 0$, so this is impossible. Hence $x - c$ is not a factor.
31. (a) Many possible answers, including: if $n = 3$ and $c = 1$, then $x + 1 = x - (-1)$ is not a factor of $x^3 - 1$ since -1 is not a solution of $x^3 - 1 = 0$.   (b) Since n is odd, $(-c)^n = -c^n$ and hence $-c$ is a solution of $x^n + c^n = 0$. Thus $x - (-c) = x + c$ is a factor of $x^n + c^n$ by the Factor Theorem.

Section 2.7, page 107

1. $(-\infty, 3/2]$
3. $(-2, \infty)$
5. $(-\infty, -8/5]$
7. $(1, \infty)$
9. $(2, 4)$
11. $[-3, 5/2)$
13. $(-\infty, 4/7)$
15. $[-7/17, \infty)$
17. $[-1, 1/8)$
19. $[5, \infty)$
21. $[4, \infty)$
23. approximately $(.602, \infty)$
25. approximately $(-\infty, -1.053)$
27. $x < \dfrac{b+c}{a}$
29. $c < x < a + c$
31. $-4/3 \leq x \leq 0$
33. $-5/2 < x < -1/2$
35. $x < -2$ or $x > -1$
37. $x \leq -11/20$ or $x \geq -1/4$
39. $x < 3/7$ or $x > 5/7$
41. $x < -53/40$ or $x > -43/40$
43. $x < -7/4$ or $x > 13/4$
45. $0 \leq C \leq 100$
47. approximately 8.608 cents per kwh
49. more than $12,500
51. between $4000 and $5400
53. The given inequality is equivalent to

$$-\frac{d-c}{2} < x - \frac{c+d}{2} < \frac{d-c}{2}$$

Adding $\dfrac{c+d}{2}$ to each part of the inequality and simplifying, we have:

$$\frac{c+d}{2} - \frac{d-c}{2} < x < \frac{c+d}{2} + \frac{d-c}{2}$$

$$c < x < d$$

Section 2.8, page 114

1. $1 \leq x \leq 3$
3. $x \leq -7$ or $x \geq -2$
5. $x \leq -3$ or $x \geq 3$
7. $x \leq -2$ or $x \geq 3$
9. $x \leq -1/2$ or $x \geq 2/3$
11. $-\sqrt{3} < x < \sqrt{2}$
13. $x < -3$ or $x > 1/2$

CHAPTER 2   A-11

15. $-1 \le x \le 0$ or $x \ge 1$ **17.** $x < -1$ or $0 < x < 3$
19. $-2 < x < -1$ or $1 < x < 2$ **21.** $x < -1/3$ or $x > 2$
23. $-2 < x < -1$ or $1 < x < 3$ **25.** $x > 1$ **27.** $x \le -9/2$ or $x > -3$
29. $-3 < x < 1$ or $x \ge 5$ **31.** $x < -3$ or $x > 5$ **33.** $x \le 1$
35. $x \le -3$ or $-2 \le x < 0$ or $x > 1$ **37.** $x \le 1 - \sqrt{5}$ or $x \ge 1 + \sqrt{5}$
39. $x < -1 - \sqrt{3}$ or $x > -1 + \sqrt{3}$ **41.** $x < -\sqrt{10}$ or $3 - \sqrt{2} < x < \sqrt{10}$ or $x > 3 + \sqrt{2}$
43. $x \le -7/2$ or $x \ge -5/4$ **45.** $x < -5$ or $-5 < x < -4/3$ or $x > 6$
47. $-\sqrt{3} < x < -1$ or $1 < x < \sqrt{3}$ **49.** $-3 < x < 3$
51. $x \le -2$ or $-1 \le x \le 0$ or $x \ge 1$ **53.** $0 < x < 2/3$ or $2 < x < 8/3$
55. $1 < x < 19$ and $y = 20 - x$ **57.** $10 < x < 35$
59. smallest 51; largest 399. The profit is 0 for 50 or 400 widgets.
61. $1 \le t \le 4$ **63.** $2 < t < 2.25$
65. (a) $x^2 < x$ when $0 < x < 1$ and $x^2 > x$ when $x < 0$ or $x > 1$.   (b) If c is nonzero and $|c| < 1$, then either $0 < c < 1$ or $-1 < c < 0$ (which is equivalent to $1 > -c > 0$). If $0 < c < 1$, then $|c| = c$ and c is a solution of $x^2 < x$ by part (a), so that $c^2 < c = |c|$. If $1 > -c > 0$, then $|c| = -c$, which is a solution of $x^2 < x$ by part (a), so that $c^2 = (-c)^2 < (-c) = |c|$.   (c) If $|c| > 1$, then either $c < -1$ or $c > 1$. In either case, c is a solution of $x^2 > x$ by part (a).

Chapter 2 Review, page 117

1. $x = 44/7$ **3.** $x = 5$ **5.** $r = \dfrac{b-a}{2Q}$ **7.** $x = \dfrac{3+2y}{1-y}$
9. 3/11 oz gold; 8/11 oz silver   **11.** $2\frac{2}{3}$ hours   **13.** 9.6 ft   **15.** $x = \dfrac{1 \pm \sqrt{14}i}{3}$
17. $y = -1$ or $5/3$ **19.** no solutions
21. No. If $x + y = 2$ and $xy = 2$, then $y = 2 - x$ and $2 = xy = x(2 - x)$. Verify that the equation $2 = x(2 - x)$ has no real solutions.
23. 4 ft   **25.** $15 - 10i$ **27.** i **29.** $16 + 11i$ **31.** $\dfrac{3}{10} - \dfrac{1}{10}i$ **33.** $\dfrac{5}{2} + \dfrac{1}{2}i$
35. $\dfrac{2}{13} + \dfrac{3}{13}i$ **37.** $x = \pm 6i$ **39.** $x = -1 \pm 2i$ **41.** $x = \dfrac{2 \pm \sqrt{11}i}{3}$
43. $x = \pm \dfrac{\sqrt{3}}{2}i$ or $\pm \dfrac{\sqrt{2}}{2}i$ **45.** $x = \pm 3$ or $\pm \sqrt{2}$ **47.** $x = \sqrt[3]{2}$
49. $x = \dfrac{5 - \sqrt{5}}{2}$ **51.** no solutions   **53.** $s = gt^2/2$ **55.** $x = -1$
57. no rational solutions   **59.** (d)   **61.** $-9/2 < x < 2$ **63.** $x \le -4/3$ or $x \ge 0$
65. (c)   **67.** $x < -5$ or $x > 4$ **69.** $x < -2$ or $x > -1/3$ **71.** $x \le -7$ or $x > -4$

CHAPTER 3

Section 3.1, page 129

1. A. $(-3, 3)$ B. $(-1.5, 3)$ C. $(-2.3, 0)$ D. $(-1.5, -3)$ E. $(0, 2)$ F. $(0, 0)$ G. $(2, 0)$ H. $(3, 1)$ I. $(3, -1)$
3. yes   5. yes   7. no   9. $13; \left(-\frac{1}{2}, -1\right)$ 11. $\sqrt{17}; \left(\frac{3}{2}, -3\right)$
13. $\sqrt{6 - 2\sqrt{6}} \approx 1.05; \left(\frac{\sqrt{2} + \sqrt{3}}{2}, \frac{3}{2}\right)$ 15. $\sqrt{2}|a - b|; \left(\frac{a+b}{2}, \frac{a+b}{2}\right)$
17. Horizontal straight line through $(0, 5)$.   19. The coordinate axes.
21.   23. $(x + 3)^2 + (y - 4)^2 = 4$ 25. $x^2 + y^2 = 2$

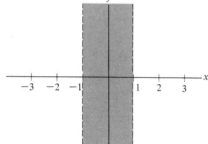

27.   29.   31. center $(-4, 3)$, radius $2\sqrt{10}$

33. center $(-3, 2)$, radius $2\sqrt{7}$ 35. center $(-12.5, -5)$, radius $\sqrt{169.25}$
37. Hypotenuse from $(1, 1)$ to $(2, -2)$ has length $\sqrt{10}$; other sides have lengths $\sqrt{2}$ and $\sqrt{8}$. Since $(\sqrt{2})^2 + (\sqrt{8})^2 = (\sqrt{10})^2$, this is a right triangle.
39. Hypotenuse from $(-2, 3)$ to $(3, -2)$ has length $\sqrt{50}$; other sides have lengths $\sqrt{5}$ and $\sqrt{45}$. Since $(\sqrt{5})^2 + (\sqrt{45})^2 = (\sqrt{50})^2$, this is a right triangle.
41. $(x - 2)^2 + (y - 2)^2 = 8$ 43. $(x - 1)^2 + (y - 2)^2 = 8$
45. $(x + 5)^2 + (y - 4)^2 = 16$ 47. $(-3, -4)$ and $(2, 1)$
49. Assume $k > d$. The other two vertices of one possible square are $(c + k - d, d)$, $(c + k - d, k)$; those of another square are $(c - (k - d), d)$, $(c - (k - d), k)$; those of a third square are $\left(c + \frac{k-d}{2}, \frac{k+d}{2}\right), \left(c - \frac{k-d}{2}, \frac{k+d}{2}\right)$.
51. $(0, 0), (6, 0)$ 53. $(3, -5 + \sqrt{11}), (3, -5 - \sqrt{11})$
55. The circle $(x - k)^2 + y^2 = k^2$ has center $(k, 0)$ and radius $|k|$ (the distance from $(k, 0)$ to $(0, 0)$). So the family consists of every circle that is tangent to the y-axis *and* has center on the x-axis.

57. (a) (0, −5) goes to (0, 0); (2, 2) goes to (2, 7); (5, 0) goes to (5, 5); (5, 5) goes to (5, 10); (4, 1) goes to (4, 6) **(b)** (0, −10) goes to (0, −5); (2, −3) goes to (2, 2); (5, −5) goes to (5, 0); (5, 0) goes to (5, 5); (4, −4) goes to (4, 1) **(c)** $(a, b+5)$ **(d)** (a, b) **(e)** $(-4a, b-5)$ **(f)** no points go to themselves.

59. If A has coordinates $(0, r)$ and C coordinates $(s, 0)$, then M has coordinates $(s/2, r/2)$ by the midpoint formula. Hence the distance from M to $(0, 0)$ is $\sqrt{\left(\frac{s}{2}-0\right)^2 + \left(\frac{r}{2}-0\right)^2} = \sqrt{\frac{s^2}{4} + \frac{r^2}{4}}$, and the distance from M to A is the same:

$$\sqrt{\left(\frac{s}{2}-0\right)^2 + \left(\frac{r}{2}-r\right)^2} = \sqrt{\left(\frac{s}{2}\right)^2 + \left(-\frac{r}{2}\right)^2} = \sqrt{\frac{s^2}{4} + \frac{r^2}{4}}$$

as is the distance from M to C:

$$\sqrt{\left(\frac{s}{2}-s\right)^2 + \left(\frac{r}{2}-0\right)^2} = \sqrt{\left(-\frac{s}{2}\right)^2 + \left(\frac{r}{2}\right)^2} = \sqrt{\frac{s^2}{4} + \frac{r^2}{4}}$$

Section 3.2, page 137

1. $\frac{5}{2}$ **3.** 4 **5.** $\frac{1-\pi}{1+\pi}$ **7.** Many correct answers, including these:

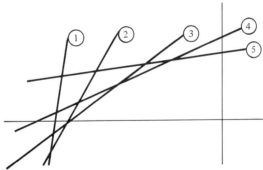

9. perpendicular   **11.** parallel

13. Many correct answers, including (5, 7), (7, 10), (9, 13), (11, 16), (13, 19).

15. 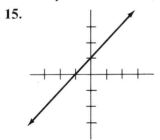   **17.**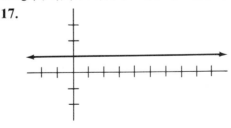

19. approximately 60.208 ft

21. The side joining $(-5, -2)$ and $(3, 0)$ has slope $(0 + 2)/(3 + 5) = 1/4$; the side joining $(-3, 1)$ and $(5, 3)$ has slope $(3 - 1)/(5 + 3) = 1/4$; so these two sides are parallel. Similarly, the side joining $(-5, -2)$ and $(-3, 1)$ and the side joining $(3, 0)$ and $(5, 3)$ are parallel since both have slope $3/2$.

23. yes   25. 22   27. -5 29. 24

31. Many correct answers, including $(0, -15)$, $(1, -20)$, $(2, -25)$.

33. P has coordinates $(1, y)$ for some y. Since $(0, 0)$ and $(1, y)$ are on L, the slope of L is $(y - 0)/(1 - 0) = y$, the second coordinate of P.

35. (a) C (b) B (c) B (d) D

Section 3.3, page 143

1. $y - 5 = 1(x - 3)$, or equivalently, $y = x + 2$ 3. $y = -x + 8$ 5. $y = -x - 5$

7. $y = -\dfrac{7x}{3} + \dfrac{34}{9}$ 9. $y = x + 2$ 11. $y = -4x + 2$

13. m = slope, b = y-intercept: $m = 2$, $b = 5$ 15. $m = -\dfrac{3}{7}$, $b = -\dfrac{11}{7}$

17. $y = 3x + 7$ 19. $y = \dfrac{3x}{2}$ 21. $y = x - 5$ 23. $y = -x + 2$ 25. $y = \dfrac{x}{3}$

27. parallel   29. perpendicular   31. $k = -\dfrac{11}{3}$ 33. $y - 4 = -\dfrac{3}{4}(x - 3)$

35. $y - 5 = -\dfrac{1}{2}(x - 2)$ 37. $y = -\dfrac{1}{2}x$

39. The equation $Ax + By + C = 0$ is equivalent to $By = -Ax - C$ and hence to $y = (-A/B)x - (C/B)$; so this line has slope $-A/B$. Similarly, $Ax + By + D = 0$ is equivalent to $y = (-A/B)x - (D/B)$ so that this line also has slope $-A/B$. Two lines with the same slopes are parallel.

41. Since L has x-intercept a and y-intercept b, the points $(a, 0)$ and $(0, b)$ are on L; also $a \neq 0$, $b \neq 0$ since $(0, 0)$ is *not* on L. So the slope of L is $(b - 0)/(0 - a) = -b/a$. The equation of L is $y = (-b/a)x + b$, which is equivalent to $ay = -bx + ab$ and hence to $bx + ay = ab$. Dividing both sides by ab shows that L has equation $(x/a) + (y/b) = 1$.

43. (a) Distance from P to Q is 9400 ft.   (b) B is 88 ft from the road.

45. (a) L has slope $\dfrac{d - 0}{c - 0} = \dfrac{d}{c}$ and equation $y = \dfrac{d}{c}x$.   (b) Since $-d = \dfrac{d}{c}(-c)$, $(-c, -d)$ lies on L.   (c) Since $c \neq 0$, c and $-c$ have opposite signs. If $d = 0$, the points (c, d) and $(-c, -d)$ lie on the x-axis, on opposite sides of the origin. If $d \neq 0$, there are four possibilities:

sign of c	+	+	−	−
sign of d	+	−	+	−
(c, d) in quadrant	I	IV	II	III
(−c, −d) in quadrant	III	II	IV	I

In each case (c, d) and $(-c, -d)$ lie on opposite sides of the origin. **(d)** The midpoint of the segment from (c, d) to $(-c, -d)$ is $((c - c)/2, (d - d)/2) = (0, 0)$.

Chapter 3 Review, page 146

1. $\sqrt{58}$ **3.** $\sqrt{c^2 + d^2}$ **5.** $\left(d, \dfrac{c + 2d}{2}\right)$ **7. (a)** $\sqrt{17}$
(b) $(x - 2)^2 + (y + 3)^2 = 17$ **9.** **11. (a)**

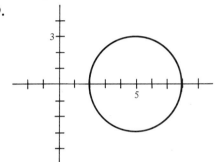

13. (c) **15.** $y = 3x - 7$ **17.** $y = -2x + 1$ **19.** $x - 5y = -29$ **21.** 25,000 ft
23. false **25.** false **27.** false **29.** false **31.** (d) **33.** (e) **35.** 5/3

CHAPTER 4

Section 4.1, page 153

1. 6 **3.** −2 **5.** 0 **7.** −17

9. Many correct answers, including: true for $u = 2$, $v = \dfrac{3}{4}$; false for $u = 2$, $v = 1$.

11. The area is a function of the radius; domain: all radii (that is, all positive real numbers); range: all areas (all positive real numbers); rule: area equals πr^2, where r is the radius.

13. Sales are a function of (money spent on) advertising; domain: all amounts (in dollars) to be spent on advertising; range: all total sales amounts (in dollars); rule: insufficient information.

15. Tax is 0 on $500 and $1509. Tax is $35.08 on $3754. Tax is $119.15 on $6783; $405 on $12,500; and $2547.10 on $55,342.

17. *Each* of the different numbers 175, 560, 1120, 1800 in the domain is assigned to the *one* number 0 in the range. The definition of the rule can be contradicted only if one number in the domain is assigned to several different numbers in the range.

19. Postage is a function of weight since each weight determines one and only one postage amount. But weight is *not* a function of postage since a given postage amount may apply to several different weights. For instance, *all* letters under 1 ounce use just one first-class stamp.

21. domain: $[-3, 3]$; range: approximately $[-4, 3.9]$ 23. 2 is assigned to $\frac{1}{2}$; 0 to $\frac{5}{2}$; and -3 to $-\frac{5}{2}$.

25. -0.2 (approximately) is assigned to -2; 3 to 0; 2 to 1; -1 to 2.5; and 1 to -1.5.

27. 1 is assigned to -2; -3 to -1; -1 to 0; 0.2 (approximately) to $\frac{1}{2}$; and 1.5 to 1.

29. (a) All positive numbers that can be entered in your calculator.   (b) All numbers between -1 and 1 (inclusive) that can be displayed on your calculator.

Section 4.2, page 160

1. $\sqrt{3} + 1$ 3. $\sqrt{11/2} - \frac{3}{2}$ 5. $\sqrt{\sqrt{2} + 3} - \sqrt{2} + 1$ 7. 4   9. $\frac{34}{3}$ 11. $\frac{59}{12}$

13. $(a + k)^2 + \frac{1}{a + k} + 2$ 15. $(2 - x)^2 + \frac{1}{2 - x} + 2 = 6 - 4x + x^2 + \frac{1}{2 - x}$

17. 8   19. -1 21. $(s + 1)^2 - 1 = s^2 + 2s$ 23. $t^2 - 1$

	$f(r)$	$f(r) - f(x)$	$\dfrac{f(r) - f(x)}{r - x}$
25.	r	$r - x$	1
27.	$3r + 7$	$3(r - x)$	3
29.	$r - r^2$	$r - r^2 - x + x^2$	$1 - r - x$
31.	$\sqrt{r}$	$\sqrt{r} - \sqrt{x}$	$\dfrac{1}{\sqrt{r} + \sqrt{x}}$

33. $f(-3) = 1.1$ (approximately); $f\left(-\frac{3}{2}\right) = 1.5$ (approximately); $f(0) = -2.8$ (approximately); $f(1) = 0$; $f\left(\frac{5}{2}\right) = 2$; $f(4) = 1.5$

35. $f\left(-\frac{5}{2}\right) = -1.2$ (approximately); $f\left(-\frac{3}{2}\right) = 0$; $f(0) = 1$; $f(3) = 3$; $f(4) = 1$

37. (iii) or (v)   39. All real numbers.   41. All real numbers.   43. All real numbers.
45. All nonnegative real numbers.   47. All nonzero real numbers.
49. All real numbers.   51. All real numbers.   53. All real numbers except -2 and 3.
55. $[6, 12]$ 57. Many possible answers, including $f(x) = x^2$ and $g(x) = |x|$.

59. $f(x) = 0$ for all x 61. 1   63. $2x + h$ 65. $2x + 3 + h$ 67. $\dfrac{-1}{x(x + h)}$

69. (i), (ii), and (iv) are true; (iii) is false since $f(-2) = -2$ and $f(2) = 2$.
71. All four are true.   **73.** $f(r) = 2\pi r$, where r = radius and $f(r)$ = circumference.
75. $f(s) = s^2$, where s = side of square and $f(s)$ = area.
77. $d(t) = \begin{cases} 55t \text{ if } 0 \le t \le 2 \\ 110 + 45(t - 2) \text{ if } t > 2 \end{cases}$, where t = time (in hours) and $d(t)$ = distance (in miles).
79. $d(t) = 2000 - 475t$, where t = time after noon (in hours) and $d(t)$ = distance from city S (in miles).
81. (a) $p(x) = \begin{cases} 12 \text{ for } 1 \le x \le 10 \\ 12 - .25(x - 10) \text{ for } x > 10 \end{cases}$, where $p(x)$ is the price per copy when x copies are purchased.   (b) $T(x) = \begin{cases} 12x \text{ for } 1 \le x \le 10 \\ 14.5x - .25x^2 \text{ for } x > 10 \end{cases}$, where $T(x)$ is the total cost of x copies.
83. $c(x) = 5.75x + (45,000/x)$

Section 4.3, page 170

1.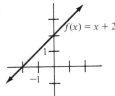

3.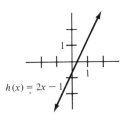

5.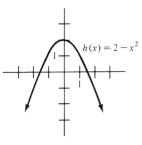

7.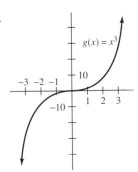

9.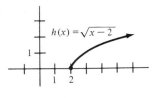

11.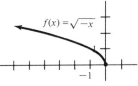

13.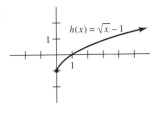

15.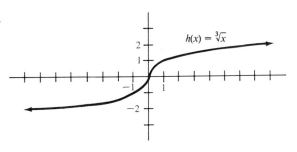

17.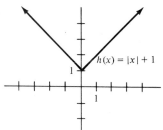

19.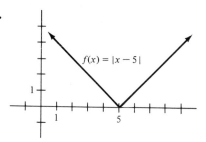

21.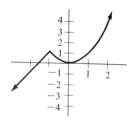

23.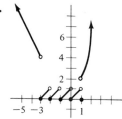

25. graph for $-2 \le x \le 3$: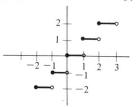

27.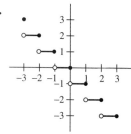

29. (a) Several possibilities, including $p(x) = \begin{cases} [x] \text{ if } x \text{ is an integer} \\ [x] + 1 \text{ if } x \text{ is not an integer} \end{cases}$ or $p(x) = -[-x]$, with $x > 0$ in all cases, where x is the weight in ounces.
(b) graph for $0 < x \le 4$: **(c)**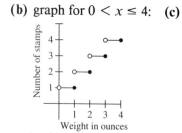

31.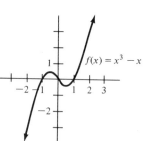

33.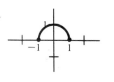

35.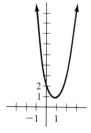

37.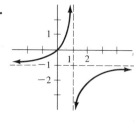

39.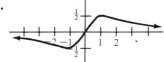

Excursion 4.3.A, page 174

1. 4 **3.** 3.5 **5.** 4.5 **7.** 4 **9.** 1, 5 **11.** [−3, 3]
13. Many correct answers, including

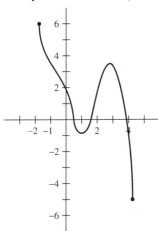

15. [−8, 9] **17.** −7, −3, 3, 7 **19.** 1, 3, 5, and others **21.** 1 and −9
23. approximately 5.5, and others **25.** domain f = [−6, 7]; domain g = [−8, 9]
27. approximately −1.5 and −.2 **29.** $x = 3$ **31.** [−2, −1] and [3, 7]
33. approximately −$13,000 (that is, a loss of $13,000) **35.** approximately 12,300
37.

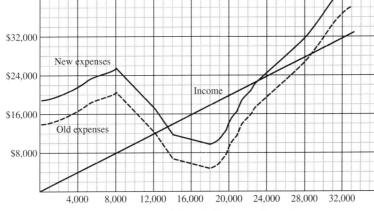

39. tem(10) ≈ 47°; tem(3 + 12) ≈ 63° **41.** approximately 11 A.M. and 8:30 P.M.

43. tem(10) ≈ 47°, so (10, 47) is on the graph; tem(16) ≈ 64°, so (16, 64) is on the graph. The point (10, 47) lies 17 units lower than (16, 64).
45. no **47.** approximately 11 A.M. and 8:30 P.M.

Section 4.4, page 185

1. odd   **3.** even   **5.** even   **7.** even   **9.** neither   **11.** yes   **13.** no
15. origin   **17.** origin   **19.** y-axis

21. 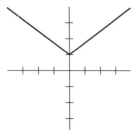   **23.** 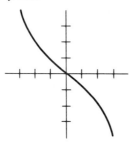   **25.**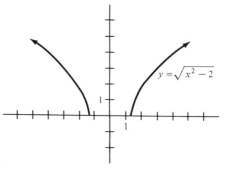

27. 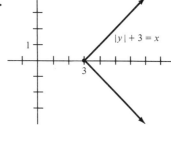   **29.** 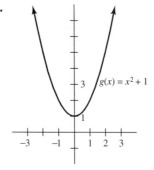   **31.**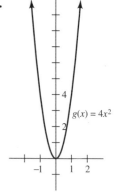

33. 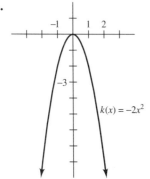   **35.**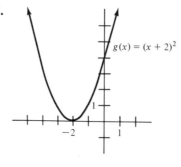

37.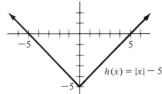

39.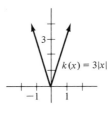

41.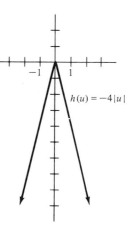

43.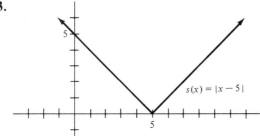

45.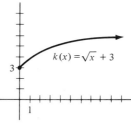

47.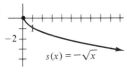

49.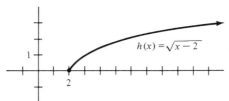

51.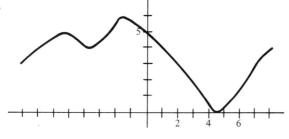

53.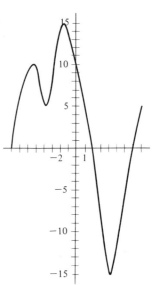

55.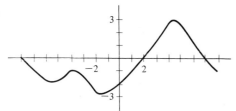

57.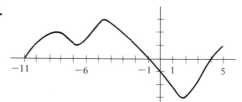

59.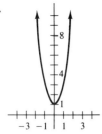

61.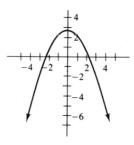

63.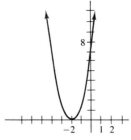

65.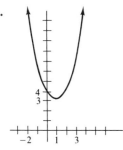

67.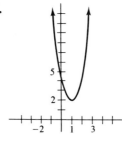

69.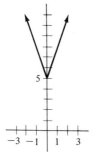

71.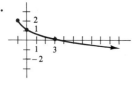

73.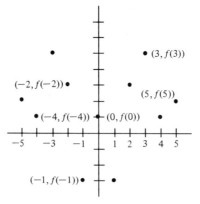

75. Many correct graphs, including the one shown here:

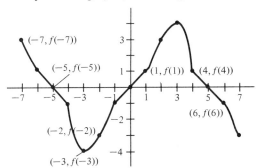

77. Suppose the graph is symmetric to the x-axis and the y-axis. If (x, y) is on the graph, then $(x, -y)$ is on the graph by x-axis symmetry. Hence $(-x, -y)$ is on the graph by y-axis symmetry. Therefore (x, y) on the graph implies that $(-x, -y)$ is on the graph, so the graph is symmetric with respect to the origin. Next suppose that the graph is symmetric to the y-axis and the origin. If (x, y) is on the graph, then $(-x, y)$ is on the graph by y-axis symmetry. Hence $(-(-x), -y) = (x, -y)$ is on the graph by origin symmetry. Therefore (x, y) on the graph implies that $(x, -y)$ is on the graph, so the graph is symmetric with respect to the x-axis. The proof of the third case is similar to that of the second case.

79. If c is in the domain of f and $f(c) = d$, then (c, d) is on the graph of f. Hence $(c, -d)$ is also on the graph of f since it is symmetric with respect to the x-axis. But the presence of $(c, -d)$ on the graph means that $f(c) = -d$. Therefore $d = f(c) = -d$. Since 0 is the only number that is its own negative, we must have $d = 0$. Hence $f(c) = 0$.

Excursion 4.4.A, page 189

1. Increasing on $[-2.5, 0]$ and $[1.5, 4]$; decreasing on $[-6, -2.5]$ and $[0, 1.5]$.
3. Increasing on $[-5.5, -3.5]$, $[-2, 0]$, and $[1, 3]$; decreasing on $[-6, -5.5]$, $[-3.5, -2]$, and $[0, 1]$.
5. **(i)** no; **(ii)** no; **(iii)** yes
7. If $0 < c < d \leq 10$, then $c^2 < d^2$. Hence $c^2 + 3 < d^2 + 3$. But this says $f(c) < f(d)$. We have shown that if $0 < c < d \leq 10$, then $f(c) < f(d)$. Therefore f is increasing on $(0, 10]$.
9. If $0 < c < d \leq 10$, then $c^2 < d^2$. But $c^2 < d^2$ and $c < d$ imply $c^2 + c < d^2 + d$. Hence $c^2 + c + 5 < d^2 + d + 5$, that is, $h(c) < h(d)$. Therefore h is increasing on $(0, 10]$.
11. graph 4   13. graph 2   15. All five; for many values of x.
17. $g =$ graph 1 and $f =$ graph 1 or 4; $g =$ graph 2 and $f =$ graph 1, 2, or 4; $g =$ graph 3 and $f =$ graph 1, 2, 4, or 5; $g =$ graph 5 and $f =$ graph 1, 2, 4, or 5.
19. Any pair of distinct functions, except $f =$ graph 3 and $g =$ graph 5.
21. $f =$ graph 1 and $g =$ graph 1, 2, 3, or 5; $f =$ graph 2 and $g =$ graph 1, 2, 3, 5; $f =$ graph 4 and $g =$ graph 1, 2, 3, 4, or 5; $f =$ graph 5 and $g =$ graph 3.

Section 4.5, page 195

1. 0   3. 30   5. 49; 1; -8 7. $-3; -3; 0$
9. $(f \circ g)(x) = (x + 3)^2; (g \circ f)(x) = x^2 + 3$ 11. $(f \circ g)(x) = 1/\sqrt{x}; (g \circ f)(x) = 1/\sqrt{x}$
13. $(f \circ g)(x) = \sqrt[3]{x^2 + 1}; (g \circ f)(x) = (\sqrt[3]{x})^2 + 1$
15. $(f \circ g)(x) = f\left(\dfrac{x - 2}{9}\right) = 9\left(\dfrac{x - 2}{9}\right) + 2 = x$ and $(g \circ f)(x) = g(9x + 2) = \dfrac{(9x + 2) - 2}{9} = x$
17. $(f \circ g)(x) = f((x - 2)^3) = \sqrt[3]{(x - 2)^3} + 2 = x$ and $(g \circ f)(x) = g(\sqrt[3]{x} + 2) = (\sqrt[3]{x} + 2 - 2)^3 = x$
19. $(f + g)(x) = x^3 - 3x + 2; (f - g)(x) = -x^3 - 3x + 2; (g - f)(x) = x^3 + 3x - 2$
21. $(f + g)(x) = \dfrac{1}{x} + x^2 + 2x - 5; (f - g)(x) = \dfrac{1}{x} - x^2 - 2x + 5;$
$(g - f)(x) = x^2 + 2x - 5 - \dfrac{1}{x}$
23. $(fg)(x) = -3x^4 + 2x^3; \left(\dfrac{f}{g}\right)(x) = \dfrac{-3x + 2}{x^3}; \left(\dfrac{g}{f}\right)(x) = \dfrac{x^3}{-3x + 2}$

25.

x	-4	-3	-2	-1	0	1	2	3	4
$f(x)$	-2.9	$-.9$	0	.6	1	1.3	1	-2	-2
$g(x) = f(f(x))$	$-.8$	.7	1	1.2	1.3	1.4	1.3	0	0

27.

x	1	2	3	4	5
$(g \circ f)(x)$	4	2	5	4	4

29.

x	1	2	3	4	5
$(f \circ f)(x)$	1	3	3	5	1

31–35. The given function is $B \circ A$, where A and B are the functions listed here. In some cases other correct answers are possible.

31. $A(x) = x^2 + 2, B(x) = \sqrt[3]{x}$ 33. $A(x) = 7x^3 - 10x + 17, B(x) = x^7$
35. $A(x) = 3x^2 + 5x - 7, B(x) = \dfrac{1}{x}$
37. Several possible answers, including $h(x) = x^2 + 2x; k(t) = t + 1$.
39. $(f \circ g)(x) = (\sqrt{x})^3$, domain $[0, \infty); (g \circ f)(x) = \sqrt{x^3}$, domain $[0, \infty)$.
41. $(f \circ g)(x) = \sqrt{5x + 10}$, domain $[-2, \infty); (g \circ f)(x) = 5\sqrt{x + 10}$, domain $[-10, \infty)$.
43. (a) $f(x^2) = 2x^6 + 5x^2 - 1$ (b) $(f(x))^2 = (2x^3 + 5x - 1)^2 = 4x^6 + 20x^4 - 4x^3 + 25x^2 - 10x + 1$ (c) no; $f(x^2) \neq (f(x))^2$ in general

Section 4.6, page 202

1. $(f \circ g)(x) = f(g(x)) = f(x - 1) = (x - 1) + 1 = x$ and $(g \circ f)(x) = g(f(x)) = g(x + 1) = (x + 1) - 1 = x$

3. $(f \circ g)(x) = f\left(\dfrac{1-x}{x}\right) = \dfrac{1}{\left(\dfrac{1-x}{x}\right)+1} = \dfrac{1}{\dfrac{(1-x)+x}{x}} = x$ and

$(g \circ f)(x) = g\left(\dfrac{1}{x+1}\right) = \dfrac{1 - \dfrac{1}{x+1}}{\dfrac{1}{x+1}} = \dfrac{\dfrac{(x+1)-1}{x+1}}{\dfrac{1}{x+1}} = x$

5. $(f \circ g)(x) = f(\sqrt[5]{x}) = (\sqrt[5]{x})^5 = x$ and $(g \circ f)(x) = g(x^5) = \sqrt[5]{x^5} = x$

7. The rule of f is "multiply by 5, then add 1," so the rule of the inverse g is "subtract 1, then divide by 5." Hence $g(x) = \dfrac{x-1}{5}$.

9. Rule of g is "take the cube root," so $g(x) = \sqrt[3]{x}$.

11. Rule of g is "add 1, then divide by 2, then take the cube root, and finally subtract 4," so

$g(x) = \sqrt[3]{\dfrac{x+1}{2}} - 4$.

13. $g(x) = -x$ 15. $g(x) = \dfrac{x+4}{5}$ 17. $g(x) = \sqrt[3]{\dfrac{5-x}{2}}$

19. $g(x) = \dfrac{x^2+7}{4}, (x \geq 0)$ 21. $g(x) = \dfrac{1}{x}$ 23. $g(x) = \dfrac{1}{2x} - \dfrac{1}{2}$

25. $g(x) = \sqrt[3]{\dfrac{5x+1}{1-x}}$ 27.

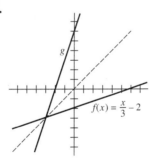

29.

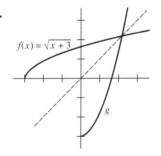

31.

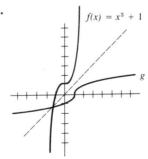

33. Many possible correct answers, including $f(x) = x$; $f(x) = -x$; $f(x) = 1/x$; $f(x) = -x + 5$; $f(x) = -x - 10$.

35. (a) Slope $L = \dfrac{2-7}{7-2} = \dfrac{-5}{5} = -1.$ (b) The slope of $y = x$ is 1 (why?). This line is perpendicular to L since the product of their slopes is -1 (see the box on page 137).
(c) Midpoint $= \left(\dfrac{2+7}{2}, \dfrac{7+2}{2}\right) = \left(\dfrac{9}{2}, \dfrac{9}{2}\right)$; it is on $y = x$.

37. (a) $f^{-1}(x) = \dfrac{x-2}{3}$ (b) $f^{-1}(1) = \dfrac{1-2}{3} = -\dfrac{1}{3}$, but $\dfrac{1}{f(1)} = \dfrac{1}{3 \cdot 1 + 2} = \dfrac{1}{5}$. Therefore $f^{-1} \neq \dfrac{1}{f}$.

39–45. There are several correct answers for each, including these:

39. One restricted function is $h(x) = |x|$ with $x \geq 0$ (so that $h(x) = x$); inverse function; $g(y) = y$ with $y \geq 0$.

41. One restricted function is $h(x) = -x^2$ with $x \leq 0$; inverse function $g(y) = -\sqrt{-y}$ with $y \leq 0$. Another restricted function is $h(x) = -x^2$ with $x \geq 0$; inverse function $g(y) = \sqrt{-y}$ with $y \leq 0$.

43. One restricted function is $h(x) = \dfrac{x^2 + 6}{2}$ with $x \geq 0$; inverse function $g(y) = \sqrt{2y - 6}$ with $y \geq 3$.

45. One restricted function is $f(x) = \dfrac{1}{x^2 + 1}$ with $x \leq 0$; inverse function $g(y) = -\sqrt{\dfrac{1}{y} - 1} = -\sqrt{\dfrac{1-y}{y}}$ with $0 < y \leq 1$.

47. Let $y = f(x) = mx + b$. Since $m \neq 0$, we can solve for x and obtain $x = \dfrac{y-b}{m}$. Hence the rule of the inverse function g is $g(y) = \dfrac{y-b}{m}$, and we have: $(f \circ g)(y) = f(g(y)) = f\left(\dfrac{y-b}{m}\right) = m\left(\dfrac{y-b}{m}\right) + b = y$ and $(g \circ f)(x) = g(f(x)) = g(mx + b) = \dfrac{(mx+b) - b}{m} = x$.

Section 4.7, page 205

1. $a = k/b$ 3. $z = kxyw$ 5. $d = k\sqrt{h}$ 7. $v = ku$; $k = 4$ 9. $v = k/u$; $k = 16$
11. $t = krs$; $k = 4$ 13. $w = kxy^2$; $k = 2$ 15. $T = kpv^3/u^2$; $k = 16$ 17. $r = 4$
19. $b = 9/4$ 21. $u = 50$ 23. $r = 3$ 25. $c = 200/3$
27. (a) 14 lb per square inch;   (b) 1.5 gal per minute   29. .064 ohm   31. 3750 kg

Chapter 4 Review, page 208

1. (a) -3 (b) 1755 (c) 2 (d) -14

3.
x	0	1	2	-4	t	k	$b-1$	$1-b$	$6-2u$
$f(x)$	7	5	3	15	$7-2t$	$7-2k$	$9-2b$	$5+2b$	$-5+4u$

5. Many possible answers, including: (a) $f(x) = x^2$, $a = 2$, $b = 3$; $f(a+b) = f(2+3) = 5^2 = 25$, but $f(a) + f(b) = f(2) + f(3) = 2^2 + 3^2 = 13$, so the statement is false.   (b) $f(x) = x + 1$, $a = 0$, $b = 1$; $f(ab) = f(0) = 1$, but $f(a)f(b) = f(0)f(1) = 1 \cdot 2 = 2$, so the statement is false.

7. $[2, \infty)$ 9. $(t+2)^2 - 3(t+2) = t^2 + t - 2$

11. $2\left(\dfrac{x}{2}\right)^3 + \left(\dfrac{x}{2}\right) + 1 = \dfrac{x^3}{4} + \dfrac{x}{2} + 1$ 13.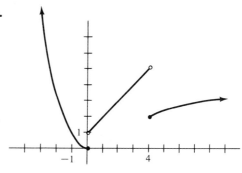

15. 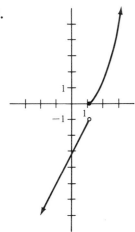   17. 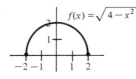   19.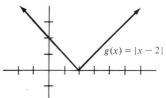

21. Approximately $[-3, 3.8]$

23. Many correct answers, including $x = -2$; all x in the interval $(2.5, 3.8)$; all x in the interval $[5, 6]$.

25. 1 27. -3 29. true   31. $x = 4$ 33. $x \le 3$ 35. $x < 3$

37. Many correct answers, including 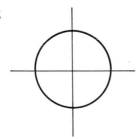   **39.** none   **41.** none

43. *x*-axis, *y*-axis, origin   **45.** even   **47.** odd   **49.** *y*-axis   **51.** $(-\infty, 3]$

53. 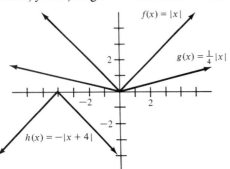   **55.** (e)

57. (a) $1/3$ (b) $(x-1)\sqrt{x^2+5}$ $(x \neq 1)$ (c) $\dfrac{\sqrt{c^2+2c+6}}{c}$

59.

x	-4	-3	-2	-1	0	1	2	3	4
$g(x)$	1	4	3	1	-1	-3	-2	-4	-3
$h(x)$	-3	-3	-4	-3	1	4	3	1	4

61. $\dfrac{82}{27}$ **63.** $\dfrac{1}{x^3} + 3$ **65.** $\dfrac{1}{4}$

67. $(f \circ g)(x) = f(x^2 - 1) = \dfrac{1}{x^2 - 1}$; $(g \circ f)(x) = g\left(\dfrac{1}{x}\right) = \dfrac{1}{x^2} - 1$

69. All nonnegative numbers except 1.   **71.**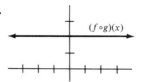

73. $g(x) = -x^2 + 14x - 44;\ x \geq 7$ **75.** 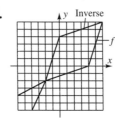   **77.** 6   **79.** 1

CHAPTER 5

Section 5.1, page 222

1. (5, 2), upward   **3.** (1, 2), downward   **5.** $\left(-\frac{3}{2}, \frac{3}{2}\right)$, downward

7. $\left(\frac{2}{3}, \frac{19}{3}\right)$, downward   **9.** $\left(\frac{1}{2}, \frac{1}{4}\right)$, downward   **11.** (3, 1), downward

13. 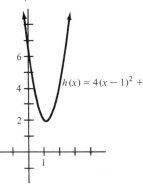   **15.** 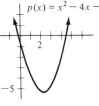   **17.**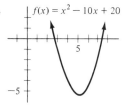

19. $f(x) = 2x^2 - 4x + 1$ **21.** $f(x) = 3x^2$ **23.** $f(x) = 2x^2 - 4x + 2$

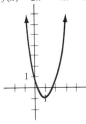

25. $f(x) = -x^2 - 6x - 6$ **27.** $f(x) = \frac{4}{25}x^2 - \frac{4}{25}x + \frac{51}{25}$ **29.** $b = 0$

31. Minimum product is -4; numbers are 2 and -2.   **33.** 9 and 18
35. $h = 15, b = 15$ **37.** Two 50-ft sides and one 100-ft side.   **39.** 30 stores
41. $3.50   **43.** $3.67 (but if tickets must be priced in multiples of .20, then $3.60 is best).
45. $t = 2.5, h = 196$ **47.** 22 ft   **49.** $t = \frac{125}{8}, h = \frac{125^2}{4} = 3906.25$

Section 5.2, page 231

1. 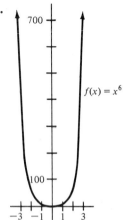   **3.** yes   **5.** yes   **7.** no

9. degree 3, yes; degree 4, no; degree 5, yes   **11.** no

13. degree 3, no; degree 4, no; degree 5, yes   **15.**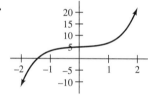

17. 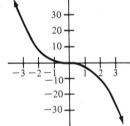   **19.** 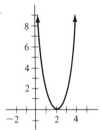   **21.**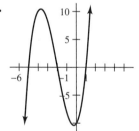

CHAPTER 5   A-31

23.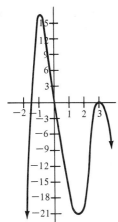

25.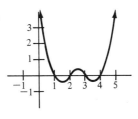

27.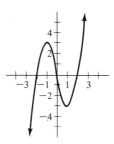

29.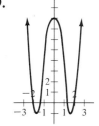

31.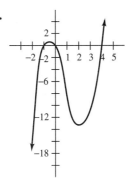

33.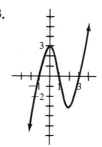

35.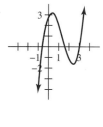

Section 5.3, page 245

1.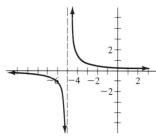

3.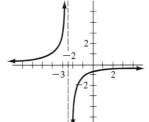

5.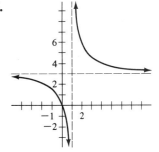

7.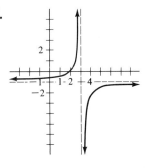

9.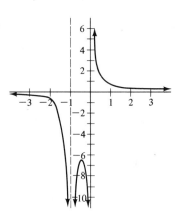

11.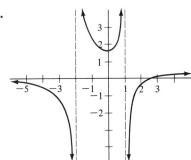

13.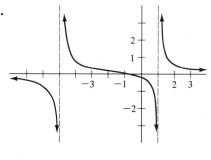

15.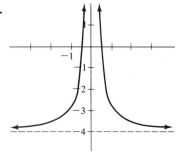

17.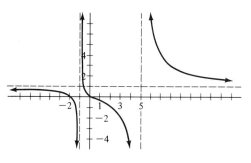

19.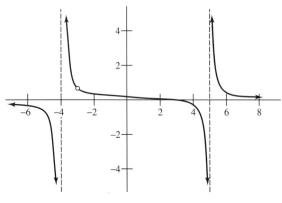

21.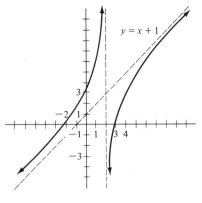

CHAPTER 5   A-33

23.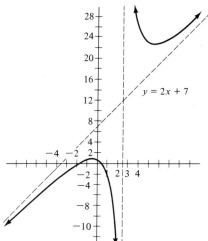

25.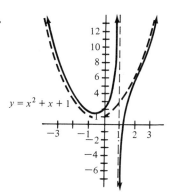

27.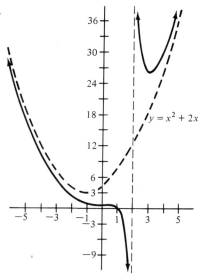

29.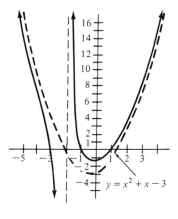

31. (a) $\dfrac{4 \cdot 10^2}{(6.4)^2} = 9.765625$ **(b)** 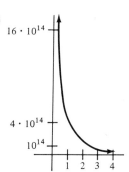 **(c)** no roots

33.

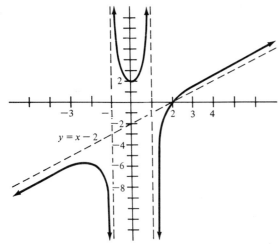

Section 5.4, page 253

1. 2 **3.** 6 **5.** −30 **7.** $x = 0$ (multiplicity 54); $x = -4/5$ (multiplicity 1)
9. $x = 0$ (multiplicity 15); $x = \pi$ (multiplicity 14); $x = \pi + 1$ (multiplicity 13)
11. $x = 1 + 2i$ or $1 - 2i$; $f(x) = (x - 1 - 2i)(x - 1 + 2i)$
13. $x = -\dfrac{1}{3} + \dfrac{2\sqrt{5}}{3}i$ or $-\dfrac{1}{3} - \dfrac{2\sqrt{5}}{3}i$; $f(x) = \left(x + \dfrac{1}{3} - \dfrac{2\sqrt{5}}{3}i\right)\left(x + \dfrac{1}{3} + \dfrac{2\sqrt{5}}{3}i\right)$
15. $x = 3$ or $-\dfrac{3}{2} + \dfrac{3\sqrt{3}}{2}i$ or $-\dfrac{3}{2} - \dfrac{3\sqrt{3}}{2}i$;

$f(x) = (x - 3)\left(x + \dfrac{3}{2} - \dfrac{3\sqrt{3}}{2}i\right)\left(x + \dfrac{3}{2} + \dfrac{3\sqrt{3}}{2}i\right)$

17. $x = -2$ or $1 + \sqrt{3}i$ or $1 - \sqrt{3}i$; $f(x) = (x + 2)(x - 1 - \sqrt{3}i)(x - 1 + \sqrt{3}i)$
19. $x = 1$ or i or -1 or $-i$; $f(x) = (x - 1)(x - i)(x + 1)(x + i)$
21. $x = \sqrt{5}$ or $-\sqrt{5}$ or $\sqrt{2}i$ or $-\sqrt{2}i$; $f(x) = (x - \sqrt{5})(x + \sqrt{5})(x - \sqrt{2}i)(x + \sqrt{2}i)$
23. Many correct answers, including $(x - 1)(x - 7)(x + 4)$.
25. Many correct answers, including $(x - 1)(x - 2)^2(x - \pi)^3$
27. $f(x) = \dfrac{17}{100}(x - 5)(x - 8)x$

29–38. Many correct answers, including the following:

29. $x^2 - 4x + 5$ **31.** $(x - 2)(x^2 - 4x + 5)$ **33.** $(x + 3)(x^2 - 2x + 2)(x^2 - 2x + 5)$
35. $x^2 - 2x + 5$ **37.** $(x - 4)^2(x^2 - 6x + 10)$ **39.** $(x^4 - 3x^3)(x^2 - 2x + 2)$
41. $3x^2 - 6x + 6$ **43.** $-2x^3 + 2x^2 - 2x + 2$
45. Many correct answers, including $x^2 - (1 - i)x + (2 + i)$.
47. Many correct answers, including $x^3 - 5x^2 + (7 + 2i)x - (3 + 6i)$.

49. $3, -\dfrac{1}{2} + \dfrac{\sqrt{3}}{2}i, -\dfrac{1}{2} - \dfrac{\sqrt{3}}{2}i$ **51.** $i, -i, -1, -2$ **53.** $1, 2i, -2i$ **55.** $i, -i, 2+i, 2-i$

57. **(a)** Since $z + w = (a + c) + (b + d)i$, $\overline{z + w} = (a + c) - (b + d)i$. Since $\bar{z} = a - bi$ and $\bar{w} = c - di$, $\bar{z} + \bar{w} = (a - bi) + (c - di) = (a + c) - (b + d)i$.   **(b)** Since $zw = (ac - bd) + (ad + bc)i$, $\overline{zw} = (ac - bd) - (ad + bc)i$. Since $\bar{z} = a - bi$ and $\bar{w} = c - di$, $\bar{z}\,\bar{w} = (a - bi)(c - di) = (ac - bd) - (ad + bc)i$.

59. **(a)** $\overline{f(z)} = \overline{az^3 + bz^2 + cz + d}$ (definition of $f(z)$)

$\phantom{\overline{f(z)}} = \overline{az^3} + \overline{bz^2} + \overline{cz} + \bar{d}$ (Exercise 57(a))

$\phantom{\overline{f(z)}} = \bar{a}\,\overline{z^3} + \bar{b}\,\overline{z^2} + \bar{c}\,\bar{z} + \bar{d}$ (Exercise 57(b))

$\phantom{\overline{f(z)}} = a\,\overline{z^3} + b\,\overline{z^2} + c\,\bar{z} + d$ ($\bar{r} = r$ for r real)

$\phantom{\overline{f(z)}} = a\bar{z}^3 + b\bar{z}^2 + c\bar{z} + d$ (Exercise 57(b))

$\phantom{\overline{f(z)}} = f(\bar{z})$ (definition of f)

(b) Since $f(z) = 0$, we have $0 = \bar{0} = \overline{f(z)} = f(\bar{z})$. Hence $\bar{z}$ is a root of $f(x)$.

61. If $f(z)$ is a polynomial with real coefficients, then $f(z)$ can be factored as $g_1(z)g_2(z)g_3(z)\ldots g_k(z)$, where each $g_i(z)$ is a polynomial with real coefficients and degree 1 or 2. The rules of polynomial multiplication show that the degree of $f(z)$ is the sum: degree $g_1(z)$ + degree $g_2(z)$ + degree $g_3(z)$ + $\cdots$ + degree $g_k(z)$. If all of the $g_i(z)$ have degree 2, then this last sum is an even number. But $f(z)$ has odd degree, so this can't occur. Therefore at least one of the $g_i(z)$ is a first-degree polynomial and hence must have a real root. This root is also a root of $f(z)$.

Excursion 5.4.A, page 257

1. 1 positive; no negative   **3.** 1 positive; 1 or 3 negative
5. 1 or 3 positive; no negative   **7.** 1 positive; 0 or 2 negative
9. 0, 2, or 4 positive; no negative

Section 5.5, page 261

1. Between 2 and 3; $x \approx 2.67$.   **3.** Between -3 and -2; $x \approx -2.43$.
5. Between -5 and -4; between -1 and 0; between 0 and 1; between 2 and 3; $x \approx -4.08$ or -0.68 or 0.14 or 2.63.
7. Lower bound: -4; upper bound: 2.   **9.** Lower bound: -4; upper bound: 2.
11. $x \approx 2.10125$

Chapter 5 Review, page 263

1. 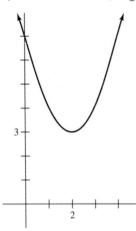 **3.**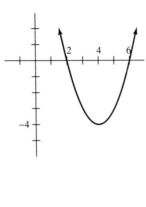

5. (c)   **7.** (f)   **9.** (b)   **11.** (d)   **13.** 30, 60, 30 ft   **15.** (a), (c), (e), (f)   **17.** 0

19. $\underline{2|}\ \ 1\ \ -5\ \ \ \ 8\ \ \ \ 1\ \ -17\ \ \ \ 16\ \ -4$

$$\begin{array}{r|rrrrrr} & & 2 & -6 & 4 & 10 & -14 & 4 \\ \hline & 1 & -3 & 2 & 5 & -7 & 2 & \underline{|\,0} \end{array}$$

other factor: $x^5 - 3x^4 + 2x^3 + 5x^2 - 7x + 2$

21. Many correct answers, including $f(x) = 5(x - 1)^2(x + 1) = 5x^3 - 5x^2 - 5x + 5$.

23. 1   **25.** (a)   **27.** Many possible answers.   **29.** 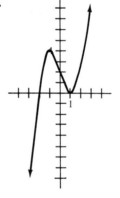   **31.** (c)

33. (e)   **35.**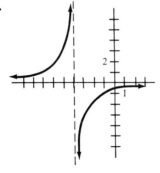

37. 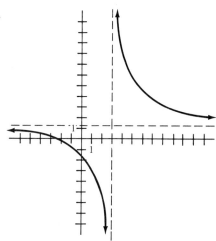   39.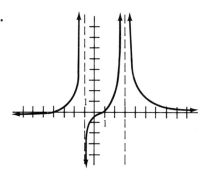

41. (d)   **43.** (d)   **45.** $x = \dfrac{-3 \pm \sqrt{31}i}{2}$ **47.** $x = \dfrac{3 \pm \sqrt{31}i}{10}$

49. $x = \sqrt{2/3}$ or $-\sqrt{2/3}$ or i or $-i$ **51.** $x = -2$ or $1 + \sqrt{3}i$ or $1 - \sqrt{3}i$ **53.** $i, -i, 2, -1$

55. Many correct answers, including $x^4 - 2x^3 + 2x^2$.

57. $f(x)$ has one positive and one negative real root.   **59.** (d)

61. $(x^4 - 4x^3 + 16x - 16) \div (x - 5)$ is $x^3 + x^2 + 5x + 41$ with remainder 189. Since all coefficients and the remainder are positive, 5 is an upper bound for the roots.

CHAPTER 6

Section 6.1, page 276

1. 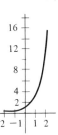   3. 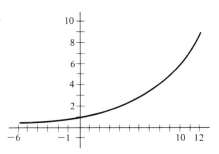   5.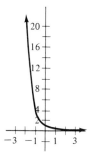

7.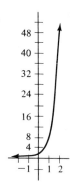

9.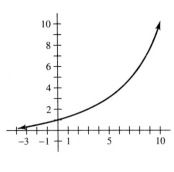

11.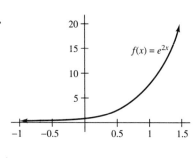

13.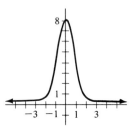

15.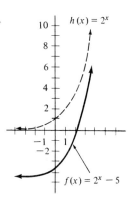

17.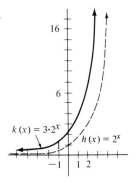

19.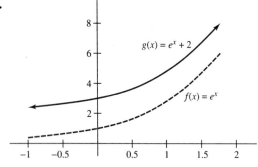

21.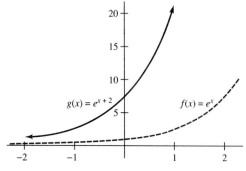

23. $\dfrac{10^x(10^h - 1)}{h}$

25. $2^x\left(\dfrac{2^h-1}{h}\right) + 2^{-x}\left(\dfrac{2^{-h}-1}{h}\right)$ 27. neither   29. even   31. even

33.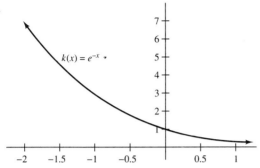

35.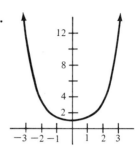

37.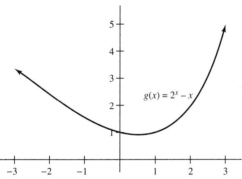

39.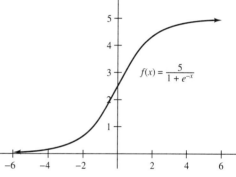

41. $878.04 after 2 years, $1007.89 after 3 years and 9 months.
43. 3.082 grams after 4000 years, 1.900 grams after 8000 years.
45. About 520 in 15 days, about 1559 in 25 days.
47. About 273,433,000 in the year 2000. The answer to the second question depends on your age. For example, if you were 18 in 1990, the U.S. population would be about 380,912,000 when you turned 65.
49. (a) 100 kg;   (b) approximately 139.3 kg, 163.2 kg, and 186.5 kg
51. (a) 100,000 now; 83,527 in 2 months; 58,275 in 6 months.   (b) No. The graph continues to decrease toward zero.
53. (a) The current population is 10, and in 5 years it will be about 149.   (b) After about 9.55 years.
55. Many correct answers: $f(x) = a^x$ for any nonnegative constant a.
57. (a) $P(x) = 2^x/100$ (b) no!

Excursion 6.1.A, page 280

1. annually: $1469.33, quarterly: $1485.95, monthly: $1489.85, weekly: $1491.37
3. $738.73   5. $821.81   7. $1065.30   9. $568.59   11. about 164 dandelions
13. about 946,984

Section 6.2, page 288

1. $e^{1.0986} = 3$ 3. $e^{6.9078} = 1000$ 5. $e^{-4.6052} = .01$ 7. $\ln 25.79 = 3.25$
9. $\ln 5.5527 = 12/7$ 11. $\ln w = 2/r$ 13. 15 15. 1/2 17. 931 19. $\sqrt{37}$
21. $x + y$ 23. x^2 25. $\ln(x^2y^3)$ 27. $\ln(x - 3)$ 29. $\ln(x^{-7})$ 31. $2u + 5v$
33. $\frac{1}{2}u + 2v$ 35. $\frac{2}{3}u + \frac{1}{6}v$ 37. $(-1, \infty)$ 39. $(-\infty, 0)$
41. All real numbers except those in $[-1, 2]$. 43.

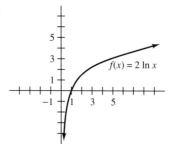

45.

47.

49.

51.

53.

55. $A = -9, B = 10$

57. $(f \circ g)(x) = \dfrac{1}{1+e^{-\ln(x/(1-x))}} = \dfrac{1}{1+\dfrac{1}{e^{\ln(x/(1-x))}}} = \dfrac{1}{1+\dfrac{1}{\dfrac{x}{1-x}}} = \dfrac{1}{1+\dfrac{1-x}{x}} = \dfrac{1}{\dfrac{1}{x}} = x;$

$(g \circ f)(x) = \ln\left(\dfrac{\dfrac{1}{1+e^{-x}}}{1-\dfrac{1}{1+e^{-x}}}\right) =$

$\ln\left(\dfrac{\dfrac{1}{1+e^{-x}}}{1-\dfrac{1}{1+e^{-x}}} \cdot \dfrac{1+e^{-x}}{1+e^{-x}}\right) = \ln\left(\dfrac{1}{1+e^{-x}-1}\right) = \ln(e^x) = x.$

59. The domain of $f(x) = \ln x^2$ is all nonzero real numbers. The domain of $g(x) = 2 \ln x$ is all positive real numbers. Since the domains are different, f and g are different functions.

61. About 4392 meters.

63. Original exam: 77; after 2 weeks: approximately 66.01; after 5 weeks: approximately 59.08; after 10 weeks: approximately 53.02.

65. About 9.9 days.

67. (a) No advertising: about 120 bikes; $1000: about 299 bikes; $10,000: about 513 bikes.
(b) $1000, yes; $10,000; no.   (c) yes; yes.

Section 6.3, page 296

1. 4   **3.** −2/3   **5.** 4   **7.** −3/2   **9.** $\log .01 = -2$ **11.** $\log \sqrt[3]{10} = 1/3$
13. $\log r = 7k$ **15.** $\log_7 5{,}764{,}801 = 8$ **17.** $\log_3 (1/9) = -2$ **19.** $10^4 = 10{,}000$
21. $10^{2.88} \approx 750$ **23.** $5^3 = 125$ **25.** $2^{-2} = \dfrac{1}{4}$ **27.** $10^{z+w} = x^2 + 2y$ **29.** $\sqrt{43}$
31. $\sqrt{x^2 + y^2}$ **33.** $\dfrac{1}{2}$ **35.** 6   **37.** $b = 10$ **39.** $b = 2$ **41.** 5   **43.** 3   **45.** 4
47. .43   **49.** .26   **51.** .1   **53.** .7   **55.** 3.3219   **57.** .8271   **59.** 1.1115
61. 1.6199   **63.** 2   **65.** approximately 2.54   **67.** 20 decibels
69. approximately 66 decibels   **71.** true   **73.** true   **75.** false   **77.** 97^{98}
79. $\log_b u = \dfrac{\log_a u}{\log_a b}$ **81.** $\log_{10} u = 2 \log_{100} u$
83. If $\log_b x = \dfrac{1}{2} \log_b v + 3$, then $b^{\log_b x} = b^{(1/2)\log_b v + 3}$, or $x = b^{(1/2)\log_b v} b^3 = (b^{\log_b v})^{1/2} b^3 = b^3 \sqrt{v}.$

Section 6.4, page 303

1. $x = 4$ **3.** $x = 1/9$ **5.** $x = \frac{1}{2}$ or -3

7. $x = -2$ or $-\frac{1}{2}$ **9.** $x = \ln 5/\ln 3 \approx 1.465$ **11.** $x = \ln 3/\ln 1.5 \approx 2.7095$

13. $x = \dfrac{\ln 3 - 5 \ln 5}{\ln 5 + 2 \ln 3} \approx -1.825$ **15.** $x = \dfrac{\ln 2 - \ln 3}{3 \ln 2 + \ln 3} \approx -.1276$

17. $x = (\ln 5)/2 \approx .805$ **19.** $x = (-\ln 3.5)/1.4 \approx -.895$

21. $x = 2 \ln(5/2.1)/\ln 3 \approx 1.579$ **23.** $x = 0$ or 1

25. $x = \ln 2 \approx .693$ or $x = \ln 3 \approx 1.099$ **27.** $x = \ln 3 \approx 1.099$

29. $x = \ln 2/\ln 4 = \frac{1}{2}$ or $x = \ln 3/\ln 4 \approx .792$ **31.** $x = \ln(t + \sqrt{t^2 + 1})$

33. $x = \frac{1}{2} \ln \left(\dfrac{1 + t}{1 - t} \right) = \frac{1}{2} [\ln(1 + t) - \ln(1 - t)]$ **35.** $x = 9$ **37.** $x = 5$

39. $x = 6$ **41.** $x = 3$ **43.** $x = \dfrac{-5 + \sqrt{37}}{2}$ **45.** $x = 9/(e - 1)$ **47.** $x = 5$

49. $x = 2$ **51.** $x = \pm\sqrt{10001}$ **53.** $x = \sqrt{\dfrac{e + 1}{e - 1}}$

55. (a) There were 20 bacteria at the beginning of the experiment and 2500 three hours later. **(b)** The time required to double the number of bacteria is $\ln 2/\ln 5 \approx 0.43$.

57. approximately 3,689 years old   **59.** approximately 2534 years ago

61. approximately 444,000,000 years   **63.** approximately 10.413 years

65. approximately 9.853 days   **67.** approximately 6.99%

69. (a) approximately 22.5 years   **(b)** approximately 22.1 years   **71.** $3197.05

73. $t = \dfrac{\ln 2}{\ln(1 + r)}$, where r is written in decimal form [for instance, 8% = .08].

75. (a) At the outbreak: 200 people; after 3 weeks: about 2795 people.   **(b)** In about 6.02 weeks.

77. (a) $k \approx 21.459$ **(b)** $t \approx .182$ **79. (a)** $k \approx .229$, $c \approx 83.3$ **(b)** 12.43 weeks

Chapter 6 Review, page 306

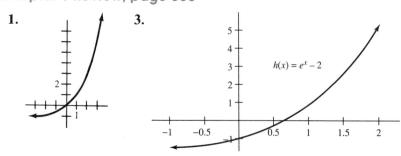

5. (d) **7.** (c) **9.** ln 756 = 6.628 **11.** ln($u + v$) = $r^2 - 1$
13. log 756 = 2.8785 **15.** $e^{7.118}$ = 1234 **17.** $e^t = rs$ **19.** $5^u = cd - k$ **21.** 3
23. 3/4 **25.** 2 ln x **27.** ln($9y/x^2$) **29.** undefined **31.** 3/2 **33.** .18
35. .70 **37.** (c) **39.** (c) **41.** $x = \pm\sqrt{2}$ **43.** $x = \dfrac{3 \pm \sqrt{57}}{4}$ **45.** $x = -\dfrac{1}{2}$
47. $x = e^{(u-c)/d}$ **49.** $x = 2$ **51.** $x = 101$ **53.** about 1.64 mg
55. approximately 12 years **57.** $452.89 **59.** 7.6

CHAPTER 7

Section 7.1, page 321

1.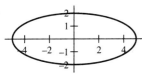

3.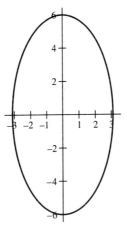

5.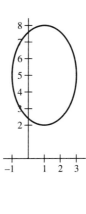

7.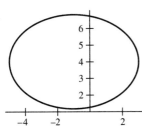

9. area = 8π **11.** area = $2\sqrt{3}\pi$ **13.** area = $7\pi/\sqrt{3}$

15. $\dfrac{x^2}{49} + \dfrac{y^2}{4} = 1$ **17.** $\dfrac{x^2}{36} + \dfrac{y^2}{16} = 1$ **19.** $\dfrac{x^2}{9} + \dfrac{y^2}{49} = 1$ **21.** $\dfrac{(x-2)^2}{4} + \dfrac{(y-3)^2}{16} = 1$

23. $\dfrac{4(x-7)^2}{25} + \dfrac{(y+4)^2}{36} = 1$ **25.** $\dfrac{x^2}{16} + \dfrac{y^2}{8} = 1$ **27.** $\dfrac{(x-3)^2}{36} + \dfrac{(y+2)^2}{16} = 1$

29. 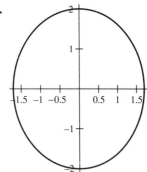 **31.** 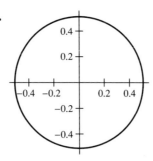 **33.**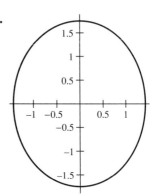

35. 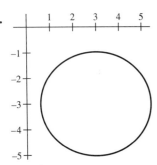 **37.** 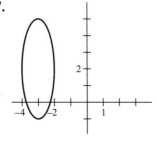 **39.**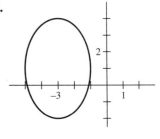

41. $\dfrac{(x+5)^2}{49} + \dfrac{(y-3)^2}{16} = 1$ or $\dfrac{(x+5)^2}{16} + \dfrac{(y-3)^2}{49} = 1$

43. If $a = b$, then $\dfrac{x^2}{a^2} + \dfrac{y^2}{a^2} = 1$. Multiplying both sides by a^2 gives $x^2 + y^2 = a^2$, the equation of a circle of radius a with center at the origin.

45. eccentricity $= .1$ **47.** eccentricity $= \sqrt{3}/2 \approx .87$ **49.** eccentricity $\approx .55$

51. approximately 226,335 miles and 251,401 miles.

53. Let P denote the punch bowl and Q the table. In the longest possible trip starting at point X, the sum of the distance from X to Q and the distance from X to P must be 100 (since the distance from Q to P is 50). Thus the fence should be an ellipse with foci P and Q and $r = 100$, as described in the introduction to this section (with $c = 25$). Verify that the length of its major axis is 100 ft and the length of its minor axis is approximately 86.6 ft.

Section 7.2, page 332

1.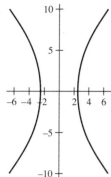

3.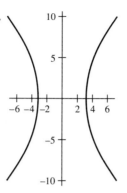

5.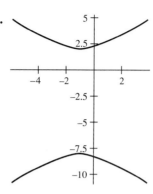

7.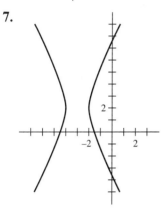

9. $\dfrac{x^2}{9} - \dfrac{y^2}{36} = 1$

11. $\dfrac{x^2}{4} - y^2 = 1$

13. $\dfrac{(y-3)^2}{4} - \dfrac{(x+2)^2}{6} = 1$

15. $\dfrac{(x-4)^2}{9} - \dfrac{(y-2)^2}{16} = 1$

17. The asymptotes of $\dfrac{x^2}{a^2} - \dfrac{y^2}{a^2} = 1$ are $y = \pm \dfrac{a}{a} x$ or $y = \pm x$, with slopes $+1$ and -1. Since $(+1)(-1) = -1$, these lines are perpendicular.

19.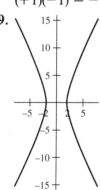

21.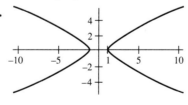

23.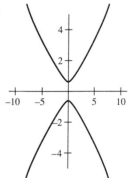

25. 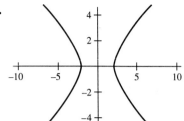   **27.**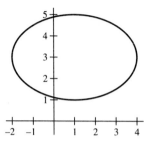

29. 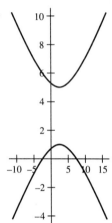   **31.**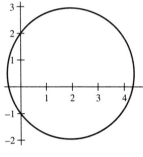

33. $\dfrac{x^2}{1{,}210{,}000} - \dfrac{y^2}{5{,}759{,}600} = 1$ (measurement in feet). The exact location cannot be determined from the given information.

Section 7.3, page 339

1. focus: (0, 1/12), directrix: $y = -1/12$.   3. focus: (0, 1), directrix: $y = -1$.
5. vertex: (2, 0), focus: (9/4, 0), directrix: $x = 7/4$, opens right
7. vertex: (2, −1), focus: (7/4, −1), directrix: $x = 9/4$, opens left
9. vertex: (2/3, −3), focus: (17/12, −3), directrix: $x = -1/12$, opens right
11. vertex: (−81/4, 9/2), focus: (−20, 9/2), directrix: $x = -41/2$, opens right
13. vertex: (−1/6, −49/12), focus: (−1/6, −4), directrix: $y = -25/6$, opens up
15. vertex: (2/3, 19/3), focus: (2/3, 25/4), directrix: $y = 77/12$, opens down

17. 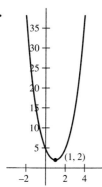   **19.** 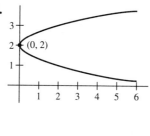   **21.**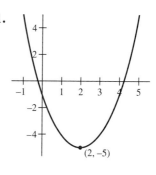

23. 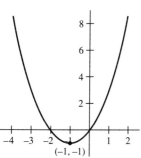   **25.** 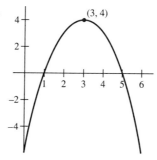   **27.**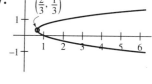

29. $y = 3x^2$ **31.** $y = 13(x - 1)^2$ **33.** $x - 2 = 3(y - 1)^2$ **35.** $x + 3 = 4(y + 2)^2$

37. $2(x - 1)^2 = y - 1$ **39.** $(y - 3)^2 = x + 1$ **41.** $b = 0$ **43.** $\left(9, -\dfrac{1}{2} \pm \dfrac{1}{2}\sqrt{34}\right)$

Chapter 7 Review, page 341

1. ellipse, foci: $(0, 2)$, $(0, -2)$, vertices: $(0, 2\sqrt{5})$, $(0, -2\sqrt{5})$

3. ellipse, foci: $(1, 6)$, $(1, 0)$, vertices: $(1, 7)$, $(1, -1)$

5. focus: $(0, 5/14)$, directrix: $y = -5/14$ **7.** eccentricity $= \sqrt{\dfrac{2}{3}} \approx 0.8165$

9. 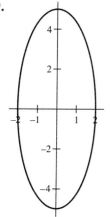   **11.**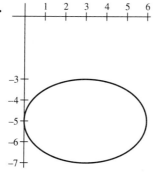

A-48 ANSWER SECTION

13. asymptotes: $y + 4 = \pm\frac{5}{2}(x - 1)$

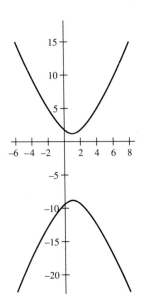

15.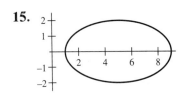

17.

19.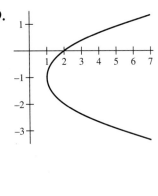

21. center: $(4, -6)$ **23.** $\frac{(x-3)^2}{4} + \frac{(y-1)^2}{2} = 1$ **25.** $\frac{y^2}{4} - \frac{(x-3)^2}{16} = 1$

27. $\left(y + \frac{1}{2}\right)^2 = -\frac{1}{2}\left(x - \frac{3}{2}\right)$

CHAPTER 8

Section 8.1, page 350

1. yes   **3.** yes   **5.** no   **7.** $x = \frac{11}{5}, y = -\frac{7}{5}$ **9.** $x = \frac{2}{7}, y = -\frac{11}{7}$

11. $x = \frac{11}{7}, y = -\frac{13}{7}$ **13.** $r = \frac{5}{2}, s = -\frac{5}{2}$ **15.** $p = -1.75, q = 3.125$

17. $x = \dfrac{3c}{2}, y = \dfrac{-c + 2d}{2}$ 19. $x = 28, y = 22$ 21. $x = 2, y = -1$

23. inconsistent   25. $x = b, y = \dfrac{3b - 4}{2}$, where b is any real number.

27. $x = b, y = \dfrac{3b - 2}{4}$ where b is any real number.   29. inconsistent

31. $x = -6, y = 2$ 33. $x = \dfrac{66}{5}, y = \dfrac{18}{5}$ 35. $x = 10.6, y = .84$

37. $x = .185, y = -.624$ 39. $x = \dfrac{rd - sb}{ad - bc}, y = \dfrac{as - cr}{ad - bc}$

41. $x = \dfrac{3c + 8}{15}, y = \dfrac{6c - 4}{15}$ is the only solution.   43. $c = -3, d = \dfrac{1}{2}$

45. 140 adults, 60 children   47. 17 and 23   49. $14,450 at 9% and $6450 at 11%
51. 3/4 lb cashews and $2\tfrac{1}{4}$ lb peanuts   53. plane 550 mph, wind 50 mph
55. boat speed 18 mph; current speed 2 mph   57. 12 liters of the 18%; 18 liters of the 8%
59. 24 g of 50% alloy; 16 g of 75% alloy

Excursion 8.1.A, page 356

1. $x = 3, y = 9$ or $x = -1, y = 1$
3. $x = \dfrac{-1 + \sqrt{73}}{6}, y = \dfrac{37 - \sqrt{73}}{18}$ or $x = \dfrac{-1 - \sqrt{73}}{6}, y = \dfrac{37 + \sqrt{73}}{18}$
5. $x = 7, y = 3$ or $x = 3, y = 7$ 7. $x = 0, y = -2$ or $x = 6, y = 1$
9. $x = 2, y = 0$ or $x = 4, y = 2$
11. $x = 4, y = 3$ or $x = -4, y = 3$ or $x = \sqrt{21}, y = -2$ or $x = -\sqrt{21}, y = -2$
13. no solutions   15. $x = \dfrac{7}{11}, y = -7$ 17. $x = 1, y = \dfrac{1}{2}$ 19. $x = -\dfrac{3}{7}, y = \dfrac{42}{13}$
21. -4 and -12 23. 1.6 and 2.6   25. 15 and -12 27. 12 ft by 17 ft
29. 8×15 inches
31. two possibilities: $2 \times 2 \times 4$ meters or approximately $3.12 \times 3.12 \times 1.64$ meters

Section 8.2, page 364

1. $x = 4, y = 1, z = 2$ 3. $x = -\dfrac{17}{3}, y = \dfrac{68}{3}, z = \dfrac{22}{3}$ 5. $x = \dfrac{3}{2}, y = \dfrac{3}{2}, z = -\dfrac{3}{2}$

7. $z = t, y = -1 + \dfrac{1}{3}t, x = 2 - \dfrac{4}{3}t$, where t is any real number.

9. $z = t, y = \dfrac{1}{2} - 2t, x = t$, where t is any real number.   11. $x = 1, y = 2$

13. no solutions   15. no solutions   17. $x = 5, y = 6, z = 0, w = -1$
19. no solutions   21. $x = \frac{1}{2}, y = \frac{1}{3}, z = -\frac{1}{4}$ 23. infinitely many solutions
25. inconsistent   27. inconsistent
29. $x = -s - t, y = s + 2t, z = s, w = t$, where t is any real number.
31. $x = t, y = 0, z = t, w = 2t$, where t is any real number.
33. $a = -\frac{7}{20}, b = -\frac{19}{20}, c = \frac{23}{10}$ 35. $a = 2, b = -3, c = 1$
37. $a = 3, b = -1, c = 2$ 39. $a = 1, b = -4, c = 1$ 41. 10 quarters, 28 dimes, 14 nickels
43. $15,000 in the mutual fund, $30,000 in bonds, $25,000 in food franchise
45. 8 hours for Tom; 24 hours for Dick; 12 hours for Harry

Excursion 8.2.A, page 370

1. $A = -1, B = 2$ 3. $A = -3/25, B = 3/25, C = 7/5$ 5. $A = 2, B = 3, C = -1$

7. $\frac{A}{x} + \frac{B}{x-1} + \frac{C}{x-2} + \frac{D}{(x-2)^2} + \frac{E}{x+3} + \frac{F}{(x+3)^2} + \frac{G}{(x+3)^3}$

9. $\frac{Ax+B}{x^2+1} + \frac{Cx+D}{(x^2+1)^2} + \frac{Ex+F}{2x^2+2x+1} + \frac{Gx+H}{(2x^2+2x+1)^2} + \frac{Ix+J}{(2x^2+2x+1)^3}$

11. $\frac{1}{2x+1} + \frac{2}{x-3}$ 13. $\frac{2}{x-3} - \frac{1}{x+2}$ 15. $\frac{-3}{x} + \frac{5}{x+1}$

17. $\frac{\frac{1}{2}}{x} + \frac{-\frac{3}{2}}{x+2} + \frac{1}{x-2}$ 19. $\frac{-1}{x-1} + \frac{2}{(x-1)^2} + \frac{6}{(x-1)^3}$ 21. $\frac{2}{(x-1)^2} + \frac{2}{(x-1)^3}$

23. $\frac{6}{x} - \frac{1}{x+1} + \frac{9}{(x+1)^2}$ 25. $\frac{1}{x-1} + \frac{2}{(x-1)^2} + \frac{3}{x+2}$

27. $\frac{3}{x-1} - \frac{2}{(x-1)^2} + \frac{1}{x+2}$ 29. $\frac{5x-3}{x^2+1} + \frac{2x}{(x^2+1)^2}$

31. $\frac{x-3}{x^2+x+1} + \frac{4x-1}{(x^2+x+1)^2}$ 33. $\frac{1}{x} + \frac{2x+4}{x^2+4}$ 35. $\frac{3}{x-1} + \frac{2x+1}{x^2+x+1}$

37. $\frac{-\frac{1}{2}x - \frac{1}{2}}{x^2+1} + \frac{-x+1}{(x^2+1)^2} + \frac{\frac{1}{2}}{x-1}$ 39. $2x + 1 + \frac{1}{x-2} + \frac{2}{3x-1}$

41. $1 + \frac{2}{x} + \frac{1}{x+1}$ 43. $\frac{1}{x-a} + \frac{1}{x+a}$ 45. $\frac{b}{x} + \frac{c}{x+a}$

Section 8.3, page 375

1. coefficient $\begin{pmatrix} 2 & -3 & 4 \\ 1 & 2 & -6 \\ 3 & -7 & 4 \end{pmatrix}$ augmented $\left(\begin{array}{ccc|c} 2 & -3 & 4 & 1 \\ 1 & 2 & -6 & 0 \\ 3 & -7 & 4 & -3 \end{array} \right)$

3. coefficient $\begin{pmatrix} 1 & -\frac{1}{2} & \frac{7}{4} \\ 2 & -\frac{3}{2} & 5 \\ 0 & -2 & \frac{1}{3} \end{pmatrix}$ augmented $\left(\begin{array}{ccc|c} 1 & -\frac{1}{2} & \frac{7}{4} & 0 \\ 2 & -\frac{3}{2} & 5 & 0 \\ 0 & -2 & \frac{1}{3} & 0 \end{array} \right)$

5. coefficient $\begin{pmatrix} 17 & -107 & 5 \\ 7 & -306 & 194 \end{pmatrix}$ augmented $\left(\begin{array}{ccc|c} 17 & -107 & 5 & \sqrt{2} \\ 7 & -306 & 194 & 0 \end{array} \right)$

7. $2x - 3y = 1$
$4x + 7y = 2$

9. $x + z = 1$
$x - y + 4z - 2w = 3$
$4x + 2y + 5z = 2$

11. $x + 2y + 3z - 4w = 1$
$ 3y + 5z + 7w = 9$
$8x - 7y = 2$
$4x + 3y + 2z - w = 0$

13. $x = 2, y = 3, z = -1$ 15. $x = -2, y = -1, z = 4$ 17. $x = \frac{13}{3}, y = \frac{5}{3}, z = -3$

19. $x = -7, y = 5$ 21. no solutions

23. $z = t, y = t - 1, x = -t + 2$, for any real number t.   25. $x = 0, y = 0, z = 0$

27. $x = -1, y = 1, z = -3, w = -2$ 29. $x = \frac{7}{31}, y = \frac{6}{31}, z = \frac{1}{31}, w = \frac{29}{31}$

31. $w = t, z = -4t, y = 3t, x = 8t$, where t is any real number.

33. 40 lbs of peanuts; 20 lbs of almonds; 80 lbs of cashews

35. 2000 chairs; 1600 chests; 2500 tables

Excursion 8.3.A, page 386

1. $A + B = \begin{pmatrix} 10 & -3 \\ 3 & 7 \end{pmatrix}$; $AB = \begin{pmatrix} 17 & -3 \\ 33 & -19 \end{pmatrix}$; $BA = \begin{pmatrix} -4 & 9 \\ 24 & 2 \end{pmatrix}$; $2A - 3B = \begin{pmatrix} -15 & 19 \\ 16 & -16 \end{pmatrix}$

3. $A + B = \begin{pmatrix} 2 & 1/2 \\ 13/2 & 9/2 \end{pmatrix}$; $AB = \begin{pmatrix} 23/4 & -1/4 \\ 43/4 & -5/2 \end{pmatrix}$; $BA = \begin{pmatrix} -21/4 & -17/4 \\ 31/4 & 17/2 \end{pmatrix}$;
$2A - 3B = \begin{pmatrix} 3/2 & 17/2 \\ 1/2 & 4 \end{pmatrix}$

5. $A + B = \begin{pmatrix} 3 & 6 \\ 8 & 9 \end{pmatrix}$; $AB = \begin{pmatrix} 7 & 9 \\ 3 & 5 \end{pmatrix}$; $BA = \begin{pmatrix} 5 & 3 \\ 9 & 7 \end{pmatrix}$; $2A - 3B = \begin{pmatrix} -9 & -13 \\ -19 & -27 \end{pmatrix}$

7. $A + B = \begin{pmatrix} 4 & 4 \\ 6 & -5 \end{pmatrix}$; $AB = \begin{pmatrix} 1 & 2 \\ -6 & 10 \end{pmatrix}$; $BA = \begin{pmatrix} 1 & -4 \\ 3 & 10 \end{pmatrix}$; $2A - 3B = \begin{pmatrix} 13 & -2 \\ -3 & 10 \end{pmatrix}$

9. AB defined, 2×4; BA not defined   11. AB defined, 3×3; BA defined, 2×2

13. AB defined, 3×2; BA not defined

15. $\begin{pmatrix} 3 & 0 & 11 \\ 2 & 8 & 10 \end{pmatrix}$

17. $\begin{pmatrix} 1 & -3 \\ 2 & -1 \\ 5 & 6 \end{pmatrix}$

19. $\begin{pmatrix} 1 & -1 & 1 & 2 \\ 4 & 3 & 3 & 2 \\ -1 & -1 & -3 & 2 \\ 5 & 3 & 2 & 5 \end{pmatrix}$

21. $\begin{pmatrix} -2 & 0 \\ 1 & 1 \end{pmatrix}$

23. $\begin{pmatrix} 7 & 0 \\ -3 & -2 \end{pmatrix}$

25. $\begin{pmatrix} 0 & -8 & -2 \\ 4 & 16 & 18 \end{pmatrix}$

27. $\begin{pmatrix} 8 & -8 \\ 5 & 2 \\ -3 & 25 \end{pmatrix}$

29. $\begin{pmatrix} -6 & 11 \\ 10 & 3 \end{pmatrix}$

31. $\begin{pmatrix} 4 & -7/2 \\ 4 & 21/2 \end{pmatrix}$

33. $\begin{pmatrix} 5/2 & 1 \\ -6 & 11/2 \end{pmatrix}$

35. $\begin{pmatrix} -3 & -4 \\ 1 & 1 \end{pmatrix}$

37. $\begin{pmatrix} -2 & 1 \\ 3/2 & -1/2 \end{pmatrix}$

39. no inverse

41. no inverse

43. $\begin{pmatrix} -3 & 2 & -4 \\ -1 & 1 & -1 \\ 8 & -5 & 10 \end{pmatrix}$

45. $A + B = \begin{pmatrix} 1 & 2 \\ 3 & 0 \end{pmatrix} + \begin{pmatrix} -1 & 2 \\ 3 & 4 \end{pmatrix} = \begin{pmatrix} 0 & 4 \\ 6 & 4 \end{pmatrix}$ and $B + A = \begin{pmatrix} -1 & 2 \\ 3 & 4 \end{pmatrix} + \begin{pmatrix} 1 & 2 \\ 3 & 0 \end{pmatrix} = \begin{pmatrix} 0 & 4 \\ 6 & 4 \end{pmatrix}$

47. By Exercise 45, $c(A + B) = 2\begin{pmatrix} 0 & 4 \\ 6 & 4 \end{pmatrix} = \begin{pmatrix} 0 & 8 \\ 12 & 8 \end{pmatrix}$ and $cA + cB = 2\begin{pmatrix} 1 & 2 \\ 3 & 0 \end{pmatrix} + 2\begin{pmatrix} -1 & 2 \\ 3 & 4 \end{pmatrix} = \begin{pmatrix} 2 & 4 \\ 6 & 0 \end{pmatrix} + \begin{pmatrix} -2 & 4 \\ 6 & 8 \end{pmatrix} = \begin{pmatrix} 0 & 8 \\ 12 & 8 \end{pmatrix}$

49. $cdA = 2 \cdot 3 \begin{pmatrix} 1 & 2 \\ 3 & 0 \end{pmatrix} = 6\begin{pmatrix} 1 & 2 \\ 3 & 0 \end{pmatrix} = \begin{pmatrix} 6 & 12 \\ 18 & 0 \end{pmatrix}$ and $c[dA] = 2\left[3\begin{pmatrix} 1 & 2 \\ 3 & 0 \end{pmatrix}\right] = 2\begin{pmatrix} 3 & 6 \\ 9 & 0 \end{pmatrix} = \begin{pmatrix} 6 & 12 \\ 18 & 0 \end{pmatrix}$

51. $A(B + C) = \begin{pmatrix} 1 & 2 \\ 3 & 0 \end{pmatrix}\left[\begin{pmatrix} -1 & 2 \\ 3 & 4 \end{pmatrix} + \begin{pmatrix} 2 & 3 \\ 1 & 2 \end{pmatrix}\right] = \begin{pmatrix} 1 & 2 \\ 3 & 0 \end{pmatrix}\begin{pmatrix} 1 & 5 \\ 4 & 6 \end{pmatrix} = \begin{pmatrix} 9 & 17 \\ 3 & 15 \end{pmatrix}$ and
$AB + AC = \begin{pmatrix} 1 & 2 \\ 3 & 0 \end{pmatrix}\begin{pmatrix} -1 & 2 \\ 3 & 4 \end{pmatrix} + \begin{pmatrix} 1 & 2 \\ 3 & 0 \end{pmatrix}\begin{pmatrix} 2 & 3 \\ 1 & 2 \end{pmatrix} = \begin{pmatrix} 5 & 10 \\ -3 & 6 \end{pmatrix} + \begin{pmatrix} 4 & 7 \\ 6 & 9 \end{pmatrix} = \begin{pmatrix} 9 & 17 \\ 3 & 15 \end{pmatrix}$

53. $k = r$ and $n = t$

55. $AB = \begin{pmatrix} 0 & 0 \\ 0 & 0 \end{pmatrix} = AC$, but $B \neq C$

57. $(A + B)(A - B) = \begin{pmatrix} 5 & 0 \\ 7 & -1 \end{pmatrix}\begin{pmatrix} 1 & 2 \\ -3 & -7 \end{pmatrix} = \begin{pmatrix} 5 & 10 \\ 10 & 21 \end{pmatrix}$, but
$A^2 - B^2 = \begin{pmatrix} 11 & -1 \\ -2 & 18 \end{pmatrix} - \begin{pmatrix} -1 & -5 \\ 25 & 4 \end{pmatrix} = \begin{pmatrix} 12 & 4 \\ -27 & 14 \end{pmatrix}$

59. $A^t = \begin{pmatrix} a & c \\ b & d \end{pmatrix}$ and $B^t = \begin{pmatrix} r & u \\ s & v \end{pmatrix}$

61. $(A^t)^t = \begin{pmatrix} a & c \\ b & d \end{pmatrix}^t = \begin{pmatrix} a & b \\ c & d \end{pmatrix} = A$

63. AX is the 3×1 matrix

$$\begin{pmatrix} a_1 & b_1 & c_1 \\ a_2 & b_2 & c_2 \\ a_3 & b_3 & c_3 \end{pmatrix} \begin{pmatrix} x \\ y \\ z \end{pmatrix} = \begin{pmatrix} a_1x + b_1y + c_1z \\ a_2x + b_2y + c_2z \\ a_3x + b_3y + c_3z \end{pmatrix}$$

If $AX = D$, then

$$a_1x + b_1y + c_1z = d_1$$
$$a_2x + b_2y + c_2z = d_2$$
$$a_3x + b_3y + c_3z = d_3$$

65. $x = -1, y = 0, z = -3$ **67.** $x = -8, y = 16, z = 5$

Section 8.4, page 395

1. -29 **3.** $\dfrac{17}{2}$ **5.** $ru - dt$ **7.** 13 **9.** -13 **11.** -26 **13.** 16

15. $2x^4 + 4x - 3$ **17.** 3 **19.** -3

21. $\begin{vmatrix} 1 & x & x^2 \\ 1 & y & y^2 \\ 1 & z & z^2 \end{vmatrix} = 1 \begin{vmatrix} y & y^2 \\ z & z^2 \end{vmatrix} - x \begin{vmatrix} 1 & y^2 \\ 1 & z^2 \end{vmatrix} + x^2 \begin{vmatrix} 1 & y \\ 1 & z \end{vmatrix} =$

$(yz^2 - y^2z) - x(z^2 - y^2) + x^2(z - y) = yz^2 - y^2z - xz^2 + xy^2 + x^2z - x^2y$ and
$(x - y)(y - z)(z - x) = (x - y)(yz - z^2 - xy + xz) =$
$-xz^2 - x^2y + x^2z - y^2z + yz^2 + xy^2$

23. (a) $ad - bc - dx + cy + bx - ay$ (b) The equation is $y - d = \dfrac{d - b}{c - a}(x - c)$, which is equivalent to $(c - a)y - d(c - a) - (d - b)(x - c) = 0$. The left side is the same as in part (a).

25. $x = -1$ or 3 **27.** $x = 0$ or 3 **29.** $x = 1$ or 5 **31.** $x = -\dfrac{9}{31}, y = \dfrac{7}{31}$

33. $x = \dfrac{39}{41}, y = \dfrac{22}{41}$ **35.** $x = 4, y = 2$ **37.** $x = -\dfrac{23}{26}, y = -\dfrac{21}{26}, z = \dfrac{17}{26}$

39. $x = \dfrac{5}{3}, y = -3, z = -\dfrac{2}{3}$ **41.** $x = 1, y = 1, z = -2$ **43.** one solution

45. infinitely many solutions **47.** 9 **49.** $\tfrac{1}{2}$ **51.** collinear **53.** not collinear

55. $\begin{vmatrix} a + r & b \\ c + s & d \end{vmatrix} = (a + r)d - b(c + s) = ad + rd - bc - bs = (ad - bc) + (rd - bs) = \begin{vmatrix} a & b \\ c & d \end{vmatrix} + \begin{vmatrix} r & b \\ s & d \end{vmatrix}$

57. $\begin{vmatrix} a + kb & b \\ c + kd & d \end{vmatrix} = (a + kb)d - b(c + kd) = ad + kbd - bc - bkd = ad - bc = \begin{vmatrix} a & b \\ c & d \end{vmatrix}$

59. $AB = \begin{pmatrix} ar + bt & as + bu \\ cr + dt & cs + du \end{pmatrix}$; $\det AB = adru + bcst - bcru - adst$

Excursion 8.4.A, page 400

1. 13 **3.** −8 **5.** Property 2: The second and fourth columns are identical.

7. Property 5: Three times the second column in the right-hand determinant is equal to the second column in the left-hand determinant.

9. Property 4: The right-hand determinant is obtained by replacing row 3 of the left-hand determinant by the sum of itself and 3 times row 1.

11. 81 **13.** 32

Section 8.5, page 410

1.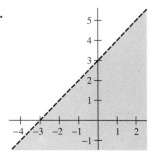

3.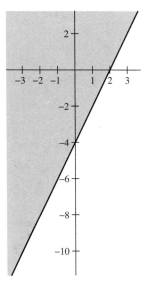

5.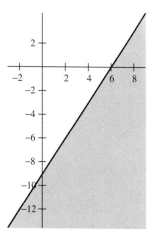

7.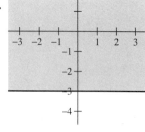

9.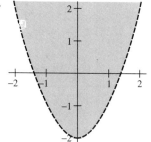

11.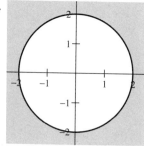

13.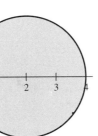

15.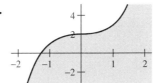

17.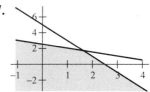

19.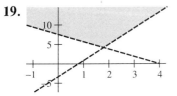

21.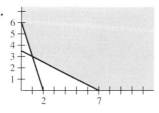

23.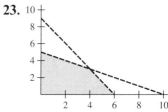

25.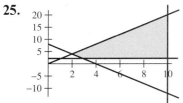

27.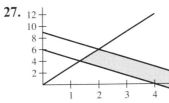

29.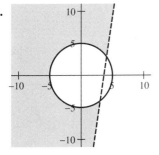

31.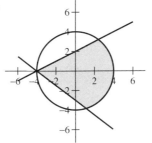

33.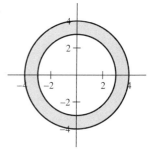

35.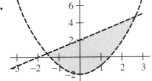

37.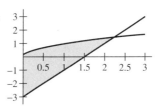

39. $x \geq 2, x \leq 7, y \geq -1, y \leq 3$ **41.** $x \leq 2, -2x + 3y \leq 8, 6y + x \geq -4$

43.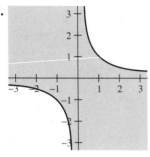

45. $2r + 3p \leq 12; 3r + 3.5p \leq 16$

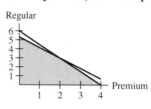

47. $4t + 2s \leq 20; 3t + 4s \leq 20$

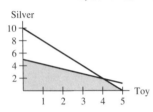

49. $2x + 3y \leq 12; 2x + y \geq 2; y \geq 1$

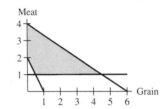

Section 8.6, page 416

1. Minimum is $F = 0$ at $(0, 0)$; maximum is $F = 15$ at $(0, 5)$.
3. Minimum is $F = 23$ at $(5, 4)$; no maximum.
5. Minimum is $F = 15$ at $(3, 6)$; maximum is $F = 28$ at $(8, 4)$.
7. Minimum is $F = 5$ at $(5, 0)$; maximum is $F = 27$ at $(3, 8)$.
9. 6 of the 4-person rafts and 18 of the 2-person rafts should be made each week for a maximum profit of $780.
11. 30 boxes of Minto-Mix (and none of Choco-Mix) should be made, which will generate the maximum revenue of $480.
13. 30 Supervites and 120 Vitahealth pills will meet the requirements at a minimum cost of $11.40.
15. 8 pounds of feed A and 3 pounds of feed B will meet the requirements at a minimum cost of $2.80.
17. Simply using 28 tons of oats (and no bran) will minimize the cost at $6300.
19. If she takes 5 humanities courses and 12 science courses, she will earn the maximum of 235 quality points.
21. Because the inequalities have no common solution (that is, there are no feasible solutions): Adding $2x + y \leq 10$ and $x + 2y \leq 8$ gives $3x + 3y \leq 18$, or $x + y \leq 6$, which contradicts $x + y \geq 8$.
23. Taking 3 humanities and 4 science courses gives an expected grade point average of 2.91. Corresponding figures for the other vertices are: $(3, 12)$: 2.96; $(5, 12)$: 2.94; $(15, 4)$: 2.81. Thus $(3, 12)$ earns the highest grade point average, whereas $(5, 12)$ earns the greatest number of quality points. There is no contradiction; the grade point average and quality point functions are different, though related, so there is no reason why they should have their maxima at the same vertex. Since the grade point average function is not linear, we cannot even be sure that the vertex $(3, 12)$ gives the maximum grade point average!

Chapter 8 Review, page 419

1. $x = -5, y = -7$ 3. $x = 0, y = -2$,   5. $x = -35, y = 140, z = 22$
7. $x = 2, y = 4, z = 6$ 9. $x = -t + 1, y = -2t, z = t$ for any real number t.
11. 37 and -19 13. (c)   15. (e)
17. $x = 4, y = 3$ or $x = -4, y = 3$ or $x = \sqrt{21}, y = -2$ or $x = -\sqrt{21}, y = -2$
19. 100   21. $\dfrac{A}{x} + \dfrac{B}{x+1} + \dfrac{C}{x-1} + \dfrac{Dx+E}{x^2+1} + \dfrac{Fx+G}{(x^2+1)^2}$ 23. $\dfrac{3}{x+2} + \dfrac{1}{x-3}$
25. $\dfrac{1}{x-2} + \dfrac{3}{(x-2)^2}$ 27. $\begin{pmatrix} 1 & -2 & 3 & | & 4 \\ 2 & 1 & -4 & | & 3 \\ -3 & 4 & -1 & | & -2 \end{pmatrix}$ 29. $\begin{pmatrix} 2 & -1 & -2 & 2 \\ 1 & 3 & -2 & 1 \\ -1 & 4 & 2 & -3 \end{pmatrix}$
31. $\begin{pmatrix} 4 & 5 \\ 24 & 1 \\ 18 & -7 \end{pmatrix}$ 33. $\begin{pmatrix} -5 & 1 \\ -6 & -5 \end{pmatrix}$ 35. DA not defined   37. FE not defined
39. $\begin{pmatrix} -9 & 7 \\ -4 & 3 \end{pmatrix}$ 41. $\begin{pmatrix} 1 & 2 & -2 \\ -1 & 3 & 0 \\ 0 & -2 & 1 \end{pmatrix}$ 43. 27   45. 24   47. (e)   49. 0
51. -54 53. $x = -25/29, y = 22/29$ 55. $x = 5/3, y = -3, z = -2/3$

57. 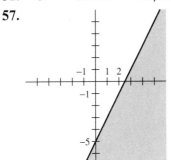   59. 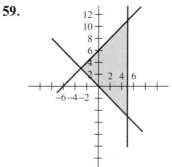   61.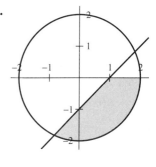

63. Minimum value of F is .72 at $x = 4, y = 8$; there is no maximum value of F.
65. 15 containers, 35 barrels; no unused space on the truck

CHAPTER 9

Section 9.1, page 430

1. 8, 10, 12, 14, 16   3. $1, \dfrac{1}{8}, \dfrac{1}{27}, \dfrac{1}{64}, \dfrac{1}{125}$ 5. $-\sqrt{3}, 2, -\sqrt{5}, \sqrt{6}, -\sqrt{7}$
7. 3.9, 4.01, 3.999, 4.0001, 3.99999   9. 2, 7, 8, 13, 14   11. $\displaystyle\sum_{i=1}^{11} i$
13. $\displaystyle\sum_{i=7}^{13} \dfrac{1}{2^i}$ (other answers possible)   15. 45   17. 24   19. 54   21. $a_n = (-1)^n$

23. $a_n = \dfrac{n}{n+1}$ 25. $a_n = 5n - 3$ 27. 4, 11, 25, 53, 109   29. 1, −2, 3, 2, 3

31. 2, 3, 3, $\dfrac{9}{2}$, $\dfrac{27}{4}$ 33. 3, 1, 4, 1, 5   35. third −10; sixth −2   37. third 5; sixth 0

39. $\sum_{n=1}^{6} \dfrac{1}{2n+1}$ (other answers possible)   41. $\sum_{n=8}^{12} (-1)^n \left(\dfrac{n-7}{n}\right)$ (other answers possible)

43. 2.613035   45. $\dfrac{1051314}{31415} \approx 33.465$ 47. 1.5759958   49. (b) 59, 61, 67, 71

51. 4, 9, 25, 49, 121   53. 3, 7, 13, 19, 23
55. (a) 1, 1, 2, 3, 5, 8, 13, 21, 34, 55   (b) 1, 2, 4, 7, 12, 20, 33, 54, 88, 143
(c) nth partial sum $= a_{n+2} - 1$
57. $n = 1$: $5(1)^2 + 4(-1)^1 = 1 = 1^2$; $n = 2$: $5(1)^2 + 4(-1)^2 = 9 = 3^2$; $n = 3$: $5(2)^2 + 4(-1)^3 = 16 = 4^2$; $n = 4$: $5(3)^2 + 4(-1)^4 = 49 = 7^2$; etc.
59. We have $a_1 = a_3 - a_2$; $a_2 = a_4 - a_3$; $a_3 = a_5 - a_4$; ... $a_{k-1} = a_{k+1} - a_k$; $a_k = a_{k+2} - a_{k+1}$. If these equations are listed vertically, the sum of the left-side terms is $\sum_{n=1}^{k} a_k$. On the right side, one term in each line is the same as a term in the next line, except for sign. So the sum of the right-side terms is $a_{k+2} - a_2$. Since $a_2 = 1$ we conclude that $\sum_{n=1}^{k} a_n = a_{k+2} - 1$.

Section 9.2, page 436

1. 13; $a_n = 2n + 3$ 3. 5; $a_n = n/4 + 15/4$ 5. 8; $-n/2 + 21/2$ 7. 87
9. −21/4   11. 30
13. $a_n - a_{n-1} = (3 - 2n) - (3 - 2(n-1)) = -2$; arithmetic with $d = -2$
15. $a_n - a_{n-1} = \dfrac{5 + 3n}{2} - \dfrac{5 + 3(n-1)}{2} = \dfrac{3}{2}$; arithmetic with $d = \dfrac{3}{2}$
17. $a_n - a_{n-1} = (c + 2n) - (c + 2(n-1)) = 2$; arithmetic with $d = 2$
19. $a_5 = 14$; $a_n = 2n + 4$ 21. $a_5 = 25$; $a_n = 7n - 10$
23. $a_5 = 0$; $a_n = -15/2 + 3n/2$ 25. 710   27. $156\tfrac{2}{3}$ 29. 2550   31. 20,100
33. 4, $5\tfrac{2}{3}$, $7\tfrac{1}{3}$, 9, $10\tfrac{2}{3}$, $12\tfrac{1}{3}$, 14; $d = 5/3$ 35. \$77,500 in tenth year; \$437,500 over ten years
37. 428   39. 18.75, 19.5, 20.25, 21, 21.75, 22.5, 23.25

Section 9.3, page 441

1. arithmetic   3. geometric   5. arithmetic   7. geometric
9. $a_6 = 160$; $a_n = 2^{n-1} \cdot 5$ 11. $a_6 = \dfrac{1}{256}$; $a_n = \dfrac{1}{4^{n-2}}$
13. $a_6 = -\dfrac{5}{16}$; $a_n = \dfrac{(-1)^{n-1} \cdot 5}{2^{n-2}}$ 15. 315/32   17. 381

19. $\dfrac{a_n}{a_{n-1}} = \dfrac{\left(-\frac{1}{2}\right)^n}{\left(-\frac{1}{2}\right)^{n-1}} = -\dfrac{1}{2}$; geometric with $r = -\dfrac{1}{2}$

21. $\dfrac{a_n}{a_{n-1}} = \dfrac{5^{n+2}}{5^{(n-1)+2}} = 5$; geometric with $r = 5$ **23.** $a_5 = 1$; $a_n = \dfrac{(-1)^{n-1}64}{4^{n-2}} = \dfrac{(-1)^{n-1}}{4^{n-5}}$

25. $a_5 = \dfrac{1}{16}$; $a_n = \dfrac{1}{4^{n-3}}$ **27.** $a_5 = -\dfrac{8}{25}$; $a_n = -\dfrac{2^{n-2}}{5^{n-3}}$ **29.** 254   **31.** $-\dfrac{4921}{19683}$

33. $\dfrac{665}{8}$ **35.** 16, 64   **37.** $25\sqrt[4]{2}, 25\sqrt{2}, 25(\sqrt[4]{2})^3$

39. $\log a_n - \log a_{n-1} = \log \dfrac{a_n}{a_{n-1}} = \log r$ **41.** 23.75 ft

43. The sequence is $\{2^{n-1}\}$ and $r = 2$. So for any k, the kth term is 2^{k-1}, and the sum of the preceding terms is the $(k-1)$-partial sum of the sequence, $\sum_{n=1}^{k-1} 2^{n-1} = \dfrac{1 - 2^{k-1}}{1 - 2} = 2^{k-1} - 1$.

45. $\sum_{n=1}^{31} 2^{n-1} = \dfrac{1 - 2^{31}}{1 - 2} = (2^{31} - 1)$ cents $= \$21{,}474{,}836.47$ **47.** $\$1898.44$

49. 92 payments

Excursion 9.3.A, page 446

1. 1   **3.** $.06/.94 = 3/47$ **5.** $\dfrac{500}{.6} = 833\dfrac{1}{3}$ **7.** $4 + 2\sqrt{2}$

9. $2/9$ **11.** $597/110$ **13.** $10{,}702/4995$ **15.** $174{,}067/99{,}900$

Section 9.4, page 452

1. 720   **3.** 220   **5.** 0   **7.** 64   **9.** 3,921,225
11. $x^5 + 5x^4y + 10x^3y^2 + 10x^2y^3 + 5xy^4 + y^5$
13. $a^5 - 5a^4b + 10a^3b^2 - 10a^2b^3 + 5ab^4 - b^5$
15. $32x^5 + 80x^4y^2 + 80x^3y^4 + 40x^2y^6 + 10xy^8 + y^{10}$
17. $x^3 + 6x^2\sqrt{x} + 15x^2 + 20x\sqrt{x} + 15x + 6\sqrt{x} + 1$
19. $1 - 10c + 45c^2 - 120c^3 + 210c^4 - 252c^5 + 210c^6 - 120c^7 + 45c^8 - 10c^9 + c^{10}$
21. $x^{-12} + 4x^{-8} + 6x^{-4} + 4 + x^4$ **23.** 56   **25.** $-8i$ **27.** $10x^3y^2$
29. $35c^3d^4$ **31.** $\dfrac{35}{8}u^{-5}$ **33.** 4032   **35.** 160

37. (a) $\binom{9}{1} = \dfrac{9!}{1!8!} = 9$; $\binom{9}{8} = \dfrac{9!}{8!1!} = 9$ (b) $\binom{n}{1} = \dfrac{n!}{1!(n-1)!} = \dfrac{n(n-1)!}{(n-1)!} = n$

39. $2^n = (1+1)^n = 1^n + \binom{n}{1} 1^{n-1} \cdot 1 + \binom{n}{2} 1^{n-2} \cdot 1^2 + \binom{n}{3} 1^{n-3} \cdot 1^3 + \cdots +$
$\binom{n}{n-1} 1^1 \cdot 1^{n-1} + 1^n = \binom{n}{0} + \binom{n}{1} + \binom{n}{2} + \binom{n}{3} + \cdots + \binom{n}{n-1} + \binom{n}{n}$

41. (a) $f(x+h) - f(x) = (x+h)^5 - x^5 =$
$\left(x^5 + \binom{5}{1} x^4 h + \binom{5}{2} x^3 h^2 + \binom{5}{3} x^2 h^3 + \binom{5}{4} xh^4 + h^5 \right) - x^5 =$
$\binom{5}{1} x^4 h + \binom{5}{2} x^3 h^2 + \binom{5}{3} x^2 h^3 + \binom{5}{4} xh^4 + h^5$

(b) $\dfrac{f(x+h) - f(x)}{h} = \binom{5}{1} x^4 + \binom{5}{2} x^3 h + \binom{5}{3} x^2 h^2 + \binom{5}{4} xh^3 + h^4$ (c) When h is very close to 0, so are the last four terms in part (b), so $\dfrac{f(x+h) - f(x)}{h} \approx \binom{5}{1} x^4 = 5x^4$.

43. $\dfrac{f(x+h) - f(x)}{h} = \dfrac{(x+h)^{12} - x^{12}}{h} = \binom{12}{1} x^{11} + \binom{12}{2} x^{10} h + \binom{12}{3} x^9 h^2 +$
$\binom{12}{4} x^8 h^3 + \cdots + \binom{12}{10} x^2 h^9 + \binom{12}{11} xh^{10} + h^{11} \approx \binom{12}{1} x^{11} = 12 x^{11}$, when h is very close to 0.

45. (a) $(n-r)! = (n-r)(n-r-1)(n-r-2)(n-r-3) \cdots 2 \cdot 1 =$
$(n-r)(n-(r+1))(n-(r+1)-1)(n-(r+1)-2) \cdots 2 \cdot 1 =$
$(n-r)(n-(r+1))!$ (b) Since $(n+1) - (r+1) = n - r$,
$((n+1) - (r+1))! = (n-r)!$

(c) $\binom{n}{r+1} + \binom{n}{r} = \dfrac{n!}{(r+1)!(n-(r+1))!} + \dfrac{n!}{r!(n-r)!} = \dfrac{n!(n-r) + n!(r+1)}{(r+1)!(n-r)!} =$
$\dfrac{n!(n+1)}{(r+1)!(n-r)!} = \dfrac{(n+1)!}{(r+1)!((n+1)-(r+1))!} = \binom{n+1}{r+1}$ (d) For example, rows 2 and 3 of Pascal's triangle are:

$\qquad$ 1 $\quad$ 2 $\quad$ 1 $\qquad\qquad \binom{2}{0} \binom{2}{1} \binom{2}{2}$

that is,

$\qquad$ 1 $\quad$ ③ $\quad$ 3 $\quad$ 1 $\qquad \binom{3}{0} \binom{3}{1} \binom{3}{2} \binom{3}{3}$

The circled 3 is the sum of the two closest entries in the row above: $1 + 2$. But this just says that $\binom{3}{1} = \binom{2}{0} + \binom{2}{1}$, which is part (c) with $n = 2$ and $r = 0$. Similarly, in the general case, verify that the two closest entries in the row above $\binom{n+1}{r+1}$ are $\binom{n}{r}$ and $\binom{n}{r+1}$ and use part (c).

Section 9.5, page 461

1. *Step 1:* For $n = 1$ the statement is $1 = 2^1 - 1$, which is true. *Step 2:* Assume the statement is true for $n = k$: that is, $1 + 2 + 2^2 + 2^3 + \cdots + 2^{k-1} = 2^k - 1$. Add 2^k to both sides, and rearrange terms:

$$1 + 2 + 2^2 + 2^3 + \cdots + 2^{k-1} + 2^k = 2^k - 1 + 2^k$$
$$1 + 2 + 2^2 + 2^3 + \cdots + 2^{k-1} + 2^{(k+1)-1} = 2^k + 2^k - 1 = 2(2^k) - 1$$
$$1 + 2 + 2^2 + 2^3 + \cdots + 2^{k-1} + 2^{(k+1)-1} = 2^{k+1} - 1$$

But this last line says that the statement is true for $n = k + 1$. Therefore by the Principle of Mathematical Induction the statement is true for every positive integer n.

[*Note:* Hereafter, in these answers, step 1 will be omitted if it is trivial (as in Exercise 1), and only the essential parts of step 2 will be given.]

3. Assume that the statement is true for $n = k$: $1 + 3 + 5 + \cdots + 2(k - 1) = k^2$. Add $2(k + 1) - 1$ to both sides: $1 + 3 + 5 + \cdots + (2k - 1) + (2(k + 1) - 1) = k^2 + 2(k + 1) - 1 = k^2 + 2k + 1 = (k + 1)^2$. The first and last parts of this equation say that the statement is true for $n = k + 1$.

5. Assume that the statement is true for $n = k$:

$$1^2 + 2^2 + 3^2 + \cdots + k^2 = \frac{k(k + 1)(2k + 1)}{6}$$

Add $(k + 1)^2$ to both sides: $1^2 + 2^2 + 3^2 + \cdots + k^2 + (k + 1)^2$

$$= \frac{k(k + 1)(2k + 1)}{6} + (k + 1)^2$$

$$= \frac{k(k + 1)(2k + 1) + 6(k + 1)^2}{6} = \frac{(k + 1)[k(2k + 1) + 6(k + 1)]}{6}$$

$$= \frac{(k + 1)[2k^2 + 7k + 6]}{6} = \frac{(k + 1)(k + 2)(2k + 3)}{6}$$

$$= \frac{(k + 1)[(k + 1) + 1][2(k + 1) + 1]}{6}$$

The first and last parts of this equation say that the statement is true for $n = k + 1$.

7. Assume the statement is true for $n = k$:

$$\frac{1}{1 \cdot 2} + \frac{1}{2 \cdot 3} + \cdots + \frac{1}{k(k + 1)} = \frac{k}{k + 1}$$

Adding $\frac{1}{(k + 1)((k + 1) + 1)} = \frac{1}{(k + 1)(k + 2)}$ to both sides yields:

$$\frac{1}{1\cdot 2} + \frac{1}{2\cdot 3} + \cdots + \frac{1}{k(k+1)} + \frac{1}{(k+1)(k+2)} = \frac{k}{k+1} + \frac{1}{(k+1)(k+2)}$$

$$= \frac{k(k+2)+1}{(k+1)(k+2)} = \frac{k^2+2k+1}{(k+1)(k+2)} = \frac{(k+1)^2}{(k+1)(k+2)} = \frac{k+1}{k+2}$$

$$= \frac{k+1}{(k+1)+1}$$

The first and last parts of this equation show that the statement is true for $n = k+1$.

9. Assume the statement is true for $n = k$: $k + 2 > k$. Adding 1 to both sides, we have: $k + 2 + 1 > k + 1$, or equivalently, $(k+1) + 2 > (k+1)$. Therefore the statement is true for $n = k+1$.

11. Assume the statement is true for $n = k$: $3^k \geq 3k$. Multiplying both sides by 3 yields: $3 \cdot 3^k \geq 3 \cdot 3k$, or equivalently, $3^{k+1} \geq 3 \cdot 3k$. Now since $k \geq 1$, we know that $3k \geq 3$ and hence that $2 \cdot 3k \geq 3$. Therefore $2 \cdot 3k + 3k \geq 3 + 3k$, or equivalently, $3 \cdot 3k \geq 3k + 3$. Combining this last inequality with the fact that $3^{k+1} \geq 3 \cdot 3k$, we see that $3^{k+1} \geq 3k + 3$, or equivalently, $3^{k+1} \geq 3(k+1)$. Therefore the statement is true for $n = k+1$.

13. Assume the statement is true for $n = k$: $3k > k + 1$. Adding 3 to both sides yields: $3k + 3 > k + 1 + 3$, or equivalently, $3(k+1) > (k+1) + 3$. Since $(k+1) + 3$ is certainly greater than $(k+1) + 1$, we conclude that $3(k+1) > (k+1) + 1$. Therefore the statement is true for $n = k+1$.

15. Assume the statement is true for $n = k$; then 3 is a factor of $2^{2k+1} + 1$; that is, $2^{2k+1} + 1 = 3M$ for some integer M. Thus $2^{2k+1} = 3M - 1$. Now $2^{2(k+1)+1} = 2^{2k+2+1} = 2^{2+2k+1} = 2^2 \cdot 2^{2k+1} = 4(3M - 1) = 12M - 4 = 3(4M) - 3 - 1 = 3(4M - 1) - 1$. From the first and last terms of this equation we see that $2^{2(k+1)+1} + 1 = 3(4M - 1)$. Hence 3 is a factor of $2^{2(k+1)+1} + 1$. Therefore the statement is true for $n = k+1$.

17. Assume the statement is true for $n = k$: 64 is a factor of $3^{2k+2} - 8k - 9$. Then $3^{2k+2} - 8k - 9 = 64N$ for some integer N so that $3^{2k+2} = 8k + 9 + 64N$. Now $3^{2(k+1)+2} = 3^{2k+2+2} = 3^{2+(2k+2)} = 3^2 \cdot 3^{2k+2} = 9(8k + 9 + 64N)$. Consequently, $3^{2(k+1)+2} - 8(k+1) - 9 = 3^{2(k+1)+2} - 8k - 8 - 9 = 3^{2(k+1)+2} - 8k - 17 = [9(8k + 9 + 64N)] - 8k - 17 = 72k + 81 + 9 \cdot 64N - 8k - 17 = 64k + 64 + 9 \cdot 64N = 64(k + 1 + 9N)$. From the first and last parts of this equation we see that 64 is a factor of $3^{2(k+1)+2} - 8(k+1) - 9$. Therefore the statement is true for $n = k+1$.

19. Assume that the statement is true for $n = k$: $c + (c + d) + (c + 2d) + \cdots + (c + (k-1)d) = \frac{k(2c + (k-1)d)}{2}$. Adding $c + kd$ to both sides, we have

$$c + (c+d) + (c+2d) + \cdots + (c + (k-1)d) + (c + kd)$$
$$= \frac{k(2c + (k-1)d)}{2} + c + kd$$
$$= \frac{k(2c + (k-1)d) + 2(c + kd)}{2} = \frac{2ck + k(k-1)d + 2c + 2kd}{2}$$
$$= \frac{2ck + 2c + kd(k-1) + 2kd}{2} = \frac{(k+1)2c + kd(k-1+2)}{2}$$

$$= \frac{(k+1)2c + kd(k+1)}{2} = \frac{(k+1)(2c+kd)}{2}$$

$$= \frac{(k+1)(2c + ((k+1)-1)d)}{2}$$

Therefore the statement is true for $n = k+1$.

21. (a) $x^2 - y^2 = (x-y)(x+y)$; $x^3 - y^3 = (x-y)(x^2 + xy + y^2)$; $x^4 - y^4 = (x-y)(x^3 + x^2y + xy^2 + y^3)$. (b) *Conjecture:* $x^n - y^n = (x-y)(x^{n-1} + x^{n-2}y + x^{n-3}y^2 + \cdots + x^2y^{n-3} + xy^{n-2} + y^{n-1})$. *Proof:* The statement is true for $n = 2, 3, 4$, by part (a). Assume the statement is true for $n = k$: $x^k - y^k = (x-y)(x^{k-1} + x^{k-2}y + x^{k-3}y^2 + \cdots + xy^{k-2} + y^{k-1})$. Now use the fact that $-yx^k + yx^k = 0$ to write $x^{k+1} - y^{k+1}$ as follows:

$$\begin{aligned} x^{k+1} - y^{k+1} &= x^{k+1} - yx^k + yx^k - y^{k+1} \\ &= (x^{k+1} - yx^k) + (yx^k - y^{k+1}) \\ &= (x-y)x^k + y(x^k - y^k) \\ &= (x-y)x^k + y(x-y)(x^{k-1} + x^{k-2}y + x^{k-3}y^2 + \cdots + xy^{k-2} + y^{k-1}) \\ &= (x-y)x^k + (x-y)(x^{k-1}y + x^{k-2}y^2 + x^{k-3}y^3 + \cdots + xy^{k-1} + y^k) \\ &= (x-y)[x^k + x^{k-1}y + x^{k-2}y^2 + x^{k-3}y^3 + \cdots + xy^{k-1} + y^k] \end{aligned}$$

The first and last parts of this equation show that the conjecture is true for $n = k+1$. Therefore by mathematical induction, the conjecture is true for every integer $n \geq 2$.

23. False; counterexample: $n = 9$.

25. True: *Proof:* Since $(1+1)^2 > 1^2 + 1$, the statement is true for $n = 1$. Assume the statement is true for $n = k$: $(k+1)^2 > k^2 + 1$. Then $[(k+1)+1]^2 = (k+1)^2 + 2(k+1) + 1 > k^2 + 1 + 2(k+1) + 1 = k^2 + 2k + 2 + 2 > k^2 + 2k + 2 = k^2 + 2k + 1 + 1 = (k+1)^2 + 1$. The first and last terms of this inequality say that the statement is true for $n = k+1$. Therefore by induction the statement is true for every positive integer n.

27. False; counterexample: $n = 3$.

29. Since $2 \cdot 5 - 4 > 5$, the statement is true for $n = 5$. Assume the statement is true for $n = k$ (with $k \geq 5$): $2k - 4 > k$. Adding 2 to both sides shows that $2k - 4 + 2 > k + 2$, or equivalently, $2(k+1) - 4 > k + 2$. Since $k + 2 > k + 1$, we see that $2(k+1) - 4 > k + 1$. So the statement is true for $n = k + 1$. Therefore by the Extended Principle of Mathematical Induction, the statement is true for all $n \geq 5$.

31. Since $2^2 > 2$, the statement is true for $n = 2$. Assume that $k \geq 2$ and that the statement is true for $n = k$: $k^2 > k$. Then $(k+1)^2 = k^2 + 2k + 1 > k^2 + 1 > k + 1$. The first and last terms of this inequality show that the statement is true for $n = k + 1$. Therefore by induction, the statement is true for all $n \geq 2$.

33. Since $3^4 = 81$ and $2^4 + 10 \cdot 4 = 16 + 40 = 56$, we see that $3^4 > 2^4 + 10 \cdot 4$. So the statement is true for $n = 4$. Assume that $k \geq 4$ and that the statement is true for $n = k$: $3^k > 2^k + 10k$. Multiplying both sides by 3 yields: $3 \cdot 3^k > 3(2^k + 10k)$, or equivalently, $3^{k+1} > 3 \cdot 2^k + 30k$. But $3 \cdot 2^k + 30k > 2 \cdot 2^k + 30k = 2^{k+1} + 30k$. Therefore $3^{k+1} > 2^{k+1} + 30k$. Now we shall show that $30k > 10(k+1)$. Since $k \geq 4$, we have $20k \geq 20 \cdot 4$, so that $20k > 80 > 10$. Adding $10k$ to both sides of $20k > 10$ yields: $30k > 10k + 10$, or equivalently, $30k >$

$10(k + 1)$. Consequently, $3^{k+1} > 2^{k+1} + 30k > 2^{k+1} + 10(k + 1)$. The first and last terms of this inequality show that the statement is true for $n = k + 1$. Therefore the statement is true for all $n \geq 4$ by induction.

35. (a) 3 (that is, $2^2 - 1$) for $n = 2$; 7 (that is, $2^3 - 1$) for $n = 3$; 15 (that is, $2^4 - 1$) for $n = 4$. (b) *Conjecture:* the smallest possible number of moves for n rings is $2^n - 1$. *Proof:* This conjecture is easily seen to be true for $n = 1$ or $n = 2$. Assume it is true for $n = k$ and that we have $k + 1$ rings to move. In order to move the *bottom* ring from the first peg to another peg (say, the second one), it is first necessary to move the top k rings off the first peg *and* leave the second peg vacant at the end (the second peg will have to be used *during* this moving process). If this is to be done according to the rules, we will end up with the top k rings on the third peg in the *same* order they were on the first peg. According to the induction assumption, the least possible number of moves needed to do this is $2^k - 1$. It now takes one move to transfer the bottom ring (the $(k + 1)$st) from the first to the second peg. Finally, the top k rings now on the third peg must be moved to the second peg. Once again by the induction hypothesis, the least number of moves for doing this is $2^k - 1$. Therefore, the smallest total number of moves needed to transfer all $k + 1$ rings from the first to the second peg is $(2^k - 1) + 1 + (2^k - 1) = (2^k + 2^k) - 1 = 2 \cdot 2^k - 1 = 2^{k+1} - 1$. Hence, the conjecture is true for $n = k + 1$. Therefore, by induction it is true for all positive integers n.

37. *De Moivre's Theorem:* For any complex number $z = r(\cos \theta + i \sin \theta)$ and any positive integer n, $z^n = r^n(\cos(n\theta) + i \sin(n\theta))$. *Proof:* The theorem is obviously true when $n = 1$. Assume that the theorem is true for $n = k$, that is, $z^k = r^k(\cos(k\theta) + i \sin(k\theta))$. Then $z^{k+1} = z \cdot z^k = [r(\cos \theta + i \sin \theta)][r^k(\cos(k\theta) + i \sin(k\theta))]$. According to the multiplication rule for complex numbers in polar form (multiply the moduli and add the arguments) we have: $z^{k+1} = r \cdot r^k(\cos(\theta + k\theta) + i \sin(\theta + k\theta)) = r^{k+1}[\cos((k + 1)\theta) + i \sin((k + 1)\theta)]$. This statement says the theorem is true for $n = k + 1$. Therefore, by induction, the theorem is true for every positive integer n.

Section 9.6, page 469

1. 24 3. 40,320 5. 132 7. 91 9. 3,764,376 11. $n(n - 1) = n^2 - n$
13. n 15. $(n^2 - n)/2$ 17. $n = 11$ 19. $n = 9$ 21. $n = 7$ 23. 40,320
25. 24 27. 3024 29. 1,860,480 31. 7,290,000 33. 192 35. 1,048,576
37. 43,545,600 39. 28 41. 20,358,520 43. 56 45. 11,550
47. (a) 630 (b) 840 49. (a) 2 (b) 6 (c) 24 (d) $(n - 1)!$
51. 479,001,600 53. 76,204,800

Excursion 9.6.A, page 472

1. 12 3. 50,400 5. 1,396,755,360

Section 9.7, page 480

1. {HHH, HHT, HTH, HTT, THH, THT, TTH, TTT} 3. {r, b}
5. (a) 1/2 (b) 1/2 (c) 5/6 (d) 2/3 7. 5/11 9. 2/11 11. 13/18

13. 1/12   **15.** 1/5   **17.** 4/5   **19.** 7/15   **21.** 1/13   **23.** 1/26   **25.** 3/26
27. 5/13   **29.** 1/54,145   **31.** 1/108,290
33. The probability of a hand with no aces is $35673/54145 \approx .66$; so the probability of a hand with at least one ace (the complement of an aceless hand) is approximately $1 - .66 = .34$.
35. (a) 1/1024   (b) $252/1024 \approx .246$ (c) $176/1024 \approx .172$ **37.** approximately .29
39. 1/7,059,052   **41.** .325   **43.** .5375

Chapter 9 Review, page 483

1. $-3, -1, 1, 3$ **3.** $1, \dfrac{1}{4}, \dfrac{1}{9}, \dfrac{1}{16}$ **5.** -368 **7.** 129   **9.** $a_n = 9 - 6n$
11. $a_n = 6n - 11$ **13.** $a_n = 2 \cdot 3^{n-1}$ **15.** $a_n = \dfrac{3}{2^{n-8}}$ **17.** -55 **19.** $\dfrac{121}{4}$
21. 8.75, 13.5, 18.25   **23.** Second method is better.   **25.** 2
27. $(1.02)^{51} = (1 + .02)^{51} = 1^{51} + \dbinom{51}{1} 1^{50}(.02) + \dbinom{51}{2} 1^{49}(.02)^2 +$ other positive terms $= 2.53 +$ other positive terms > 2.5.
29. 455   **31.** $n + 1$ **33.** 1140   **35.** 80
37. True for $n = 1$. If the statement is true for $n = k$, then $1 + 5 + \cdots + 5^{k-1} = \dfrac{5^k - 1}{4}$ so that

$$1 + 5 + \cdots + 5^{k-1} + 5^k = \dfrac{5^k - 1}{4} + 5^k = \dfrac{5^k - 1 + 4 \cdot 5^k}{4} = \dfrac{5 \cdot 5^k - 1}{4} = \dfrac{5^{k+1} - 1}{4}.$$

Hence the statement is true for $n = k + 1$ and therefore true for all n by induction.
39. Since the statement is obviously true for $x = 0$, assume $x \neq 0$. Then the statement is true for $n = 1$. If the statement is true for $n = k$, then $|x^k| < 1$. Then $|x^k| \cdot |x| < |x|$. Thus $|x^{k+1}| = |x^k| \cdot |x| < |x| < 1$. Hence the statement is true for $n = k + 1$ and therefore true for all n by induction.
41. True for $n = 1$. If the statement is true for $n = k$, then $1 + 4 + \cdots + 4^{k-1} = \dfrac{1}{3}(4^k - 1)$.
Hence $1 + 4 + \cdots + 4^{k-1} + 4^k = \dfrac{1}{3}(4^k - 1) + 4^k = \dfrac{1}{3}(4^k - 1) + \dfrac{3 \cdot 4^k}{3} = \dfrac{4^k - 1 + 3 \cdot 4^k}{3} = \dfrac{4 \cdot 4^k - 1}{3} = \dfrac{1}{3}(4^{k+1} - 1)$. Hence the statement is true for $n = k + 1$ and therefore for all n by induction.
43. If $n = 1$, then $9^n - 8n - 1 = 0$. Since $0 = 0 \cdot 8$, the statement is true for $n = 1$. If the statement is true for $n = k$, then $9^k - 8k - 1 = 8D$, so that $9^k - 1 = 8k + 8D = 8(k + D)$. Consequently, $9^{k+1} - 8(k + 1) - 1 = 9^{k+1} - 8k - 8 - 1 = 9^{k+1} - 9 - 8k = 9(9^k - 1) - 8k = 9[8(k + D)] - 8k = 8[9(k + D) - k]$. Thus 8 is a factor of $9^{k+1} - 8(k + 1) - 1$ and the statement is true for $n = k + 1$. Therefore it is true for all n induction.
45. 96,909,120   **47.** 10,626   **49.** 200   **51.** 24   **53.** 280
55. 5/36   **57.** 1/4   **59.** approximately .38

Index

absolute value, 25, 26
   equations 28, 98
   functions 152
   inequalities 29, 105, 106, 114
addition
   of algebraic expressions 33
   of fractions 51
   matrix 378
algebra, fundamental theorem 247
algebraic
   expression 33
   tests for symmetry 179
algorithm, division 39, 247
analytic geometry 121
antilogarithm 495
applications
   of first-degree equations 68
   of quadratic equations 82
applied problems, guidelines 68
approximation techniques 257
area of ellipse 321
arithmetic
   means 435
   sequence 431
associative laws 3
asymptote 236, 239, 240, 244, 323
augmented matrix 362
average 435
axis 121
   major—of ellipse 313
   minor—of ellipse 313
   of parabola 216, 333
   transverse 323

back substitution 358
bad point 235
base
   change of—formula 294
   for logarithms 290

Big-Little Principle 235
binomial
   coefficient 447, 467
   expansion 451
   theorem 36, 449, 459
bisection method 257
bounds 260

calculator 271
cancellation property 50
Cartesian coordinate system 121
center
   of circle 127
   of ellipse 313
   of hyperbola 323
change of base 294
characteristic 495
Charlie 87
circle 127, 311
   unit 128
closed interval 30
coefficient
   binomial 447, 467
   leading 37
   matrix 374, 392
   of polynomial 37
cofactor 397
column 377
   expanding determinant along 390
combination 467
common
   denominator 50
   difference 432
   logarithm 291
   ratio 437
commutative laws 3
complement of event 478
completing the square 77

I-1

complex
   conjugate 89
   numbers 87
   construction of the—numbers 93
   factorization of polynomials over the—numbers 249
composite function 192
   domain 193
composition of functions 191
compound interest 273, 278
conic
   degenerate 311
   section 311
conjugate 89
   roots theorem 251
   solutions 91
constant 33
   polynomial 37
   sequence 424
   term 37
   of variation 203, 204
constraints 411, 413
continuity 226
contraction of graph 183
convention, domain 157
convergent series 444
coordinate(s)
   axes 121
   plane 121
   of point 121
   tests for symmetry 179
counting principle 464
Cramer's rule 392, 393
cube(s)
   difference of 46
   perfect 46
   root 20, 91
   sum of 46
cubic
   factoring patterns 46
   polynomial 37

decay
   exponential 273
   radioactive 274
decimal(s) 1, 9
   rounding 11
decreasing
   function 188, 201
   on interval 188

degenerate conic 311
degree of polynomial 37
denominator 50
   rationalizing the 54
dependent system 347, 349, 363
Descarte's Rule of Signs 255
determinant 388, 389, 398
   expanding along row or column 390
   higher-order 397
   properties 399
diagonal, main 383
difference
   common 432
   of cubes 46
   of functions 191
   of squares 46
direct variation 203
directrix 333
discriminant 79
distance formula 123
distance on number line 28
distinguishable permutation 472
distribution, probability 479
distributive laws 3
dividend 37
division
   algorithm 39, 247
   of fractions 52
   of polynomials 37
   synthetic 42
   by zero 8
divisor 37
domain 151
   convention 157

e(irrational number) 274, 279
eccentricity
   of ellipse 318
   of hyperbola 328
echelon form 358
   matrix 373
elimination
   Gaussian 358
   method 344, 359
ellipse 311, 313
   area 321
   eccentricity 318
   equation 314–316

endpoints of interval 30
equality rule for fractions 50
equation(s) 62
  absolute value 98
  of circle 127
  of ellipse 314–316
  exponential 298
  first-degree 62, 68, 142
  graph of 127
  of hyperbola 324, 327
  of line 139
  linear 343, 357, 392
  logarithmic 302
  of parabola 335, 336, 338
  point-slope form 139
  principles for solving 62
  quadratic 76, 82
  quadratic type 94
  radical 95
  slope-intercept 141
  solution of 126
  system of 343, 352, 357, 392
equivalent
  equations 62
  inequalities 103
even function 179
event(s) 670
  complement of 478
  mutually exclusive 477
expansion
  binomial 451
  of determinant 390, 398
  of graph 182, 183
experiment 473
exponent(s)
  decimal 22
  integer 13
  irrational 271
  laws 14, 23, 272
  negative 14
  rational 22
exponential
  bank 279
  decay 273
  equation 298
  function 272
  graph 274
  growth 273
extended principle of mathematical
    induction 462

Factor Theorem 101, 248
factorial 447
factoring 45
  method for quadratic equations 77
factorization
  over the complex numbers 249
  over the real numbers 252
feasible solution 411, 413
Fibonacci sequence 426, 431
figures, significant 12
first-degree equations 62, 142
  applications 68
foci
  of ellipse 313
  of hyperbola 323
focus of parabola 333
FOIL 35
formula
  compound interest 279
  distance 123
  midpoint 125
  quadratic 79
fraction(s) 50
  partial 366
fractional expressions 49
function(s)
  absolute value 152
  composite 192
  decreasing 188, 201
  difference of 191
  domain convention 157
  even 179
  exponential 272
  graphs 163
  greatest integer 152
  identity 194
  increasing 187, 201
  inverse 197, 199, 286, 295
  inverse—theorem 200
  linear 167
  linear rational 237
  logarithmic 286, 294
  odd 180
  one-to-one 200
  operations on 190
  polynomial 224
  product of 191
  quadratic 216, 219
  quotient of 191
  range 151

functions *(Continued)*
   rational 234
   square root 164
   step 171
   sum 190
functional notation 155
   mistakes with 159
fundamental
   counting principle 464
   theorem of algebra 249

Gaussian elimination 358
geometric
   means 441
   sequence 437
   series 445
geometry review 487
graph(s)
   contraction 183
   of equation 127
   expansion 183
   of exponential functions 274
   of first-degree equations 143
   of functions 163
   of inequality 403
   of inverse function 199
   of logarithmic functions 287, 295
   of polynomial functions 224, 232
   of quadratic functions 219
   of rational functions 235
   reading 171
   reflection 183
   shifting 180
   shrinking 182
   stretching 182
graphing calculator 164, 170
   method 258
graphing techniques 177
greater than 2
greatest integer function 152
growth
   exponential 273
   logarithmic 287
   polynomial 232, 273
   population 275
guidelines for applied problems 68

half-life 274
hierarchy of operations 5

higher-order determinants 397
hole 239
homogeneous system of equations 362, 395
horizontal
   asymptote 236, 240
   line test 200
   shifts 182, 501
How Green Was My Valley 75
hyperbola 311, 323
   eccentricity 328
   equation 324, 327
hypotenuse 487

i(complex number) 87
identity 3, 67
   function 194
   matrix 383
imaginary
   number 80
   part 93
inconsistent system 347, 349, 363
increasing
   function 187, 201
   on interval 187
index, summation 427
induction 454
   hypothesis 457
inductive step 457
inequalities 7, 103
   absolute value 105, 106, 114
   graphs of 599
   linear 103
   polynomial 108
   principles for solving 103
   rational 111
   system of 343, 405
infinite
   sequence 424
   series 443
infinity 31
integer(s) 1
   exponent 13
   greatest—function 152
   prime 431, 458
intercepts 141
interest
   compound 273, 278
   simple 69

Intermediate Value Theorem 227
interpolation 492
interval
  decreasing on 188
  increasing on 187
  notation 30
inverse 3
  function 197, 199, 286, 295
  of matrix 383
  variation 204
inversely proportional 204
invertible matrix 383
irrational
  exponents 271
  numbers 1, 10

joint variation 205

laws
  exponent 14, 23, 272
  logarithm 283–285, 293
leading coefficient 37
less than 2
line(s)
  equation of 139, 141
  horizontal—test 200
  number 1
  parallel 136
  perpendicular 137
  slope of 132
  vertical 139
  vertical—test 163
linear
  equations 343, 357
  function 167
  inequalities 103
  interpolation 492
  polynomial 37
  programming 413
  rational function 237
  systems of—equations 343, 357, 371
logarithm(s)
  to base b 290
  common 291
  natural 282, 291
  power law 285, 293
  product law 283, 293
  quotient law 284, 293
  tables 492

logarithmic
  equations 302
  functions 286, 294
  growth 287
LORAN 329
lower bound 260
lowest terms 50

main diagonal 383
major axis of ellipse 313
mantissa 495
mathematical induction 454
matrix 377
  addition 378
  augmented 372
  coefficient 374, 392
  determinant 388, 389, 398
  echelon form 373
  equality 378
  identity 383
  inverse 383
  invertible 383
  main diagonal of 383
  methods 371
  minor 389
  multiplication 380
  nonsingular 383
  scalar multiplication 379
  zero 382
maximum
  relative 226
  value 221, 412, 413
means
  arithmetic 435
  geometric 441
midpoint formula 125
minimum
  relative 226
  value 221, 413
minor
  axis of ellipse 313
  of determinant 389
mistakes
  with functional notation 159
  with induction 458
Morgan, Mrs. 75
multiplication
  matrix 380
  patterns 35

multiplication *(Continued)*
   scalar 379
multiplicity of roots 250
mutually exclusive events 477

n factorial 447
natural
   logarithm 282, 291
   logarithmic function 286
   number 1
negative 3, 4
   exponent 14
   number 2
nonlinear system 352
nonnegative numbers 2
nonrepeating decimal 1, 10
nonsingular matrix 383
number
   complex 87
   Fibonacci 426
   imaginary 88
   irrational 1, 10
   line 1
      distance on 28
   natural 1
   negative 2
   positive 2
   rational 1, 10
   real 1, 10, 88
   test 108
numerator 50
   rationalizing the 54

oblique asymptote 240
odd function 180
one-to-one function 200
open interval 30
operation(s)
   on equations 359
   on functions 190
   row 373
opposite signs 3
order 2
   of operations 5
ordered pair 123
origin 121
   symmetry 178
outcome 473

parabola 216, 311, 333
   equation 335, 336, 338
parabolic asymptote 246
parallel lines 136
parentheses 6
   omitting 34
partial
   fractions 366
   sum 428, 433, 439
Pascal's triangle 448
peak 226
perfect
   cube 46
   square 46
permutation 465
   distinguishable 472
perpendicular lines 137
plane, coordinate 121
point
   coordinates 121
   -slope form 139
polynomial(s) 36
   bounds for roots 261
   coefficient 37
   constant (term) 37
   degree 37
   division 37
   equations 62, 76, 91, 94, 247
   factoring 45
   function 224
   growth 232
   inequalities 108
   linear 37
   quadratic 37
   root 226
   zero 37, 226
population growth 275
positive number 2
postage-stamp function 171
power
   law for logarithms 285, 293
   principle 95
prime 431, 458
principle(s)
   Big-Little 235
   fundamental counting 464
   of mathematical induction 456, 462
   for solving equations 62
   for solving inequalities 103

probability 473
   distribution 479
product
   of functions 191
   law for logarithms 283, 293
   matrix 380
progression
   arithmetic 431
   geometric 437
proportional 203
Pythagorean Theorem 487

quadrants 121
quadratic
   applications of—equations 82
   equations 76, 80
   factoring patterns 46
   formula 79
   function 216, 219
   polynomial 37
   type equations 94
quotient 37
   of functions 191
   law for logarithms 284, 293

radical 20, 21
   equation 95
radioactive decay 274
radiocarbon dating 299
radius 127
range 151
ratio, common 437
rational
   exponents 22
   expression 49
   functions 234
   inequalities 111
   numbers 1, 10
   solutions test 100
rationalizing denominators and numerators 54
real
   numbers 1
      factorization of polynomials over 252
   part 93
rectangular coordinate system 121

recursively defined sequence 426
reflection
   in line $y = x$ 199
   in x-axis 183
relative maxima/minima 226
remainder 38, 40
   theorem 247
repeating decimal 1, 10
Richter scale 296
right
   angle 487
   triangle 487
rise 131
roots 250
   approximation 257
   conjugate 251
   cube 20, 91
   nth 20
   multiplicity 250
   of polynomials 226
   square 20
   of unity 91
rounding 11
row 377
   expanding determinant along 390
   operations 373
rule
   of function 151
   of 72 289
run 131

sample space 473
scalar multiplication 379
scientific notation 17
section, conic 311
sequence 424
   arithmetic 431
   constant 426
   Fibonacci 426, 431
   geometric 437
   partial sum 428, 433, 439
   recursively defined 426
series
   convergent 444
   geometric 445
   infinite 443
shifts 180, 313
shrinking graphs 182

sigma notation 427
sign(s) 5
   Descartes' Rule of 255
   opposite 3
   same 2
significant figures 12
similar triangles 490
simple interest 69
slope(s) 132
   -intercept form 141
   of parallel lines 136
   of perpendicular lines 137
   point—form 139
   properties 135
   theorem 133
solution 62
   conjugate 91
   of equation in two variables 126
   feasible 411, 413
   of inequality 401
   optimal 413
   of quadratic equation 80
   rational 100
   of system 343, 349
   trivial 362
solving
   equations 62
   inequalities 103
square(s)
   completing the 7
   difference of 46
   perfect 46
   root 20
   root function 164
   root of negative number 90
standard
   form for complex numbers 87
   form for logarithms 495
step function 171
stretching graphs 182
substitution
   back 358
   method 343
subtraction, matrix 574
sum
   of cubes 46
   of functions 190
   matrix 378
   partial 428, 433, 439

sum *(Continued)*
   of series 444, 445
   of squares 46
summation
   index 427
   notation 427
symmetry
   origin 178
   tests 179
   x-axis 178
   y-axis 177
synthetic division 42
system
   dependent 347, 349, 363
   of equations 343, 352, 357, 392
   homogeneous 362, 395
   inconsistent 347, 349, 363
   of inequalities 343, 405
   of linear equations 343, 357
   of nonlinear equations 352
   solution of 343

table, logarithm 492
term of sequence 424
terminating decimal 10
Test
   Number Theorem 108
   for symmetry 179
triangle(s) 487
   inequality 27
   Pascal's 448
   right 487
   similar 490
trivial solution 362
turning point 226

unit circle 128
unity, roots of 91
upper bound 260

valley 226
value
   absolute 25, 26
   of function 157
variable 33
variation 203

vertex 412
  of ellipse 313
  of hyperbola 323
  of parabola 216, 220, 333
vertical
  asymptote 236, 239
  line 139
  line test 163
  shift 180, 313

x-axis 121
  reflection in 183
  symmetry 178
x-coordinate 121

x-intercept 141, 226

y-axis 121
  symmetry 177
y-coordinate 121
y-intercept 141

zero
  division by 4, 8
  exponent 14
  matrix 382
  polynomial 37
  of a polynomial 226
  products 4